2020 ACVIM Forum On Demand

Online
10 June 2020

ISBN: 978-1-7138-1351-4

Printed from e-media with permission by:
Curran Associates, Inc.
57 Morehouse Lane
Red Hook, NY 12571

Some format issues inherent in the e-media version may also appear in this print version.

Address correspondence to:
American College of Veterinary Internal Medicine (ACVIM)
8301 E. Prentice Avenue, Suite 300
Greenwood Village, CO 80111
www.acvim.org

Additional copies of this publication are available from:

Curran Associates, Inc.
57 Morehouse Lane
Red Hook, NY 12571 USA
Phone: 845-758-0400
Fax: 845-758-2633
Email: curran@proceedings.com
Web: www.proceedings.com

TABLE OF CONTENTS

CARDIOLOGY

EQUINE

FOOD ANIMAL

NEUROLOGY

ONCOLOGY

SMALL ANIMAL INTERNAL MEDICINE

EQUINE & FOOD ANIMAL INTERNAL MEDICINE & SMALL ANIMAL INTERNAL MEDICINE & CARDIOLOGY

Abstract C01: Effects of Pimobendan on Left Atrial Transport Function in Cats with Hypertrophic Cardiomyopathy

Samantha L. Kochie[1]; Karsten Schober[1]; Jaylyn Rhinehart[1]; Randolph Winter[1]; John Bonagura[2]; Annie Showers[1]; Vedat Yildiz[1]
[1]The Ohio State University; [2]North Carolina State University

Arterial thromboembolism (ATE) is a sequelae of hypertrophic cardiomyopathy (HCM) in cats related to left atrial (LA) enlargement and dysfunction. This study addresses the general hypothesis that pimobendan, an inodilator, improves LA transport function in healthy cats and cats with HCM. Prospective, double-blind, randomized, placebo-controlled clinical cohort study. 21 cats with HCM and 10 healthy control cats underwent two examinations 4–7 days apart. Cats were randomized to receive either pimobendan (0.25 mg/kg body weight q 12 h) or placebo. Two-dimensional and Doppler echocardiographic variables of LA function (LA shortening fraction and fractional area change, velocities of A, AR, and A′ waves, LA ejection force, LA kinetic energy, and peak left auricular flow velocity) were evaluated. Data were statistically compared using standard test procedures for before–after comparisons and linear mixed model analysis. Several echocardiographic variables characterizing LA size and function were increased after pimobendan: Left auricular flow velocity (0.85±0.20 m/s vs. 0.71±0.22 m/s, p=0.009), estimated total LA volume (p=0.030), LA emptying area (p=0.036) and LA emptying volume (p=0.032), and A velocity (0.77±0.12 vs. 0.62±0.17, p=0.049). LA kinetic energy (p=0.19) and LA ejection force (p=0.27) were not different after pimobendan. Status (HCM vs. healthy) and LA size (normal vs. increased) were not identified as independent predictors of drug effect in the multivariate model. This study identified positive, albeit minor, effects of pimobendan on LA reservoir and pump function in cats with pre-clinical HCM. Whether or not chronic therapy with pimobendan can reduce the risk of cardiogenic embolism deserves further study.

Abstract C02: Canine Oblique Angiographic Projections in Dorsal, Left Lateral, and Right Lateral Recumbency

Caroline Q. Sloan; Brian Scansen; Kursten Pierce; I-Jung Chi; Christopher Orton
Colorado State University

In humans, oblique angiographic projections help optimize cardiac interventions and assume dorsal recumbency. Interventions in animals are performed in both dorsal and lateral recumbency without standardized veterinary terminology. The study aim was to develop and evaluate an angiographic nomenclature for oblique fluoroscopic projections in dogs that is consistent regardless of patient recumbency. Twelve canine cadavers underwent left (n=6) and right (n=6) heart catheterization. Cardiac structures were opacified with iodinated contrast. Oblique projections were acquired every 30 degrees across 180 degrees of detector arc. Projections were obtained in variable recumbency (dorsal, left lateral, right lateral) for each dog with staggered sequence order. A rotational angiogram was then acquired, allowing reconstruction of all projection angles. Evaluators were blinded to patient positioning and a scoring system was used to evaluate similarity of each projection obtained in different states of recumbency. Angiographic projections were labeled according to beam-to-detector path relative to the dog, regardless of C-arm or table position. For example, a 30 degree right ventral oblique projection was defined as the detector angled 30 degrees from sternum toward the dog's right shoulder. Contrast was gravitationally dependent resulting in subjective differences in chamber opacity between recumbent states. All projections were graded as equivalent irrespective of recumbency. We propose a unified terminology for canine angiography that can be applied to oblique projections and is agnostic to patient recumbency and C-arm position. This nomenclature may improve consistency in future studies and as structural heart interventions begin to adopt unique angulations to guide transcatheter therapies in animals.

Abstract C03: Effect of Pimobendan on Cardiac and Renal Function in Canine Subclinical Myxomatous Mitral Valve Disease

Joanna L. Kaplan; Lance Visser; Catherine Gunther-Harrington; Eric Ontiveros; Luke Wittenburg; Carrie Palm; Joshua Stern
University of California, Davis

We sought to determine the effects of standard-dose and high-dose pimobendan on cardiac size and function and renal function estimates in dogs with subclinical myxomatous mitral valve disease (MMVD). Dogs <15 kg diagnosed with stage B2 MMVD were enrolled in this prospective, randomized, blinded, placebo-controlled, clinical trial. Azotemic dogs and dogs taking cardiac medications were excluded. N-terminal pro-brain natriuretic peptide (NT-proBNP), quality-of-life (QOL) scoring, renal function (iohexol clearance ELISA-estimated glomerular filtration rate, symmetric dimethylarginine, and creatinine concentration), and echocardiographic cardiac size and function were evaluated before and 7–10 days after placebo (n=6), standard-dose pimobendan 0.2–0.3 mg/kg q 12 (SD_pimo; n=12), or high-dose pimobendan 0.5–0.6 mg/kg q 12 h (HD_pimo; n=12). Paired t-tests (or nonparametric equivalent) were used for comparisons within group. ANOVAs and Tukey's post-hoc tests (or nonparametric equivalent) were used for comparisons among groups. There were no significant differences in renal function indices within the placebo, SD_pimo, and HD_pimo groups ($p \geq 0.13$). There were no significant differences ($p \geq 0.48$) of QOL scores within the placebo, SD_pimo, and HD_pimo groups. Significant differences ($p < 0.001$) in percent change of NT-proBNP were identified among the placebo (0.7%), SD_pimo (−36.7%), and HD_pimo (−46.0%) groups. Echocardiographic data are presented in Table 1. Results suggest pimobendan possesses a dose-dependent benefit on cardiac function but does not significantly affect (increase or decrease) QOL or tests that estimate renal function in dogs with subclinical MMVD. Further study of the dose-dependent effects of pimobendan is warranted, particularly in dogs with MMVD and renal dysfunction, heart failure, or both.

Table 1. Mean (SD) of echocardiographic variables before and 7–10 days after placebo, standard-dose pimobendan (SD_pimo, 0.2–0.3 mg/kg PO q 12 h), and high-dose pimobendan (HD_pimo, 0.5–0.6 mg/kg PO q 12 h) in dogs with subclinical myxomatous mitral valve disease

Variables	Group	Before	After	P-value (Before vs After)	% change	P-value (% change among groups)
LAV (mL/kg)	Placebo	3.0 (1.0)	3.1 (1.2)	0.68	1.3 (15.6)	**0.004**
	SD_pimo	2.8 (0.8)	2.1 (0.6)	**0.002**	-22.7 (14.9)[a]	
	HD_pimo	3.3 (1.4)	2.4 (1.1)	**<0.001**	-27.1 (16.9)[a]	
LVVd (mL/kg)	Placebo	4.0 (1.1)	3.9 (0.8)	0.1	-0.2 (8.2)	**0.009**
	SD_pimo	3.8 (0.7)	3.2 (0.7)	**<0.001**	-16.7 (12.5)[a]	
	HD_pimo	3.7 (0.8)	2.9 (0.8)	**0.001**	-21.8 (15.0)[a]	

LVVs (mL/kg)	Placebo	0.86 (0.32)	0.77 (0.29)	0.47	-7.3 (35.6)	**<0.001**
	SD_pimo	0.99 (0.41)	0.57 (0.28)	**<0.001**	-41.6 (14.8)[a]	
	HD_pimo	1.06 (0.30)	0.48 (0.29)	**<0.001**	-55.0 (20.7)[a]	
LV EF (%)	Placebo	78.5 (4.3)	80.3 (6.7)	0.49	2.3 (7.3)	**0.007**
	SD_pimo	74.4 (7.2)	82.5 (5.4)	**<0.001**	11.3 (7.8)	
	HD_pimo	71.3 (8.3)	83.5 (8.9)	**<0.001**	17.6 (10.4)[a]	

LAV, left atrial volume; LVVd, left ventricular volume at end-diastole; LVVs, left ventricular volume at end-systole; LV EF, left ventricular ejection fraction; SD_pimo, standard-dose pimobendan; HD_pimo, high-dose pimobendan.
[a]Significantly different ($p<0.05$) compared to percent change of the placebo group.
Significant differences are in bold typeface.

Abstract C04: Complications and Outcomes Associated with Epicardial Pacemakers in Cats

Eva W. Frantz; Sonja Tjostheim; Alexandra Palumbo; Heidi Kellihan; Rebecca Stepien
School of Veterinary Medicine, University of Wisconsin

Epicardial pacemaker implantation is the treatment of choice for cats with clinical signs secondary to bradyarrhythmias. Complication rates and long-term outcomes, however, have only been reported in a small number of cats. The objectives of this study were to describe the major and minor complications in a larger population of cats (n=20) that received epicardial pacemakers and to report their survival rates. Medical records (2003–2019) were reviewed. Presenting complaint, indication for pacing, presence of structural heart disease, presence of congestive heart failure, presence of a major or minor complication, and survival time were recorded and assessed in the statistical analysis. A complication was determined to be major if it was life-threatening or required replacement of the pacemaker system, while minor complications were not life-threatening and generally self-limiting or required minimal intervention. None of the variables evaluated were associated with a significant increase in the incidence of major or minor complications. The most common major complication was loss of ventricular capture (n=6), which was successfully treated in all cases by increasing pacemaker output and/or replacing the pacemaker lead. Lead dislodgement, the most common major complication reported previously in dogs with epicardial pacemakers, was not documented in this population of cats. The overall survival time (MST 948 days) was similar to that previously reported in dogs with epicardial pacemakers (Figure 1). Fifty percent of cats had one or more major complication, which is comparable to major complication rates previously documented in cats. No statistical difference in survival time was identified between cats that experienced a major complication and those that did not ($p=0.773$; Figure 2). This study provides relevant insights into the complications and outcomes of epicardial pacemakers in cats. Major complications were not associated with reduced survival time in this population.

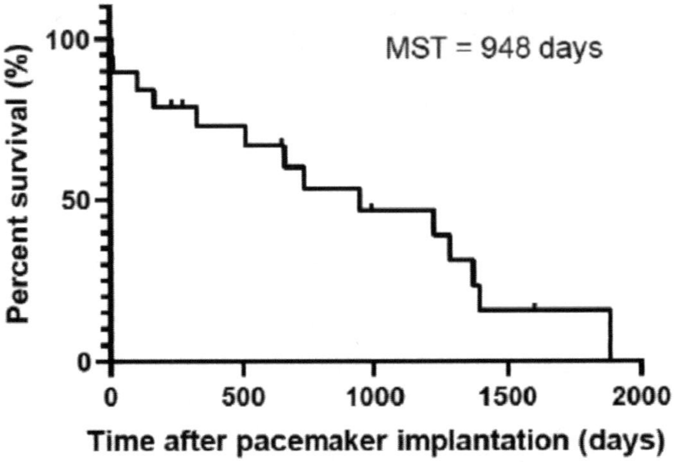

Figure 1. Kaplan-Meier survival curve for 19 cats. Tick marks represent censored data for cats still alive at study end or lost to follow-up. One cat died in the perioperative period and was excluded from the analysis.

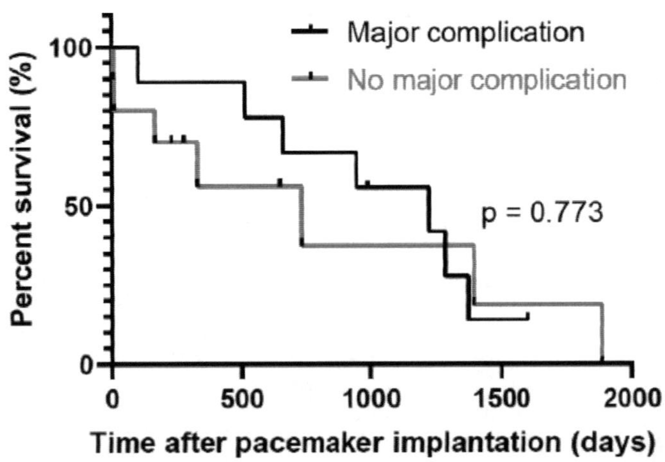

Figure 2. Kaplan-Meier survival curve comparing 19 cats with (n = 9) and without (n = 10) major complications. Tick marks represent censored data for animals still alive at study end or lost to follow-up. One cat died in the perioperative period and was excluded from the analysis.

Abstract C05: The Use of Activity Response Settings in Rate Responsive Pacemakers in Dogs: A Retrospective Study

Ananda Pires[1]; Michelle Colpitts[1]; Shari Raheb[1]; Lynne O'Sullivan[2]; Sonja Fonfara[1]
[1]University of Guelph; [2]University of Prince Edward Island

Third degree atrioventricular block (3rdAVB) is a common bradyarrhythmia in dogs for which pacemaker implantation is the gold-standard treatment. Pacemakers with rate responsive mode (VVIR) capability are usually implanted, as these provide the option of heart rate adjustment in response to activity. The aim of the present study was to determine the efficacy of the rate responsive setting of VVIR pacemakers in dogs that have pacemakers for 3rdAVB. Medical records of the Ontario Veterinary College were retrospectively searched and a total of 28 dogs with VVIR pacemakers for 3rdAVB were identified. Pacemaker interrogations, 6–12 months after placement, detected that the dogs spent the most time at the programmed lower rate with only minor heart rate variation. In 9/28 (32%) dogs, the activity daily living (ADL) and exertion responses were increased one or two levels to improve pacemaker response to activity. In 5/9 (55%) dogs, this did not elicit heart rate variation; in one dog this resulted in heart rate variation (this dog had an epicardial lead with abdominal generator, contrary to the other dogs); 3/9 (33%) dogs died before re-evaluation. In the majority of dogs with VVIR pacemakers for 3rdAVB, the default rate responsive mode did not result in heart rate variation. An increase of the activity response settings did not improve pacemaker response to activity, and it is possible that further adjustments are required. Furthermore, positioning of the pacemaker generator might influence rate responsiveness of a VVIR pacemaker.

Abstract C06: Use of Omics Technologies in the Investigation of Diet-Associated Dilated Cardiomyopathy (DCM) in Dogs

Lisa M. Freeman[1]; Caren Smith[2]; John Rush[1]
[1]Cummings School of Veterinary Medicine, Tufts University; [2]Human Nutrition Research Center on Aging, Tufts University

The objective of this study was to use metabolomic and lipidomic data from diets and dogs to identify possible dietary factors associated with canine diet-associated DCM. Eighteen diets (9 non-traditional and 9 traditional) were analyzed in duplicate using metabolomic and lipidomic platforms (i.e., foodomics) at a commercial laboratory. Metabolomic analysis was also performed on 20 canine plasma samples: Dogs with DCM and congestive heart failure (CHF) eating traditional diets (n=5), dogs with DCM and CHF eating non-traditional diets (n=5), dogs with DCM but no CHF eating non-traditional diets (n=5), and healthy control dogs eating traditional diets (n=5). Samples were analyzed using Ultrahigh Performance Liquid Chromatography-Tandem Mass Spectroscopy. Of 830 biochemicals identified in the metabolomics profile of the dog foods, 261 were significantly increased and 153 were significantly decreased in non-traditional vs. traditional diets associated with multiple metabolic pathways. Of 966 biochemicals in the dietary lipidomic profile, 140 were significantly increased and 277 were significantly decreased in non-traditional vs. traditional diets. A number of metabolites distinguished the 2 diet groups including 3 unnamed biochemicals that were identified in 100% of non-traditional diets and 0% of traditional diets, but none of the 3 biochemicals was found in the plasma of affected dogs. Plasma metabolomic profiles identified multiple significant differences between dogs with CHF eating non-traditional vs. traditional diets. Omics technologies, such as metabolomics, can provide functional information on both individual diet constituents and blood biochemicals that may assist in drawing mechanistic connections between diet and complex, multifactorial diseases, such as diet-associated DCM.

Abstract C07: Evaluation of Myocardial Fibrosis in Cats with Hypertrophic Cardiomyopathy Using Cardiac Magnetic Resonance Imaging

Ryan Fries; Saki Kadotani; Stephanie Keating; David Schaeffer; Jonathan Stack
College of Veterinary Medicine, University of Illinois at Urbana–Champaign

Hypertrophic cardiomyopathy (HCM) is the most common heart disease in cats. Echocardiography is the non-invasive reference standard for diagnosing HCM; however, it has little to no value in assessing myocardial composition. Recent advances in cardiac magnetic resonance imaging (CMR) have allowed for detection of diffuse and focal fibrosis in a variety of heart conditions, including HCM. The objective of this study was to quantify myocardial fibrosis, namely T1 mapping and extracellular volume fraction (ECV) in healthy and HCM cats using CMR. Seventeen healthy and twelve HCM, age-matched, client-owned cats were prospectively enrolled. Tests performed included physical examination, indirect blood pressure, complete blood count, biochemical analysis including total thyroid, urinalysis, transthoracic echocardiogram, and CMR with contrast. Cats were considered healthy if all diagnostic tests were within normal limits and a diagnosis of HCM was determined by the presence of either focal or generalized left ventricular concentric hypertrophy >6 mm on echocardiography. Between groups, there were no statistical differences in regard to age, indirect blood pressure, or laboratory results. Statistically significant CMR parameters included left ventricular mass (healthy = 5.87 g, HCM = 10.3 g, $p<0.0001$), T1 native mapping (healthy = 1122 ms, HCM = 1209 ms, $p=0.004$), and ECV (healthy = 26.0%, HCM = 32.6%, $p<0.0001$). Results of this study confirm that CMR is useful at detecting myocardial fibrosis in cats with HCM and that myocardial composition is different between normal and HCM cats.

Abstract C08: Renin-Angiotensin-Aldosterone System Activation in Hypertensive Cats Receiving Amlodipine

Clarke E. Atkins[1]; Darcy Adin[2]; Catherine Glahn[3]; Oliver Domenig[4]; Teresa DeFrancesco[1]; Kate Meurs[1]
[1]North Carolina State University; [2]University of Florida; [3]Coastal Cat Clinic; [4]Attoquant Diagnostics

Chronic Renin-Angiotensin-Aldosterone-System (RAAS) activation is harmful. Vasodilators, including amlodipine in man and the dog, activate RAAS. Contradictory data exist, however, for the hypertensive cat. We sought to clarify this, with a novel method of global circulating RAAS analysis. We hypothesized that naturally hypertensive cats, receiving chronic amlodipine therapy experience RAAS activation, as compared to normal, untreated subjects. We enrolled 9 client-owned, unmedicated, normotensive cats, normal by physical examination, basic laboratory tests, and echocardiography and 5 client-owned cats, with proven hypertension, IRIS Stage 2–3 chronic kidney disease (2 also hyperthyroid, 1 diabetic), receiving amlodipine chronically (average 1.5 years). Average systolic blood pressure was 166 mm Hg, down from an average maximum of 188.6 mm Hg. None received RAAS-suppressant therapy. Serum was frozen (–80C) for future blinded LC-MS/MS-based RAAS equilibrium analysis quantification/calculation of RAAS components and select RAAS enzymes. Analytes, reported as mean (± SD) and median (25th–75th quartiles), were compared between groups, using non-paired student's t-test. RAAS activation was significantly greater in hypertensive cats receiving amlodipine, as compared to normal cats. Significant elevations in hypertensive cats were observed for values of Renin, Angiotensin I, Angiotensin 1-7, Angiotensin III, and Angiotensin 1-5, and Aldosterone. Mean angiotensin converting-enzyme activity was less in hypertensives ($p<0.05$), while chymase activity was 50% greater ($p>0.05$). While RAAS is chronically activated in hypertensive cats receiving amlodipine, the exact role played by amlodipine is not clarified by this study. Nevertheless, the data indicate that concurrent RAAS-suppressant therapy is indicated in hypertensive cats with hyperthyroidism and/or chronic kidney disease, receiving amlodipine.

Abstract C09: Prevalence of the PDK4 and Titin Gene Mutations in North American Doberman Pinschers

Ryan Fries[1]; Nancy Morris[2]; Amara Estrada[3]; Kathryn Meurs[4]
[1]College of Veterinary Medicine, University of Illinois at Urbana-Champaign; [2]Veterinary Emergency and Specialty Hospital; [3]University of Florida; [4]North Carolina State University

Dilated cardiomyopathy (DCM) is the most common cardiomyopathy in dogs and in Doberman Pinschers (DP), the estimated lifetime risk is 25-50%. In DP of North America, two genetic mutations are reportedly associated with developing DCM. One encoding for the pyruvate dehydrogenase kinase 4 (PDK4) gene and one encoding for the titin gene. The objective of our prospective, longitudinal cohort study is to determine the prevalence of these two mutations in a population of North American DP and evaluate the influence of these mutations on the development of DCM in this population. All dogs are screened annually including physical examination, echocardiogram, Holter monitor, and client questionnaire. Diagnosis of DCM is based on criteria from the European Society of Veterinary Cardiology screening guidelines. A total of 385 dogs are currently enrolled. Regarding the PDK4 mutation, 52.3% are negative, 38.1% are heterozygous positive, and 9.6% are homozygous positive. Regarding the titin mutation, 20.6% are negative, 51.1% are heterozygous positive, and 28.3% are homozygous positive. Only 13.1% are negative for both mutations. At the time of initial screening, 73.4% of DP had a normal echocardiogram, 15.9% had occult DCM, 6.9% were considered equivocal, and 3.1% were diagnosed with myxomatous mitral valve disease. In the occult DCM subset, 89.1% of dogs had at least one mutation. With an incidence of at least one mutation in 86.9% of DP in this population, ongoing screening and outcome analysis is essential for understanding the influence of these genetic mutations on the development of DCM in DP.

Abstract C10: Diagnostic Utility of Caudal Vena Cava Measurements in Dogs with Cavitary Effusions or Heart Failure

Yen Yu Chou[1]; Jessica Ward[2]; Lara Barron[3]; Melissa Tropf[2]; Shane Murphy[2]; Gregory Lisciandro[4]; John Bonagura[3]; Anna McManamey[3]; Teresa DeFrancesco[3]
[1]Iowa State University; [2]College of Veterinary Medicine, Iowa State University; [3]College of Veterinary Medicine, North Carolina State University; [4]Hill Country Veterinary Specialists

Thoracic ultrasound has proven effective for the diagnosis of left-sided congestive heart failure (L-CHF) dogs and cats with respiratory distress. The objective of this study was to determine whether ultrasonographic indices of the caudal vena cava (CVC) could be used to diagnose right-sided CHF (R-CHF) in dogs with cavitary effusions. Dogs were prospectively enrolled in four groups: R-CHF (n=34), L-CHF (39), cavitary effusions of noncardiac etiology (NC, 41), and pericardial effusion with tamponade (PCEFF, 17). Ultrasonographic indices included right ventricular to left ventricular ratio (RV:LV) and 2D and M-mode subxiphoid measures of CVC maximal and minimal size (CVCmax and CVCmin), CVCmax indexed to aortic dimension (CVC:Ao), and CVC collapsibility index (CVC-CI). Variables were compared between study groups using Kruskal-Wallis and Dunn's-Bonferroni testing. All indices (RV:LV, CVCmax, CVCmin, CVC:Ao, and CVC-CI) were significantly different between R-CHF and NC dogs ($p<0.005$). A CVC-CI less than 30% (in either M-mode or 2D) was 97% sensitive and 93% specific for diagnosis of R-CHF versus NC effusion. All CVC indices, but not RV:LV ratio, also differed between PCEFF and NC dogs ($p<0.005$). Compared to NC, L-CHF dogs had higher CVC:Ao in both 2D ($p=0.017$) and M-mode ($p=0.014$); compared to R-CHF, L-CHF dogs had higher CVC-CI in both 2D and M-mode ($p<0.005$). Ultrasonographic indices of CVC size and collapsibility are useful to differentiate R-CHF versus NC disease as causes of cavitary effusions. Dogs with L-CHF demonstrate CVC measurements intermediate between R-CHF and NC dogs.

Abstract C11: Tricuspid Valve Dysplasia is Associated with a PLA2G4F Variant in the Labrador Retriever

Kathryn Meurs[1]; Sandy Tou[2]; Teresa DeFrancesco[1]; Bruce Keene[1]; John Bonagura[1]; Steven Friedenberg[3]
[1]North Carolina State University; [2]College of Veterinary Medicine, North Carolina State University; [3]University of Minnesota

Tricuspid valve dysplasia (TVD) is a congenital heart defect observed most commonly in the Labrador Retriever. Some dogs have only a mild form of the defect and can live comfortably, often with no outward signs, while others have significant valvular insufficiency and congestive heart failure. TVD has been shown to be a heritable trait in the Labrador Retriever. We hypothesized that we could identify a genetic marker for the trait in the Labrador Retriever by performing Whole Genome Sequencing with affected Labrador Retrievers and filtering them against 478 unaffected non-Labrador Retriever dogs. We identified a variant in a candidate gene, PLA2G4F (phospholipase A2 group IVF) that contains a malignant variant that changes a conserved amino acid from glutamic acid to aspartic acid. The variant is strongly associated with TVD (p=0.0001) and was found in 63 of 64 affected Labrador Retrievers and was not in 12 echocardiographically confirmed unaffected Labrador Retrievers. It was also identified in 2% of the general canine population of 478 non-Labrador Retriever dogs and 56 of 70 Labrador Retrievers of unconfirmed phenotype. These findings are suggestive of incomplete penetrance of the variant. In silico predictions suggest that this variant could change the protein structure of PLA2G4F. PLA2G4F is responsible for facilitating the breakdown of phospholipase to arachidonic acid and is ultimately involved in the development of prostaglandins thought to play a role in heart disease and cardiac development. We conclude that PLA2G4F may be associated with the development of tricuspid valve dysplasia in the Labrador Retriever.

Abstract C12: Evaluation of Galectin-3 as a Novel Biomarker in Feline Hypertrophic Cardiomyopathy

Jonathan Stack; Ryan Fries; Saki Kadotani; Leah Kruckman; Gabrielle Wallace
College of Veterinary Medicine, University of Illinois at Urbana–Champaign

Galectin-3, a circulating biomarker, correlates with myocardial fibrosis in humans. It independently predicts mortality and adverse outcomes in cardiovascular disease, including hypertrophic cardiomyopathy (HCM). In veterinary medicine, risk-stratification of feline HCM patients is difficult and largely based on specific echocardiographic findings such as left atrial enlargement. The purpose of this study was to evaluate galectin-3 as a potential clinical biomarker in feline HCM. Thirty healthy cats, nineteen cats with occult HCM (left ventricular concentric hypertrophy >6 mm), and twenty-six cats with HCM and past or present congestive heart failure (CHF) were enrolled. Tests performed included echocardiography, complete blood count, serum chemistry, total T4, thoracic radiographs, and blood pressure to categorize each cat and ensure no concurrent diseases. When possible, quantitative NT-proBNP levels, cardiac troponin I levels, and urinalysis were performed. Serum galectin-3 levels were determined using a validated, commercially available colorimetric quantitative ELISA. Galectin-3 was significantly elevated in all cats with HCM compared to healthy cats ($p<0.009$). Subgroup analysis found galectin-3 was significantly elevated in CHF compared to healthy cats ($p=0.0140$), but was not significantly different between healthy and occult HCM cats ($p=0.1094$) or between occult and CHF cats ($p=0.8318$). Cats with a history of CHF had significantly higher levels of blood urea nitrogen and creatinine, and lower levels of potassium and chloride compared with all other groups. Results of this study confirm that galectin-3 is increased in cats with HCM and warrants further investigation as a clinical biomarker.

Abstract C13: Use of Three-Dimensional Models of Echocardiographic Imaging Planes in the Teaching of Echocardiography

Marlos Gonçalves Sousa; Bruna Costa; Marcela Wolf; Simone Stedile
Federal University of Paraná

Heart disease in dogs and cats has a high incidence in clinical routine, with echocardiography being the most used test for its diagnosis. The main difficulty presented by beginners in echocardiography is the orientation of the echocardiographic imaging plane with the three-dimensional (3D) anatomy of the heart. As 3D models are advantageous in understanding two-dimensional images, the hypothesis of this study is that 3D echocardiographic models would be useful in teaching echocardiography. Thus, the objective of this work was to develop and test 3D models of healthy and sick hearts of dogs and cats, to evaluate its effectiveness in undergraduate vet students understanding echocardiographic imaging planes. Model hearts made out resin with the main normal echocardiographic imaging planes of dogs and cats were produced, as well as models with mitral endocardiosis in dogs and hypertrophic cardiomyopathy in cats. Students from the 4th year of the Veterinary Medicine degree were invited to participate, and after a theoretical class they were randomly assigned to two groups, the model group and the control group. The model group had access to 3D models, along with a self-explanatory text on echocardiographic imaging planes, and the control group had access only to the self-explanatory text. Both groups had two weeks to study. The students answered a quiz to evaluate their learning, and also answered an anonymous questionnaire to give their opinion about the experience with the learning method used. A total of 39 students participated in the study, 19 of them in the model group and 20 in the control group. The dedicated study time was longer in the model group ($p=0.0027$); being that 7 of these students (36.8%) exceeded 2 hours of study. The proportion of students who achieved a satisfactory grade in the quiz was 89.5% in the model group and 60% in the control group ($p=0.0449$). Students with no previous experience who studied on the model were better at the quiz when compared to students in the control group who also had no previous experience ($p=0.0516$). The present study shows that the three-dimensional models of echocardiographic imaging planes of dogs and cats facilitated and significantly improved learning and the identification of cardiac structures and abnormalities, and provided greater motivation for students through the study of echocardiography.

Abstract C14: Clinical Relevance of Serum Electrolytes in Small Animals with Acute Heart Failure: A Retrospective Study

Marine Roche-Catholy[1]; Iris Van Cappellen[2]; Laurent Locquet[2]; Bart Broecks[2]; Dominique Paepe[2]; Pascale Smets[2]
[1]Faculty of Veterinary Medicine, Ghent University; [2]Ghent University

Electrolyte abnormalities frequently occur in congestive heart failure (CHF) patients. Hypochloremia has been identified as a marker of diuretic response and as a strong negative prognostic factor in people with CHF. However, publications on prognostic implications of electrolyte abnormalities in small animals with CHF are lacking. We aimed to document electrolyte abnormalities upon admission of dogs and cats with acute CHF, and to evaluate the associations between electrolyte concentrations and diuretic dose, and between electrolyte concentrations and prognostic variables. Dogs and cats with first onset of acute CHF, confirmed by medical imaging, and a complete electrolyte panel performed upon admission were retrospectively included. A total of 46 dogs and 34 cats met the inclusion criteria. The most commonly encountered electrolyte anomaly was hypochloremia observed in 24% (9/46 dogs and 10/34 cats) of cases. In dogs only, a significant negative correlation was identified between serum chloride levels at admission and furosemide doses both at discharge and at end-stage heart failure ($r=-0.59$; $p<0.001$ and $r=-0.62$; $p=0.004$ respectively). With regard to prognostic variables, no significant hazard ratios were found for duration of hospitalization nor survival time for any of the electrolyte concentrations. The present study provides new information supporting hypochloremia as a potential marker of impaired diuretic response in dogs presented with a first episode of acute CHF. Further prospective studies are needed to clarify the relationship between hypochloremia and diuretic response in dogs and cats with acute CHF, and its long-term implications.

Abstract C15: Rapid Arteriotomy Site Closure using Preplaced Purse-String Suture Technique in Dogs

Kursten V. Pierce; Brian Scansen; E Christopher Orton
Colorado State University

Ligation following arterial catheterization after image-guided interventions is common practice in veterinary medicine, which may affect vascular supply or limit future access. This prospective case series investigated the utility of a preplaced purse-string suture technique for rapid closure of arterial catheterization sites in dogs. An arterial cutdown was performed and a purse-string suture either in a square or diamond-shaped pattern was preplaced around the intended arterial puncture site using 6-0 monofilament, nonabsorbable polypropylene suture in 8 dogs with a median body weight of 15.5 kg (range, 4.3 to 32 kg) undergoing femoral (n=7) and carotid (n=1) arterial catheterization. Within the loose preplaced suture, arterial puncture was performed with an over-the-needle catheter and an introducer sheath was placed. When the procedure was finished, the introducer sheath was removed while the preplaced suture was tightened to achieve hemostasis at the arteriotomy site. Catheterizations included ductal occlusion (n=3), diagnostic left heart catheterization (3), transarterial chemoembolization of hepatocellular carcinoma (1), and stent graft implantation for aortoiliac thrombosis (1). Complete hemostasis was achieved in 2 dogs at time of purse-string closure, mild bleeding occurred in 4 dogs with complete hemostasis after placement of an addition single inverting stich, and the suture tore through the vessel wall in 2 dogs during tightening of the purse-string leading to ligation. Human suture-mediated vascular closure devices provided the basis for this direct suture technique, which appears to be a low-cost, viable option to avoid ligation during arterial catheterization in dogs. Success appears dependent on operator experience and surgical skill.

Abstract C16: Correlation of Hepatic Venous Doppler with Right Ventricular Morphofunctional Parameters in Dogs with Pulmonary Hypertension

Marlos Gonçalves Sousa; Vinícius Silva; Tilde Froes; Stephany Lucina; Marcela Wolf
Federal University of Paraná

The aims of this study were three-fold: to investigate, prospectively, the reliability of the hepatic vein Doppler waveform to diagnose and characterize the magnitude of pulmonary hypertension in dogs; to evaluate whether a correlation exists between hepatic venous flow waves and the structural and functional characteristics of the right ventricle; and to determine whether age, gender, body weight, heart rate, heart rhythm and systolic blood pressure play a role in the hepatic venous waveform pattern. The cross-sectional observational study included 43 dogs with varying degrees of pulmonary hypertension and a control group composed of 15 healthy dogs. From the pulsed spectrum of the hepatic veins, the direction and velocity of each of the four components was established. The wave forms were classified into three different patterns: type 1, type 2, and type 3 (Figure 1). The velocities of the hepatic A, S, V and D spectral waves and the phasic pattern varied according to the severity of pulmonary hypertension (Table 1). The pattern of hepatic veins waveforms varied in patients with pulmonary hypertension as compared to control dogs ($p<0.05$). There were no differences between the results obtained in males and females. In mild pulmonary hypertension, 8 dogs showed pattern 1 and 9 dogs, pattern 2. For moderate pulmonary hypertension, the majority of the dogs (n=15) had pattern 2, while only 2 dogs showed pattern 1. In animals with severe pulmonary hypertension, there was a substantial qualitative change in the documented pattern: 7 dogs had pattern 3 whereas the other 2 showed pattern 2. Correlation were documented between hepatic vein waves and age, heart rate and the right ventricle structural and functional variables. Wave D had a sensitivity of 100% when a cut-off of 11.6 cm s-1 was used to differentiate dogs with pulmonary hypertension from healthy dogs. The hepatic flow pattern was most affected in patients with moderate and severe pulmonary hypertension. This study shows the feasibility of this methodology as a supplementary technique for the recognition and staging of pulmonary hypertension. Also, Doppler hepatic venous flow was correlated with parameters that describe right ventricle function and structure.

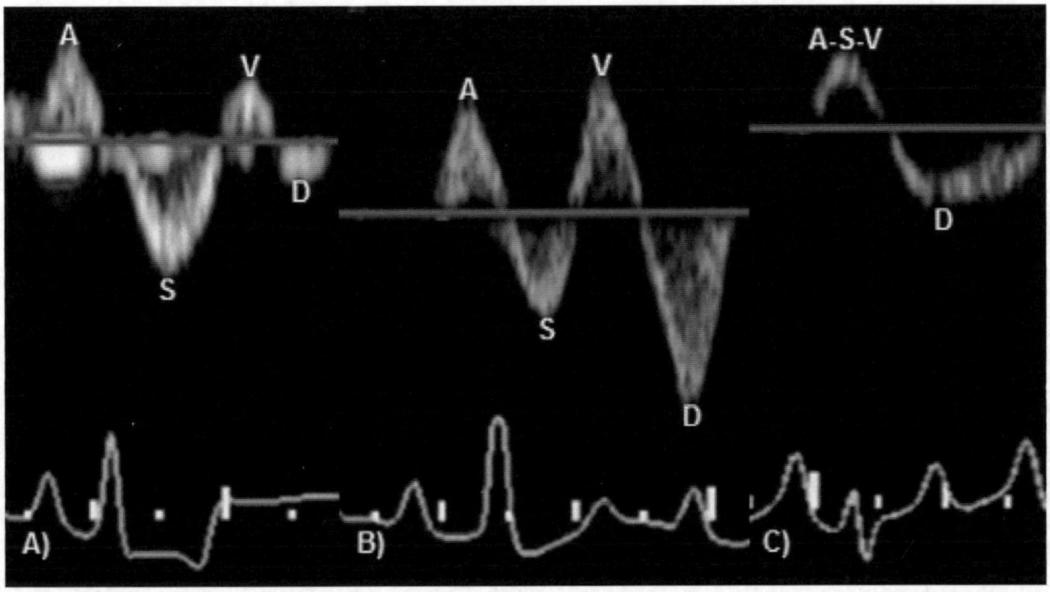

Figure 1 - Patterns of the hepatic Doppler waveform in dogs with pulmonary hypertension. A) Type 1, three-phasic, represented by a wave S larger than wave D and S/D ratio ≥1. B) Type 2, three-phasic, characterized by a wave S smaller than wave D and S/D ratio <1. C) Type 3, two-phasic, with the merging of wave S and waves A and V during ventricular systole and S/D ratio <1.

Table 1 - Comparison of the hepatic venous flow waves assessed through triplex Doppler in accordance with pulmonary hypertension severity.

	Pulmonary hypertension severity								*P*
	Absent		Mild		Moderate		Severe		
	Mean	SD	Mean	SD	Mean	SD	Mean	SD	
Wave A (cm · s^{-1})	7.7 ^{A}	2.4	7.1 ^{A}	1.9	11.6 ^{B}	4.3	13.9 ^{C}	3.9	< 0.001
Wave S (cm · s^{-1})	12.1	3.3	11.8	4.9	11.3	5	13.3	3	0.317
Wave V (cm · s^{-1})	-	-	3.9 ^{A}	1.5	4.5 ^{A}	0.9	14.8 ^{B}	6.5	0.002
Wave D (cm · s^{-1})	9.2 ^{A}	3.1	11.5 ^{AC}	6.9	14.1 ^{BC}	5.6	17.9 ^{B}	7.5	0.001
S/D	1.3 ^{A}	0.2	1.2 ^{AC}	0.7	0.8 ^{BC}	0.3	0.8 ^{B}	0.2	< 0.001

SD, standard deviation; S/D, wave S/wave D ratio. Values with different superscript letters indicate significant statistical difference between the groups.

Abstract C17: Ultrasonographic Assessment of Canine Femoral Vessels: Relationship to Age, Body Size, Sex, and Conformational Measurements

Katrina Cusack; Sonya Wesselowski; Ashley Saunders; Nicholas Jeffery
Texas A&M University

Minimally invasive transvascular procedures have advanced in veterinary medicine, but basic knowledge about expected vascular size and vascular imaging requires further exploration. Ultrasonographic imaging of the femoral artery (FA) and vein (FV) in a large population of dogs to describe relationships between vessel diameter and various patient characteristics. Animals were 230 client-owned dogs. Body measurements were obtained: body weight, withers height, body length, thigh circumference, head length and head width. A linear ultrasound probe was used to measure the internal diameter of the FA and FV bilaterally. Measurement repeatability was assessed. Allometrically scaled body weight had the strongest correlation with diameter of the FA and FV (correlation coefficients: 0.92, 0.80, respectively), although thigh circumference (FA:0.89, FV:0.78) and withers height (FA:0.84, FV:0.76) were also strongly correlated. Within the entire population, males had a smaller FA ($p=0.005$), but not FV ($p=0.278$), than females and age was negatively associated with FA ($p=0.031$) and FV ($p\leq0.001$) diameter. Comparison of left and right FA and FV diameter revealed minimal mean differences, though limits of agreement could encompass multiple French gauge sizes: [Mean difference (limits of agreement): FA=0 mm (0.85 to 0.84 mm), FV=0.035 mm (1.92 to 1.99 mm)]. Coefficients of variation for intra and interoperator repeatability were <9%. The diameter of the canine FA and FV is influenced by multiple factors, with potential for clinically relevant differences between right and left-sided vessels. Ultrasound measurement of the FA and FV was fast, repeatable, and could improve pre-procedural planning.

Abstract C18: Presence of Gut Dysbiosis in Dogs with Heart Failure: A Pilot Study

Joonbum Seo; Linda Matthewman; Dong Xia; Yu-Mei Chang; Jenny Wilshaw; David Connolly
Royal Veterinary College, University of London

The contribution of gut microbiota in the development and maintenance of heart failure known as the "gut hypothesis" describes impaired gut health in heart failure patients, resulting in dysbiosis and bacterial translocations across the edematous mucosal layer. This results in immune-stimulation via endotoxins, contributing to a chronic inflammatory state, malnutrition and cachexia. While this has been the focus of evaluation of novel therapies in people with advanced heart failure, no such research has been described in veterinary medicine. We hypothesized that dogs in congestive heart failure (CHF) have quantifiable alterations in gut microbiota compared to healthy control dogs. This is a cross-sectional pilot study. Pre-study power calculation showed 15 dogs were required in each group. Fecal samples were collected from healthy dogs and dogs affected by left CHF or right CHF. Fecal DNA was extracted, and the samples were transported to an external molecular Lab (Mr DNA, Texas, USA) for 16S rRNA sequencing (Illumina®). QIIME2 pipeline was used to analyze the microbial communities, and R, SPSS and Galaxy (Huttenhower Lab, harvard.edu) were used for statistical analysis. A total of 15 healthy dogs, 16 left CHF dogs, and 15 right CHF dogs were enrolled. Results showed no difference in alpha- (p=0.68) and beta-diversity (p=0.33) among the control, left CHF and right CHF groups. However, taxa analysis showed an increase in abundance of the Proteobacteria line in dogs with CHF (p=0.01). In conclusion, our pilot study demonstrated an alteration in the gut microbial communities in dogs with CHF without a change in the overall biodiversity and richness. The gut health is likely compromised in dogs with CHF.

Abstract C19: Combination Radiation Therapy and Chemotherapy for Right Atrial Masses with Pericardial Effusion in Dogs

Kimberly N. Cook; Thaibinh Nguyenba; Ioannis Giatis; Heather Lasher; Lisa Fulton; Deborah Prescott
MedVet Medical and Cancer Center for Pets

Pericardial effusion (PE) secondary to right atrial (RA) masses in dogs carries a poor long-term prognosis with limited treatment options. Our objective was to assess combination radiation and chemotherapy outcome in dogs with PE and RA masses. Medical records between January 2010 and June 2019 were reviewed retrospectively to identify treatment group dogs diagnosed with PE and RA masses via echocardiography that received both radiation and chemotherapy. A retrospective disease matched control group without radiation and chemotherapy was also sought. Data collected included presenting temperature, heart rate, blood pressure, ascites, arrhythmias, need for pericardiocentesis, and metastasis along with time to initial radiation treatment, radiation protocol, chemotherapy protocol, survival time, and cause of death. Eighteen treatment group and 22 control group dogs were identified. Treatment group median survival time was 124.5 days (range 4–358 days), and survival was unaffected by chemotherapy protocol or presenting heart rate, temperature, blood pressure, ascites, arrhythmias, pericardiocentesis, or metastasis. Control group median survival time following pericardiocentesis was 7.5 days (range 1–77 days). For survival comparison, dogs that received radiation after 6 days (mean time to initial radiation treatment) and control dogs that died before 6 days were excluded. Baseline characteristics were similar between groups. Treatment group survival was longer than the control group ($p<0.01$). Radiation and chemotherapy were tolerated. This study suggests combination radiation and chemotherapy is viable and extends survival time in dogs with PE secondary to RA masses, compared to control dogs. Further research is needed to determine independent effects of radiation therapy.

Abstract C20: Effect of Multi-Dose Oral Trazodone on Arterial Blood Pressure in Normal Beagles

Sienna R. Drizin; Kathryn Wotman; Kyle Kline; Michael Lappin; Rebecca Ruch Gallie
Colorado State University

Trazodone is prescribed by veterinarians for reducing stress in hospitalized dogs, treating behavioral disorders, and facilitating confinement in post-surgery recovery. However, there are unknowns about how this drug affects canine physiology and no studies have evaluated the peripheral cardiovascular effects of multi-dose oral trazodone in awake patients. The objective of this randomized double-blinded crossover study was to evaluate the effect of multi-dose oral trazodone on blood pressure in normal dogs. Eight research beagles were randomly assigned to treatment and control groups, then switched after a 3-day washout. Control groups received placebo and treatment groups received 50 mg of trazodone orally once per day. Systolic blood pressure was measured via Doppler prior to and at 1, 4, 8 and 12 hours post-treatment for a total of three days. A sedation score was recorded for each dog at every time-point. A mixed model was fit to the data, and Dunnett's method was used to compare downstream time points versus baseline for each treatment. Wilcoxon signed-rank test was used to compare sedation scores between treatments at each time point. Average blood pressure in the treated dogs compared to the untreated dogs did not change significantly throughout the measurement period. Sedation scores did not significantly change between the treated and untreated groups apart from T1 on day 1 ($p=0.031$). Based on the findings of this study, oral administration of clinically useful doses of trazodone in healthy dogs does not cause a significant change in systolic blood pressure.

Abstract C21: Prevalence of the Angiotensin-Converting Enzyme Gene Polymorphism in Dogs

Darcy B. Adin[1]; Clarke Atkins[2]; Steven Friedenberg[3]; Joshua Stern[4]; Kathryn Meurs[2]
[1]University of Florida; [2]North Carolina State University; [3]University of Minnesota; [4]University of California, Davis

The angiotensin-converting enzyme (ACE) gene intronic polymorphism at canine chromosome 9:11507816:G >A has recently been shown to increase the magnitude of aldosterone breakthrough, despite adequate suppression of angiotensin II by ACE-inhibitors. Genotype influence on non-angiotensin mediated aldosterone production holds implications for dogs with advanced heart disease, but polymorphism prevalence is unknown. We hypothesized that this polymorphism is more common in some breeds than others and therefore sought to determine prevalence and breed distributions in a large number of dogs. A database of 497 canine whole genome sequences from 54 breeds sequenced for various disease-specific studies was utilized to genotype dogs at chromosome 9:11507816:G >A. Allele frequency of the polymorphism within the studied population was 20.8%, with 19.9% of dogs heterozygous and 10.8% homozygous. The polymorphism was absent in 22 breeds and present in 32 breeds. Breed predisposition was evaluated for breeds with ≥10 dogs. The polymorphism was overrepresented in Irish Wolfhounds, Dachshunds, Cavalier King Charles Spaniels, Great Danes, and Bull Mastiffs. Irish Wolfhounds were more commonly homozygous than heterozygous. The ACE gene polymorphism was found in nearly a third of dogs, with unequal breed distribution. Some breeds demonstrated high polymorphism prevalence, which may be functionally important for breeds predisposed to heart disease. However, it is possible that the polymorphism prevalence was linked to the disease for which the dogs were sequenced; prevalence within a larger population could vary. Genotyping for this polymorphism may be a step towards personalized medicine and targeted clinical trials in the future of canine medicine.

Abstract C22: Left Atrial Function Assessment Using Tissue Motion Annular Displacement in Chronic Mitral Valve Diseased Dogs

Ana Paula Sarraff[1]; Vinícius Silva[2]; Giovana Tuleski[2]; Marcela Wolf[2]; Marconi Farias[1]; Marlos Souza[2]
[1]Pontifícia Universidade Católica do Paraná (PUCPR), Curitiba, PR, Brazil; [2]Federal University of Paraná (UFPR)

A strong correlation between left atrial disfunction, heart disease severity and prognosis are observed in dogs with myxomatous mitral valve disease (MMVD). The methods used to evaluate left atrial function are highly dependent on technical factors, are difficult and time consuming to use in clinical practice. The aims of this study were to study longitudinal left atrial function in dogs with chronic mitral valve disease by Tissue Motion Annular Displacement (TMAD), correlate TMAD with other echocardiographic methods of left atrial function and compare sensibility and specificity of this new tool to differentiate MMVD stages. In this prospective cross-sectional study, one hundred fifty-five client-owned dogs, including 95 dogs with myxomatous mitral valve disease (MMVD) and 60 healthy control dogs underwent echocardiogram. The animals were divided into A (control), B1, B2, C and D groups, according to ACVIM consensus. Left atrial function was assessed by biplanar area-length method, LA global strain (by speckle tracking), tissue Doppler imaging (mitral A') and TMAD (global and systolic) (by speckle tracking). Left atrial reservoir function was evaluated by LA global TMAD, LA global strain and LA emptying fraction and LA systolic function by LA systolic TMAD, LA ejection fraction and mitral A'. Data underwent the Shapiro-Wilk test to check for a normal distribution. Left atrial global TMAD was greater in B2, C and D groups ($p<0.001$) and LA systolic TMAD was greater in B1, B2 and C and lower in D group ($p=0.002$). Left atrial global TMAD was an excellent method to differentiate dogs without (B1) from with remodeling (B2, C and D) ($p<0.05$), whereas global strain and ejection fraction were better to discriminate the asymptomatic (B1 and B2) from the symptomatic (C and D) dogs ($p<0.05$) (Table 1). According to ROC analysis, the best discriminator between the asymptomatic from symptomatic and between animals with and without cardiac remodeling was LA global TMAD (AUC=0.715; AUC=0.756). Left atrial global TMAD value with optimal sensitivity and specificity to distinguish asymptomatic from symptomatic dogs was 21.07 mm/m2 (sensitivity 61.1% and specificity 76.3%) and animals with from without cardiac remodeling was 17.59 mm/m2 (sensitivity 66% and specificity 81%). The best method to discriminate asymptomatic from symptomatic animals and with from without cardiac remodeling was LA global TMAD. This method can be used as complementary, in conjunction with traditional echocardiographic examination, to assess global atrial function in patients with MMVD. A strong correlation between left atrial disfunction, heart disease severity and prognosis are observed in dogs with myxomatous mitral valve disease (MMVD). The methods used to evaluate left atrial function are highly dependent on technical factors, are difficult and time consuming to use in clinical practice. The aims of this study were to study longitudinal left atrial function in dogs with chronic mitral valve disease by Tissue Motion Annular Displacement (TMAD), correlate TMAD with other echocardiographic methods of left atrial function and compare sensibility and specificity of this new tool to differentiate MMVD stages. In this prospective cross-sectional study, one hundred fifty-five client-owned dogs, including 95 dogs with myxomatous mitral valve disease (MMVD) and 60 healthy control dogs underwent echocardiogram. The animals were divided into A (control), B1, B2, C and D groups, according to ACVIM consensus. Left atrial function was assessed by biplanar area-length method, LA global strain (by speckle tracking), tissue Doppler imaging (mitral A') and TMAD (global and systolic) (by speckle tracking). Left atrial reservoir function was evaluated by LA global TMAD, LA global strain and LA emptying fraction and LA systolic function by LA systolic TMAD, LA ejection fraction and mitral A'. Data underwent the Shapiro-Wilk test to check for a normal distribution. Left atrial global TMAD was greater in B2, C and D groups ($p<0.001$) and LA systolic TMAD was greater in B1, B2 and C and lower in D group ($p=0.002$). Left atrial global TMAD was an excellent method to differentiate dogs without (B1) from with remodeling (B2, C and D) ($p<0.05$), whereas global strain and ejection fraction were better to discriminate the asymptomatic (B1 and B2) from the symptomatic (C and D) dogs ($p<0.05$) (Table 1). According to ROC analysis, the best discriminator between the asymptomatic from symptomatic and between animals with and without cardiac remodeling was LA global TMAD (AUC=0.715; AUC=0.756). Left atrial global TMAD value with optimal sensitivity and specificity to distinguish asymptomatic from symptomatic dogs was 21.07 mm/m2 (sensitivity 61.1% and specificity 76.3%) and animals with from without cardiac remodeling was 17.59 mm/m2 (sensitivity 66% and specificity 81%). The best method to discriminate asymptomatic from symptomatic animals and with from without cardiac remodeling was LA global TMAD. This method can be used as complementary, in conjunction with traditional echocardiographic examination, to assess global atrial function in patients with MMVD.

Table 1 Comparison of LA TMAD, LA Strain, LA Emptying, LA Ejection Fraction and Mitral A' wave in accordance with groups.

	Groups					
	Control	B1	B2	C	D	*p*
(n)	60	42	17	31	5	
LA TMAD **(mm)**						
Global AP 2,4	5.45 (1.36)^{A}	5.87 (1.61)^{A}	7.43 (1.89)^{B}	7.14 (2.17)^{B}	7.14 (1.82)^{AB}	<0,001
Systolic AP 2,4	2.76 (1.02)^{AD}	3.80 (1.31)^{ABC}	4.02 (1.92)^{B}	3.42 (2.03)^{C}	1.62 (2.26)^{D}	<0,001
LA TMAD **(mm/m2)**						
Global AP 2,4	13.25 (5.68)^{A}	13.33 (4.80)^{A}	21.03 (8.94)^{B}	22.16 (9.94)^{B}	21.09 (11.73)^{B}	<0,001
Systolic AP 2,4	7.00 (3.63)^{A}	9.24 (4.23)^{BC}	10.39 (8.83)^{B}	8.97 (8.66)^{BC}	4.65 (11.07)^{AC}	0,002
LA Strain **(%)**	33.61 (6.72)^{A}	31.17 (8.98)^{A}	29.96 (7.75)^{A}	23.41 (7.37)^{B}	17.42 (8.41)^{B}	<0,001
LA Empt Fr **(%)**	62.84 (9.15)^{A}	65.19 (12.73)^{A}	65.99 (12.59)^{A}	55.05 (8.52)^{B}	44.10 (21.96)^{B}	<0,001
LA Ej Fr **(%)**	38.61 (8.59)^{A}	49.19 (10.38)^{B}	49.78 (12.59)^{B}	31.62 (13.69)^{C}	17.59 (20.90)^{C}	<0,001
TDI A' **(cm/s)**	6.60 (1.79)^{A}	9.45 (2.10)^{B}	9.53 (2.69)^{B}	8.68 (4.47)^{B}	8.12 (3.48)^{AB}	<0,001

(n), number of animals in quartile; LA, left atrium; TMAD, tissue mitral annular displacement; AP4, apical 4-chamber; AP2, apical 2-chamber; AP 2,4, average of 2 and 4 chamber; kg, kilograms; LAL, left atrium length; Emp Fr, Emptying fraction; Ej Fr, Ejection fraction, TDI A', Peak velocity of diastolic mitral annular motion as determined by tissue Doppler imaging.
Data are expressed as means (standard deviation) or medians (interquartile range) depending on the parameter attaining a normal distribution or not on the Shapiro-Wilk normality test. Values with different superscripted letters indicate statistically significant differences between groups.

Abstract C23: Electrocardiography in Dogs with Mammary Tumors

Gabriela Bahr Arias; Mariângela Kilpp Oliveira; Josiana Schnitzer; Giovana Di Santis; Maria Isabel Martins; Fabio Gava
Londrina State University

Malignant tumors can cause paraneoplastic syndrome and consequent cardiovascular disorders. The objective of this study was to evaluate the electrocardiography (ECG) in bitches with mammary tumors in order to detect significant changes in cardiac rhythm, ECG waves, segments or cardiac axis. Twenty-nine female dogs were used in this study (dogs with mitral valve disease were excluded), allocated in three groups: G1: control group (n=10), G2: benign neoplasia group (n=6) and G3: malignant neoplasia group (n=13), being considered malignant or not after histopathological analysis. Five minutes of computerized ECG record was used and statistical analysis was performed by analysis of variance followed by Tukey's test. This study was approved by the local ethics committee. The most prevalent type of neoplasia in G2 was the benign mixed tumor (83%), followed by adenoma (17%) whereas in G3: carcinoma in a mixed tumor (61%); papillar carcinoma (23%); tubular carcinoma (8%) and carcinosarcoma (8%). Regarding cardiac rhythm, it was found sinus arrhythmia (SA) and normal sinus rhythm (NS): G1: 50% SA and 50% NS; G2: 67% SA and 33% NS; G3: 54% SA and 46% NS. No ventricular or atrial arrhythmias were detected. For other parameters in G1, G2 and G3, respectively (mean ± Std error): FC (bpm): 110±9.2, 120±8.5, 124±7,5; P (ms): 48±1.6, 51±1.8, 50±1.2; P (mV) 0.19±0.02, 0,2±0.02, 0,19±0.02; PR (ms): 94±4.3, 93±5.5, 89±3.9; QRS (ms): 56±1.54, 60±4, 62±1.2; R (mV): 1.1±0.06, 1.2±0.24, 0.9±0.13; QT (ms): 203±9.4; 204±7.9; 197±13.8; and cardiac axis (°): 66±6.2, 61±7.9, 70±7.5. There were no significant differences for all these parameters and also for ST interval and T wave morphology, concluding that ECG has no changes in benign or malignant mammary tumors in dogs.

Abstract C24: Comparison of Heart Rate Obtained using AliveCor Electrocardiography to 24-Hour Holter in Canine Atrial Fibrillation

Sumana Prabhakar; Tamilselvam Gunasekaran; Robert Sanders
Michigan State University

Canine atrial fibrillation (AF) is commonly managed through pharmacological heart rate (HR) control. Accurate HR assessment is required to determine the need for HR control therapy and to monitor drug efficacy. Heart rate obtained using in-hospital electrocardiography (ECG) has been shown to be inaccurate in dogs likely due to heightened sympathetic tone. Twenty-four-hour continuous ambulatory ECG monitoring (Holter) provides at-home HR data and valuable prognostic information. The mean 24-hour Holter HR of less than 125 bpm has been shown to be associated with survival. However, Holter monitoring is not always feasible in a primary care setting due to the need for additional equipment. The AliveCor smartphone ECG has been shown to produce accurate instantaneous HR in dogs with sinus rhythm and presents a possible alternative to Holter monitoring for at home HR assessment. The objectives of this study are 1. To determine the accuracy of the AliveCor ECG to estimate HR in AF by comparing HR obtained using AliveCor ECG to the corresponding Holter ECG. 2. To compare 24-hour Holter mean HR to the mean HR obtained using multiple one and five-minute AliveCor ECG recordings in dogs with AF. Dogs with AF underwent Holter monitoring. Simultaneously, an AliveCor electrode was adhered to the left side of each dog's chest (AliveCorattach) and an area was prepared on the right side of the dogs' chest for the owners to place an additional AliveCor electrode (AliveCorowner). The owners were instructed to record four five-minute resting ECGs on their personal smartphones (AliveCor Vet smart phone application) using both AliveCor electrodes. The accuracy of AliveCor ECG was assessed by comparing the HR obtained from each AliveCor electrode to the HR calculated from the simultaneous Holter recording. AliveCor ECG obtained HR measurements for one and five minute recording durations were compared to 24-hour mean HR using modified Bland Altman limits of agreement (LOA) plots. A total of 11 dogs were enrolled. One dog was excluded due to poor quality of the AliveCor ECG tracings and one dog was excluded due to an inability to match the timing of Holter and AliveCor ECGs. Seven out of 9 dogs were receiving drug therapy to control HR. All 9/9 dogs had at least a single one minute recording and 8/9 dogs had at least a single five minute recording for both AliveCorattach and AliveCorowner. A complete data set was obtained from 3/9 dogs. As only 3/9 dogs had complete data sets, we were only able to compare a single, one minute and five minute AliveCor HR to the 24-hour Holter mean HR. The median difference between the HR provided by all the available five minute AliveCorattach and AliveCorowner to the simultaneous Holter ECG was 4 beats per minute (bpm) (IQR = 1.4 to 14 bpm) and 7 bpm (IQR = 4.3 to 12 bpm) respectively. The median difference between the HR provided by one minute AliveCorattach and AliveCorowner to the simultaneous Holter ECG was –0.5 beats per minute (IQR = –5 to 8.5 bpm) and –4 bpm (IQR = –10 to –0.5 bpm) respectively. The median HR difference between AliveCor HR and simultaneous Holter ECG HR was not significantly different between AliveCorattach and AliveCorowner for either recording durations. No significant correlation was found between the recording duration to the median difference between AliveCor HR and simultaneous Holter ECG HR. Comparison of a single five minute AliveCorattach HR and AliveCorowner HR to 24-hour Holter mean HR using the modified Bland-Altman LOA plots revealed a mean difference of 1 bpm (LOA = –30 to 29 bpm) and 2 bpm (LOA = –36 to 39 bpm) respectively. Comparison of a single one minute AliveCorattach HR and AliveCorowner HR to 24-hour Holter mean HR using the modified Bland-Altman LOA plots revealed a mean difference of 9.5 bpm (LOA = –51 to 32 bpm) and 4.5 bpm (LOA = –41 to 50 bpm) respectively. The median HR difference between the 24-hour Holter mean HR and HR obtained using AliveCor ECG was not significantly different between one and five minute AliveCor recordings. All the HR values provided by five minute AliveCorowner recordings were within 20% range from 24-hour mean HR, while 87.5% of the HR values were within similar range for AliveCorattach recordings. One minute AliveCorattach recording provided 88% of the HR values within 20% range from 24-hour mean HR as compared to 77% HR values within similar range for AliveCorowner recordings. This study shows that the AliveCor ECG can provide reasonable estimates of simultaneous HR in dogs with AF. The wide LOA between AliveCor ECG HR and the 24-hour mean HR, makes a single, one or five minute AliveCor ECG recordings, unreliable for HR assessment in an individual dog with AF. In this study, due to poor owner compliance we were not able to compare more than a single, one or five minute AliveCor ECG recordings to 24-hour Holter mean HR.

Abstract C27: Transvenous Thromboembolic Coiling of Feline Patent Ductus Arteriosus

Brittany Stewart; Christopher Stauthammer
University of Minnesota

Minimally invasive per-catheter occlusion of Patent Ductus Arteriosus (PDA) in feline patients is limited due to femoral vascular diameter. Surgical ligation has historically been the treatment of choice. The purpose of this study is to assess the safety and efficacy of transvenous thromboembolic coiling in cats as a valid treatment option for PDA occlusion. Client-owned cats with left to right shunting PDAs underwent minimally invasive occlusion via a transvenous approach using a Flipper detachable coil (Cook Medical LLC, Bloomington, IN). Minimal ductal diameter and ampulla diameter measurements were taken via transthoracic echocardiography and angiography. Ductal occlusion was evaluated on transthoracic echocardiography performed <24 hours post procedure and during long term follow-up (recommended at 3 months with additional recheck frequency decided based on residual flow). Residual ductal flow was graded based on color Doppler. A total of 7 cats, median weight 3.5 kg (1.9–4.8) and age 259 days (105–668) were enrolled. PDA morphology types 2a (3), 2b (3), and 3 (1) were represented. Based on echocardiography, the ductal ampulla diameter was 4.1 mm (1.2–6.5) and the minimal ductal diameter was 2.9 mm (0.7–5.8). Based on angiography, the ductal ampulla diameter was 5.9 mm (4–8) and the minimal ductal diameter was 2.3 mm (1.2–5.5). Flipper coil diameters ranged from 5 to 8 mm varying from 5 to 12 cm in extended length. Coil placement was successful in 6 of the 7 cases (86%). One patient (type 3 morphology) was unable to have the coil (8 mm/12 cm) safely deployed based on size of ampulla (8 mm) and the procedure was converted to surgical ligation. Average fluoroscopy time was 32.5 minutes (16.9–57.6). There were zero incidents of device embolization or perioperative death. All patients were provided pain medication and antibiotics post procedure. Residual flow was noted on the <24-hour post echocardiogram in three of the six successful cases (50%) with all considered moderate in severity. Follow up echocardiogram (mean 235 days post procedure) revealed residual flow in two cases; one trivial and one moderate in severity. Transvenous thromboembolic coiling in cats with type 2a and 2b morphology appears to be a safe and effective approach to PDA occlusion regardless of patient's size. In cats with larger PDA measurements and/or type 3 morphology, standard surgical ligation is still likely the best approach though further investigation is warranted. Especially with measurement inconsistencies between transthoracic echocardiography and angiography, the best occlusion approach may need to be made intra-operation.

Abstract C28: Comparison of 3D Printing Segmentation Software Programs for Veterinary Cardiology Applications

Kaitlin M. Abbott-Johnson[1]; Susanne Stieger-Vanegas[2]
[1]QTest Labs; [2]Oregon State University

Three-dimensional (3D) printing is an emerging technology with a wide range of medical applications, but has only been utilized in veterinary cardiology for a few select purposes. As cardiac gated CT and cardiac MRI are more commonly performed in veterinary medicine, it is important to consider the potential ways to utilize these new imaging studies, such as for education and procedural planning purposes. Complex congenital abnormalities can vary between individual cases, and understanding disease variations and especially spatial anatomy can be challenging. Producing digital and printed 3D models can allow deeper understanding of cardiac anatomic pathology. The primary goal of this study was to describe the methods and quantify the time investment necessary for creating 3D cardiac models from diagnostic imaging. The secondary goal was to compare free open source (3D Slicer, NIH, Bethesda, MD) and proprietary 3D modeling (Mimics, Materialise, Belgium) software to create 3D digital images of the heart. The ECG-gated CT angiography studies of six New World Camelids with confirmed complex congenital abnormalities including truncus arteriosus, tetralogy of Fallot, double outlet right ventricle, coronary anomalies, and ventricular septal defects were selected for this study. 3D bloodpool and myocardial models of each heart were created in each software program. Evaluation of the software included the time needed to create each model type and documentation of the most useful tools and methods used within each program. All 3D cardiac models created had acceptable quality for printing and identification of complex cardiac abnormalities (Figure 1). Creation of 3D bloodpool models was easier and took significantly less time than creating myocardial models ($p=0.021$) (Table 1). Myocardial models took approximately 1.2 hours longer to complete in 3D Slicer than in Mimics, but this difference was not statistically significant ($p=0.0711$). A limitation of this study is the lack of objective methods to evaluate the accuracy of the final 3D cardiac models. Reported linear measurements utilizing 3D printed bone models are not well suited to evaluate cardiac models, though a recent veterinary orthopedic study showed that 3D Slicer had higher accuracy than other open source programs and the quality of 3D models created was comparable to those created with Mimics. We conclude that when proprietary segmentation software is cost prohibitive, open-source software is a practical way for veterinarians to create a 3D cardiac model, though more time investment may be needed for segmentation.

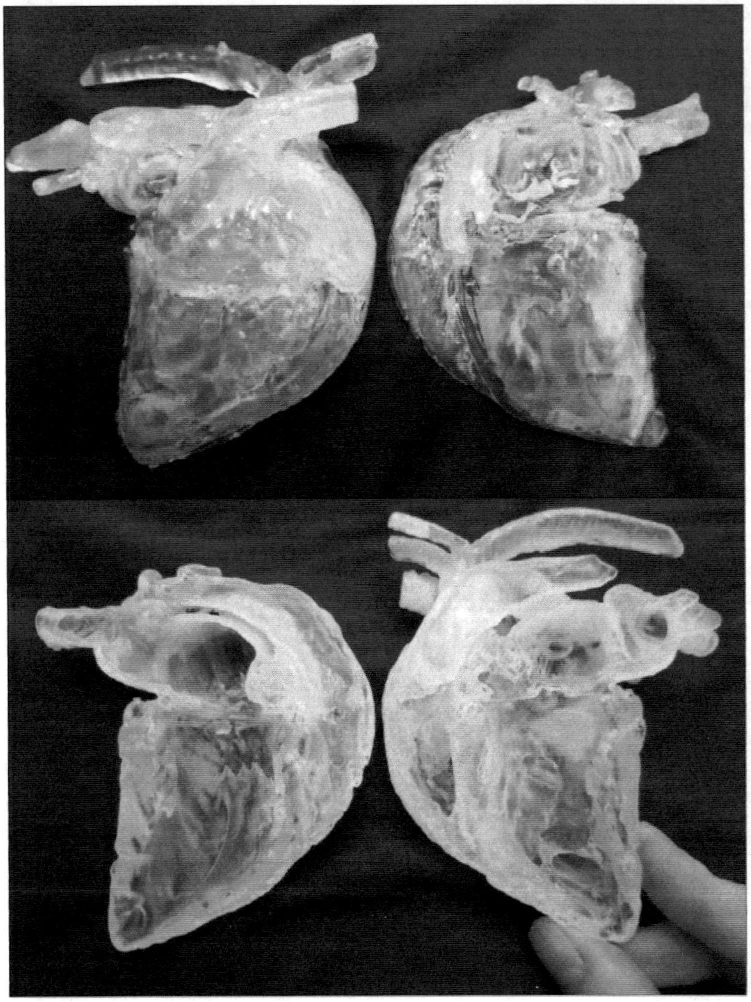

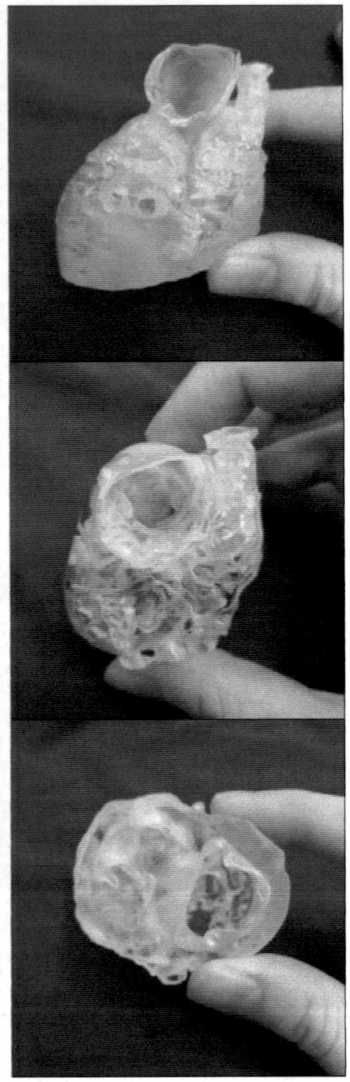

Figure 1. Left and right sides of an alpaca cria myocardial model, with a small muscular VSD. (above) Heart base of an alpaca cria with truncus arteriosus. (right) Both models segmented with 3D Slicer software, printed at 60% size.

Table 1

	3D Slicer			Mimics		
Case	Time (hrs) bloodpool segmentation	Time (hrs) myocardial model	Total Time	Time (hrs) bloodpool segmentation	Time (hrs) myocardial model	Total Time
1	1.5	4.25	5.75	1.5	2.25	3.75
2	1.5	4.5	6	1	1.75	2.75
3	0.5	3.75	4.25	2	1.5	3.5
4	2	2.75	4.75	1.25	2.25	3.5
5	1	0.75	1.75	0.5	2	2.5
6	1.5	2	3.5	2.25	1.5	3.75
Mean:	1.33	3.00	4.33	1.42	1.88	3.29

Abstract C29: Retrospective Investigation of Diet and Dilated Cardiomyopathy (DCM) in Dogs

Kim J. Freid; Lisa Freeman; John Rush; Suzanne Cunningham; Megan Davis; Emily Karlin; Vicky Yang
Cummings School of Veterinary Medicine, Tufts University

The FDA has been investigating an apparent association between certain diets and DCM in dogs. The objectives of this study were to 1) compare diets between dogs with DCM and myxomatous mitral valve disease (MMVD) to determine prevalence of non-traditional diets and 2) retrospectively review DCM cases to determine signalment, diets, echocardiographic changes over time, and outcome. Objective 1: Diet histories of dogs with DCM and those with MMVD examined between May and September, 2018 were compared. Objective 2: Medical records of all dogs diagnosed with DCM between January 1, 2014 and September 30, 2018 were reviewed. Dogs were grouped into traditional or non-traditional diet categories based on ingredients and whether the manufacturer met World Small Animal Veterinary Association recommendations. Objective 1: 30/33 (91%) of dogs with DCM were eating non-traditional commercial diets compared to 34/70 (49%) of dogs with MMVD ($p<0.001$). Objective 2: 75 dogs with DCM were identified retrospectively: 9 cases in 2014, 8 in 2015, 21 in 2016, 16 in 2017, and 21 in the first 9 months of 2018. Median age was 7.2 yrs (range 0.4–13.6 yrs) and median weight was 37.7 kg (range, 10.0–75.7 kg). The most common breeds were Doberman Pinscher (n=21) and Great Dane (n=17), but a variety of breeds were represented. 76.4% of dogs were eating non-traditional diets. Taurine was measured in 20 dogs: 3/14 had low plasma and 1/18 had low whole blood taurine concentrations. 48 dogs had follow-up echocardiography. Median survival for all dogs was 293 days (range, 1–1901 days).

Abstract C30: Circulating Cell Free DNA in Cats with Hypertrophic Cardiomyopathy and Cardiogenic Arterial Thromboembolism

Arianne F. Fabella[1]; Ronald Li[1]; Stephanie Istvan[2]; Yu Ueda[1]; Joshua Stern[1]
[1]University of California, Davis; [2]Veterinary Specialist Hospital, San Diego

Cardiogenic arterial thromboembolism (CATE) is a life-threatening complication of feline hypertrophic cardiomyopathy (HCM). Clinicians have limited diagnostic tools to identify cats at risk of this devastating outcome. Novel biomarkers may aid in the early detection cats at high risk of CATE. Circulating cell free DNA (cfDNA) has been demonstrated in humans and dogs to play a key role in thrombosis by augmenting the intrinsic pathway and impairing fibrinolysis. Using immunofluorescence, cfDNA in the form of neutrophil extracellular traps has recently been identified as a structural element of feline distal aortic thrombi. Thus, cfDNA is a potential biomarker for assessing thrombotic risk in cats with HCM. This study aimed to characterize and compare levels of circulating cfDNA in cats with CATE (n=4), in cats with HCM (n=13) and in clinically healthy cats (n=18) without HCM. Citrated whole blood was collected following cardiovascular and echocardiographic evaluations. The concentration of isolated cfDNA from plasma was measured by spectrophotometry, followed by agarose gel electrophoresis to determine the size of cfDNA fragments. The mean concentration of plasma cfDNA in cats with CATE was 19.58±20.3 ng/ul compared to those in healthy cats (10.09±4.3 ng/ul) and HCM cats (9.22±5.44 ng/ul) (p=0.06). Gel electrophoresis identified cats with CATE had longer cfDNA fragments between 150 to 400 base pairs compared to healthy cats and HCM cats with fragments of <50 base pairs. Results of this preliminary study suggest that cfDNA concentrations and cfDNA fragment size may identify HCM cats at risk of thromboembolic complications and continued studies are warranted.

Abstract C31: Cellular Localisation of Key Proteins Controlling Pro-Fibrotic Pathways in Cats with Hypertrophic Cardiomyopathy

Wan-Ching Cheng[1]; Lois Wilkie[1]; Melanie Dobromylskyj[2]; Virginia Luis Fuentes[1]; David Connolly[1]
[1]Royal Veterinary College, University of London; [2]Finn Pathologists

Myocardial fibrosis is one of the major pathological changes in feline HCM. Besides the quantity of collagen deposited, the degree of collagen crosslinking in a collagen bundle also qualitatively affects its structural characteristics and hence the quality of the extracellular matrix (ECM). ECM composed of predominately highly crosslinked collagen will have reduced compliance and pre-dispose to diastolic dysfunction. Lumican is a small leucine rich protein involved in collagen crosslinking. Recently it was shown that human patients with hypertrophic obstructive cardiomyopathy had increased myocardial lumican, which correlated with cardiac fibrosis determined by late gadolinium enhancement on cardiac magnetic resonance imaging. Furthermore, lumican can induce transcription of LOX (a collagen cross-linking enzyme), TGF-β (a potent pro-fibrotic mediator) and collagen protein in rodent cardiac fibroblasts. We have shown upregulation of Lumican and downstream proteins including LOX and TGF-β isoforms by Western Blot in the myocardium from cats with HCM. The aim of this present work is to quantify and describe the cellular location of these proteins within the myocardium of HCM cats and explore the SMAD-dependent TGF-β signalling in feline HCM. The middle section of the left ventricle from 6 normal control cats and 6 HCM cats presented for necropsy were used. All the hearts underwent histopathological examination. No difference in age and gender was detected between the control and HCM group. Immunohistochemistry was used to localise expression of lumican, LOX, LOXL2, TGF-β1, TGF-β2, phosphorylated SMAD2. Using imageJ, the area immunostained for lumican, LOX and LOXL2 (%) was quantified. In addition, activation of fibroblasts was determined by quantifying the area (excluding blood vessels) immunostained for α-SMA (%). The number of interstitial cells with nuclear staining of p-SMAD2 was quantified as an indicator for activation of TGF-β regulated SMAD signalling. Parametric or non-parametric statistical tests were used where appropriate. Cut-off for statistical significance was set at 0.05. By immunohistochemistry, lumican localised to the ECM, the cardiomyocytes, and around some infiltrative leucocytes. LOX and LOXL2 localised to cardiomyocytes. LOXL2 also localised to some interstitial cells. Lumican, LOX and LOXL2 (%) were significantly increased in the HCM group (lumican median 36.6% [IQR 26.2–55.7] (n=4) versus 1.1% [IQR 0.8–2.2] (n=5); p=0.016, LOX median 28.2% [IQR 21.3–39.6] (n=6) versus 1.7% [IQR 1.3–10.1] (n=5); p=0.004, LOXL2 median 14.1% [IQR 6.9–15.3] (n=5) versus 2.7% [IQR 2.1–4.7] (n=5); p=0.016) compared to the control group. TGF-β1 localised to the tunica media of vessels in both groups with the HCM group showing increased immunolabelling. Marked difference existed in the expression of TGF-β2 between the control and HCM group where the controls had minimal TGF-β2 expression whereas the HCM cats showed extensive immunolabelling in cardiomyocytes and the tunica media in some vessels. The number of interstitial cells with nuclear translocation of p-SMAD2, and the area occupied by α-SMA positive cells (%) were higher in the HCM cats (Interstitial cells Nuclear p-SMAD2 mean 51.5 [IQR 33.0–95.5], n=4 versus 12.0 [range 6.0–28.5], n=5; p=0.034) (α-SMA (%) mean 3.42 [IQR 0.38–5.17], n=6 versus 0.53 [IQR 0.18–0.54], n=5; p=0.054) compared to the control cats. In conclusion, lumican was elevated in feline HCM and might promote collagen cross-linking through increased expression of LOX and LOXL2. The localisation of lumican across cell populations in the diseased heart has not been widely explored but our results are in line with the limited published data. Activation of myofibroblasts via TGF-β2 regulated SMAD signalling could increase collagen production. Together this could result in a less compliant ECM, increased mechanical stiffness and reduced diastolic performance of the left ventricle.

Abstract C32: A Dose-Ranging Study of Coenzyme Q10 in Dogs with Myxomatous Mitral Valve Disease

Natalia Druzhaeva; Aleksandra Domanjko Petrič; Gabrijela Tavčar-Kalcher; Janja Babič; Alenka Nemec Svete
Veterinary Faculty, University of Ljubljana

Coenzyme Q10 (CoQ10) supplementation improves left ventricular systolic function in advanced heart failure human patients. According to human studies, at least a 3-fold rise of basal plasma CoQ10 concentration is needed to achieve the biological effects of CoQ10 supplementation. Our goal was to determine the dose sufficient to achieve a 3-fold increase in CoQ10 plasma concentration in dogs with congestive heart failure (CHF) due to myxomatous mitral valve disease (MMVD). In a randomized, double-blinded, placebo-controlled study 19 client-owned dogs with MMVD and CHF were allocated to 3 groups, receiving water-soluble CoQ10 in a daily dose either 100 mg (6 dogs), 200 mg (6 dogs) or placebo (7 dogs), divided twice daily, for two weeks as an addition to cardiac therapy. Twelve healthy dogs not receiving CoQ10 were also included. Blood was drawn five times in MMVD patients (basal, 4 hours, one and two weeks after the beginning of supplementation and one week after the last dose) and once in healthy dogs. Liquid chromatography-tandem mass spectrometry was used to measure plasma CoQ10 concentration. We found no significant difference in basal CoQ10 concentration among groups of patients and healthy dogs (Kruskal-Wallis test). After two weeks, CoQ10 concentration was significantly higher in both CoQ10 groups compared to their basal concentrations (Wilcoxon test; $p<0.05$) and compared to the placebo group (Kruskal-Wallis test). A dose of 100 mg resulted in 1.7–4.7-fold and a dose of 200 mg in a 3.2–6.8-fold increase of plasma CoQ10 concentration. One week after the last dose, CoQ10 concentration decreased in both groups; however, in the 200 mg group, it remained significantly higher in comparison to basal concentration. We found no plasma CoQ10 deficiency in our cardiac patients. A daily dose of 200 mg resulted in at least a 3-fold increase of plasma CoQ10 concentration in all patients and might be used in supplementation studies in canine CHF patients.

Abstract C33: Genetic Polymorphisms Impact Platelet Inhibition by Clopidogrel in Cats with Hypertrophic Cardiomyopathy

Yu Ueda; Maureen Oldach[1]; Ronald Li[1]; Eric Ontiveros[1]; Samantha Fousse[1]; Karen Vernau[1]; Michael Court[2]; Joshua Stern[1]
[1]University of California, Davis; [2]College of Veterinary Medicine, Washington State University

Clopidogrel inhibits adenosine diphosphate (ADP)-induced platelet activation and is frequently used in cats with hypertrophic cardiomyopathy (HCM) to prevent thromboembolic complications. However, response to clopidogrel has significant interpatient variability in cats. Previous studies reported multiple nonsynonymous single nucleotide polymorphisms (SNPs) within ADP receptor (P2RY1, P2RY12) and cytochrome P450 (CYP) 2C genes. The objective of this study was to determine the impact of the genetic polymorphisms on clopidogrel response in cats with HCM. 49 cats were enrolled after echocardiographic evaluation, baseline blood work, and Doppler blood pressure measurement. Blood samples were obtained before and after 10 to 14 days of clopidogrel therapy, 18.75 mg orally every 24 hours, to assess the degree of platelet inhibition using whole blood platelet aggregometry (Multiplate® analyzer) and flow cytometry. The post-clopidogrel blood sample was used for quantification of clopidogrel and its active metabolite. One nonsynonymous SNP in both P2RY1 and P2RY12 genes and two in the CYP2C gene were identified. Whole blood platelet aggregometry revealed significant resistance to clopidogrel-induced platelet inhibition in cats with the P2RY1 variant (P2RY1:A236G) ($p=0.012$ for area under the curve [AUC], $p=0.0084$ for aggregation, $p=0.037$ for velocity). The associations for AUC ($p=0.047$) and aggregation ($p=0.034$) remained significant after false discovery rate adjustment. No other pharmacogenetic effects were identified for the tested polymorphisms that withstood false discovery rate adjustment. This study demonstrated that a genetic polymorphism in the P2RY1 gene significantly altered response to clopidogrel therapy.

Abstract C34: Establishing Normal Reference Intervals for Radiographic, Echocardiographic, and NT-proBNP Values in Apparently Healthy Kittens

Karen M. Vernau[1]; Catherine Gunther-Harrington[2]; **Ashley Sharpe**[3]; Yu Ueda[2]; Ashley Walker[4]; Elizabeth Montgomery[3]; Jennifer Surmick[3]; Nicole Fernandez[3]; Eric Ontiveros[2]; Joshua Stern[2]
[1]University of California, Davis; [2]Department of Medicine and Epidemiology, School of Veterinary Medicine, University of California, Davis; [3]Veterinary Medical Teaching Hospital, School of Veterinary Medicine, University of California, Davis; [4]College of Veterinary Medicine, North Carolina State University

Subjective assessment of heart size in kittens is challenging and there is a need for established reference ranges to prevent misdiagnoses of cardiomegaly in this patient population. The purpose of this study was to generate reference intervals for echocardiographic and radiographic quantification of cardiac size in apparently healthy kittens. Eighty-eight kittens aged 6–16 weeks were included in the study. Radiographic and echocardiographic evaluations were performed without sedation. Echocardiographic images and thoracic radiographs were measured to establish reference intervals for multiple echocardiographic variables, vertebral heart score (VHS) and cardiac thoracic ratio. N-terminal pro-brain type natriuretic peptide (NT-proBNP) was also measured. Statistical correlations between the echocardiographic parameters and age, body weight and sex were all evaluated, and reference ranges were generated. Multiple moderate to strong correlations between body weight and age with various echocardiographic parameters were noted and allometric scaling was performed for body weight. Kittens had a VHS of 9.4±0.73 which is significantly larger ($p<0.0001$) and exceeds the previously proposed cutoff of 7.8 in adult cats. Tricuspid regurgitation was a common finding and was present in 37.5% (33/88) of kittens. Measured NT-proBNP levels were comparable to healthy adult cats with a median of 31 pmol/L (range 24–129 pmol/L). Specific reference ranges are reported in part as Table 1 below and may be utilized when evaluating heart size in this population.

Table 1

Parameter	Median	Min-Max	95% reference interval (90% CI)
Left sided heart measurements			
LVIDd$^{2DE, Sx}$ (cm)	1.12	0.82-1.43	0.83 (0.78-0.87) to 1.40 (1.37-1.45)
LVIDs$^{2DE, Lx}$ (cm)	0.60	0.26-0.92	0.34 (0.29-0.38) to 0.88 (0.83-0.91)
IVSd$^{2DE, Lx}$ (cm)	0.34	0.21-0.44	0.23 (0.22- 0.24) to 0.44 (0.42-0.45)
IVSs$^{2DE, Lx}$ (cm)	0.52	0.31-0.77	0.33 (0.31-0.36) to .71 (0.68-0.74)
LVFWd$^{2DE, Lx}$ (cm)	0.32	0.21-0.48	0.20 (0.19-0.22) to 0.45 (0.43-0.47)
LVFWs$^{2DE, Lx}$ (cm)	0.50	0.33-0.65	0.34 (0.32-0.37) to 0.65 (0.63-0.68)
LVIDd$^{MM, Sx}$ (cm)	1.19	0.75-1.62	0.86 (0.81-0.91) to 1.53 (1.48-1.57)
LVIDs$^{MM, Sx}$ (cm)	0.55	0.17-0.94	0.28 (0.23-0.32) to 0.84 (0.80-0.89)
IVSd$^{MM, Sx}$ (cm)	0.33	0.20-0.52	0.21 (0.20-0.23) to 0.45 (0.43-0.46)
IVSs$^{MM, Sx}$ (cm)	0.54	0.30-0.71	0.37 (0.34-0.39) to 0.72 (0.69-0.74)
LVFWd$^{MM, Sx}$ (cm)	0.33	0.21-0.48	0.20 (0.19-0.23) to 0.45 (0.43-0.46)
LVFWs$^{MM, Sx}$ (cm)	0.55	0.39-0.80	0.38 (0.35-0.40) to 0.74 (0.70-0.77)
LV FS$^{MM, Sx}$ (%)	53.8	35.8-80.0	36.6 (34.0-39.3) to 68.6 (65.9-71.2)
LAD$^{2DE, Lx}$ (cm)	0.90	0.64-1.25	0.65 (0.61-0.69) to 1.14 (1.10-1.18)
LA$^{2DE, Sx}$ (cm)	0.85	0.60-1.11	0.63 (0.60-0.66) to 1.06 (1.03-1.10)
Ao$^{2DE, Sx}$ (cm)	0.63	0.46-0.78	0.47 (0.45-0.50) to 0.78 (0.75-0.80)
LA/Ao$^{2DE, Sx}$	1.37	1.00-1.64	1.12 (1.08-1.59) to 1.63 (1.59-1.66)
Right sided heart measurements			
RVIDd$^{2DE, Lx}$ (cm)	0.51	0.22-0.81	0.26 (0.23-0.30) to 0.76 (0.72-0.80)
RVIDs$^{2DE, Lx}$ (cm)	0.28	0.13-0.52	0.12 (0.10-0.14) to 0.44 (0.41-0.47)
RVFW$^{2DE, Lx}$ (cm)	0.18	0.12-0.24	0.12 (0.11-0.13) to 0.23 (0.22-0.24)
RAD$^{2DE, Lx}$ (cm)	1.03	0.68-1.32	0.75 (0.71-0.80) to 1.36 (1.31-1.40)
RV FS$^{2DE, Lx}$ (%)	42.7	26.4-66.6	23.4 (20.8-26.1) to 62.8 (59.6-65.7)
Radiographic measurements			
VHS	9.50	7.74-10.80	7.96 (7.74-8.21) to 10.92 (10.70-11.11)
Cardiac thoracic ratio (%)	67.2	49.2-85.0	54.4 (52.5-56.5) to 79.8 (77.6-81.9)

Abstract C35: Clinical Usage and Tolerability of Intravenous Pimobendan in Dogs Suffering from Acute Congestive Heart Failure

Michael Aherne; John Hertzer; Darcy Adin
College of Veterinary Medicine, University of Florida

Oral pimobendan is used to treat acute and chronic canine congestive heart failure (CHF). Intravenous pimobendan is not FDA-approved and is only available in the U.S. with special import license. This retrospective study aimed to evaluate clinical usage and tolerability associated with IV pimobendan administration in dogs with acute CHF. Medical records from 7/2019–1/2020 were reviewed, identifying dogs that received IV pimobendan to treat acute CHF. Controls comprised of age- and size-matched dogs that received only oral pimobendan for acute CHF. Clinical and diagnostic variables, medication dosages, and 30-day mortality were compared between groups. Data are reported as median(range). Seven dogs received 1(1–6) dose of IV pimobendan at 0.15(0.15–0.16) mg/kg followed by oral pimobendan at 0.38(0.35–0.65) mg/kg q 8(8–12) hours. Controls received 0.34(0.26–0.42) mg/kg oral pimobendan q 12(8–12) hours. No adverse effects were noted. For study vs. control groups respectively, hospitalization time was 47(5–107) vs. 20(8–75) hours, presenting RR was 60(28–100) vs. 60(40–88) breaths/min, cumulative furosemide dose was 9(2.11–30.24) vs. 6.42(2.1–9.98) mg/kg, time to RR <40 breaths/min was 11(1–17) vs. 1(1–3) hours, presenting pulmonary edema score was 7(6–13) vs. 8(4–16). No variables significantly differed between groups. Of the 7 IV pimobendan patients, 3 survived to discharge, 1 died from cardiac arrest and 3 were euthanized due to lack of clinical improvement. Of the 7 control dogs, 5 survived to discharge, and 2 were euthanized. Thirty-day mortality did not differ between groups. In conclusion, IV pimobendan appears to be tolerated without significant adverse effects in dogs with acute CHF.

Abstract C36: Evaluation of Cardiac Troponin I in Neonate Dogs Born through Eutocia

Maria Lucia Lourenço; Keylla Helena Nobre Pacifico Pereira; Luiz Eduardo Santos Correia; Viviane Hibaru; Viviane Maria Codognoto; Fabiana Souza; João Carlos Ferreira; Simone Chiacchio
School of Veterinary Medicine and Animal Science, São Paulo State University (Unesp), Botucatu, São Paulo, Brazil

Cardiac Troponin I is a specific biomarker for cardiomyocytic injury and is used to identify myocardial damage through asphyxia in human newborns. However, there are no similar studies pertaining to neonate dogs. This study aimed at evaluating the serum levels of cardiac troponin I in dogs and establish possible correlations with modified Apgar score, heart rate and respiratory rate, blood gas levels, blood sugar levels and weight. A total of 15 neonate dogs of several breeds (Pug, Shih-tzu and Spitz), all delivered vaginally, were evaluated immediately after birth and after 60 minutes. The reference values adopted for troponin I were <0.006 to 0.05 ng/mL. The mean weight of the puppies was 137.1±38.5 grams and the mean values for the parameters at the time of birth were: Apgar score (9±0.9; 10±0), heart rate (208.6±23; 222.4±8.47 bpm), respiratory rate (39.9±8.7; 38.8±11.17 mpm), pH (7.1±0.1; 7.2±0.03), HCO3 (16±4.19; 21.7±3.44mEq/L), pCO2 (42±8.75; 47±9.03 mm Hg), pO2 (22.3±8.04; 21.6±5.10 mm Hg), sO2 (33.1±16.5; 33.5±9.2 mm Hg), blood sugar (95.9±15.86; 155.7±37.5 mg/dL) and troponin I (0.013±0.007; 0.012±0.006 ng/mL). The neonatal parameters did not diverge significantly between both moments ($p<0.05$), and there were no significant correlations between troponin I and the remaining parameters. The newborns presented mixed acidosis (respiratory PCO2 and metabolic HCO3) due to transitory physiological hypoxemia at the time of birth. However, the results obtained from the Apgar score and the blood gas assay indicate that the dogs evaluated in this study were not born asphyxiated (severe hypoxemia) and presented good vitality and favorable clinical conditions, which means the levels of troponin I were within the reference values for healthy animals eutocia delivered. Further studies evaluating asphyxiated neonates born through dystocia or cesarean sections are needed to assess if there are increases in the levels of troponin I, as happens in humans, and ascertain that the biomarker can be an early indicator of myocardial injury in births with severe hypoxemia. Fapesp grant: 2019/21366-0.

Abstract C37: Comparison of Different Protamine Doses for Heparin Reversal in Dogs Undergoing Mitral Valve Plasty (MVP)

Akane Yoshikawa[1]; Haruhiko Suzuki[2]; Masayuki Enokizono[3]; Dai Nagakubo[4]; Naoko Oshima[5]; Nobuyuki Kanno[5]
[1]Nihon University; [2]Meguro Animal Medical Center; [3]Japan Small Animal Medical Center; [4]The University of Tokyo; [5]Animal Cardiovascular and Thoracic Surgery Center

Protamine is used in patients after cardiopulmonary bypass (CPB) to reverse the anticoagulant effect of heparin. Usually, a dose of protamine equal to 1 mg for each 100 units of heparin is used to reverse the residual heparin activity after CPB. However, overdosing of protamine is known to produce a paradoxical anticoagulant effect in human recently, and may contribute to bleeding. In dogs, protamine administration will induce an anaphylactic response with hypotension as a side effect with high probability, thus determining the proper dose of protamine is necessary. Here, we compared two groups to investigate the optimum dosage of protamine. 25 dogs who underwent MVP under CPB were included. Patients were divided into two groups: High dose (4 mg/kg) protamine group (HD group; n=13) and low dose (2 mg/kg) protamine group (LD group; n=12). Activated clotting time (ACT) was used to monitor coagulation and guide management of anticoagulation control in patients undergoing MVP. We administered 400 IU/kg heparin as an initial dose and increased it if it was difficult to maintain the ACT values above 400 seconds during the CPB. Despite the different doses of heparin administrated, the ACT returned to normal values in both groups (HD group: 108.07 seconds, LD group: 106.75 seconds; p=0.513), and the postoperative total amount of bleeding from chest drainage stayed the same (HD group: 0.43 ml/kg/hour, LD group: 0.64 ml/kg/hour; p=0.437). These results suggest that it might be possible to decrease the dose of protamine.

Abstract C38: Layer-Specific Myocardial Function Assessed by Two-Dimensional Speckle-Tracking Echocardiography in Cats with Restrictive Cardiomyopathy

Ryohei Suzuki; Yunosuke Yuchi; Ayaka Niina; Takahiro Teshima; Hirotaka Matsumoto; Hidekazu Koyama
Nippon Veterinary and Life Science University

Restrictive cardiomyopathy (RCM) is one of the myocardial diseases characterized by restriction of diastolic filling in cats. Layer-specific myocardial strain obtained by two-dimensional speckle-tracking echocardiography is expected to provide additional information for the assessment of myocardial function in cats with cardiomyopathy. However few studies have evaluated the myocardial function in cats with RCM. We aimed to evaluate the layer-specific myocardial function in cats with RCM. Twelve healthy cats and six cats with RCM were enrolled in this study. Layer-specific myocardial function (global, endocardial, and epicardial) were compared using left ventricular (LV) longitudinal and circumferential strain (SL and SC), and right ventricular (RV) SL. RV-SL was measured by tracing only RV free wall or the entire RV wall (6seg). SL were significantly low in cats with RCM compared with controls in the epicardial LV-SL (13.7% vs. 18.3%), all layers of RV-SL6seg (global; 20.5% vs. 31.1%, endocardial; 24.9% vs. 35.1%, epicardial; 17.1% vs. 28.0%). Additionally, endocardial-to-epicardial LV-SL ratio was significantly high in cats with RCM (1.5 vs. 1.3) ($p<0.05$). SC were significantly low in cats with RCM compared with controls in global and endocardial SC (global; 17.6% vs. 22.2%, endocardial; 31.7% vs. 40.6%) ($p<0.05$). Depressed endocardial SC might be a characteristic in cats with RCM, which could be induced by endocardial fibrosis. Additionally, the results of RV-SL might suggest that RCM could induce the entire RV dysfunction caused by the myocardial fibrosis as with LV, or pulmonary hypertension secondary to LV dysfunction.

Abstract C39: Left and Right Myocardial Function in Dogs with Pulmonary Hypertension Secondary to Mitral Valve Disease

Ryohei Suzuki; Yunosuke Yuchi; Ayaka Niina; Takahiro Teshima; Hirotaka Matsumoto; Hidekazu Koyama
Nippon Veterinary and Life Science University

Pulmonary hypertension (PH) is a common complication in dogs with myxomatous mitral valve disease (MMVD). In dogs with MMVD, pulmonary artery pressure increases as a consequence of the increased left atrial pressure (passive condition), leading to pulmonary arterial remodeling and increased pulmonary vascular resistance (reactive condition). We hypothesized that right ventricular (RV) function diminished by PH may affect left ventricular (LV) function in dogs with MMVD and PH, especially reactive conditions. We aimed to detect the differences between mild and moderate-to-severe PH to compare LV and RV function. Eight healthy dogs, eight dogs with mild PH, and seven dogs with moderate-to-severe PH were enrolled in this study. The following indices obtained by two-dimensional speckle-tracking echocardiography were compared among controls, mild, and moderate-to-severe PH groups: LV and RV longitudinal strain (SL). RV-SL were measured by tracing only RV free wall or the entire RV wall including interventricular septum (6seg). LV-SL in mild PH group (24.9%) was higher than controls (16.6%), and lower in moderate-to-severe PH group (18.1%). RV-SL6seg was significantly lower in moderate-to-severe PH group (17.3%) than mild PH group (22.5%). In dogs with mild PH, increased LV-SL and RV-SL may reflect compensative left and right myocardial function. Whereas, the decreased RV-SL might be decompensation by advanced pressure-overload (reactive condition) in dogs with moderate-to-severe PH. The decreased LV-SL might be induced by the decreased RV-SL because of its low perfusion to LV and paradoxical motion of interventricular septum. Myocardial function may detect reactive PH with poor outcome in dogs with MMVD.

Abstract C40: Short-Term Evaluation of Clinical, Radiographic, and Echocardiographic Effects of Patent Ductus Arteriosus Closure in Dogs

Sage H. Hubert; Sonya Gordon; Ashley Saunders
Department of Small Animal Clinical Sciences, School of Veterinary Medicine, Texas A&M University

Introduction: Patent ductus arteriosus (PDA), a common congenital abnormality in dogs, results in cardiomegaly and clinical signs. **Objectives**: To describe pre- and short-term post-operative results in dogs with isolated PDA and subpopulations with concurrent congenital heart disease (CHD) and that received medical therapy only prior to PDA closure. **Methods**: Retrospective study of 57 client-owned dogs with pre- and post-operative imaging within 48 hours of PDA closure. Data collected included signalment, heart and respiratory rates, cardiac medications. Imaging parameters obtained included radiographic vertebral heart size (VHS), vertebral left atrial size (VLAS) and echocardiographic left atrial (LA) and ventricular (LV) size, fractional shortening (FS), transmitral peak E-wave, concurrent CHD, presence of residual flow. **Results**: Females represented 70% (40/57). PDA closure methods included surgical ligation (n=8) or Amplatz canine duct occluder (n=49). 30/57 were uncomplicated. 14/57 had concurrent CHD (subaortic/aortic stenosis (n=9), pulmonic stenosis (n=3), ventricular septal defect (n=1), mitral dysplasia (n=1). 19/57 received cardiac medications (furosemide, ACE inhibitor, pimobendan). Significant mean value reduction occurred for each variable ($p<0.015$): respiratory rate 38 to 32 breaths/min, heart rate 125 to 107 beats/min, VHS 11.7 to 11.3, VLAS 2.3 to 2.0, LV internal diameter normalized to body weight in diastole 2.21 to 1.82 and in systole 1.40 to 1.33, FS 34.0 to 24.0%, LA to aorta ratio short axis 1.49 to 1.26 and long axis 2.44 to 2.09, E-wave 0.98 to 0.63m/s. Residual flow in 1 dog resolved at recheck. **Conclusion**: PDA closure results in significant reductions in clinical and imaging parameters within 48 hours of closure, including uncomplicated cases.

Abstract C41: Prevalence of Maine Coon and Ragdoll Associated MYBPC3 Mutations in Feline Population with Hypertrophic Cardiomyopathy

Kerry O'Donnell[1]; Darcy Adin[2]; Clarke Atkins[1]; Teresa DeFrancesco[1]; Bruce Keene[1]; Sandy Tou[1]; Kathryn Meurs[1]
[1]College of Veterinary Medicine, North Carolina State University; [2]College of Veterinary Medicine, University of Florida

Myosin binding protein C3 (MYBPC3) is a component of the sarcomere A band that functions to bind myosin and titin, thereby facilitating sarcomere contractions. Mutations in this gene have been associated with familial hypertrophic cardiomyopathy (HCM) in both human and feline patients. Two cat breeds, the Maine Coon and the Ragdoll, have mutations in the MYBPC3 gene that have been associated with disease. In the Maine Coon cat, a single base pair variation changes the amino acid produced from Alanine to Proline (A31P) and in the Ragdoll cat, a single base pair variation changes the amino acid from Arginine to Tryptophan (R820W). Here, we analyze the prevalence of these breed-specific mutations in a greater feline population diagnosed with HCM. Samples of DNA were collected from non-Maine Coon and non-Ragdoll cats, diagnosed with HCM, through the North Carolina State University College of Veterinary Medicine, Cardiology section from April 2014 to August 2018. DNA was isolated from blood or buccal swab and endpoint genotyping was performed with previously designed probes and protocols to test for the A31P and R820W mutations. A total of 109 samples were collected from cats of the following breeds: Domestic Shorthair (80), Domestic Longhair (12), Persian (4), Siberian (4), Bengal (4), Domestic Medium Hair (2) Rex (1), Siamese (1) Norwegian (1). All cats were negative (homozygous wildtype) for both mutations. The absence of two breed-specific mutations in this general population sampling of cats with HCM reinforces the allelic heterogeneity of feline HCM, with different breed types being associated with different disease-causing mutations. This is supported by the human model of the disease, in which over 7,000 MYBPC3 mutations have been identified, and 230 have been pathogenically associated with disease. The value of genetic testing for either known MYBPC3 mutation in the overall feline HCM population is likely minimal.

Abstract C42: Role of Echocardiographic Views Adapted for Lung Evaluation in Diagnosis of Pulmonary Edema in Dogs

Marlos Gonçalves Sousa; Bruna Bruler; Amalia Giannico; Marcela Wolf
Federal University of Paraná (UFPR)

The purpose was to test a lung-ultrasound-like technique during echocardiographic examination to help identify the cause of dyspnea in dogs. 75 dogs were enrolled into 5 groups: (1) control, (2) MMVD stage B2 with mitral E peak velocity <130 cm/s and E/IVRT <2,5, (3) MMVD B2 with mitral E peak velocity >130 cm/s and E/IVRT >2,5, (4) cardiogenic edema and (5) pulmonary disease. Chest x-rays were used as gold standard tests. Loop recordings with decreased depth, placement of the heart in the upper 1/3 of the screen and change of focal point to the lung plane were used for posterior evaluation in the following standard echocardiographic views: apical 4 chambers, long axis 4 chambers, short axis at aortic level and short axis at papillary muscle level. Videos were classified as **negative** and **positive** based on the number of B-lines (0, 1 or 2 and 3 or >3). A positive classification was able to differentiate dogs with cardiogenic edema or pulmonary disease from dogs in the other three groups in all four adapted views. The best views for identification of edema or lung disease were right parasternal short axis at papillary muscle level and long axis 4 chambers view. Interobserver tests showed a kappa of 0.88 and 0.86 for right parasternal short axis at papillary muscle level and long axis 4 chambers, respectively. We've concluded that echocardiographic views adapted for lung evaluation, in addition to echocardiographic parameters, may aide in differential diagnosis of dyspnea in dogs.

Abstract C43: Myocardial Transcriptome Profiling in Cats with and without Hypertrophic Cardiomyopathy

Jessica Joshua; Jeff Caswell; Sonja Fonfara
University of Guelph

Feline hypertrophic cardiomyopathy (HCM) is the most common heart disease in cats. Although, genetic mutations have been identified for certain breeds, not all cats with mutations develop HCM, indicating incomplete penetrance. Therefore, other genes and their expression levels may be important in uncovering HCM pathophysiology. However, myocardial mRNA profiles for feline HCM are unknown. We hypothesize that the myocardial transcriptome will differ in cats with HCM compared to healthy cats. To determine mRNA profiles, RNA was isolated from the left ventricle (LV) and left atrium (LA) of 5 healthy cats and 5 cats with HCM. Expression levels were quantified using next generation sequencing and compared using bioinformatic tools for differential gene expression and principal component analysis. First, we found 355 genes were differently expressed (DE) in HCM LV compared to control LV. Second, we identified 122 DE genes in HCM LA compared to control LA, of which 11 are already known to be associated with HCM Third, we detected 1509 DE genes in HCM LA compared to HCM LV, 16 of which are currently linked to HCM. Overall, we show myocardial global mRNA profiles in healthy cats and altered expression profiles in cats with HCM and between cardiac regions. Our findings may help in detecting novel genes and pathways involved in HCM to benefit feline health.

Abstract C44: Tricuspid Annular Plane Systolic Excursion Normalized by Right Ventricular Size in Dogs with Pulmonary Hypertension

Yunosuke Yuchi; Ryohei Suzuki; Ayaka Niina; Takahiro Teshima; Hirotaka Matsumoto; Hidekazu Koyama
Nippon Veterinary and Life Science University

Tricuspid annular plane systolic excursion (TAPSE) is one of right ventricular (RV) systolic function indicators. There was no significant difference for TAPSE among severity groups of pulmonary hypertension (PH) secondary to myxomatous mitral valve disease (MMVD) because of its load dependency. We hypothesized that TAPSE normalized by RV size could assess RV function regardless of RV load. We aimed to evaluate the utility of TAPSE normalized by RV size. Twenty healthy dogs and sixty-eight dogs with MMVD were enrolled in this study. The following indices were measured as the indicators of RV size; end-diastolic RV internal dimension (RVIDd), end-diastolic and end-systolic RV area (RVEDA and RVESA), and end-diastolic RV wall thickness. TAPSE was measured by B-mode and M-mode methods, and TAPSE normalized by body weight (TAPSEn) was also calculated. TAPSE normalized by RV size were calculated as TAPSE divided by each RV size indices in the same cardiac cycles. We compared all indices among PH severity groups and presence or absence of right-sided congestive heart failure (R-CHF), and reproducibility was also assessed. There was no significant difference in TAPSE and TAPSEn. Whereas, all indices about TAPSE normalized by RV sizes were significantly decreased in dogs with severe PH and R-CHF. TAPSEB-mode/RVIDd and TAPSEB-mode/RVEDA had high reproducibility, sensitivity, and specificity to detect the R-CHF. TAPSE normalized by RV size could reflect RV dysfunction, which could not be detected by non-normalized TAPSE. Additionally, TAPSEB-mode/RVIDd and TAPSEB-mode/RVEDA might be useful to estimate the onset of R-CHF.

Abstract C45: P-Wave Terminal Force in Dogs with Naturally Occurring Myxomatous Mitral Valve Disease

Karla L. Calderón Olaguivel; Marcela Wolf; Vinícius Costa Silva; Giovana Ruviaro Tuleski; Bruna Bruler; Marlos Sousa
Federal University of Paraná

In medicine, P-wave terminal force in precordial lead V1 (PTFV1) is the most widely used electrocardiographic criteria to assess left atrial abnormalities. It is the algebraic product of the amplitude and the duration of the negative deflection at the end of P-wave in precordial lead V1. In veterinary, PTFV1 is still an unexplored electrocardiographic parameter. Myxomatous mitral valve disease (MMVD) is the most common cardiac disease in dogs. Progressive degeneration of the mitral valve predisposes to left-sided volume overload and chamber enlargement. We sought to investigate the existence of PTFV1 and its value as electrocardiographic marker of left atrial remodeling in a population of dogs with naturally occurring MMVD. Forty-seven healthy dogs (control group) and seventy-four client-owned dogs with MMVD were enrolled in this prospective investigation. All dogs underwent a three-minute electrocardiographic recording with all frontal leads and three simultaneous locations for precordial lead V1: first intercostal space at the level of the costo-chondral junction (1CCJ), third intercostal space at the level of the costo-chondral junction (3CCJ), and fifth intercostal space at the level of the esterno-chondral junction (5ECJ). A single PTFV1 value for each precordial lead location was obtained after averaging five representative PTFV1 values. Also, standard echocardiography was performed in all dogs. In 1CCJ and 3CCJ locations, P-wave was negative. In consequence, PTFV1 was found as a positive deflection at the end of P-wave. In the control group, PTFV1 was detected in 96% (45/47) of dogs at both 1CCJ and 3CCJ locations (1.36±0.28 and 1.7±0.44, respectively). Frequency of detection of PTFV1 at 5ECJ was similar to 1CCJ and 3CCJ in control dogs, but notably diminished in MMVD groups. ROC curves evaluating sensitivity and specificity of PTFV1 to differentiate controls and MMVD dogs at 3CCJ, produced an AUC = 0.6137 ($p<0.05$). In the same way, ROC analysis to differentiate controls and dogs with enlarged left atrium obtained an AUC = 0.672 at 1CCJ, while at 3CCJ we documented an AUC = 0.6933. In MMVD group, weak positive correlations were found between PTFV1 values at both 1CCJ and 3CCJ locations and left atrium-to-aorta ratio, diastolic left ventricular internal diameter, normalized diastolic left ventricular internal diameter and E-wave. In addition, a correlation existed between normalized left ventricular internal parameter and 1CCJ PTFV1 when only dogs with enlarged left atrium were considered ($r=0.42$; $p<0.05$). In conclusion, our study found that PTFV1 was best identified at 1CCJ and 3CCJ locations and showed a polarity opposite to the one in man. Differences in mean vector of atrial activation and distribution of electrical potentials on body surface might justify such contrasting finding. In the same way, PTFV1 lacks accuracy to distinguish between healthy and MMVD dogs, but its performance was a bit better when only B2, C and D MMVD dogs, i.e., dogs with enlarged left atrium, were considered. Moreover, there were positive correlations between PTFV1 values and echocardiographic surrogates of left atrium dilation and congestion. To the best of the author´s knowledge, this study is pioneer in evaluating PTFV1 as an electrocardiographic marker in dogs with MMVD. Further studies are warranted to better clarify its applicability in other cardiac diseases, and well as the role played by the thorax conformation in detecting PTFV1 in dogs.

Abstract C46: Real-time Dosimetry Monitoring in the Interventional Catherization Laboratory

Kursten V. Pierce; Brian Scansen; Nicole Maddox
Colorado State University

Exposure to low-dose ionizing radiation during fluoroscopic procedures is a health risk for operators and staff in veterinary catheterization laboratories, though data is lacking for typical exposure levels per case and by staff member. The study aim was to determine radiation exposure to all staff members for various interventional procedures. Radiation exposure was collected prospectively using a real-time dosimetry monitoring system from September 2018 to January 2020. Procedure type, fluoroscopy time, and real-time individual doses to operators and staff were collected. Complete data was obtained from 105 procedures: balloon pulmonary valvuloplasty (BPV, n=38), transvenous pacemaker implantation (27), patent arterial duct occlusion (PDA, 25), diagnostic right heart catheterization (5), percutaneous embolization for portocaval shunts (4), transarterial chemoembolization for hepatocellular carcinoma (3), and tracheal stenting (3). Median radiation dose exposure over all procedures was significantly higher for cardiology residents (4.93 μSv, IQR 1.59–15.64) compared to faculty (3.46 μSv, 0.64–13.02), anesthetist (0.53 μSv, 0.10–2.43) and transesophageal sonographer (TEE, 0.46 μSv, 0.35–2.20), while the lowest exposure was recorded by the technician (0.001 μSv, 0–0.10) ($p<0.0001$). Median doses to faculty, resident, anesthetist, TEE, and technician during BPV were 4.33 μSv, 13.80 μSv, 0.57 μSv, 0.46 μSv, and 0.001 μSv. Median doses in the same order for PDA were 0.67 μSv, 3.40 μSv, 0.67 μSv, 0.63 μSv, and 0.01 μSv. Median doses to faculty, resident, anesthetist, and technician during pacemaker were 2.48 μSv, 3.05 μSv, 0.03 μSv, and 0.01 μSv. These data provide information on dose to operator for fluoroscopic procedures performed in animals. Real-time dosimetry monitoring may enhance radiation awareness, optimizing radiation safety practices for all interventional team members.

Abstract C47: Pharmacokinetics and Relative Bioavailability of Pimobendan and O-Desmethyl Pimobendan in Healthy Dogs After Rectal Administration

Jiwoong Her[1]; Dawn Boothe[2]; Lenore Bacek[3]; Crisanta Cruz-Espindola[2]; Randolph Winter[4]; Kendon Kuo[2]
[1]Department of Clinical Sciences, College of Veterinary Medicine, Auburn University; [2]College of Veterinary Medicine, Auburn University; [3]BluePearl Veterinary Partners; [4]The Ohio State University

Pimobendan is used as therapy for acute congestive heart failure (CHF) in dogs with advanced cardiac disease. In some patients with CHF, oral administration is difficult and increases patient stress. This study aimed to determine if administration per rectum (PR) would be an acceptable alternative to administration per os (PO). Pharmacokinetics of pimobendan and ODMP were prospectively characterized in healthy dogs (n=8) using a randomized, crossover design with a 24 hr washout after a single dose of pimobendan (0.5 mg/kg) administered either PR or PO. Plasma pimobendan and ODMP were quantitated using HPLC using an assay validated in dogs. Data were subjected to non-compartmental analysis (Phoenix Winnonlin®). Key parameters (mean ± SD) after PO and PR administration, were, respectively, for pimobendan, peak concentration (Cmax, ng/ml) 49±30.9 vs. 10±1.95, time to reach Cmax (Tmax, hr) 2.07±0.93 vs. 1±0.42, disappearance half-life (T½, hr) 1.83±0.80 vs. 2.20±0.58, and area under the concentration-time curve (AUC0–∝ ng/ml/hr), 148±71 vs. 31±12, with a relative bioavailability (F) of 0.24. For ODMP, PO vs. PR, respectively, Cmax was 31±10 vs. 8.8±4.76, Tmax of 3.2±1.6 vs. 1.65±1.1, T½ was 4.96±2.74 vs. 8.34±4.81, AUC0–∝ was 168±36 vs. 50±19, F of 0.27. The F of ODMP vs. pimobendan was 1.31 (PR) and 1.51 (PO). This study demonstrates pimobendan may achieve potentially therapeutic concentrations when administered PR at 0.5 mg/kg.

Abstract C48: Repeat Balloon Valvuloplasty for Dogs with Pulmonary Valvar Restenosis

Randolph L. Winter[1]; Jaylyn Rhinehart[1]; Amara Estrada[2]; Herbert Maisenbacher[3]; Thaibinh Nguyenba[4]; Brian Scansen[5]; John Bonagura[6]; Karsten Schober[1]
[1]College of Veterinary Medicine, The Ohio State University; [2]College of Veterinary Medicine, University of Florida; [3]Veterinary Heart Care; [4]MedVet Medical and Cancer Centers for Pets; [5]College of Veterinary Medicine, Colorado State University; [6]College of Veterinary Medicine, North Carolina State University

Pulmonary stenosis (PS) is a common congenital defect in the dog, characterized by valve leaflet dysplasia, fusion, or both. Severe valvar PS can be treated with balloon valvuloplasty (BV) to reduce the obstruction severity and improve clinical signs. Typically, additional interventional procedures are unnecessary, as restenosis rates are reportedly low in both dogs (10–17%) and in humans. Repeated pulmonary BV in humans is generally successful and safe, but outcomes in dogs with pulmonary restenosis have not been reported. This retrospective study describes procedural outcomes and length of time between procedures in dogs undergoing multiple BV procedures for pulmonary valvar restenosis. Medical records and stored echocardiographic images were reviewed. Measured variables included maximum systolic ejection velocity (PVmax) and velocity time integral (VTI); maximal pressure gradient across the pulmonary valve (maxPG) was calculated as maxPG = 4V2. Twelve dogs were included; one underwent three procedures. Median time between BV procedures was 31.4 months (range: 6.0 to 120.7). One dog died during repeat BV, but no others experienced adverse effects. Mean reductions in PVmax, max PG, and VTI following initial and repeat BV were 1.9 m/sec, 79.3 mm Hg, 45.1 cm, and 1.53 m/sec, 63.8 mm Hg, 31.4 cm, respectively (all $p<0.01$). Differences in pre-BV and post-BV PVmax, max PG, and VTI weren't different between initial and repeat BV ($p=0.42$, $p=0.64$, $p=0.37$). This study suggests that repeat BV for dogs with pulmonary valvar restenosis might be effective and safe in a majority of canine patients.

Abstract C49: Congestive Heart Failure Affects Peripheral Blood Lymphocyte Subtypes in Canine Myxomatous Mitral Valve Disease

Natalia Druzhaeva[1]; Alenka Nemec Svete[1]; Alojz Ihan[2]; Katka Pohar[3]; Aleksandra Domanjko Petrič[1]
[1]Veterinary Faculty, University of Ljubljana; [2]Faculty of Medicine, University of Ljubljana; [3]Faculty of Medicine, Institute of Microbiology and Immunology, University of Ljubljana

The inflammatory response is a part of the pathophysiology of congestive heart failure (CHF), and studies show that lymphocytes play a role in this process. We aimed to compare lymphocyte and their subtype concentrations and proportions and correlate them to the stage of the disease in dogs with different stages of myxomatous mitral valve disease (MMVD) and healthy dogs. In a prospective cross-sectional study, we included 65 dogs with MMVD in 3 ACVIM stages (B2, C2, and C1+D) receiving cardiac treatment (where indicated) and 13 healthy controls aged 5 to 12 years. Fresh whole blood samples were analyzed with multicolor flow cytometry using rat and mouse anti-canine antibodies (BioRad) for T lymphocytes (CD3+), including T helper lymphocytes (CD3+CD4+), cytotoxic T lymphocytes (CD3+CD8+), and B lymphocytes (CD45+CD21+). Plasma NT-proBNP concentrations were measured using IDEXX ELISA and serum cardiac troponin I (cTnI) concentrations were measured using a high-sensitivity immunoassay (ADVIA Centaur TnI-Ultra; Siemens). Dogs in CHF (stages ACVIM C2 and ACVIM C1+D) had significantly ($p<0.05$) lower percentages of CD3+CD4+ and significantly higher percentages of CD3+CD8+ (one-way ANOVA, Tukey post hoc test) as well as significantly lower CD4+/CD8+ ratio (Kruskal-Wallis; pairwise comparisons with Bonferroni corrections) in comparison to healthy controls. We did not find any significant differences in total lymphocyte concentration and percentage, as well as concentrations and percentages of T and B lymphocytes among different groups of cardiac patients and healthy controls. Plasma NT-proBNP concentrations significantly differed among groups of patients and healthy dogs, except for healthy and ACVIM B2; however, NT-proBNP did not correlate significantly (Spearman test) with any of the lymphocyte subtypes. Cardiac troponin I concentrations significantly differed among groups and correlated significantly negatively with B lymphocyte concentration and proportion of CD3+CD4–CD8– (double negative T lymphocytes) in ACVIM B2, positively with proportion and concentration of CD3+CD4+CD8+ (double-positive T lymphocytes) in ACVIM C2, and positively with concentration of CD3+ (T lymphocytes) in ACVIM C1+D group (Spearman test). We found that the proportion of cytotoxic T lymphocytes is increased, and the proportion of T helper lymphocytes decreased in CHF in dogs with MMVD but not in the ACVIM B2 group.

Abstract C50: Mathematical Prediction of Infiniti Medical Standard and Duality Tracheal Stents Deployed Length

Eric De Madron
Central Hospital for Veterinary Medicine

Careful selection of a tracheal stent is crucial to obtain the best outcome. Currently, the predicted length of the stent once deployed in the trachea is derived from lengthening charts provided by the manufacturer and obtained by measuring the length of the stent under varying degrees of compressions in the workshop. Although useful, these tables do not cover all the possible scenarios, especially when the trachea does not have similar proximal and distal diameters and when the use of a Duality stent is considered. The goal of this study was to establish a mathematical formula allowing prediction of the deployed length of a tracheal stent and to determine whether the results would match the measurements obtained by the engineers. The hypothesis is that the lateral surface of the stent does not change with compression. For a standard cylindrical stent, its lateral surface $A = \pi DL$. Therefore, if D1 and L1 are the nominal (uncompressed) length and diameter of the stent, D2 the tracheal diameter and L2 the length of the stent once deployed, one has: $\pi D1L1 = \pi D2L2$. After simplification, one has: $D1L1 = D2L2$, therefore $L2 = L1*D1/D2$. The calculated deployed lengths obtained using this formula correlated very well with the measured lengths with a linear regression close to the identity line ($Y = 0.9958X + 4.6723$, $R2 = 0.9796$) (Figure 1). Similarly, I used the lateral surface of a truncated cone for the Duality stents: $A = 1/2(D + d)\pi s$, s being the slant height: $s = \sqrt{((D/2 - d/2)^2 + L^2)}$. If D1 and d1 are the proximal and distal diameters of the nominal stent and D2 and d2 the proximal and distal diameters of the trachea, then: $1/2(D1 + d1)\pi s1 = 1/2(D2 + d2)\pi s2$, from which $(D1 + d1)s1 = (D2 + d2)s2$, therefore: $s2= s1(D1+ d1)/ (D2+ d2)$. Again, the predicted deployed lengths obtained using this formula correlated very well with the measured lengths with a linear regression close to the identity line ($Y = 0.8968X + 10.992$, $R2 = 0.8803$) (Figure 2). In conclusion, it is possible to mathematically predict the deployed length of a standard cylindrical or asymmetric Duality tracheal stent in a cylindrical or conical trachea. This should help the clinician in the proper selection of the stent prior to deployment.

Figure 1

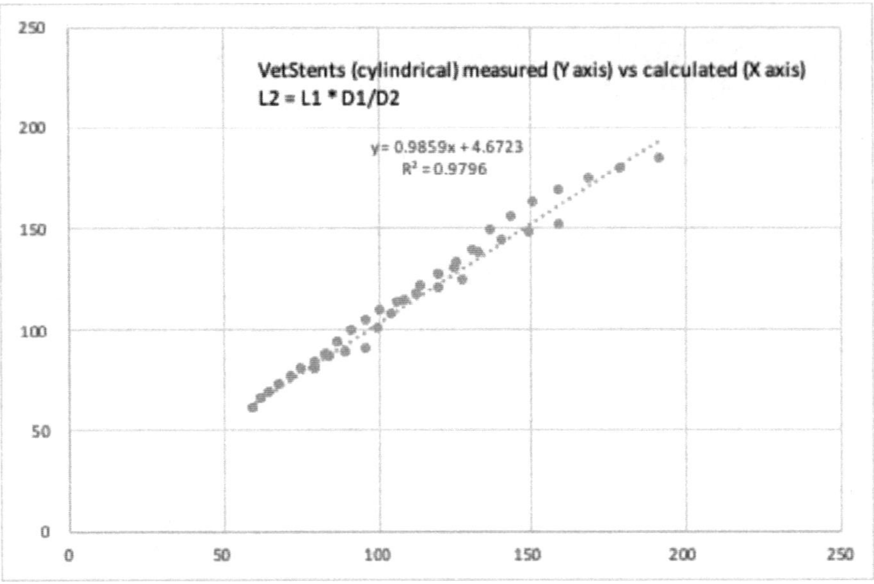

Figure 2

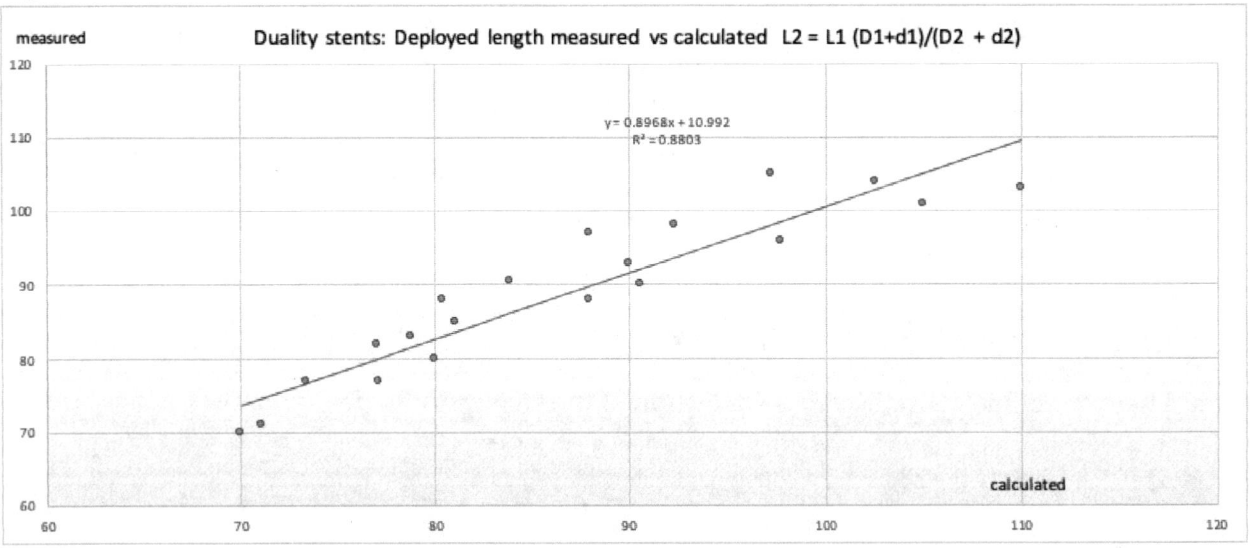

Abstract C51: Myocardial Lumican is Associated with Fibrosis in Cats with Hypertrophic Cardiomyopathy Phenotype

Wan-Ching Cheng[1]; Lois Wilkie[1]; Melanie Dobromylskyj[2]; Virginia Luis Fuentes[1]; David Connolly[1]
[1]Royal Veterinary College, University of London; [2]Finn Pathologists

Lumican is a small leucine rich protein with a well characterised function of facilitating collagen organisation. Human patients with heart failure have increased quantities of myocardial lumican, which correlate with measures of myocardial fibrosis. *In vitro* studies in rodent fibroblasts have shown that lumican up-regulates pro-fibrotic genes such as TGF-β, lysyl oxidase (LOX), and collagen I. Recently, by mass spectrometry, it was shown that human patients with hypertrophic obstructive cardiomyopathy had increased myocardial lumican, which correlated with cardiac fibrosis determined by late gadolinium enhancement on cardiac magnetic resonance imaging. Lysyl oxidases are important downstream components of the lumican pathway. They are a group of collagen cross-linking enzymes that facilitate the maturation of collagen and thereby reduced the susceptibility of collagen to degradation. The aim of this research was to investigate the role of lumican in feline HCM and assess lumican-associated pro-fibrotic signalling pathways in cats with HCM. Full thickness tissue from left ventricular free wall sized 1 x 1 x 1 (cm) was collected following necropsy. Ten cats aged between 1.6–18.9 (median 5.8) years were assigned to the normal control group and another 10 cats aged between 1.7–17.0 (median 8.7) years were assigned to the HCM group based on pre-mortem echocardiography and or/histopathology. There was no difference in the age and sex between the control and HCM group. Transcripts for lumican (LUM), lysyl oxidases (LOX and LOXL2), TGF-betas (TGFB1 and TGFB2), and collagen (COL1A1 and COL3A1) were quantified using qPCR after reverse transcription with RPS7 and RPL30 being used as reference genes. Immunoblotting was used to quantify the protein products of lumican, LOX, LOXL2, TGF-β1, and TGF-β2. The quantity of soluble and insoluble collagen (ug/mg) in 2 control and 4 HCM cats was measured using a Sircol collagen assay, a commercially available colorimetric kit for collagen quantification. Student t test, Mann-Whitney test, and Welch's t test was used to compare the transcripts level (ΔCt), protein level, and the collagen level, respectively. Correlation of the gene transcripts was assessed using Pearson's test while the association between protein products and the different collagens was assessed using Spearman's test. A p-value lower than 0.05 was deemed significant. Compared to the control group, the HCM group showed upregulation of all the pro-fibrotic transcripts except for ACTA2 (which encodes α-SMA). Lumican transcripts positively correlated with TGFB1, TGFB2, LOX, LOXL2, COL1A1, and COL3A1 transcripts. Lumican transcripts also correlated with the amount of total collagen (ug/mg). Immunoblotting confirmed the presence of lumican as a smear of band at 37–55 kDa and bands at above 100 kDa. Both isoforms of TGF-beta appeared at around 50 kDa and were elevated in the HCM cats compared to the control cats. LOX and LOXL2 were also increased in the HCM cats compared to controls and were identified at 50 and 75 kDa, respectively. The Sircol collagen assay showed that about 80% of the collagen was insoluble. The 2 controls had significantly higher amount of soluble collagen compared to the 4 HCM cats. In the HCM cats, the insoluble collagen component accounted for higher percentage of total collagen compared to the 2 controls (88–99% vs. 82–87%, p=0.054). TGF-β2, but not TGF-β1, correlated with the amount of total collagen. Moreover, only LOX, but not LOXL2, showed a non-significant positive trend to collagen cross-linking (CCL=insoluble collagen/ soluble collagen) (r=0.771,

N=6, $p=0.072$). In conclusion, we have shown that lumican is increased in the left ventricle of cats with HCM and is associated with fibrosis potentially by (1) increasing collagen production through TGF-β signalling and (2) regulating the percentage of insoluble protein through LOX. In human heart disease, increased quantities of insoluble protein are associated with reduced ventricular compliance and diastolic dysfunction.

Abstract C52: Metabolic Markers of Insulin Resistance in Hypertrophied Myocardium of Cats

Dmitrii Oleynikov
Almazov National Medical Research Center, IEM, Saint-Petersburg, Russian Federation

Objectives. Investigation of myocardial tissue concentration of ATP, glucose transporters 1 and 4, pyruvate dehydrogenase, hexokinase 2, insulin receptor, and adropin proteins, determining metabolic changes and possible insulin resistance in feline myocardium with hypertrophic phenotype. **Methods**. Eighteen cats were studied, divided into 3 groups: without cardiac disease (n=5), cats with hypertrophic cardiomyopathy (n=8), cats with chronic kidney disease and secondary myocardial hypertrophy (n=5). This is the pilot study for metabolic markers determination. Animals in the study were diagnosed for primary disease by standard methods and algorithms. Cats were euthanized due to end-stage chronic kidney disease, refractory heart failure or by owners will. The material was obtained immediately after death. Samples for the metabolic study were taken from the apical part of the left ventricular free wall and fixed in liquid nitrogen at once and were stored in -80C refrigerator. Studied protein concentrations were analyzed in a specialized research laboratory, using ELISA kits, provided by Cloud-Clone Corp. (USA), included: total ATP, pyruvate dehydrogenase, hexokinase II, Adropin, insulin receptor, GLUT1, and GLUT4. **Results**. In the group with HCM, we discovered that levels of ATP, pyruvate dehydrogenase and adropin were severely suppressed in comparison to healthy cats, while GLUT1 and GLUT4 were unchanged. The concentration of hexokinase 2 and insulin receptor proteins was significantly increased. In the group of secondary myocardial hypertrophy, suppression of most studied proteins was admitted, except insulin receptor. **Conclusion**. In conclusion, we found out that metabolic remodeling and development of insulin resistance in observed diseases with hypertrophy phenotype. We observed depression of pivotal enzymes proteins, limiting energy restoration potency for cardiomyocytes.

Abstract C53: The Effect of Timolol Ophthalmic Solution 0,5% on Systolic Function in Healthy Cats

Marlos Gonçalves Sousa[1]; Giovana Tuleski[2]; Vinicius Costa Silva[2]; Julio Santos[2]; Matheus Silveira[3]; Marcela Wolf[2]; Marlos Sousa[2]
[1]Federal University of Parana; [2]Universidade Federal do Paraná; [3]Instituto Federal Catarinense

Introduction: The ophthalmic application of timolol solution 0,5% reduces heart rate (HR) and facilitates the separation of E and A waves on the transmitral flow of cats, after a twenty minutes waiting. It is expected that this beta-blocker drug will also decrease the contraction force, being the aim of this study to verify if a single drop of a timolol ophthalmic solution decreases the systolic function significantly. **Animals**: Thirty-three healthy cats (1–13 years, median six years), 2,1 to 11 kilos (median 4,6 kg); 18 females and 15 males; 29 mixed breeds, two Persians, one Maine coon, and one Siamese. **Methods**: Each cat had a complete echocardiography examination before the administration of a single drop of timolol ophthalmic solution 0,5% at the left eye, repeating the exam 20 minutes later. The parameters used to evaluate the left ventricle (LV) systolic function were: shortening fraction (SF), ejection fraction (EF), Mitral Plane Systolic Excursion (MAPSE), longitudinal Strain (LSt), and Tissue Mitral Annular Displacement (TMAD). The Tricuspid Annular Plane Systolic Excursion (TAPSE) allowed the evaluation of the right ventricle contraction. The variables that helped to asses the left atrium (LA) function were: SF (M-mode from the LA and aorta view, right parasternal window), Complete ejection fraction (LA cEF), Active ejection fraction (LA aEF), and Passive ejection fraction (LA pEF), TMAD, and LSt. A record of apical 4-chamber (4AP) and 2-chamber (2AP) image allowed the off-line calculation of all TMAD and LSt values. The average AP4 and 2AP values resulted in the Global indexes. All the standard parameters of an echocardiographic exam were made, including Tissue Doppler (TD) of the lateral and the septal mitral annulus for diastolic valuation. For statistical evaluation, SPSS for Windows V.20.0 (SPSS Inc. Headquarters, Chicago, USA) was used. Paired t-test or Wilcoxon verified differences between animals with and without timolol. The significance level was $p<0.05$. **Results**: As expected, the heart rate (HR) decreased by 19% with timolol. Table 1 shows the results of the systolic evaluation. SF, a classical echocardiographic parameter to evaluate systolic function, reduced 20%. Other systolic variables also decreased: lateral MAPSE (12%), LA SF (16%), Global LA St (12%), Global LA cEF (9%), Global LA aEF (12%). LV St and all TMAD parameters were not affected by the beta-blocker. The assessment of the diastolic function before and after the timolol is in Table 2. From the thirty-three cats, only ten presented fused outflow mitral or TD waves before the timolol administration, having timolol succeed help in separating 2 of them. The proportion of fused outflow mitral waves was the same, being timolol effective to separate only the fused lateral TD waves. With lower HR, IRVT values increased 15% after timolol, being most of the IRVT values considered enlarged after the drug administration (>60 ms). **Conclusions**: Using timolol eye drops to assess cat diastolic function requires careful, as this procedure alters the systolic function parameters used routinely. Less than a third of cats presented fused waves before the timolol administration, being questionable the necessity of this practice on a routine daily basis.

Table 1- Echocardiographic systolic parameters before and after timolol administration

	Control M±SD	Timolol M±SD	p-value
Echo HR	190,9 ± 26,1	160,3 ± 18	<0,001
SF	51 ± 7,8	42,4 ± 5,6	<0,001
EF	87 ± 8,4	77,8 ± 2,1	0,190
Lateral MAPSE	4,6 ± 0,7	4,1 ± 0,7	<0,001
Septum MAPSE	4,4 ± 1	3,9 ± 0,9	0,058
TAPSE	8 ± 1,8	7,5 ± 2,1	0,404
Global LV LSt	33,9 ± 3,9	31,8 ± 5,4	0,092
Global LV MP TMAD	4,1 ± 0,8	4,1 ± 0,7	0,918
Global LV MP% TMAD	17,5 ± 3	16,6 ± 2,9	0,287
Global LA SF	35,2 ± 7	30,2 ± 5,3	0,022
Global LA St	36,9 ± 4,4	32,9 ± 6,1	0,004
Global LA TMAD Total Dysplacement	4 ± 0,8	4,1 ± 0,8	0,371
Global LA TMAD Systolic annulus dysplacement	2,4 ± 0,6	2,4 ± 0,7	0,879
Global LA Complete EF	57,4 ± 4,6	52,7 ± 4,2	0,001
Global LA Active EF	43,5 ± 5,9	39 ± 5,5	0,006
Global Passive EF	24,2 ± 7,4	22,3 ± 6,5	0,434

Table 2- Echocardiographic diastolic parameters before and after timolol administration

Variables		Control N	Control %	Timolol N	Timolol %	Total N	Total %	p-value
Fused E A outflow	No	25	37,9%	25	37,9%	50	75,8%	0,613
	Yes	8	12,1%	8	12,1%	16	24,2%	
Fused Septal E' A'	No	28	42,4%	29	43,9%	57	86,4%	0,500
	Yes	5	7,6%	4	6,1%	9	13,6%	
Fused Lateral E' A'	No	25	37,9%	31	47,0%	56	84,8%	0,041
	Yes	8	12,1%	2	3,0%	10	15,2%	
IVRT Interpretation	Enlarged	8	12,1%	20	30,3%	28	42,4%	0,003
	Normal	25	37,9%	13	19,7%	38	57,6%	
Diastolic function interpretation using Septal TD	Fused	9	13,6%	8	12,1%	17	25,8%	0,625
	Normal	12	18,2%	15	22,7%	27	40,9%	
	Abnormal	4	6,1%	3	4,5%	7	10,6%	
	Pseudonormal	6	9,1%	7	10,6%	13	19,7%	
	Inverted EA normal TD	2	3,0%	0	0,0%	2	3,0%	
Diastolic function interpretation using Lateral TD	Fused	11	16,7%	8	12,1%	19	28,8%	0,552
	Normal	14	21,2%	20	30,3%	34	51,5%	
	Abnormal	5	7,6%	3	4,5%	8	12,1%	
	Pseudonormal	2	3,0%	2	3,0%	4	6,1%	
	Inverted EA normal TD	1	1,5%	0	0,0%	1	1,5%	

Whole Genome Sequencing to Explore Genetic Risk Factors in Canine Myxomatous Mitral Valve Disease

Melanie J. Hezzell[1]; Marsha Wallace[2]; Jenny Wilshaw[2]; Stephen Hare[3]; Nicholas Timpson[4]; Adrian Boswood[2]; Lucy Davison[2]
[1]University of Bristol; [2]Royal Veterinary College, University of London; [3]Wellcome Centre for Human Genetics, University of Oxford; [4]Bristol Medical School, University of Bristol

Canine myxomatous mitral valve disease (MMVD) is the most common acquired cardiac disease in dogs, with cavalier King Charles spaniel (CKCS) being particularly overrepresented. Almost all CKCS >10 years of age are affected, suggesting that genetic risk factors may be 'fixed' in this breed. MMVD is believed to be a polygenic trait, and previous genetic work in canine MMVD has employed a case-control design within individual breeds. However, this design precludes the identification of disease-associated variants that are 'fixed' within high risk breeds and may result in only minimal genetic differences between cases and controls being identified. The aim of the present study was to identify new canine MMVD genes as potential future targets for prevention or treatment of disease. This was achieved by applying a new model for identification of genetic risk variants in complex disease, using whole genome sequencing (WGS) of breeds at relatively high and low risk of disease. Additionally, genetic differences between CKCS with early-onset MMVD that died due to congestive heart failure (group C) and CKCS with late-onset MMVD that died due to non-cardiac disease (group NC) were explored. Whole genome sequencing (WGS) at 30X coverage was undertaken on 12 CKCS, 6 in group C and 6 in group NC. Results were compared with WGS from 48 non-CKCS breed dogs, mainly sequenced as part of the Canine Diabetes Genetics Partnership initiative. DNA was extracted from blood samples that were surplus to requirements for clinical purposes. A custom bioinformatics pipeline was used to call, annotate and prioritise variants for follow-up, based on the Genome Analysis ToolKit, using CanFam3.1 as a reference. Variants were prioritised according to a range of criteria, including their alternate allele frequency (by breed or disease severity status), predicted impact on gene function, or location near a region with a plausible role in cardiac disease. In CKCS, 1993 breed-unique high or moderate impact variants were identified, 1393 of which were present in more than one CKCS, and 43 of which were found exclusively in group C. A total of 103 variants of high or moderate impact were identified in or near genes with a plausible role in cardiac disease and at a high frequency in all CKCS. A further 539 similar variants of potential cardiac relevance were at a higher frequency in group C compared to group NC. A follow-up genotyping study is in progress using a larger cohort and more breeds. This work demonstrates that WGS offers a promising route for investigation of complex diseases where genetic risk may be fixed within a breed.

Evaluation of Radiographic Predictors of Left Heart Enlargement in Dogs

Lance C. Visser; Laetitia Duler; Kristina Jackson
University of California, Davis

We sought to determine the diagnostic accuracy of vertebral left atrial size (VLAS), vertebral heart size (VHS), and radiologists' assessment for predicting left atrial enlargement (radLAE) in a large population of dogs with and without echocardiographic left atrial enlargement (echoLAE). Dogs that had an echocardiogram and thoracic radiographs performed within 24-hours were consecutively enrolled provided they did not have pericardial effusion or a cardiac mass. EchoLAE was defined by a left atrium to aortic ratio acquired from short-axis (LA/AoSx) and long-axis (LA/AoLx) >1.68 and >2.54, respectively. Radiographic assessments were blinded to echocardiographic assessments and vice versa. Cutoffs for VLAS and VHS were generated using receiver operating characteristic analyses. Logistic regression was utilized to identify associations with echoLAE. One-hundred and eighty-three dogs were enrolled. Dogs were diagnosed with myxomatous mitral valve disease (56%; 35 stage B1, 35 stage B2, and 33 stage C/D), other cardiovascular diseases (25%) or no cardiovascular disease (19%). Prevalence of EchoLAE was 63%. Optimal cutoffs for VLAS (>2.3 vertebrae; sensitivity [Sn] 88.9%; specificity [Sp] 73.3%) and VHS (>11.1 vertebrae; Sn=74.6%; Sp=75.8%) were generated and compared with radLAE (yes/no). VLAS >2.3 (area under the curve [AUC]=0.81) was more accurate than VHS >11.1 (AUC=0.75) and radLAE (AUC=0.76) for predicting echoLAE. VLAS >2.3 and radLAE (both p<0.001) were independently associated with echoLAE but VHS >11.1 was not (p=0.054). VLAS is a useful radiographic measurement for predicting echoLAE, particularly if echocardiography or a radiologist's assessment are unavailable.

Short-Term Outcome of a Prospective Study of Diet-Associated Dilated Cardiomyopathy (DCM) in Dogs

Lisa M. Freeman[1]; John Rush[1]; Darcy Adin[2]; Kelsey Weeks[1]; Kristen Antoon[1]; Sara Brethel[2]; Suzanne Cunningham[1]; Luis dos Santos[1]; Renee Girens[2]; Emily Karlin[1]; Katherine Lopez[1]; Camden Rouben[2]; Michelle Vereb[2]; Vicky Yang[1]
[1]Cummings School of Veterinary Medicine, Tufts University; [2]College of Veterinary Medicine, University of Florida

Objective: To evaluate serial changes in echocardiography and cardiac biomarkers in dogs with suspected diet-associated DCM, after diet change. Dogs with DCM [M-mode fractional shortening (FS) ≤25%, normalized LVIDd ≥1.8, and normalized LVIDs (nLVIDs) ≥1.2] eating either a non-traditional or traditional diet were enrolled in a 9-month study. Echocardiography, ECG, high-sensitivity cardiac troponin I (cTnI), NT-proBNP, and taurine were measured at baseline. Dogs were treated with cardiac medications as clinically indicated, diet was changed, and echocardiography and biomarker analyses were repeated after 3, 6, and 9 months. As of January, 2020, 59 dogs have been enrolled (36 M/23 F). Median age=7.7 yrs (2.6–15.9 yrs) and weight was 32.0 kg (3.8–97.8 kg). Enrolled breeds included Doberman Pinscher (n=14), Golden Retriever (n=7), Pit Bull (n=6), and Boxer (n=6), but multiple other breeds were represented. Dogs were ACVIM Stage C (n=45) or Stage B (n=14). Fifty-one of 59 dogs (86.4%) were eating non-traditional diets. Six of 59 dogs had mildly reduced plasma (n=3) or whole blood (n=3) taurine concentrations. To date, 41 of 42 dogs still alive were re-evaluated at 3 months (23 at 6 months, 16 at 9 months). At 3 months, dogs originally eating non-traditional diets had increased FS ($p<0.001$) and decreased nLVIDs ($p=0.003$), left atrium:aorta ($p=0.002$), NT-proBNP ($p=0.04$), and cTnI ($p=0.003$) compared to baseline. 3/9 traditional diet dogs were alive at 3 months. Overall, 27 dogs died or were euthanized [12/27 (44.4%) due to sudden death].

Insights from Corrected Chloride Calculation in Dogs with Congestive Heart Failure

Darcy B. Adin[1]; Clarke Atkins[2]; Bruna Del Nero[1]
[1]University of Florida; [2]North Carolina State University

Hypochloremia associated with congestive heart failure (CHF) in dogs is likely multifactorial. Loop diuresis causes 1:2 sodium (Na+):chloride (Cl-) loss, whereas free water retention (FWR) causes 1:1 Na+:Cl- dilution. We hypothesized that serum Cl- corrected for Na+ would approximate measured Cl- (mCl-) in most CHF dogs because of loop diuresis, but that Cl- would normalize after correction in some CHF dogs, indicating FWR. Dogs with preclinical heart disease served as controls (no diuretics, no FWR expected) for comparison to CHF dogs. The corrected Cl- (cCl-), mCl-, and their difference were compared between 71 preclinical and 100 CHF dogs, and between Stage C (76) and Stage D dogs (24), using the formula: cCl-=(mid-reference range Na+/measured Na+) x mCl-. Both mCl- and cCl- were lower in CHF vs. preclinical dogs, and Stage D vs. Stage C dogs (both $p<0.0001$). The cCl- was higher than mCl- in CHF dogs ($p=0.014$), but not preclinical dogs ($p=0.2$). Median (10th–90th percentile) difference between cCl- and mCl- was 0.69 mmol/L (-2.16 to 3.4) for CHF dogs. This difference was greater for Stage D vs. Stage C dogs ($p=0.0006$). A diuretic resistance indicator (uNa:uK<1.0) was found in 5 CHF dogs with >90th percentile increase in Cl- after correction. No hypochloremic CHF dogs had normal cCl-. Serum Cl- remained low after correction in hypochloremic CHF dogs, consistent with loop diuretic effects on electrolytes; however some showed notable increases in Cl- after correction, supportive of concurrent FWR. Future study correlating cCl- to serum antidiuretic hormone concentrations is warranted.

Current Understanding of the Pathobiology of Canine Myxomatous Mitral Valve Disease

Brendan M. Corcoran, MVB, PhD, DPharm, MRCVS
Edinburgh, Scotland, UK

"A doctor, eh?" cried he, much excited. "Have you your stethoscope? Might I ask you—would you have the kindness? I have grave doubts as to my mitral valve, if you would be so very good. The aortic I may rely upon, but I should value your opinion upon the mitral."

—A. Conan Doyle (Sherlock Holmes), *The Sign of the Four*

INTRODUCTION

Mitral valve disease has long been recognised in both the dog and human, and the history of canine mitral valve disease was extensively reviewed by Borgarelli and Buchanan in 2012.[1] A wide range of synonyms has been used to describe the disease, but basically the disease is best described as myxomatous mitral valve disease (MMVD), as that describes the characteristic pathology, features of which are shared with the two most common forms of the disease in humans, Barlow's disease (BD) and fibroelastic deficiency (FED). While some descriptions of canine MMVD can be found in the early literature, it was John Whitney's paper in 1967 that gave the first comprehensive pathological description of the disease, and writing with Ariel Pomerance in 1970 the first credible suggestion of the comparative elements between dog and human.[2,3] In James Buchanan's paper of 1977, the total body of knowledge about the disease was summarised, and then reports on the pathobiology of the disease declined.[4] For much of the intervening period, MMVD research concentrated on understanding disease prevalence, breed susceptibility, improving therapeutics, and (with the advances in echocardiography) understanding the structural and haemodynamic consequences of mitral valve regurgitation. In tandem with the development of echocardiography have been advances in the use of cardiac biomarkers, with both targeting disease diagnosis, monitoring therapeutic response, and attempting to predict risk and outcome.

With increasing understanding of the clinical consequences of MMVD, a parallel research stream looking again at the pathobiology of the disease emerged, with identifiable themes including reappraisal of the pathology of MMVD, cellular changes, genetics, genomics, and molecular events. Much of this later work was only possible once a credible canine molecular reagent toolbox became available.

PATHOLOGY APPRAISAL FROM GROSS TO ULTRASTRUCTURAL CHANGES

The pathology descriptions of canine MMVD from the 1960s and 1970s need little revision and are still used today.[2-4] The gradual development of free-edge nodular lesions as the disease progresses and the association with age were well recognised, but the makeup of these nodules was less understood and they were often classified as fibroelastic nodules. Changes to the chordae tendineae were also reported, as were occasional ruptured cords. The myxomatous nature of the pathology, with increased expression of proteoglycans resulting in expansion of the spongiosa layer and disruption of the tight collagen bundles of the fibrosa, are recognised as cardinal features of the disease and dominate the pathological profile of the canine disease.[5] In humans, FED has a similar profile, but in BD there are additional layers of fibrosis on both the ventricular and atrial sides of the valve, which are never seen in the dog.[6] The disruption of the fibrosa and general myxoid degeneration of the valve can be better appreciated on electron microscopy, including the increased proteoglycan content and the production of immature fibrillar collagen by valve interstitial cells.[7] Other imaging modalities such as x-ray and nuclear diffraction show there is collagen loss, and collagen bundles lose their structural integrity and orientation necessary for the valve to withstand chamber pressures.[8,9] Lastly, electron microscopy has identified endothelial damage, detachment, and dysfunction in diseased valves.[10,11]

VALVE CELL CHANGES

The original pathology reports are limited in their description of the cell changes, not least because the molecular reagents were not available. The primary cellular abnormality involves transition of quiescent valve interstitial cells (qVICs) to an activated myofibroblast phenotype (aVICs).[12-14] This transition is recognised by increased expression of alpha smooth-muscle actin (αSMA) by VICs, and it is presumed they then have altered synthetic and catabolic activity. However, in other tissues (including myocardium), activated myofibroblasts typically result in tissue fibrosis, and that occurs in the latter stages of human BD, but not in canine MMVD. Activated myofibroblasts regulate their activity through a combination of apoptosis and/or senescence so avoiding hypertrophic scarring, but there is evidence of aVICs persistence in canine MMVD valves, with aVICs first appearing in young adults.[15] In canine MMVD, there is no evidence of inflammatory cell infiltration, but there are small changes in resident macrophage and mast cell numbers.[12-14] In agreement with what has been identified on electron microscopy, histopathology identifies loss of valve endothelial cells and evidence of endothelial-to-mesenchymal transition (EndoMT), which may contribute to the valve interstitial cell population and overall endothelial dysfunction.[16,17] Unlike in human valves, there is no evidence of recruitment of fibrocytes from the blood pool.[18]

THE GENETICS OF MMVD

The heritability of MMVD in certain breeds, with the Cavalier King Charles Spaniel (CKCS) being the best known example, is well recognised, and application of breeding strategies has had some success in reducing incidence in certain breeds.[19-23] Studies have had partial success in identifying loci of interest on chromosome 2 and 15 in the whippet and 13 and 14 in the CKCS, but the complexity of the inheritance pattern in susceptible breeds is a confounding factor in genetic studies and the identification of important gene mutations.[22,23] Familial mitral valvulopathies are recognised in human medicine, and some gene defects have been identified, but none have been found to have analogous representation in dogs.[24,25] It is likely that non-syndromic mitral valvulopathies in human patients lack a well-defined inheritance pattern, and this might match what is seen in the general non-pedigree dog population. Gene expression in CKCS diseased valve tissue does segregate away from that seen in age-matched non-CKCSs with similar severe disease, and further studies are needed to understand this difference (Corcoran *et al.*, unpublished observation).

VALVE TISSUE GENOMICS

While there are a limited number of studies looking at valve gene expression in canine and human MMVD, the data do give valuable insight into the mechanisms that are involved in disease pathogenesis and development of the eventual pathology phenotype.[26-28] Confounding factors in such studies are where the data has been generated from one specific breed without comparison to disease-free valves from the same breed, and has involved only dogs with severe end-stage disease. Similar problems arise with human studies where familial studies have looked at close-knit kindreds and/or have only examined severely affected valves from non-syndromic BD and FED patients.[29-31] Gene expression changes do not necessarily equate with changes in protein expression, which will control the eventual biological outcome. The valve proteome has identified protein changes shared between early and late-stage disease, but also differences and only some agreement with the published transcriptomic data.[32] Single gene changes can give some insight, but transcriptomics is more informative if multiple genes are examined and the associated network connections are identified using data mining and bioinformatics tools. Gene changes have been identified for a large number of gene families including genes associated with extracellular matrix (ECM), cell signalling and inflammation, peptidases, cytoskeletal filaments and EndoMT.[26-28] While some of these changes (such as ECM) make sense in the context of what is known about valve pathology, others (such as inflammation) are less obvious. However, by applying gene enrichment analysis to what are very large and complicated data sets, clear themes linking all these gene families together can be identified. Gene ontology terms are still dominated by ECM and inflammation, but further analysis indicates that the top upstream regulators are TGFβ1, TNF and IFγ, and that it is TGFβ signalling which is the dominant pathway at all stages of disease (Corcoran *et al.*, unpublished observation). The importance of TNF and IFγ signalling likely relates not to a role of inflammation (see **Valve Cell Changes**), but to the downstream signals they share with the TGFβ pathway.

Serotonergic (5HT) Signalling

Interest in the potential role of 5HT in MMVD stems from the observation that drug-induced, enhanced-serotonergic signalling in human patients treated with appetite suppressants or various anti-Parkinsonian drugs, or affected by carcinoid syndrome results in myxomatous valvulopathies.[33,34] The prevalence of such effects are relatively rare and seem to require very high levels of circulating 5HT or 5HT-agonists for valve pathology to be initiated. Nevertheless, several clinical studies have identified changes in serotoninergic and platelet function in MMVD dogs, suggesting possible contribution to disease pathogenesis. In both the human and dog mitral valve transcriptome, there is increased expression of 5HT receptor genes, and in the dog this expression increases as disease progresses.[26,28,34] However, downregulation of proteins important in tryptophan hydroxylase function, important in synthesis of 5HT, might mean this pathway is not activated.[34] 5HT receptor gene expression is highly controlled by TGFβ; 5HT has not been shown to induce VIC activation, and 5HT blockade has limited effects in transitioning aVICs to a quiescent phenotype, not doing so in canine cultured aVICs but doing so in human mechanically activated valve explants.[34,35] While 5HT can affect downstream signalling pathways that are implicated in MMVD, such as ERK1/2, the data could also suggest serotonergic pathway changes are a consequence rather than a cause of disease.[34-37]

Transforming Growth Factor Beta (TGFB) Signalling

There is compelling evidence that TGFβ signalling is involved in MMVD pathogenesis and increased TGFβ protein expression has been reported in diseased mitral valves.[38] While there is no clear upregulation of tissue *TGFβ* gene expression in canine mitral valves (see **Serotonergic Signalling**), genomic analysis identifies TGFβ signalling as the dominant activated pathway in disease valves and that this activation is present at all stages of the disease and in aVICs cultured from diseased valves.[35] A wide range of genes associated with TGFβ/5HT/BMP signalling are altered in canine and mitral valve MMVD, but the complex mechanisms that control these changes are unknown, and there are shared gene changes across all these pathways.[26-28,34,39] The TGFβs can transition quiescent VICs to aVICs, and aVICs treated with TGFβRI kinase inhibitors or pan-anti-TGFβ antibody are transitioned to the normal phenotype.[35] In Barlow's disease, there is increased differential expression of *TGFβ* genes, but this likely reflects the dominance of fibrosis characteristic of the later stages of disease in human patients.[29] To what extent TGFβ is implicated in pathogenesis of the earlier stages of human MMVD is unknown, as examining valve tissue from such patients is not readily feasible.

The Future and What Next?

The preferred outcome is that veterinary science continues to examine the molecular aspects of MMVD in order to identify the cause of the disease (basic science) and then the therapeutic strategies to prevent, arrest or reverse valve pathology (translation to the clinical patient). The initial stage will be to examine valve biology in more complex cell culture systems than used to date, see if that can be replicated in whole valves, and then identify targets that can be pharmacologically manipulated in a safe and effective way in affected patients. Such an approach takes time and is costly, and the concern is that veterinary medicine alone will not have the resources to do so effectively or to completion. However, an interesting clinical conundrum can be posed about MMVD. All dogs, if they live long enough, will have evidence of valve pathology, but only a proportion will be adversely affected, and many with severe valve damage can live long and healthy lives. MMVD is only important if it causes harm; for those in most danger of developing heart failure, any medication that has been shown to have significant beneficial effect in delaying time to heart failure and time to death can be used. A problem for the veterinary pharmaceutical industry is the potential market, and return on their investment for any new drug is compromised by the fact that many dogs live happily with their leaky valves, and those that do not can rely on drugs such as pimobendan to give them an acceptable lifespan.[39] It could be argued that the main challenge for MMVD research for veterinary cardiology is refining techniques that allow identification of those dogs at genuine risk of developing heart failure, and those that can be left alone.

References

1. Borgarelli, Buchanan. *J Vet Cardiol*. 2012;14:93.
2. Whitney. *J Small Anim Pract*. 1967;8:459.
3. Pomerance, Whitney. *Cardiovasc Res*. 1970;4:61.
4. Buchanan. *Adv Vet Sci Comp Med*. 1977;21:75.
5. Han, *et al. J Heart Valve Dis*. 2010;19:60.
6. Roberts, *et al. J Am Coll Cardiol*. 2015;63:1667.
7. Black, *et al. Am J Vet Res*. 2005;66:1408.
8. Hadian, *et al. Biophys J*. 2007;93:2472.
9. Hadian, *et al. Cardiovasc Pathol*. 2010;19:141.
10. Corcoran, *et al. Am J Vet Res*. 2004;65:198.
11. Han, *et al. Vet J*. 2013;197:388.
12. Han, *et al. Am J Vet Res*. 2008;69:763.
13. Disatian, *et al. J Heart Valve Dis*. 2008;17:402.
14. Lu, *et al. J Vet Cardiol*. 2016;18:100.
15. Blake, *et al. Res Vet Sci*. 2020;128:99.
16. Lu, *et al. Vet J*. 2015;206:377.
17. Olsen. *J Comp Path*. 2003;129:120.
18. Barth, *et al. Path Res Pract*. 2005;201:301.
19. French, *et al. Vet J*. 2012;193:283.
20. Lewis, *et al. Vet J*. 2011;188:73.
21. Birkegard, *et al. J Vet Intern Med*. 2016;30:63.
22. Stern, *et al. Plos ONE*. 2015; DOI:10.1371/journal.pone.0141234.
23. Madsen, *et al. J Hered*. 2011;102:S62.
24. Gasser, *et al. BMC Cardiovasc Disord*. 2018; DOI 10.1186/s12872-018-0755-y.
25. Meurs, *et al. Vet J*. 2018;232:16.

26. Oyama, Chittur. *Am J Vet Res*. 2006;67:1307.
27. Li, *et al. OMICS*. 2015;19:461.
28. Lu, *et al. Vet J*. 2015;204:23.
29. Thalji, *et al. Circ Cardiovasc Genet*. 2015;8:516.
30. Greenhouse, *et al. Gene*. 2016;586:56.
31. Sainger, *et al. J Cell Physiol*. 2012;227:2595.
32. Lacerda, *et al. Proteomics Clin App*. 2009;3:1422.
33. Connnoly, *et al. Am J Path*. 2009;175:988.
34. Dreisburgh, *et al. J Mol Cell Cardiol*. 2018;115:94.
35. Tan, *et al. Plos ONE*. 2019; DOI: 10.1371/journal.pone.0221126.
36. Arndt, *et al. J Vet Intern Med*. 2009;23:1208.
37. Disatian, Orton. *J Heart Valve Dis*. 2009;18:44.
38. Aupperle, *et al. J Comp Pathol*. 2009;139:97.
39. Boswood, *et al. J Vet Intern Med*. 2016;30:1765.

Rational Dose Selection for ACE Inhibitors in Canine Heart Disease

Jessica L. Ward, DVM, DACVIM (Cardiology)
Ames, IA, USA

Introduction to Angiotensin-Converting-Enzyme Inhibitors

Angiotensin-converting enzyme inhibitors (ACEi) are common medications used in both human and veterinary medicine for treatment of cardiac disease,[1-7] systemic hypertension,[8-11] and proteinuria[8,12,13]. ACEi mitigate the renin-angiotensin-aldosterone system (RAAS) by inhibiting the action of ACE, the enzyme that catalyzes the conversion of angiotensin I into angiotensin II. Decreased production of angiotensin II (and consequently decreased release of aldosterone) results in balanced vasodilation, reduced sodium and water retention, dilation of the glomerular efferent arteriole, and decreased cardiac fibrosis and remodeling.[14,15] Enalapril and benazepril are the most commonly prescribed ACEi in veterinary medicine.[16] Although ACEi are frequently prescribed in dogs, no consensus exists regarding the ideal dose of ACEi in this species.

Pharmacokinetic and Pharmacodynamic Studies in Healthy Dogs

Pharmacokinetic and pharmacodynamic studies comparing various doses of ACEi in healthy dogs have not provided consistent recommendations regarding ideal dosing. One study in dogs showed that single or repeated oral doses of benazepril resulted in long-lasting suppression of plasma ACE activity (75–85% enzyme suppression at 24 hours), and that ACE inhibition was independent of doses at or above 0.25 mg/kg.[17] These results formed the basis of the minimum recommended label dosage for benazepril in Europe as 0.25 mg/kg PO once daily (Fortekor, Elanco Animal Health; Cardalis, Ceva Santé Animale). However, subsequent reanalysis of these data using pharmacokinetic modeling suggested that q 12 h dosing (as opposed to q 24 h dosing) would achieve greater inhibition of plasma ACE with the same total daily dose.[18] Furthermore, a different study of single-dose enalapril and benazepril at 0.5 mg/kg demonstrated much shorter duration of effect, with suppression of ACE activity falling below 50% only 12 hours after administration.[19]

These pharmacokinetic and pharmacodynamic studies of ACEi in dogs are limited by the use of plasma ACE activity as the surrogate marker for RAAS suppression. ACE activity is now considered a suboptimal endpoint to characterize the effects of ACEi on RAAS for several reasons, including ACE-independent mechanisms of angiotensin II and aldosterone production[20-23] and the potential for renin activation and accumulation of angiotensin I during ACEi therapy[24]. A more recent study evaluating the pharmacodynamic effect of benazepril (0.34 mg/kg PO q 24 h) in a canine low-sodium diet model of RAAS activation used circulating angiotensin II and aldosterone levels as outcome measures, and reported a marked but transitory decrease (<10 hours) in these biomarkers.[25]

Pharmacokinetic results are also complicated by chronobiology of the RAAS itself. Studies in healthy dogs demonstrate that the circadian rhythm of the RAAS is influenced by time of feeding (i.e., sodium intake); for dogs fed in the morning, peak RAAS activation occurs approximately 12 hours later during the evening and overnight.[26-28] These results suggest that within a q 24 h dosing schedule, evening dosing would be more effective than morning dosing, a phenomenon that has been demonstrated in humans receiving ACEi for systemic hypertension.[29-31]

Doses of ACE Inhibitors in Clinical Trials

Beyond questions of pharmacokinetics and pharmacodynamics, the degree and duration of RAAS suppression needed to achieve **clinical** benefit (vasodilation, decreased proteinuria, or long-term cardioprotection) is unknown.

Varying doses of ACEi have been used in clinical trials of dogs with naturally occurring cardiovascular disease. In early studies of dogs with CHF secondary to myxomatous mitral valve disease (MMVD) or dilated cardiomyopathy (DCM), enalapril was initially prescribed at 0.5 mg/kg PO q 24 h, with clinicians having the option to increase dose frequency to q 12 h at their discretion.[1,32,33] In the first study of benazepril in this context, the dose range was wide (0.25–1.0 mg/kg PO q 24 h), with a final mean dose of 0.45 mg/kg. These studies demonstrated a variety of hemodynamic and clinical benefits with ACEi compared to placebo, including decreased pulmonary wedge pressures, improved clinical signs and quality of life, and prolonged survival. However, these studies offer little insight into an ideal ACEi dose regime, particularly because the decision to increase ACEi dose frequency to q 12 h could be made by individual clinicians for any reason. More recently, a short-term clinical trial of dogs with CHF secondary to MMVD was performed to assess non-inferiority of Fortekor Plus (combined benazepril + pimobendan, Elanco Animal Health) compared to the combination of Fortekor (Elanco) + Vetmedin (Boehringer Ingelheim Vetmedica).[34] No clinical or clinicopathologic differences were noted between dogs who received benazepril q 12 h or q 24 h as part of various drug product formulations in this study; however, the study period was limited to only 56 days and was not designed to evaluate effects of benazepril independent of pimobendan.

Two large-scale clinical trials have investigated the use of ACEi in dogs with **preclinical** MMVD, prescribing enalapril at doses of 0.25–0.5 mg/kg PO q 24 h (mean dose 0.37 mg/kg)[35] or 0.23–0.66 mg/kg PO q 24 h (mean dose 0.46 mg/kg)[36]. Results of these two studies are somewhat discordant and have been widely debated and compared. The first study suggested no benefit of enalapril in delaying onset of CHF in a population of Cavalier King Charles spaniels in Scandinavia, while the second study suggested a modest (5-month) benefit of enalapril in terms of number of CHF-free days in a more heterogeneous population of dogs in the United States. The difference in total daily dose of enalapril used between studies (as well as the limitation to q 24 h dosing for both studies) is among the methodological differences cited as possible reasons for discrepancy in results.

ACE inhibitors at varying doses have also been investigated in laboratory studies of dogs with experimentally induced mitral regurgitation. In this context, enalapril (0.5 mg/kg PO q 24 h for 9 days, then q 12 h thereafter) led to improved exercise tolerance compared to placebo,[5] while captopril (2.0 mg/kg PO q 8 h) led to significant echocardiographic and hemodynamic improvements[3]. However, these studies did not attempt to compare differences between dose or administration schedules.

As a treatment for systemic hypertension, ACEi are listed in the ACVIM Consensus Statement with a wide dose range of 0.5–2.0 mg/kg PO q 12 h,[37] but no studies have investigated a dose-response effect of ACEi on systemic blood pressure (BP) in dogs. In contrast, a dose-responsive effect of ACEi on BP has been demonstrated in cats; degree and duration of enalapril's BP-lowering effect increased in a dose-dependent manner at doses of 0.25, 0.5, and 1.0 mg/kg PO, and the most effective antihypertensive regimen involved q 12 h dosing.[38] For treatment of proteinuria in dogs, similarly wide doses of ACEi are cited in consensus documents (0.5–2.0 mg/kg/day),[39] although one study has shown that over half of dogs require dose escalation to q 12 h to achieve reduction in UPC >50% of baseline[40].

Ongoing Research and Future Directions

While different studies have assessed the effect of various ACE inhibitor doses in a variety of disease states, relatively few have compared ACE inhibitor dosage **within** the same study, and cross-study comparisons are inherently difficult. Additional research is needed to determine whether the RAAS-mitigating (and thereby antihypertensive, renoprotective, or cardioprotective) properties of ACE inhibitors are dose-dependent.

Our research group has begun two investigations to explore these questions. The first is a retrospective study to assess a dose-response effect of ACE inhibitors on short-term outcome variables (BP, urine protein:creatinine ratio, and renal values) in client-owned dogs receiving ACEi for any indication. The second study is a prospective trial in healthy laboratory beagle dogs to assess dose-exposure-response effect of benazepril on a comprehensive profile of RAAS biomarkers.

Preliminary results of the retrospective study suggest a mild dose-response effect of ACEi on BP in dogs, with higher doses (and/or q 12 h dosing) resulting in an average ~15 mm Hg decrease in BP, compared to ~5 mm Hg decrease for lower doses (and/or q 24 h dosing). This effect was most pronounced in dogs being treated with ACEi for proteinuria (with or without concurrent systemic hypertension); in this subgroup, q 12 h dosing of ACEi led to a 7.4% decrease in BP, while q 24 h dosing caused essentially no antihypertensive effect (actually led to a 2% **increase** in BP). In the subgroup of dogs receiving ACEi for systemic hypertension (with or without concurrent proteinuria), q 12 h dosing was associated with a greater absolute and percent decrease in urine protein:creatinine ratio compared to q 24 h dosing. Overall, these results suggest that ACEi may have mild dose-dependent effects on target short-term outcome variables (BP, proteinuria) and that dose frequency (q 12 h vs. q 24 h) may be a more important predictor of outcome compared to total daily dose. This retrospective study also demonstrates the safety and tolerability of ACEi across a wide dosing range. No statistically significant nor clinically relevant effects of ACEi—dose-dependent or otherwise—were noted in terms of renal values or electrolytes. Furthermore, in this population of 326 dogs receiving an initial mean total daily ACEi dose of 0.69 mg/kg/day (with 58% of dogs receiving q 12 h dosing), side effects related to ACEi were noted in only 14 dogs (4.3%), and only 6 dogs (1.8%) had ACEi discontinued due to adverse effects.

How to Choose a Dose?

In the United States, ACEi are widely available and very inexpensive (generally $3–4/month when utilizing promotional programs for generic drugs offered at many commercial pharmacies). Numerous studies have demonstrated that ACEi are generally safe and well tolerated across a wide range of doses. Based on currently available pharmacokinetic and pharmacodynamic data, effect of single-dose administration of ACEi on ACE activity may be independent of doses >0.25 mg/kg, but several lines of evidence suggest that q 12 h dosing may be preferable to q 24 h dosing. Higher dose of ACEi (and particularly q 12 h dosing) may be particularly useful when treating systemic hypertension or proteinuria. For these reasons, I currently advocate doses of ACEi (enalapril or benazepril) of 0.25–0.5 mg/kg PO q 12 h.

References

1. The COVE Study Group. Controlled clinical evaluation of enalapril in dogs with heart failure: results of the Cooperative Veterinary Enalapril Study Group. *J Vet Intern Med.* 1995;9:243–252.
2. The BENCH Study Group. The effect of benazepril on survival times and clinical signs of dogs with congestive heart failure: results of a multicenter, prospective, randomized, double-blinded, placebo-controlled, long-term clinical trial. *J Vet Cardiol.* 1999;1:7–18.

3. Blackford LW, Golden AL, Bright JM, *et al*. Captopril provides sustained hemodynamic benefits in dogs with experimentally induced mitral regurgitation. *Vet Surg*. 1990;19:237–242.
4. Flood JA, Hoover JP. Improvement in myocardial dysfunction in a hypothyroid dog. *Can Vet J*. 2009;50:828–834.
5. Hamlin RL, Benitz AM, Ericsson GF, *et al*. Effects of enalapril on exercise tolerance and longevity in dogs with heart failure produced by iatrogenic mitral regurgitation. *J Vet Intern Med*. 1996;10:85–87.
6. Swedberg K, Eneroth P, Kjekshus J, Wilhelmsen L. Hormones regulating cardiovascular function in patients with severe congestive heart failure and their relation to mortality. CONSENSUS Trial Study Group. *Circulation*. 1990;82:1730–1736.
7. Swedberg K, Kjekshus J, Snapinn S. Long-term survival in severe heart failure in patients treated with enalapril. Ten year follow-up of CONSENSUS I. *Eur Heart J*. 1999;20:136–139.
8. Lefebvre HP, Toutain PL. Angiotensin-converting enzyme inhibitors in the therapy of renal diseases. *J Vet Pharmacol Ther*. 2004;27:265–281.
9. Labato MA. Antihypertensives. In: Silverstein DC, Hopper K, eds. *Small Animal Critical Care Medicine*. St. Louis, MO: Elsevier; 2009:763–767.
10. Li EC, Heran BS, Wright JM. Angiotensin converting enzyme (ACE) inhibitors versus angiotensin receptor blockers for primary hypertension (review). *Cochrane Libr*. 2014;8:1–49.
11. Lenz T, Hoyer J, Schulte KL. Clinical pharmacokinetics of angiotensin converting enzyme (ACE) inhibitors in renal failure. *Clin Pharmacokinet*. 1993;24:230–254.
12. Vaden SL, Elliott J. Management of proteinuria in dogs and cats with chronic kidney disease. *Vet Clin North Am Small Anim Pract*. 2016;46:1115–1130.
13. King JN, Font A, Rousselot JF, *et al*. Effects of benazepril on survival of dogs with chronic kidney disease: a multicenter, randomized, blinded, placebo-controlled clinical trial. *J Vet Intern Med*. 2017;31:1113–1122.
14. Regulski M,, Regulska K, Stanisz B, *et al*. Chemistry and pharmacology of angiotensin-converting enzyme inhibitors. *Curr Pharm Des*. 2015;21:1764–1775.
15. Toutain PL, Lefèbvre HP. Pharmacokinetics and pharmacokinetic/pharmacodynamic relationships for angiotensin-converting enzyme inhibitors. *J Vet Pharmacol Ther*. 2004;27:515–525.
16. Lefebvre HP, Brown SA, Chetboul V, *et al*. Angiotensin-converting enzyme inhibitors in veterinary medicine. *Curr Pharm Des*. 2007;13:1347–1361.
17. King JN, Mauron C, Kaiser G. Pharmacokinetics of the active metabolite of benazepril, benazeprilat, and inhibition of plasma angiotensin-converting enzyme activity after single and repeated administrations to dogs. *Am J Vet Res*. 1995;56:1620–1628.
18. Toutain PL, Lefebvre HP, King JN. Benazeprilat disposition and effect in dogs revisited with a pharmacokinetic/pharmacodynamic modeling approach. *J Pharmacol Exp Ther*. 2000;292:1087–1093.
19. Hamlin RL, Nakayama T. Comparison of some pharmacokinetic parameters of 5 angiotensin-converting enzyme inhibitors in normal beagles. *J Vet Intern Med*. 1998;12:93–95.
20. Fyhrquist F, Saijonmaa O. Renin-angiotensin system revisited. *J Intern Med*. 2008;264:224–236.
21. Roig E, Perez-Villa F, Morales M, *et al*. Clinical implications of increased plasma angiotensin II despite ACE inhibitor therapy in patients with congestive heart failure. *Eur Heart J*. 2000;21:53–57.
22. Ames MK, Atkins CE, Eriksson A, Hess AM. Aldosterone breakthrough in dogs with naturally occurring myxomatous mitral valve disease. *J Vet Cardiol*. 2017;19:218–227.
23. van De Wal RMA, Plokker HWM, Lok DJA, *et al*. Determinants of increased angiotensin II levels in severe chronic heart failure patients despite ACE inhibition. *Int J Cardiol*. 2006;106:367–372.
24. Geary K, Hunt M, Peach M, *et al*. Effects of angiotensin converting enzyme inhibition, sodium depletion, calcium, isoproterenol, and angiotensin II on renin secretion by individual renocortical cells. *Endocrinology*. 1992;131:1588–1594.
25. Mochel JP, Fink M, Peyrou M, *et al*. Pharmacokinetic/pharmacodynamic modeling of renin-angiotensin aldosterone biomarkers following angiotensin-converting enzyme (ACE) inhibition therapy with benazepril in dogs. *Pharm Res*. 2015;32:1931–1946.
26. Mochel J, Danhof M. Chronobiology and pharmacologic modulation of the renin-angiotensin-aldosterone system in dogs: what have we learned? *Rev Physiol Biochem Pharmacol*. 2015;169:43–69.
27. Mochel J, Fink M, Peyrou M, *et al*. Chronobiology of the renin-angiotensin-aldosterone system in dogs: relation to blood pressure and renal physiology. *Chronobiol Int*. 2013;30:1144–1159.
28. Mochel J, Fink M, Bon C, *et al*. Influence of feeding schedules on the chronobiology of renin activity, urinary electrolytes and blood pressure in dogs. *Chronobiol Int*. 2014;31:715–730.
29. Hermida RC, Ayala DE, Fernández JR, *et al*. Administration-time differences in effects of hypertension medications on ambulatory blood pressure regulation. *Chronobiol Int*. 2012:30:280–314.

30. Hermida RC, Ayala DE, Mojón A, *et al*. Sleep-time ambulatory blood pressure as a novel therapeutic target for cardiovascular risk reduction. *J Hum Hypertens*. 2014;28:567–574.
31. Hermida RC, Crespo JJ, Domínguez-Sardiña M, *et al*. Bedtime hypertension treatment improves cardiovascular risk reduction: the Hygia Chronotherapy Trial. *Eur Heart J*. 2019:1–12.
32. Temers F, Temers B. Acute and short-term hemodynamic, echocardiographic, and clinical effects of enalapril maleate in dogs with naturally acquired heart failure: results of the Invasive Multicenter PROspective Veterinary Evaluation of Enalapril study. The IMPROVE Study Group. *J Vet Intern Med*. 1995;9:234–242.
33. Ettinger S, Benitz A, Ericsson G, *et al*. Effects of enalapril maleate on survival of dogs with naturally acquired heart failure. The Long-Term Investigation of Veterinary Enalapril (LIVE) Study Group. *J Am Vet Med Assoc*. 1998;213:1573–1577.
34. King JN, Hirakawa A, Sonobe J, *et al*. Evaluation of a fixed-dose combination of benazepril and pimobendan in dogs with congestive heart failure: A randomized non-inferiority clinical trial. *J Vet Sci*. 2018;19:117–128.
35. Kvart C, Häggström J, Pedersen HD, *et al*. Efficacy of enalapril for prevention of congestive heart failure in dogs with myxomatous valve disease and asymptomatic mitral regurgitation. *J Vet Intern Med*. 2002;16:80–88.
36. Atkins CE, Keene BW, Brown WA, *et al*. Results of the veterinary enalapril trial to prove reduction in onset of heart failure in dogs chronically treated with enalapril alone for compensated, naturally occurring mitral valve insufficiency. *J Am Vet Med Assoc*. 2007;231:1061–1069.
37. Acierno MJ, Brown S, Coleman AE, *et al*. ACVIM consensus statement: guidelines for the identification, evaluation, and management of systemic hypertension in dogs and cats. *J Vet Intern Med*. 2018;32:1803–1822.
38. Uechi M, Imamoto S, Ishikawa Y. Dose-dependent inhibition of angiotensin converting enzyme by enalapril in cats. *J Vet Med Sci*. 2002;64:385–387.
39. IRIS Canine GN Study Group Standard Therapy Subgroup. Consensus recommendations for standard therapy of glomerular disease in dogs. *J Vet Intern Med*. 2013;27:S27–43.
40. Grauer GF, Greco DS, Getzy DM, *et al*. Effects of enalapril versus placebo as a treatment for canine idiopathic glomerulonephritis. *J Vet Intern Med*. 2000;14:526–523.

Abstract E02: Heart Rate Variability During Exercise and Recovery in Thoroughbred Racehorses Presented for Poor Performance

Anna N. Hammond[1]; Melanie Hezzell[1]; Sarah Smith[1]; Samantha Franklin[2]; Kate Allen[1]
[1]University of Bristol; [2]University of Adelaide

Heart rate variability (HRV) measures the beat-to-beat variation of successive cardiac cycles. There are few studies investigating HRV at exercise, although assessment of HRV may facilitate identification of arrhythmias. This study explored relationships between HRV during exercise and recovery, and pathologies (arrhythmias, lameness, gastric ulceration, airway inflammation and upper respiratory tract obstructions) in poorly performing Thoroughbred (TB) racehorses. One hundred and eighty-three TBs underwent a high-speed treadmill exercise test with an exercising electrocardiogram (ECG) recorded. ECGs were analysed using Televet 100™, divided into peak exercise and recovery portions and examined for rhythm abnormalities. HRV data were derived using specialized software (Kubios™). The derived HRV variables included the standard deviation of normal to normal intervals (SDNN), the root mean square of successive differences (RMSSD), the R-R triangular index (RRTI), the triangular interpolation of R-R intervals (TINN) and the stress index (SI) which is a square root derivation from the histogram of R-R intervals. Horses were categorised according to arrhythmia during exercise (yes/no), arrhythmia during recovery (yes/no), lameness (yes/no), upper respiratory tract obstruction (yes/no) and airway inflammation (yes/no). Proportions were compared using the Chi squared test. Correlations were assessed using Spearman's Rho. Univariate and multivariable general linear models were constructed to assess associations between each HRV measure during recovery, the same HRV measure during exercise, and clinical variables. Variables were logarithmically transformed if the residuals were not normally distributed. If no suitable transformation was possible, differences in HRV variables between clinical categories were assessed using Mann-Whitney U tests. P values <0.05 were considered significant. More horses had arrhythmias during recovery than exercise (104/183 vs. 60/183, $P<0.001$) and the proportion of complex arrhythmias was higher during recovery (39/104 vs. 3/60, $P<0.001$). Arrhythmia during exercise (AEx) was not correlated with the presence of arrhythmia during recovery (ARec, P=0.446). During exercise, none of the HRV variables were associated with any of the clinical variables ($P>0.05$ in all cases) except RRTI with AEx (P=0.023). During recovery, AEx and ARec were the only independent predictors of any HRV measure: Log SDNN (P=0.030 exercise, P=0.027 recovery), log RRTI (P=0.028 exercise, P=0.025 recovery), log TINN (P=0.038 exercise, P=0.005 recovery) and SI (P=0.031 exercise, P=0.001 recovery). During recovery, RMSSD was significantly different between categories for AEx (P=0.010) and ARec (P=0.039); no other differences between clinical categories were detected. In conclusion, AEx, gastric ulceration, respiratory disorders and lameness have little impact on HRV during exercise in poorly performing TB racehorses. However, during recovery HRV is influenced by AEx and ARec, but not by gastric ulceration, respiratory disorders or lameness.

Abstract E03: Repeatability and Reproducibility of Collapsibility Index of Jugular Veins in Healthy Adult Horses: Pilot Study

Ludovic A. Tanquerel; Aude Giraudet
Clinique Equine, Ecole Nationale Vétérinaire d'Alfort, Maisons-Alfort, France

Collapsibility Index (CI) is a noninvasive ultrasound method used in human medicine to estimate volemia, and has not been described in adult horses. The objectives were to investigate the repeatability, and reproducibility of CI of Jugular Veins (JV) in healthy adult horses. Ultrasound was performed on 26 JV of 13 healthy adult horses (9 WB, 2 SB, 2 others). Mean weight was 540 kg (SD=45), and mean age was 11 yo (SD=5). Ultrasound of both JV was performed with a 5 MHz linear probe by positioning the probe on the brachiocephalicus muscle (to avoid compression of the vein), with the head of the horse in neutral position. Anatomic M-mode were performed and CI was calculated as: CI (%)=([maximum JV diameter] - [minimum JV diameter])/[maximum JV diameter]. Ultrasounds were performed by 2 operators, and measures were repeated twice. Bland-Altman plots were produced. Acceptable agreement was considered to be present if bias were less than 20% and if the Limits Of Agreement (LOA) were included in [-20%; +20%]. For repeatability, bias were acceptable in 6/6 plots, and LOA were not acceptable in 6/6. For reproducibility, bias were acceptable in 6/6 plots, and LOA were not acceptable in 6/6. Based on this preliminary results, CI of the external jugular vein in adult horses has a poor repeatability and reproducibility. Further studies with more horses are needed before using CI as a clinical tool in horses.

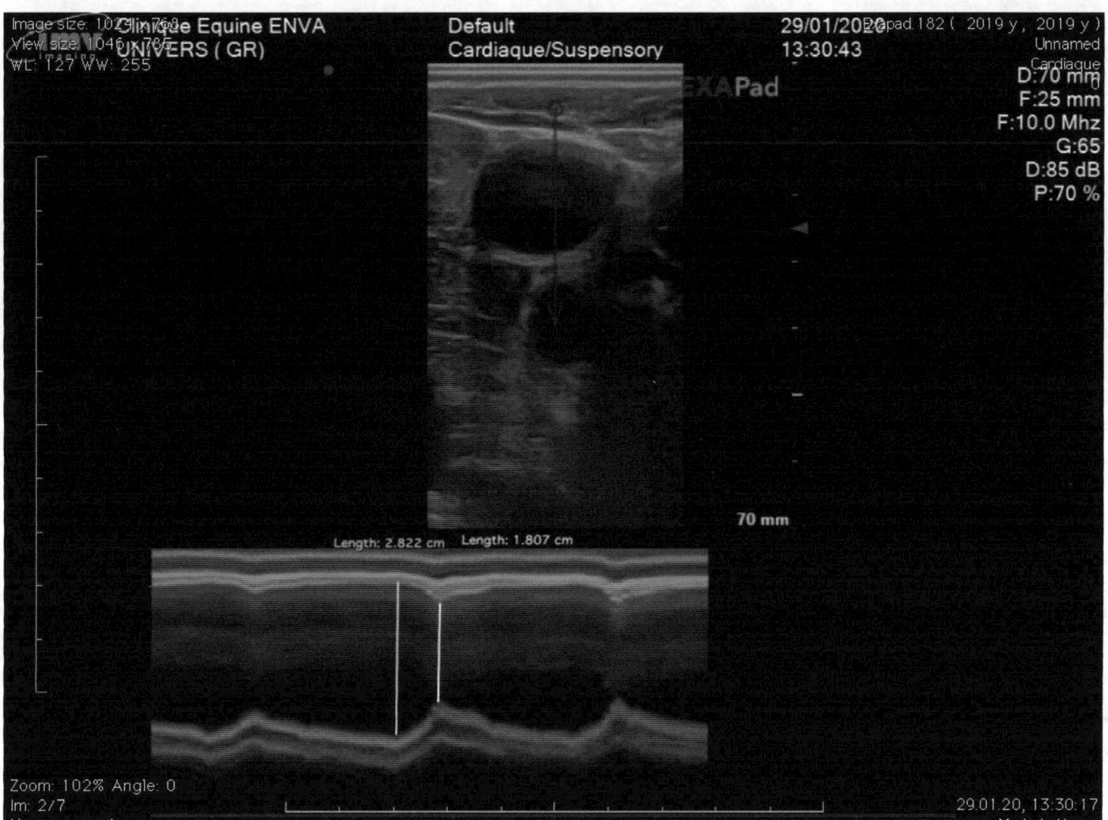

	Site	Biais (%)	LOA
operator 1 vs operator 1	left jugular vein	-0,5	-25; 24
operator 1 vs operator 1	right jugular vein	10,3	-17 ; 38
operator 1 vs operator 1	left jugular vein (transverse view)	7,8	-16 ; 32
operator 2 vs operator 2	left jugular vein	-2,3	-38 ; 33
operator 2 vs operator 2	right jugular vein	1,9	-42 ; 46
operator 2 vs operator 2	left jugular vein (transverse view)	1,7	-48 ; 51
operator 1 vs operator 2	left jugular vein	-15,3	-52 ; 2
operator 1 vs operator 2	right jugular vein	-14,5	-62 ; 3
operator 1 vs operator 2	left jugular vein (transverse view)	-3,6	-40 ; 33
operator 1 vs operator 2	left jugular vein	-13,6	-73 ; 46
operator 1 vs operator 2	right jugular vein	-12,8	-65 ; 39
operator 1 vs operator 2	left jugular vein (transverse view)	-13,6	-36 ; 9

Abstract E04: Factors Affecting Cardiac Auscultation in the Horse

Gayle D. Hallowell; Paul Lewis; James Heale; James Bailey; Adam Redpath; Steffi Pratt; I Mark Bowen
School of Veterinary Medicine and Science, University of Nottingham

The main objective of this study was to evaluate factors affecting the heart sounds that could be auscultated in the adult. 248 horses were auscultated by at least 5 individuals. Records were made regarding operator experience, body condition score, hair length, ambient noise level, breed type, heart rate, which heart sounds could be auscultated in the 4th and 5th left and right intercostal spaces. Any murmurs and dysrhythmias were also recorded. The standard stethoscope used by the individual was used for the entire study, which included Littman electronic, cardiology and classic. Data was evaluated for normality. Various statistical tests were used to evaluate the data including Chi-squared and Fisher's exact test and Wilcoxon matched pairs signed rank tests using Graphpad Prism 8.3, Significance was determined at $p<0.05$. Horses were aged between 5–23 years (13.8±4.9) years. The distribution was 32% mares and 68% geldings. They were equally distributed between TB type (38%), ISH (33%) and cob (29%). The median weight was 620 kg (IQR600–650). Median body condition score was 6 (IQR5–6) and median heart rate was 36 beats per minute (IQR32–36). 20% of horses were found to have second degree atrio-ventricular blockade and 17% had murmurs (50% systolic murmurs associated with the mitral valve and 50% physiological aortic flow murmurs). S4 was audible in 94% of horses and S3 in 78% of horses. S4 was most likely to heard in the left fourth intercostal space and S3 very ventral in the fifth left intercostal space. There was no effect of breed, body condition score, ambient noise level or hair length as to whether S3 or S4 were audible (Table 1). Lower heart rates increased the likelihood that S4 was audible ($p=0.01$; S4 audible - 32(32–36) bpm and not audible 36(36–40)). Level of experience of cardiac auscultation improved the likelihood of hearing S3 ($p=0.03$), but not S4 ($p=0.16$). In summary, S3 is audible in many horses and not just thin, fit Thoroughbreds if auscultated in the optimal position and experience improves auscultation skills as has been previously reported.

Factor	**P value for S4**	**P value for S3**
Breed	0.92	0.37
BCS	0.50	0.67
Heart rate	**0.01**	0.30
Noise level	0.49	0.69
Hair length	0.46	0.17

Table 1: Factors influencing heart sounds audible. Significance P<0.05.

Abstract E05: Comparison of Two-Dimensional versus Three-Dimensional Echocardiography for Assessment of Left Atrial Volume in Thoroughbred Racehorses

Francesca C. Worsman; Karen Blissitt; Darren Shaw; John Keen
The Royal (Dick) School of Veterinary Studies, University of Edinburgh

Left atrial size predicts morbidity and mortality in human and small animal cardiology. In horses, assessing left atrial volume may be of particular interest for volume overload associated with conditions such as mitral valve regurgitation, atrial fibrillation, and ventricular septal defects. The aim of this study was to determine the intra-observer measurement variability of equine left atrial volume using the modified Simpson's method of discs, and compare this to volumes obtained using a real-time three-dimensional software analysis package (4D LVQ, EchoPAC v 202, GE Healthcare). The study used graded datasets of the left atrium of athletic Thoroughbreds horses (n=24; 4–9 yrs; 411–534 kg), obtained from right parasternal views using a Vivid E9 with 3V transducer (GE Healthcare). Horses with grade >3/6 cardiac murmurs assessed by auscultation were excluded from this study. Random generated order measurements were obtained by a single observer on 4 occasions. Two-dimensional end-systolic (2DE-ESV) and end-diastolic (2DE-EDV) left atrial volumes were measured by a modified Simpson's method of discs, using the 2DE Volume Biplane function (EchoPAC v 202, GE Healthcare). Intra-observer variability was assessed via calculation of 1 – the intra-class correlation coefficient (ICC) from random-effect linear models on EDV and ESV with horse added as the random effect (1 - ICChorse) in R (v 3.5.2, R) using the lmer and sjPlot packages. Intra-observer variability for the three-dimensional measurements has been previously presented (Worsman *et al.* 2019) and was assessed via calculation of 1 – the intra-class correlation coefficient. The same datasets were used for both measurements and comparison between two-dimensional and three-dimensional volumes were assessed using Bland Altman plots in R (v 3.5.2) using the BlandAltmanLeh package. Average 2DE-EDV was 684 ml (range 440–1144 ml) while 2DE-ESV was 404.4 ml (range 212–786 ml), n=24. Lower observer variation (1 - ICChorse) for 2DE-EDV measurements was observed (25%) compared to 2DE-ESV (38%). This suggests agreement between measurements (1 - ICChorse <50%). Bland Altman analysis of two-dimensional versus three-dimensional volumes suggested a mean difference of -23.4 ml for 3DE-ESV (lower limit CI -33, upper limit CI 290) (fig. 1), and -90.9ml for 3DE-EDV (lower limit CI -469.5, upper limit CI 287.8) (fig. 2). Although volumes assessed from two-dimensional and three-dimensional datasets were comparable, assessment of left atrial volume using the modified Simpson's method showed greater intra-observer variability and only moderate agreement between measurements, in comparison to the previously reported results for three-dimensional volume assessment which suggested good agreement (Worsman *et al.* 2019). Three-dimensional techniques for assessing volume are likely to be preferable for longitudinal evaluation of left atrial volume owing to less variability. Worsman, F.C.F., Blissitt, K., Shaw, D.J. and Keen, J.A. (2019) Assessment and intra-observer variability of equine left atrial volume using 4D Manual LVQ algorithm (GE Healthcare). ECEIM Congress Valencia 2019 Proceedings Abstract 74, page 85.

Figure 1. Bland Altman Plot ESV 3DE + 2DE agreement

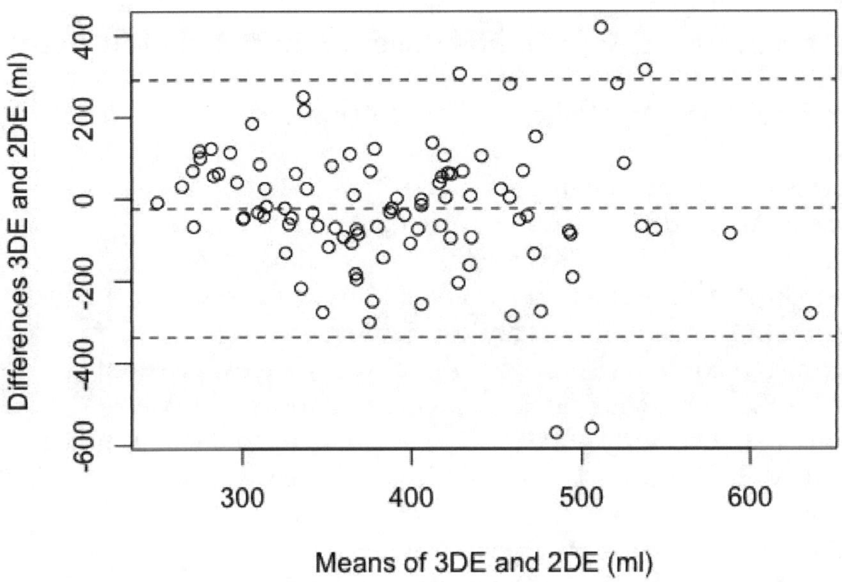

Figure 2. Bland Altman Plot EDV 3DE + 2DE agreement

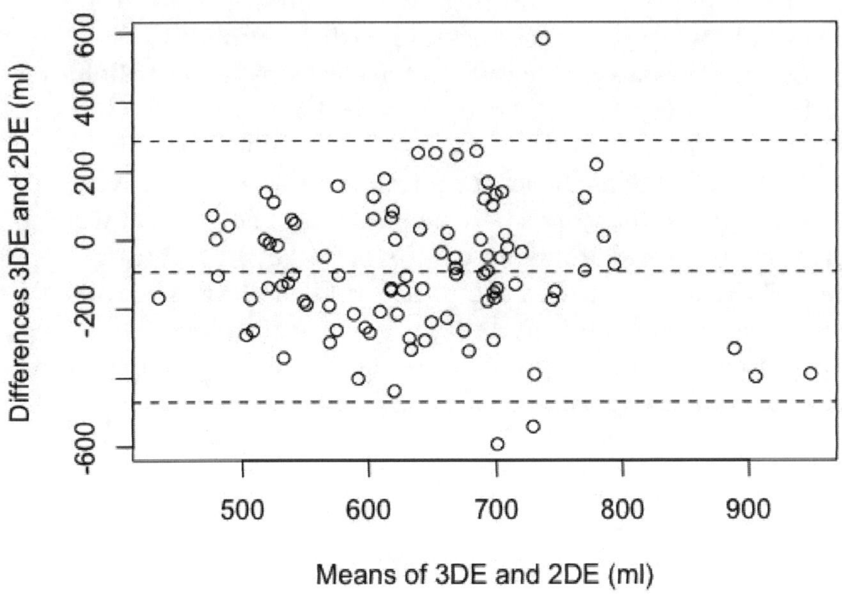

Abstract E06: Influence of Stethoscope on Cardiac Auscultation in the Horse

Gayle D. Hallowell; James Heale; Paul Lewis; James Bailey; Adam Redpath; Steffi Pratt
School of Veterinary Medicine and Science, University of Nottingham

The main objective of this study was to evaluate the impact that different stethoscopes have on the ability to auscultate the adult equine heart. 20 horses were auscultated by at least 5 individuals using nine different stethoscopes. Records were made regarding operator experience, body condition score, hair length, ambient noise level, breed type, heart rate, which heart sounds could be auscultated in the 4th and 5th left and right intercostal spaces. Any murmurs and dysrhythmias were also recorded. Users were also asked to give a preference score for each stethoscope per horse. Stethoscopes evaluated included various Thinklabs, Kruuse, Littman and MDF. Various statistical tests were used to evaluate the data including Chi-squared and Fisher's exact test, Mann Whitney U test, Friedman test and Wilcoxon matched pairs signed rank tests using Graphpad Prism 8.3. Significance was determined at $p<0.05$. Horses were aged between 5-23 years (13.8±4.9) years. The distribution was 32% mares and 68% geldings. They were equally distributed between TB type (38%), ISH (33%) and cob (29%), The median weight was 620 kg (IQR600–650). Median body condition score was 6(IQR5–6) and median heart rate was 36 beats per minute (IQR32–36). 20% of horses were found to have second degree atrio-ventricular blockade and 17% had murmurs (50% systolic murmurs associated with the mitral valve and 50% physiological aortic flow murmurs). The electronic stethoscopes and one of the Littman stethoscopes were superior regarding audibility of S4 ($p=0.04$) and S3 ($p=0.03$). Grading of murmurs was higher when the electronic and cardiology stethoscopes were used, compared to others ($p=0.02$). All users reported that their preference was influenced by how easy it was to hear heart sounds on the right. All stethoscopes had similar preferences, except for the MDF, which was the least preferred ($p=0.04$). There was a wide range in preferability of the electronic stethoscopes between users and horses that was related to how long the haircoat was ($p=0.01$) and ambient noise level ($p=0.03$). Preferred stethoscope within users was related to what they were used to using. Overall the electronic stethoscopes increased the ability to hear heart sounds and those and the cardiology stethoscopes allowed for improved auscultation of murmurs. The electronic stethoscopes were less good in noisier environments and on hairier animals.

Abstract E07: Effects of Short-Term Induced Atrial Fibrillation on Atrial Function After Cardioversion

Charlotte Hopster-Iversen[1]; Rikke Buhl[2]; Helena Carstensen[2]; Sanni Hansen[2]; Eva Hesselkilde[2]; Sarah Nissen[2]
[1]Faculty of Health and Medical Sciences, University of Copenhagen; [2]University of Copenhagen

Atrial fibrillation (AF) causes electrical and contractile remodeling in horses. Aim of this study was to evaluate the effects of short-term induced AF on left atrial contractile function after spontaneous cardioversion. Nine healthy standardbred horses were included. The right atrium was tachy-paced (600 bpm) for 20 hours to simulate AF. Echocardiographic evaluation of the left atrium (LA) was performed before induction of AF (T0) and immediately after cessation of the atrial pacing and spontaneous conversion to sinus rhythm (T1). The echocardiographic measurements included LA diameter, area and ejection phases indices. Atrial TDI (Tissue Doppler Imaging) measurements included peak myocardial velocity during atrial contraction (A), time to onset A (onset A), time to peak A (tA) and duration of A (durA). Short-term simulated AF resulted in significantly reduced atrial contractile function (active fractional area change of the LA and ratio of active-to-total LA area change) and prolonged TDI-based conduction time (onset A) at T1 compared to T0. Previous studies in horses with naturally occurring AF have shown atrial electrical and contractile remodeling compared to horses in sinus rhythm with a gradual improvement over time after cardioversion. The present study demonstrated reduced atrial function when the horses returned to sinus rhythm already after just 20 hours of tachy-paced AF. This highlights the importance of rapid treatment of AF in order to prevent contractile remodeling of the atria and avoid possible negative effects on the long-term prognosis for sustained sinus rhythm.

Abstract E08: Can the Frequency of Vagally-Mediated Arrhythmias After Conversion Predict Atrial Fibrillation Reoccurrence in Horses?

Melodie J. Schneider; Katharyn Mitchell; Colin Schwarzwald
University of Zurich

Reoccurrence of atrial fibrillation (AF) after cardioversion is a common problem in horses, with recurrence rates of 16–50% reported. Known predictors of reoccurrence include previous episodes of AF, left atrial (LA) enlargement, atrial stunning after conversion, mild or greater mitral regurgitation and >4 mth duration of AF. In human medicine, highly vagal individuals (particularly extremely fit athletes) have a 3-fold increase in the risk of developing AF through vagally-mediated electrophysiology mechanisms such as shortened atrial effective refractory period and action potential duration. This study aimed to examine the relationship between atrioventricular blocks (AVB) detected in 24 hr ECG recordings after successful cardioversion of AF and the rate of reoccurrence of AF. The medical records for all AF cardioversions between 2007–2019 were examined and information about AF reoccurrence collected. Three 24 hr ECG recordings (0–24, 24–48 and 48–72 hr periods after conversion) were analysed for the frequency of arrhythmias. Two-way, repeated measures ANOVAs were performed to compare the number of arrhythmias in horses with and without AF reoccurrence. A Chi-squared test was used to evaluate the effect of LA enlargement and treatment on AF reoccurrence. Twenty-one horses underwent 24 successful conversions (12 with quinidine sulfate, 12 with transvenous electrical cardioversion) during this period. At follow-up, 10 horses were in normal sinus rhythm (48%), 9 horses were in AF (43%) and 2 horses were lost to follow up. The median number of AVB per 24 hr recording period was 60 (min 0–max 4905). There was no difference in the number of AVB/24 hr period in horses with and without reoccurrence of AF (P=0.51). There was a median of 27 atrial premature complexes (APCs) per 24 hr recording period (min 2–max 1417). No difference between the number of APCs/24 hr period in horses with or without AF reoccurrence was observed (P=0.27). Horses with LA enlargement had a higher rate of reoccurrence (reoccurrence rate 80% vs. 36% with no LA enlargement, P=0.04). There was no difference in the rate of reoccurrence between the two treatments (P=1). In this cohort of horses with AF, there was no association between the frequency of AVBs or APCs in the immediate (0–72 hr) post conversion period and the reoccurrence of AF. This study is limited by the small sample size.

Abstract E09: New Diagnostic Opportunities for Diagnosing Paroxysmal Atrial Fibrillation in Horses

Rikke Buhl; Sarah Nissen; Helena Carstensen; Sofie Poulsen; Marie Louise Winther; Charlotte Hopster-Iversen; Thomas Jespersen; Eva Hesselkilde
University of Copenhagen

In equine cardiology, only limited information about paroxysmal atrial fibrillation (PAF) is available and undiagnosed PAF could be related to poor performance in horses. Due to the intermittent occurrence, PAF is difficult to diagnose, however by implanting a small ECG device (Implantable loop recorder, ILR) subcutaneously automatic detection of PAF can be obtained. The aim of the study was to investigate if ILRs were able to detect PAF in poor performing horses and how reliable the method was. Twelve Standardbreds (mean age 6 years (range 3–8 years)) with a history of reduced performance were recruited from the racetracks. After cardiac examination, an ILR (Reveal Linq, Medtronic) was inserted subcutaneously at the sixth left intercostal space after local anesthesia. After implantation, the horses were trained and raced with no restrictions. The ILRs were read at various time points. The mean period of ILR implantation was 6.5 months (range 2–24 months). In three horses, PAF was detected with a mean duration of 7 hours (range 1 hour 36 min–10 hours 5 min, in total five episodes). Furthermore, one horse had persistent AF detected by the ILR. The ILR detected marked sinus arrhythmia in three horses. Overall, the ILRs were able to detect correct heart beats in 95% with 5% misclassifications. The algorithm had a positive detection in 85% and a false detection in 15% of the AF episodes. No episodes were recorded during exercise, due to movement artefacts. Surprisingly, PAF was detected in three out of 12 horses, suggesting that PAF may occur more frequently in the horse population than previously expected. Therefore, ILRs can be a useful ECG tool for horses presented with poor performance. As no episodes were recorded during exercise, we cannot conclude whether PAF also occurs during physical activity.

Abstract E10: Phenotypic and Functional Characterisation of Equine Endothelial Cells

Elizabeth J. Finding; Robert Purcell; Nicola Menzies-Gow; Jonathan Elliott; Caroline Wheeler-Jones
Royal Veterinary College

Endothelial cells are ubiquitous and universally useful, but isolation without proper identification can lead to data misinterpretation. Studies making use of equine endothelial cells (EC) include those exploring viruses (e.g., EHV), laminitis and inflammation but investigation of equine EC is also relevant to wound healing, regenerative medicine and exercise physiology, areas that have received little attention to date. Although equine EC have been used in scientific research since 1985 there is limited information on the *in vitro* biology of these cells. This study characterises equine aortic endothelial cells (EAEC), providing the essential information for researchers to study these cells with confidence. EAEC were isolated from mixed breed adult horses euthanased at a commercial abattoir using collagenase digestion (0.25 mg/ml; 10–20 mins) or mechanical scraping. Cells were cultured on gelatin-coated tissue culture flasks in Medium 199 (M199) supplemented with 20% equine serum, endothelial cell growth factor, heparin and antibiotics, or in Endothelial Cell Growth Medium 2 (ECGM2; Promocell) supplemented with 20% equine serum and antibiotics. Cells were visualised using phase contrast light microscopy and fluorescent immunocytochemistry (ICC). ICC results were validated using en face imaging of equine intercostal arteries. EAEC angiogenic potential was investigated using a thin layer 2-dimensional angiogenesis assay (TLA). Intracellular signalling responses to treatment with vascular endothelial growth factor (VEGF; 1–100 ng/ml; 1–60 min) were investigated using Western blotting to detect phosphorylation of extracellular signal-regulated kinase 1 and 2 (ERK 1/2). Equine blood outgrowth cells (eBOC) from healthy horses were isolated by culturing fresh peripheral blood monocytic cells (PBMC) on collagen using equine-specific reagents and a standard protocol in use for human blood outgrowth EC isolation (also known as endothelial cell colony forming cells; ECFC). eBOC colonies were observed 11–15 days after initiation of PBMC culture. Colonies of more than 100 cells were passaged and propagated on collagen-coated plates prior to analysis by ICC, flow cytometry and immunoblotting. Human ECFC were isolated and characterised in parallel. Mechanical scraping yielded a higher proportion (80%; n=30) of EC-rich cultures (>60% vWF-positive cells) than collagenase digestion (33%; n=27; p=0.0005). EAEC growth rate was greater in ECGM2 than M199 (2.5 ± 1.2 cells/h vs. 17.3 ± 1.2 cells/h; n=1; p<0.0001). The endothelial nature of EAEC was confirmed by immunostaining for von Willebrand factor (vWF), with cells displaying punctate staining suggestive of vWF storage within Weibel-Palade bodies. This distribution was also seen in en face preparations. Other antibodies raised against EC-specific or EC expressed proteins did not cross react with EAEC antigens in ICC in unstimulated cells or in cells exposed to inflammatory cytokines; these include CD31, CD34, CD105, VE-cadherin, VEGFR2, ICAM-1, ICAM-2, VCAM-1, claudin 5 and ZO-1. EAEC tubulogenesis, assessed using the TLA, was increased in response to VEGF but performance of the assay varied between cells cultured in M199 or ECGM2. Treatment with VEGF resulted in phosphorylation of ERK1/2 with a maximal response at 25 ng/ml and after 5, 10 and 30 minutes incubation. Human ECFC displayed typical EC cobblestone morphology, expressed EC markers (e.g., CD31; vWF, VE-cadherin) and displayed VEGF-stimulated tubulogenesis in the TLA assay. In marked contrast, eBOC cultures (n=8) had a fibroblastic morphology, expressed typical mesenchymal stromal cell (MSC) markers (CD29, CD44, CD90, CD105) and failed to form tube-like structures in the TLA assay. This study provides the initial essential information necessary for detailed cell biology studies of EAEC. Not all previous studies using equine vessel EC or purporting to have isolated equine ECFC have confirmed the endothelial identity of the cell population isolated. The level of contamination evident in some EAEC cultures raises concerns that results from studies without EC identification are representative of a mixed vascular cell population, rather than a specific endothelial population. One major challenge facing equine EC research is the lack of commercially available equine-specific antibodies for endothelial antigens. Despite thorough investigations, the only EC-specific protein reliably detected was vWF. This is an essential antibody to identify EC but it cannot be used to purify cells from a mixed population; this would require an antibody targeted to a cell surface antigen. Preliminary investigations of the angiogenic functions of EAEC have been performed, including assessment of VEGF signalling, to provide a framework for further studies. The difficulties associated with isolation of equine ECFC from PBMC cultures means that rigorous characterisation of equine vascular EC models is critical for progressing understanding of equine endothelial cell biology.

Abstract E17: Poincaré Plots as a Visual Measure of Heart Rate Variability in Healthy Horses

Katharyn J. Mitchell; Michaela Kuehni; Colin Schwarzwald
University of Zurich

Poincaré plots are graphical displays of heart rate variability (HRV) that are used in ECG analysis. Although regularly utilised in human and small animal cardiology, these graphical analyses are not well described in horses. This study aimed to define Poincaré patterns in overnight ECG recordings from healthy horses and assess between-day variability in Poincaré patterns in a subset of horses. Overnight ECG recordings were obtained from 30 healthy horses. Five horses had three ECGs recorded. The ECGs were corrected for artefacts. Arrhythmias were noted but not removed from the data set. Percent instantaneous beat-to-beat cycle length variation (% R-R variation) was calculated, Poincaré plots were created and traditional HRV analyses performed. The mean (min, max) recording length was 11 hrs 32 min (10 hrs 18 min, 13 hrs 21 min). Atrioventricular (AV) blocks, sinus pauses or both were present in 27 horses. Five horses showed frequent (>1/hr) atrial or ventricular premature complexes and were separated from the remaining analyses. The maximum % R-R variation was between -57.9 and 133.2%, with the median variation -0.2%. Evaluation of the Poincaré plots revealed healthy horses to have a comet or wedge pattern with RR intervals sequences containing second-degree AV blocks and sinus pauses/blocks appearing as clusters in the upper and lower right quadrants. Premature complexes appear as clouds in the lower left, upper left and middle-lower right quadrants. Repeatability of the Poincaré patterns was excellent, with most variables showing very low-low variability (coefficients of variation between 3.8 and 28.5%). Poincaré plots as graphical representations of HRV allow the observer an immediate overview of the cardiac rhythm patterns during the period of recording. The pattern of overnight equine Poincaré plots appears similar to those of resting humans but quite different to those reported in dogs. This difference is postulated to relate to the marked respiratory sinus arrhythmia observed in canines. Overnight ECG recordings have excellent between-day repeatability and provide very similar Poincaré patterns.

Abstract E18: Novel Mobile Applications Enable Wearable Devices to Yield Accurate Exercising ECG and Heart Rate Data in Horses

Joanne E. Haughan[1]; Noah Cohen[2]; Mary Robinson[3]; Cristobal Navas de Solis[3]
[1]Department of Clinical Studies, University of Pennsylvania School of Veterinary Medicine, New Bolton Center, Kennett Square, Pennsylvania; [2]College of Veterinary Medicine & Biomedical Sciences, Texas A&M University; [3]University of Pennsylvania School of Veterinary Medicine, New Bolton Center

User-friendly tools are needed to investigate sudden cardiac death in horses. We hypothesized wearable devices would provide diagnostic-quality electrocardiograms (ECGs). ECGs obtained with 2 wearable devices (W2nd™and Polar Equine™) were compared to simultaneous recordings with a telemetry unit (Televet™) in 5 Thoroughbreds completing 43 separate submaximal exercise tests on a treadmill. Voltage data were generated by mobile applications and analyzed using Kubios software by 1 blinded investigator. Televet recordings were analyzed routinely. Maximal heart rate (HR) generated by mobile applications (HRmaxapp), HR after manual correction (HRmaxcorr), percentage diagnostic ECGs (%diag) at the gallop, and overall quality assessed by visual analogue scale (VAS) were compared. Statistical analysis was performed using linear mixed-effects modeling, chi-squared tests, and Bland-Altman plots. Significance was set at $P<0.05$. Acceptable %diag was defined as >95%/ECG and a priori limits of agreement for HRmax were set as ±2.5% maximal HR. HRmaxcorr did not differ between groups. HRmaxapp was significantly lower for W2nd (166.8/min, 95% CI: 160.5–173.1/min) but did not differ significantly between Televet (178.8/min 95% CI:165.8–191.1/min) and Polar (181.3/min, 95% CI:174.5–188.1/min). HRmaxcorr was accurate and precise in all runs but HRmaxapp was within a priori limits of agreement in 16/23 W2nd and 18/19 Polar recordings. %diag was significantly lower (77.1%, 95% CI:67.4–86.8) for W2nd than Polar (100%, 95% CI:89.9–110.3). VAS was lower for W2nd (46.2, 95% CI:35.5–57.0) than Polar (90.6, 25% CI:79.4–101.9) and Televet (82.8, 95%CI:74.0–91.5). Wearable devices are promising tools for large scale studies of exercising arrhythmias in horses.

Abstract E19: Plasma and Urine Aldosterone in Normal Horses and Horses with Subclinical Valve Disease

Babetta A. Breuhaus
North Carolina State University College of Veterinary Medicine

The renin-angiotensin-aldosterone system (RAAS) is activated in heart failure to support arterial blood pressure. However, chronic activation of RAAS becomes maladaptive, potentially causing myocardial injury and fibrosis. Horses with heart disease are sometimes treated with angiotensin-converting inhibitors; however, benefits of therapy are theoretical and cost-effectiveness uncertain. Accordingly, we attempted to measure when RAAS is activated in horses with cardiac disease. A commercially-available assay for plasma and urine aldosterone concentrations was evaluated in 22 healthy horses, and in 30 horses with chronic, valvular heart disease (VHD) due to mitral or aortic regurgitation. Diagnoses were based on clinical examination and Doppler echocardiography. Blood and urine samples were obtained from all horses and stored for quantitation of aldosterone and creatinine concentrations. Percent recovery, intra- and inter-assay coefficients of variation were acceptable. Preliminary reference intervals were established in normal horses using CLSI robust statistical methods. Median (upper confidence limit) concentrations of plasma and urine aldosterone were 254 (1058) and 7890 (18007) pmol/L, while urine aldosterone:creatinine ratio (UA:C) was 745.3 (UCL=1672.6) pmol/mmol. Accepting the limitations of a small sample for generating relatively wide reference ranges, aldosterone concentrations were not increased in horses with VHD, including subanalysis of 16 with increased left heart size. Plasma aldosterone was weakly correlated to UA:C ($r2=0.231$, $p<0.001$). These preliminary results suggest that RAAS is not continually active in some horses with subclinical VHD, even when there is cardiac enlargement. Further studies in horses should focus on those with heart failure and specific preclinical stages of VHD.

Abstract E20: Seasonal Variation of Endogenous Adrenocorticotropic Hormone (ACTH) in Healthy Non-Geriatric Donkeys in Northern California

Sarah Schale[1]; Erin Goodrich[2]; Philip Kass[1]; Emily Berryhill[1]
[1]University of California Davis; [2]Cornell University

Background: Elevated plasma adrenocorticotropic hormone (ACTH) is often used to diagnose pituitary pars intermedia dysfunction (PPID) in horses. The hormone is known to naturally increase in the fall season in horses. Donkeys have been found to have higher ACTH concentrations than horses, however, variation of ACTH concentrations over a 12-month period has not been assessed. **Hypothesis**: We hypothesized that donkey ACTH concentrations would be higher than horses in all seasons, and that, similarly to horses, ACTH concentrations would further increase in the fall. **Animals**: 25 healthy donkeys (10 standards, 15 miniatures), a median (range) of 6 (2–13) years of age, and at the same location under the same management. **Methods**: Prospective observational study. Serial plasma samples obtained bimonthly from June through November and monthly from December through May. Plasma ACTH concentrations were determined by a validated assay. Data are presented as medians with lower and upper 95% confidence intervals (95% CI). **Results**: Donkey ACTH concentrations were lowest in the winter and spring (12.6 [9.7–15.5] pg/mL and 19.0 [16.5–21.5] pg/mL, respectively), with an increase in the summer (52.9 [45.5–60.3] pg/mL), and peak in the fall (77.2 [64.5–89.7] pg/mL). Fall concentrations were higher than those in spring and winter (P-value <0.0001). ACTH concentrations were highest in September (123.0 [95.4–150.4] pg/mL). **Conclusion and clinical importance**: Donkey ACTH concentrations were similar to normal concentrations reported in horses in winter and spring but were higher in summer and fall, with a marked increase in fall. Donkey-specific reference ranges are required for accurate endocrinopathy screenings.

Abstract E21: Androgens and Estrogens in Healthy and Hospitalized Neonatal Foals

Jacob M. Swink[1]; Lindsey Rings[2]; Hailey Snyder[1]; Katarzyna Dembek[3]; Ahmed Kamr[1]; Nimet Browne[4]; Katherine Christie[2]; Ramiro Toribio[1]
[1]The Ohio State University; [2]Rood and Riddle Equine Hospital; [3]Iowa State University; [4]Hagyard Equine Medical Institute

Sepsis and neonatal maladjustment syndrome (NMS) are major causes of morbidity and mortality in neonatal foals. Dysfunction of the hypothalamic-pituitary-adrenal axis is frequent in critically ill newborn foals. Some information is available on glucocorticoids, mineralocorticoids, and progestogens in healthy and sick foals, but little is known about androgens and estrogens. High progestogens have been associated with sepsis and NMS. The goal of the study was to measure androgens and estrogens in hospitalized foals. Blood was collected on admission (0) and at 24, 48, and 72 hours from 62 healthy, 56 septic, and 41 sick non-septic (SNS) foals of <3 days of age. Serum androgens and estrogens were measured using immunoassays. At admission, estrone sulfate, androstenedione, and testosterone concentrations were higher in septic compared to healthy and SNS foals ($P<0.05$). Estradiol, estrone sulfate, testosterone, dehydroepiandrosterone (DHEA), and androstenedione were higher in foals with evidence of NMS ($P<0.05$). Testosterone, estrone sulfate, and estradiol were higher in non-surviving foals ($P<0.05$). All hormones decreased over time in healthy foals but remained elevated in septic foals. Foals with higher testosterone concentrations were more likely to die. The areas under the curve for estradiol, estrone, dihydrotestosterone (DHT), androstenedione, and DHEA were good predictors of non-survival. This study showed that androgen and estrogen concentrations are altered in sick foals and their dynamics during hospitalization are associated with disease severity and outcome. Increased sex steroid concentrations likely reflect delayed clearance and organ dysfunction. Some of these steroids may have prognostic value.

Abstract E22: Effect of Altrenogest Administration to Pregnant Mares on the Endocrine Profile of Their Foals

Jacob M. Swink[1]; Lindsey Rings[2]; Hailey Snyder[1]; Rachel McAuley[1]; Katarzyna Dembek[3]; Michele Frazer[4]; William Gilsenan[2]; Ramiro Toribio[1]
[1]The Ohio State University; [2]Rood and Riddle Equine Hospital; [3]Iowa State University; [4]Hagyard Equine Medical Institute

Sepsis remains the leading cause of mortality in newborn foals. Critical illness alters the hypothalamic-pituitary-adrenal axis in newborn foals. Administration of exogenous progestogens (altrenogest, progesterone) is common practice in the equine industry. Exogenous steroids cross the placental barrier and elevated endogenous progesterone concentrations have been associated with illness in newborn foals. Therefore, the goal of this study was to determine the association between altrenogest administration to pregnant mares and the steroid profile/disease status of their foals. Blood samples were collected on admission from 62 healthy, 56 septic, and 41 sick non-septic (SNS) foals of <3 days of age. Foals within groups were further divided into born to altrenogest-treated or -untreated mares. Serum progesterone, 17α-hydroxyprogesterone, pregnenolone, dehydroepiandrosterone (DHEA), allopregnanolone, and cortisol, and plasma adrenocorticotropic hormone (ACTH) were measured using immunoassays. Altrenogest-exposed septic foals had lower ACTH and 17α-hydroxyprogesterone, but higher progesterone than unexposed foals ($P<0.05$). Altrenogest-exposed non-surviving foals had higher pregnenolone and progesterone concentrations than those from untreated-mares ($P<0.05$). Allopregnanolone, cortisol, DHEA, or ACTH:steroid ratios were not different between exposed and unexposed foals. Hormone levels between exposed and unexposed foals within healthy, SNS, surviving, or NMS groups were not different. Altrenogest exposure was not associated with likelihood of mortality. This study demonstrates that exogenous progestogens could influence the steroid profile of septic newborn foals. Progestogen difference between septic altrenogest-exposed and unexposed foals does not imply a cause of disease, but an association, which due to its implications deserves additional controlled studies before these findings are extrapolated.

Abstract E23: The Effect of Trailering and Dentistry on Resting Adrenocorticotropic Hormone Concentration in Horses

John C. Haffner[1]; Steve Grubbs[2]; Rhonda Hoffman[3]
[1]Middle Tennessee State University Horse Science Center; [2]Boehringer Ingelheim; [3]Middle Tennessee State University

Pituitary pars intermedia dysfunction (PPID) may affect >20% of horses aged ≥15 years. The measurement of adrenocorticotropic hormone (ACTH) is the most commonly used diagnostic test used for the diagnosis of horses with PPID. The diagnosis of PPID is supported by a plasma ACTH concentration greater than the seasonally adjusted reference range. However, several studies have concluded that pain, stress and concurrent illness were only likely to affect diagnostic usefulness of resting ACTH when severe. The objective of this study was to identify if trailering or teeth floating (common stressful situations/procedures) increased plasma ACTH concentrations in horses. Twelve horses were enrolled and randomized into 3 groups of 4 horses/group. Each horse group was randomly assigned to the initial treatment group, dentistry (DN), trailered (TR) or stabled controls (CN). Following initial treatment, each horse group was randomly assigned to each of the two remaining treatment groups; therefore, each horse group underwent all 3 treatments. Plasma was collected from all horses prior to each treatment (baseline). The DN horses were placed in stocks, sedated with 0.1 to 0.3 mg/lb xylazine IV and following mouth speculum placement, teeth were floated with a PowerFloat®. The TR group was loaded on a six-horse slant trailer and hauled for 40 minutes. Immediately following the dental procedure and trailer ride, post-procedure (P0) plasma samples were collected. Plasma samples were then collected from all horses at 15, 30, 60 and 120 minutes post-procedure. Plasma samples from the CN horses were taken when the trailered horses returned. Plasma samples were frozen (-80C) until analysis at Cornell Animal Health Diagnostic Center. Data were analyzed using a mixed model with repeated measures (i.e., each horse as its own control), with main effects of treatment (CN, DN, TR) and time, and day x time as the repeated effect. Statistical significance was designated at $P<0.05$, and $0.05<P<0.10$ was considered a trend. Data were summarized as mean ± SE. No change occurred in ACTH over time in the CN or DN horses (P=0.14). ACTH was higher in TR compared to CN (P=0.026) and DN (P=0.016) horses. In TR horses, ACTH was higher than baseline (PRE) immediately after (T0; P=0.0003) and tended to be higher (P=0.066) at 15 min after trailering. By 30 min post-trailering, there were no differences in mean resting ACTH compared to PRE concentrations (P=0.55). No significant difference in resting ACTH concentrations over time was observed in horses undergoing dentistry procedure compared to baseline. A forty-minute trailer ride resulted in significantly increased resting ACTH concentrations in horses up to 30 min post-unloading. Based on results of this study, collecting blood from horses within 30 minutes from trailer unloading may result in elevated resting ACTH concentrations.

Abstract E37: Diagnostic Evaluation of Insulin and Glucose Dynamics in Light Breed Horses with Experimentally-Induced Insulin Dysregulation

Kathryn Timko; Laura Hostnik; Mauria Watts; Chiaming Chen; Adam Bercz; Ramiro Toribio; James Belknap; Teresa Burns
The Ohio State University

Insulin dysregulation (ID) increases the risk for developing laminitis and other metabolic derangements. Therefore, proper identification of horses with ID is important to facilitate preventive treatment strategies. This study compared three diagnostic tests for ID: baseline insulin (BI) and glucose concentrations, oral sugar test (OST), and combined glucose and insulin test (CGIT) in horses before and after induction of ID. Fourteen adult light breed horses received dexamethasone (0.08 mg/kg PO q 24 h; 7 days) to induce ID. BI, OST, and CGIT were evaluated before and after ID. Observed agreement and kappa coefficients were calculated for BI, OST and CGIT for the diagnosis of ID. Proxy measurements of ID (reciprocal of the square root of insulin [RISQI], quantitative insulin sensitivity check index [QUICKI], homeostasis model assessment [HOMA] and insulin-to-glucose ratio [IG]) were calculated. Dexamethasone administration significantly changed the classification of horses from normal to ID based on BI, OST, and CGIT (P=0.0013, 0.0128, 0.0006 respectively). RISQI, QUICKI, HOMA and IG were all significantly different after dexamethasone administration. There was moderate agreement between BI and OST [ins]60 >60 µIU/mL, BI and [ins]60 >45 µIU/mL and between BI evaluations performed 24 hours apart (k=0.512, 0.435, 0.444 respectively). There was poor agreement between BI and CGIT using diagnostic criteria for insulin and glucose (k=-0.026, 0.122 respectively). All tests and proxy measurements supported induction/exacerbation of ID after dexamethasone treatment. The variability in diagnostic performance of common tests for equine ID may affect clinical decisions; performing multiple tests may improve accuracy.

Abstract E38: Effect of Early or Late Sampling on Thyrotropin-Releasing Hormone (TRH) Stimulation Test Results

Kristen Thane[1]; Nicholas Frank[1]; Amy Rubin[2]; Cassandra Uricchio[2]
[1]Tufts University; [2]University of Massachusetts

Diagnosis of pituitary pars intermedia dysfunction (PPID) using the thyrotropin-releasing hormone (TRH) stimulation test requires precise timing from injection of TRH to blood sample collection. The objective of this study was to determine whether early or late sampling would result in a significant (>10%) difference in plasma adrenocorticotropic hormone (ACTH) concentration compared to standard 10-minute sampling. Sixteen healthy adult horses with unknown PPID status underwent a single TRH stimulation test, with blood aseptically collected into EDTA vacuum tubes immediately prior to TRH injection and exactly 9 minutes (early), 10 minutes (standard), and 11 minutes (late) post-injection. Blood was processed within 4 hours of collection; plasma was stored at -20C until analysis and ACTH was measured by chemiluminescent immunoassay. Two aliquots of each 10-minute plasma sample were analyzed separately to assess intra-assay variability. Minor variability was observed between the paired 10-minute sample aliquots (range 0.0% to 5.0%; median 2.3%). Both early and late samples exhibited greater variability compared to the average of the paired 10-minute samples, and overall variability of early or late samples compared to the corresponding 10-minute mean concentration ranged from 0.1% to 62.7% (median 10.3%). Seventy-five percent of horses tested had at least one early or late reading that differed by >10% from its corresponding 10-minute mean value. Incidence of >10% variability was independent of PPID status. Precise timing of sample collection is critical to ensure accurate assessment of PPID status given the observation of significant variability associated with minor alterations in sample collection timing.

Abstract E39: Cartilage Oligomeric Matrix Protein Differential Expression in Lamellar Tissue from Prolonged Euglycemic Hyperinsulinemic Clamp Model

Sarah F. Colmer[1]; Andrew Van Eps[2]; Hannah Galantino-Homer[1]
[1]University of Pennsylvania School of Veterinary Medicine; [2]University of Pennsylvania School of Veterinary Medicine, New Bolton Center

Endocrinopathic laminitis (EL) is one of the most common forms of equine laminitis, although its precise pathogenesis remains to be discovered. The euglycemic hyperinsulinemic clamp (pEHC), implemented by intravenous continuous rate infusion of insulin while plasma glucose concentrations are held constant at basal levels by glucose infusion, induces laminitis and epidermal lamellar lesions in both fore and hind limbs. In addition, distal limb cooling prevents pEHC laminitis and epidermal lesions in the cooled limb. However, little is known about dermal tissue damage in EL. Cartilage oligomeric matrix protein (COMP) serves as a marker of damage or degradation of collagen in connective tissue (cartilage, tendon, or dermis). The objective of this study was to determine if COMP is reduced in lamellar tissue from the pEHC model and if cryotherapy prevents COMP loss and dermal lamellar damage. COMP was quantified through immunoblotting and densimetry of protein extracts from snap-frozen lamellar tissue from a cooled forelimb and ambient fore and hind limbs from 8 young Standardbred horses subjected to 48 hours of pEHC. A statistically significant decrease in COMP was identified in extracts from both the ambient forelimbs ($P<0.001$) and hind limbs ($P<0.001$) compared with cooled forelimbs. These results suggest that dermal damage contributes to EL pathogenesis in both forelimbs and hind limbs and that it can be prevented by cryotherapy.

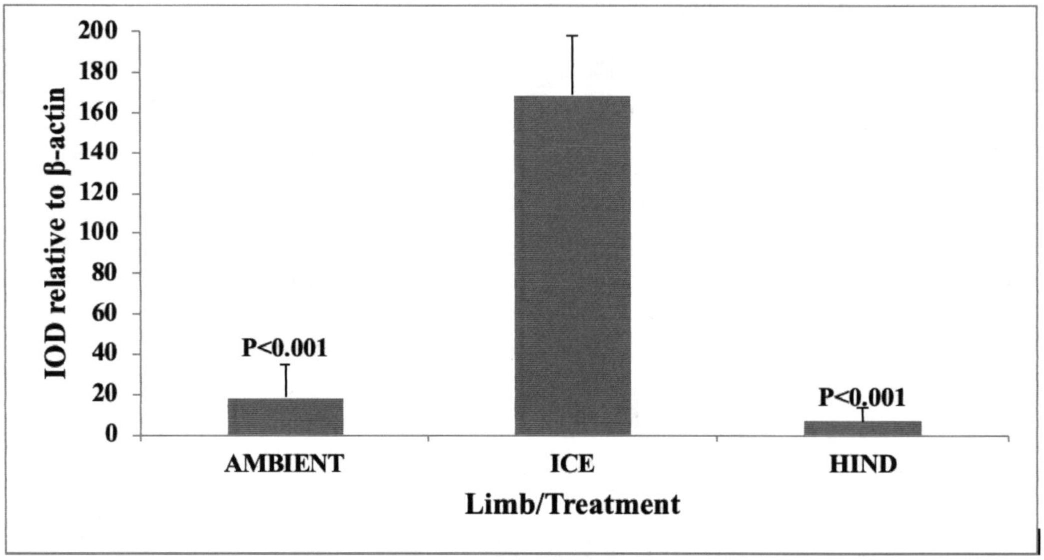

Figure 1: Concentrations of lamellar COMP in pEHC-model protein extracts. AMBIENT = ambient temperature forelimb, ICE = cryotherapy-treated forelimb, HIND = ambient temperature hind limb. Statistically significant decreases in the integrated optical density (IOD) of lamellar COMP concentrations relative to β-actin controls were observed in ambient temperature forelimbs and hind limbs compared with cryotherapy-treated forelimbs (N=8)

Abstract E80: The Value of Transabdominal Ultrasound in Horses Presenting with Medical Gastro-Intestinal Disease

Gayle D. Hallowell[1]; Carolyn Bates[2]; Emma Peal[3]; Caroline Bullard[3]; James Bailey[4]; Adam Redpath[4]; Mark Bowen[4]
[1]School of Veterinary Medicine and Science, University of Nottingham; [2]Defence Animal Training Regiment; [3]Royal Army Veterinary Corps, Household Cavalry Mounted Regiment; [4]SVMS, University of Nottingham

To evaluate the frequency with which abnormalities are detected using transabdominal ultrasound in horses that present with weight loss, chronic diarrhoea, pyrexia of unknown origin or recurrent bouts of abdominal pain. Clinical and ultrasound case records were searched for abdominal evaluations over the period of 2015–2020 in horses >1 year of age. Signalment, ultrasound findings and final diagnoses were recorded as was outcome. Abnormalities recorded from abdominal ultrasound included: increased small (SI) or large intestinal (LI) wall thickness defined as >4 mm, abnormal intestine, abnormal appearance or volume of peritoneal fluid or abnormal abdominal organ or other. Other information recorded included specific intestinal wall thickness, other tests performed, treatments administered and any further findings. Recurrent colic was classified as >3 episodes in 3 months and chronic diarrhoea of >7 days duration. Images were obtained using a Vivid Q ultrasound machine (GE Healthcare) using both a 3 MHz convex and 7.5 MHz microconvex probes and saved as cineloops that were evaluated offline. 224 evaluations were found that had both abdominal ultrasound images and clinical reports or records. Horses ages ranged from 3–28 (13.5±4.8 years), were predominantly TB/part-breds (29%) or ID/part-breds (29%) with 59% geldings and 41% mares. 61% of evaluations were for horses with weight loss, 24% with recurrent colic, 9% with diarrhoea and 6% with other signs. Abnormalities were identified in 80% of examinations. Of those with abnormalities, 19% had thickened SI, 40% thickened LI, 21% thickened SI and LI, 4% abnormal motility, 8% abnormal peritoneal fluid and 8% with other abnormalities. Transabdominal ultrasound can identify abnormalities in a large percentage of horses that present with these clinical signs, making the time spent performing this technique worthwhile.

Abstract E42: The Bioavailability and Efficacy of a Novel Omeprazole Product in Horses

Jessica Wise; Kristopher Hughes; Sharanne Raidal
Veterinary Clinical Centre, Charles Sturt University

Omeprazole facilitates gastric ulcer healing through the suppression of gastric acid secretion. Enteric-coated omeprazole formulations have increased bioavailability; however, optimal dose rates of these preparations remain uncertain. The current study compared the bioavailability and efficacy pharmacodynamics of a novel, in-feed, enteric-coated granule formulation of omeprazole (NOV) with an existing, registered enteric-coated paste formulation (REF). Bioavailability was assessed by comparison of plasma concentrations of omeprazole following administration of a single dose of NOV and REF, with intravenous administration of omeprazole. The study included 9 horses randomly assigned to a three sequence, three period Latin square design, with a 7 day wash out period. Efficacy was assessed by comparison of plasma omeprazole concentrations, gastric pH and ulcer scores following repeated administration of NOV or REF to 12 horses for seven days in a prospective two period, blinded and randomised trial with sequential cross over. The median bioavailability of NOV was 58.2% (range: 11.0–91.1) and for REF was 21.6% (range: 10.1–82.6); there was no significant difference between products ($p=0.312$). Treatment was associated with increased gastric pH ($p< 0.001$) and reduced squamous ulcer scores ($p< 0.001$). There was no difference in gastric pH or squamous ulcer scores between omeprazole products. For both omeprazole products, treatment resulted in increased gastric pH ($p\leq0.001$). There was no difference in gastric pH profiles between products. Omeprazole treatment was associated with decreased gastric emptying. These results demonstrate good bioavailability and efficacy of the in-feed omeprazole formulation indicating, supporting the use of this method of drug administration in horses. Our results also indicate that plasma concentrations of omeprazole and drug dose used do not predict drug efficacy.

Abstract E43: Inter-Observer Agreement and Intra-Observer Repeatability of Two Scoring Systems for Equine Gastric Ulcer Syndrome

Jessica Wise; Sharanne Raidal; Kristopher Hughes
Veterinary Clinical Centre, Charles Sturt University

Clinical grading systems require good inter-observer agreement and intra-observer repeatability. A reliable scoring system for equine gastric ulcer syndrome (EGUS) would optimise ulcer scoring, including the effects of treatment and management protocols on ulcer healing, and comparison of results between studies. Six observers (three equine medicine specialists, three residents in equine disciplines) graded 60 pre-recorded de-identified equine gastroscopy videos, using two grading systems applied three times: the EGUS Council grading system and a novel visual analogue scale (VAS). There was a minimum period of one week between each grading occasion. The order of videos was randomised for each grading. Agreement coefficients were calculated using Gwet's agreement coefficient, with ordinal weights applied (AC2), for the EGUS Council grading system and the Intraclass Correlation Coefficient (ICC) for the VAS. Inter-observer agreement was poor for both squamous (mean AC2= 0.687) and glandular mucosa (mean AC2 0.721) using the EGUS grading system. The intra-observer repeatability of the squamous (mean AC2 0.801) and glandular mucosa (mean ICC=0.804) was acceptable for five observers. Inter-observer agreement was poor for both squamous (mean ICC=0.353) and glandular mucosa (mean ICC=0.635) grading, using the VAS. The intra-observer repeatability for squamous mucosa was acceptable for all experienced observers (mean ICC=0.828. The intra-observer repeatability for glandular mucosa was poor for all observers (mean ICC 0.555). These results indicate that agreement between observers is poor for grading of both squamous and glandular mucosal appearance, irrespective of the grading system used. Intra-observer repeatability is influenced by grading system used and observer experience. These findings have implications for comparison of results between observers and studies and selection of grading system for monitoring of horses with EGUS.

Abstract E45: Comparison of Blood Gas Analysis, Electrolytes and Plasma Protein Concentrations in Horses at Different Altitudes

Camilo Jaramillo[1]; Maria Arias[2]; David Renaud[1]; Diego Gomez[1]
[1]University of Guelph; [2]Universidad CES

The objective of this study was to determine the venous blood gas values, electrolyte and plasma protein concentrations in adult horses residing at 5 different altitudes. One hundred and forty-eight healthy Paso Fino horses >1.5 year-old were enrolled. The horses lived at five different altitudes: 0 masl (n=23), 1000 mamsl (3280 famsl, n=25), 1500 mamsl (4921 famsl, n=30), 2000 mamsl (6561 famsl n=30) and 3000 mamsl (9842 famsl, n=30). An EDAN i15® (Edan Instruments, Inc – Guangzhou – China, 2016), portable system was used to determine pH, PvCO2 (mm Hg) and HCO3- (mmol/L), hemoglobin (Hb, mg/dL), BE (mmol/L)] and whole blood electrolyte concentrations (mmol/L) (Na++, K+, Ca++, Cl- and L-Lactate). Packed cell volume (PCV, %) was measured by blood centrifugation in a microhematocrit tube in a microhematocrit centrifuge. Total plasma protein was measured as total solids using a refractometer. A One-way ANOVA with Tukey's HSD post hoc test or a Wilcoxon signed-rank test, depending of the normality of the data, were conducted to compare groups. Horses at 0 mamsl had lower PCV (25 ± 2%) and Hb (8.3 ± 0.8 mg/dL) concentration than those horses living above 1000 mamsl (P<0.05, for all comparisons). No differences in PCV and Hb were identified in horses living between 1000 and 3000 mamsl (P>0.05, for all comparisons). The venous PvCO2 was significantly lower in horses residing at altitudes above 1000 mamsl (37.8 to 39 mm Hg) than those living at 0 mamsl (42 ± 4.8 mm Hg) (P=0.001, for all comparisons). No differences in PvCO2 were identified in horses living between 1000 and 3000 mamsl. No differences in the concentration of whole blood electrolytes or plasma protein were observed in horses residing at any altitude. This study confirms the observation that red blood cell numbers increase with altitude. However, these differences were not identified when comparing horses living between 1000 to 3000 mamsl. Differences in PvCO2 also confirm hyperventilation as a respiratory adaptation to altitude.

Abstract E46: Imidocarb Dipropionate Fails to Clear Both *Theileria equi* and *Theileria haneyi* in Superinfected Horses

Kelly Sears[1]; Lindsay Fry[2]; Don Knowles[1]
[1]Washington State University; [2]USDA-ADRU

Equine theileriosis continues to present challenges for the international movement of horses. A newly identified *Theileria* sp. of equids, *Theileria haneyi*, further complicates regulatory efforts as it is not detected by current diagnostic assays for *Theileria equi*. Mounting prevalence data supports the distribution of this species, along with *T. equi*, in multiple regions around the world, and recent publications have verified that superinfection with both species is possible. Prevention of equine theileriosis is largely reliant on chemosterilization with imidocarb dipropionate (ID). However, a previous report suggested that complete clearance is not achievable when a horse is infected with multiple species. The purpose of this study was to determine if the currently accepted ID treatment regimen (4 mg/kg q 72h IM for 4 doses) would clear *T. equi* and *T. haneyi* in persistently, superinfected horses. Six experimentally infected horses were utilized. Three horses were infected with *T. equi* first, followed by *T. haneyi*, and three horses were infected with *T. haneyi* first, followed by *T. equi*. *T. equi* nested PCR was performed after each round of treatment, and suggested chemosterilization had not been achieved in a subset of horses. To verify the infection status following treatment, each horse was subsequently splenectomized. 2/3 of the horses first infected with *T. haneyi* developed severe *T. equi* recrudescence, requiring euthanasia, and verifying lack of successful treatment. In contrast, 3/3 of the horses first infected with *T. equi* failed to develop severe disease following splenectomy, confirming partial treatment success in this group. However, all horses remained persistently infected with *T. haneyi*. In conclusion, ID does not appear to be effective for complete chemosterilization of superinfected horses.

Abstract E47: Bile Acids, GGT, and Direct Bilirubin as Prognostic Indicators for Horses with Liver Disease

Barbara Delvescovo[1]; Joy Tomlinson[1]; Sally Anne DeNotta[2]; Elizabeth Hodge[3]; Lauren Bookbinder[4]; Hussni Mohammed[1]; Thomas Divers[1]
[1]Cornell University; [2]University of Florida; [3]University of Pennsylvania; [4]College of Veterinary Medicine Michigan State University

Bile Acids (BA) are markers of hepatic function. In equine practice they are commonly used to establish severity of liver disease and monitor response to treatment. Recent published data from Europe supports the use of BA as prognostic indicator of short and long-term outcome in horses. Similar studies are not available in the United States. Etiopathology of hepatic diseases in horses from United States might differ from horses included in the European study causing predictive outcome value of BA to differ. Determining the prognostic value of BA in a population of horses with liver disease from Eastern United States might therefore provide a more accurate prognostic information for this marker in horses in this region. Our hypothesis was that moderately to severely increased BA values are not correlated with outcome in the population considered. The objective of the study was to retrospectively evaluate the predictive value of elevated BA on outcome in a population of horses presented to four referral hospitals in the Eastern United States. To test our hypothesis, we determined if BA values differed between survivors and non-survivors. Moreover, GGT, and direct bilirubin (DB) were also evaluated in the studied population. Records of horses 1 or more years of age admitted to four referral centers between 1997 and 2018 with a BA measurement >30 µmol/L and at least 6 months follow up information were included in the study. Four of these horses were research horses. The highest single BA, GGT and DB value recorded for each horse was used for statistical analysis. BA, GGT, and direct bilirubin values were compared between survivors and non-survivors. Eighty-two horses met the inclusion criteria. These horses presented for liver disease. 55% of the horses survived at least 6 months post-discharge. BA values were not significantly different between survivors and non-survivors (P=0.45). BA and GGT or Direct Bilirubin were not correlated. In the population evaluated GGT values were also not correlated with outcome (P=0.15), instead direct bilirubin values were negatively correlated with outcome (P=0.0). According to these results, BA values do not represent a reliable long-term prognostic indicator across the different liver pathologies for this population of horses with liver dysfunction from Eastern United States. Instead DB values may be more reliable indicator of prognosis in a similar population of horses.

Abstract E48: Glucose Attenuates AMPK Signaling in the Absence of Insulin in Equine Digital Lamellae

Erin F. Pinnell[1]; James Belknap[2]; Teresa Burns[2]; Mauria Watts[2]; Paty Weber[3]
[1]The Ohio State University College of Veterinary Medicine; [2]The Ohio State University; [3]Michigan State University

Equine insulin dysregulation (ID) is a known risk factor for the development of laminitis. 5′ adenosine monophosphate activated protein kinase (AMPK) is an important enzymatic regulator of energy homeostasis. AMPK signaling influences epithelial differentiation and adhesiveness, which are relevant to equine laminitis; therefore, AMPK is an important target to evaluate, particularly given that agonists are commercially available. The objective of this study was to determine the effects of insulin and glucose on AMPK activation (pAMPK) in keratinocytes and lamellar explants. Lamellar explants and keratinocytes from 2 different horses were collected and assigned to 3 incubation conditions: media control, media + glucose (0, 5, and 25 mg/dL), and media + glucose (0, 5, and 25 mg/dL) + insulin (10,000 mIU units/L). The concentration of pAMPK in treated cells and explants was determined using Western blot and compared between the groups. Glucose, in the absence of insulin at both the 5 and 25 mg/dL concentrations, decreased pAMPK concentration by ~80% compared to control conditions; the addition of insulin to the incubation did not appreciably change this response (pAMPK decreased by ~75% compared to the control). Glucose appears to attenuate AMPK signaling independently of insulin. In the digital lamellae of equids with ID, decreased concentrations of pAMPK could adversely influence differentiation and adhesiveness of keratinocytes in a way that contributes to lamellar failure following enhanced glucose delivery to the foot. AMPK agonists should be evaluated as potential therapeutic targets for equine ID and endocrinopathic laminitis.

Abstract E49: FluAvert Stimulates Immunity and Reduces Equine Herpesvirus 1 Replication in Equine Respiratory Epithelial Cells

Gisela Soboll Hussey[1]; Lila Zarski[1]; Wendy Vaala[2]; Craig Barnett[2]; Fairfield Bain[2]
[1]Michigan State University; [2]Merck Animal Health;

Equine herpesvirus 1 (EHV-1) is ubiquitous in horses worldwide and the cause of respiratory disease, late term abortion, and equine herpesvirus myeloencephalopathy (EHM). Currently, available vaccines are limited in their efficacy and prevention of viremia or EHM. This is likely related to the immunomodulatory properties of EHV-1. Multiple EHV-1 proteins have shown to interfere with induction of interferons, cytokines, chemokines, antigen presentation, and cellular immunity. We hypothesized that the mucosal influenza vaccine FluAvert, which is known to stimulate strong antiviral responses, will enhance antiviral innate immunity and that these responses would also provide protection from EHV-1 infection. Primary equine respiratory epithelial cells (ERECs) were treated with FluAvert or Mock 1, 2, 5, or 7 days prior to inoculation with EHV-1. Induction of interferons, cytokine and chemokine mRNA expression, and protein secretion was evaluated by high throughput qPCR and multiplex protein analysis. Further, intracellular and extracellular EHV-1 titers were determined by qPCR. FluAvert treatment resulted in modulation of IL-8, CCL2 and CXCL9 starting on day 5 post treatment. Coinciding with the timing of optimal chemokine/cytokine induction, EHV-1 replication was most efficiently reduced in ERECs treated with FluAvert 5 days prior to EHV-1 inoculation. In conclusion, our results suggest that FluAvert may be effective at counteracting some of the immune-modulatory properties of EHV-1 at the airway epithelium, and the peak for this response occurs 5–8 days post FluAvert treatment. Future *in vivo* studies are needed to investigate FluAvert as a prophylactic in situations where EHV-1 exposure may occur.

Abstract E50: Effect of AMPK Agonists on Insulin and Glucose Dynamics in Experimentally-Induced Insulin Dysregulation in Horses

Kathryn Timko; Laura Hostnik; Mauria Watts; Chiaming Chen; Adam Bercz; Ramiro Toribio; James Belknap; Teresa Burns
The Ohio State University

Insulin dysregulation (ID), a hallmark component of equine metabolic syndrome, is linked to the development of laminitis, a condition for which there are few pharmacologic treatment options. 5′-Adenosine-monophosphate-activated protein kinase (AMPK) is a highly conserved heterotrimeric enzyme essential to cellular energy regulation and an important therapeutic target in humans with metabolic syndrome. The objective of this study was to assess the effects of AMPK agonist administration (metformin [MET], aspirin [ASP], and combination [MET/ASP]) on insulin and glucose dynamics in experimentally-induced ID in light breed horses. Insulin dysregulation was induced in 14 adult light breed horses with dexamethasone (0.08 mg/kg PO q 24 h); horses were randomly assigned to groups and received either ASP (10 mg/kg PO q 24 h; n=7) or MET (30 mg/kg PO q 12 h; n=7) for 7 days. Seven horses then received MET/ASP for an additional 7 days. The oral sugar test (OST) and combined glucose and insulin test (CGIT) were used to assess ID at four time points: baseline, ID, ID + monotherapy, and ID + combination therapy. The OST area under the curve for glucose (AUCgluc0-240) after combination treatment was significantly lower than at the ID and ID + monotherapy timepoints (P=0.0052). The CGIT area under the curve for glucose (AUCgluc0-150) was not significantly affected by AMPK agonist treatment (P=0.0779). These findings suggest that MET/ASP combination therapy can synergistically improve insulin and glucose dynamics in horses and may have therapeutic value.

Abstract E51: Effect of AMPK Agonists on Hepatic AMPK-Related Gene Expression in Horses with Experimentally-Induced Insulin Dysregulation

Kathryn Timko; Laura Hostnik; Mauria Watts; Chiaming Chen; Adam Bercz; Ramiro Toribio; James Belknap; Teresa Burns
The Ohio State University

Insulin dysregulation (ID), a hallmark of equine metabolic syndrome (EMS), is intrinsically related to the development of laminitis. 5′-Adenosine-monophoshpate-activated protein kinase (AMPK) agonists are used in humans to promote euglycemia and enhance systemic insulin sensitivity, representing a promising therapy for horses with EMS. This study investigated hepatic AMPK signaling in response to AMPK agonists (metformin [MET], aspirin [ASP], combination [MET/ASP]) in horses with experimentally-induced ID. Insulin dysregulation was induced in 14 adult light breed horses with dexamethasone (0.08 mg/kg PO q 24 h). The horses were assigned to groups, and each horse received either ASP (10 mg/kg PO q 24 h; n=7) or MET (30 mg/kg PO q 12 h; n=7) for 7 days. Seven horses then received MET/ASP for an additional 7 days. To assess AMPK regulation, liver biopsy samples were obtained at the following time points: baseline, ID, ID + monotherapy, and ID + combination therapy. Real-time polymerase chain reaction (qPCR) was performed to evaluate expression of AMPKα, peroxisome proliferator-activated receptor-γ (PPARγ), peroxisome proliferator-activated receptor- γ coactivator 1-α (PGC1α), and phosphoenolpyruvate carboxykinase (PEPCK) at all timepoints. The expression of AMPKα was significantly increased with combination therapy compared to baseline (P=0.0114). PPAR γ expression was significantly decreased at ID + monotherapy (P=0.0026) but not with combination therapy, suggesting a partial response to combination treatment. The expression of PGC1α was not significantly affected by any treatment. PEPCK expression was significantly decreased by administration of dexamethasone (P=0.0114). These findings suggest that AMPK agonist therapy can alter hepatic AMPK-related gene expression in adult horses.

Abstract E52: Prevalence of *Streptococcus equi* in Healthy Horses in Colombia

Camilo Jaramillo[1]; Diego Gomez[1]; Brayan Fonseca[2]; Gabriela Mandiola[1]; Luis Arroyo[1]
[1]University of Guelph; [2]Iowa State University

Establishing the prevalence of *Streptococcus equi* in different locations is necessary since this information is underreported in the literature. The objective of this study was to determine the prevalence of *S. equi* in healthy horses in Colombia and the antimicrobial susceptibility of the recovered isolates. One hundred and thirty-seven horses, older than six months of age, from 15 farms were enrolled. The sample size was estimated based on the horse population of the municipality sampled (3000 – government population census, 2017). The error rate was set at 5%, for an expected prevalence of 10% and a confidence interval of 95%. The sampling was carried out in a stratified manner such that it was proportional to the population size of each farm. The guttural pouch was swabbed via endoscopic guidance and culture was performed on the samples in 5% sheep blood agar. Gram positive cocci, β hemolytic and C Lancefield positive colonies were subcultured and DNA was extracted for PCR analysis. The primers SodA F (5′- CAG CAT TCC TGC TGA CAT TCG TCA GG -3′) and SodA R (5′- CTG ACC AGC CTT ATT CAC AAC CAG CC -3′) were used for *Streptococcus equi* bacterial identification. To confirm the genus and species of *S. equi*, the DNA PCR products were sequenced. An antibiotic disc diffusion test was performed with penicillin, ceftiofur, trimethoprim-sulfamethoxazole (TMS), enrofloxacin and oxytetracycline. It was found that 15% (21/137) of the horses cultured positive, with all of the isolates reporting positive on PCR. Sequencing determined that *S. equi* ssp. *equi* accounted for 90% (19/21) of the isolates, while 10% were found to be *S. equi* ssp. *zooepidemicus* (2/21). In conclusion, this study confirms the presence of *S. equi* ssp. *equi* and *S. equi* ssp. *zooepidemicus* in healthy horses in Colombia. The calculated prevalence for individual animals was 15%. Lastly, all isolates were sensitive to TMS, penicillin, and ceftiofur and resistant to enrofloxacin and oxytetracycline.

Abstract E53: A Prospective Study of Common Laboratory Variables in Neonatal Foals in Texas

Laszlo Hunyadi[1]; Emily Sundman[2]
[1]Texas Tech University; [2]Equine Sports Medicine & Surgery

This study was to evaluate inflammatory biomarkers in neonatal Quarter Horse foals with normal parturition, illness (including sepsis), and foals born to mares with complicated parturitions and placental abnormalities (abnormal mares). A prospective study was conducted with 64 foals at a private practice in northern Texas. Forty-three normal foals, 7 ill/septic foals, and 14 clinically normal foals from abnormal mares were included in the study. Blood samples from foals were assessed for C-reactive protein (CRP), haptoglobin, cortisol, adrenocorticotropic hormone (ACTH), and serum amyloid A (SAA) within 24 hours of birth. Statistically significant differences were found for haptoglobin, ACTH, and SAA. Evaluation of haptoglobin values demonstrated that 1 foal in the ill/septic category was above the normal range and all foals in the abnormal mare category were within the range for normal foals (30–107 mg/dL). Evaluation of ACTH values demonstrated that 2 ill/septic foals were above the normal range (5–91 pg/mL) and 1 foal in the abnormal mare category was below the range for normal foals. Evaluation of SAA values demonstrated that 5 ill/septic foals were above the range for normal foals (0–31 μg/mL), while all foals in the abnormal mare category were with the range for normal foals. The study showed a statistically significant difference in haptoglobin, ACTH, and SAA concentrations in neonatal foals; however, neither haptoglobin nor ACTH values showed clinically useful differences between the groups. SAA was clinically useful to identify ill/septic foals in comparison to normal foals. Evaluation of neonatal cortisol, ACTH, haptoglobin, or CRP did not provide clinically useful information to differentiate normal foals from sick foals. Birth to a mare with subsequent placental abnormalities did not result in clinically abnormal neonatal foals in this study.

Abstract E54: Comparison of Different Methods for Measurement of Electrolytes and Detection of Acid-Base Disorders in Horses

Diego E. Gomez[1]; Sebastien Buczinski[2]; Shannon Darby[3]; Megan Palmisano[3]; Rob Mackay[3]
[1]University of Guelph; [2]University of Montreal; [3]University of Florida

The concentration of strong electrolytes (Na, K and Cl) can be determined using ion selective electrode technology based on direct or indirect potentiometric measuring. Significant differences in the measured concentrations of Na or Cl have been reported between analyzers, which can lead to disparities in the calculated values of the simplified strong ion difference (sSID) approach. The aims of this study were to determine the level of agreement between 2 analyzers for determining concentrations of strong electrolytes (Na and Cl) and the calculated variables of sSID approach, and to determine the level of agreement of both analyzers in identifying acid-base disorders in sick horses. This study was conducted on 124 paired samples from sick horses. Samples for both analyzers were collected simultaneously by venipuncture of the jugular vein. Electrolytes were measured using a Beckman Coulter AU480 Chemistry analyser® (AA) and a Nova Biomedical Stat Profile® (BGA), respectively. The sSID variables were calculated as SID4= (Na++K+)-(Cl-+L-Lactate-); A-=(0.22 x TP)/(1+10(6.65-pH)) (g/dL) and USI= SID4 - HCO3- - A-. The acid-base disorders were described when the following variables were outside of the following reference ranges: SID4 (38 to 47 mM/L), Atot (12 to 16 mM/L) and USI (-2 to 2 mM/L). The SID4 and USI concentrations were calculated with the values obtained from each analyzer. Agreement between the 2 analyzers was explored using Passing-Bablok regression and Bland–Altman analysis. Intraclass correlation coefficient (ICC) was also used to determine the reliability between the SID and USI measurements from the two analysers. Kappa coefficient test was used to evaluate the level of agreement between the two analysers in detecting acid-base disorders using prespecified cut-offs. The results of Passing-Bablok regression indicated significant methodologic differences in measuring Na and Cl, and the calculated values SID4 and USI (the Intercept (I) 95% CI did not include 0 and the slope (S) was different to 1 for all regressions). The Bland-Altman plot indicated that the mean bias (95% limits of agreement) for Na, Cl, SID4 and USI between the 2 analyzers were: -1.2 mmol/L (-9.2 to 6.8), 4.4 mmol/L (-4.4 to 13), -5.4 mmol/L (-13 to 2) and -6.2 mmol/L (-14 to 1.7), respectively. The ICC of the measurement of the SID4 and USI for both analyzers was 0.55 (95%CI: -0.2 to 0.8) and 0.2 (95%CI: -0.15 to 0.48), respectively. Kappa coefficient analysis showed no agreement between BGA and AA for detection of SID4 (K=0.20, 95%CI, 0.1 to 0.31) or USI abnormalities (K= -0.04, 95%CI, -0.11 to 0.02). Comparison between both analyzers showed methodologic differences in measuring Na and Cl, and calculated SID4 and USI, which resulted in a poor agreement between the analyzers to diagnose acid-base disorders in sick horses.

Abstract E55: Urinary L-Lactate Measures in Adult Horses

Ilana Glasberg; Scott Austin; Pamela Wilkins
University of Illinois College of Veterinary Medicine

Urinary L-lactate (ULAC) and ULAC:creatinine ratios are altered in human renal disease and neonatal Hypoxic-Ischemic Encephalopathy, but have not been assessed in the horse. Urine (U) and blood sodium (Na), LAC and creatinine concentrations were measured in 6 healthy adult horses. Na and creatinine were measured by the clinical laboratory. LAC was measured using 2 point-of-care (POC) devices, Lactate Scout and Lactate Plus. LAC measurements were performed in triplicate with the mean used in analysis. USG was measured by refractometer and dipstick. ULAC:creatinine, UNa:creatinine and fractional excretions (Fx) of LAC and Na were calculated. Data are reported as mean ± SD, associations examined by linear regression. Results are from the Lactate Scout as the Lactate Plus did not report values for urine. Respectively, blood LAC and Na were 0.9±0.3 mmol/L and 134.7±2.9 mEq/L, ULAC and UNa 5.1±1.0 mmol/L and 20±9 mEq/L, UCreatinine and creatinine 178.4±89.4 and 1.00±0.14. ULAC:creatinine and UNa:creatinine were 0.27±0.03 mmol/L and 0.24±0.32 respectively while FxLAC and FxNa were 3.03±1.06 and 0.07±0.03. Refractometer USG was 1.041±0.01, dipstick USG was always >1.005. Blood LAC and Na were positively associated (R2=62.5%, P<0.001), as were ULAC and UNa, (R2=59.9%, P<0.001). ULAC:creatinine and UNa:creatinine were more strongly associated (R2=81.7%, P<0.001) as were FxNa and FxLAC (R2=84.4%, P<0.001). Urinary LAC measurements obtained using the Lactate Scout POC are reported here for the first time in the horse. LAC and Na appear related with that relationship being stronger in urine than in blood, suggesting that both undergo modification by the kidney.

Abstract E56: Evaluation of Three Different Doses of Thyrotropin Releasing Hormone in Miniature Horses

Alfredo Sanchez-Londono[1]; Nicholas Frank[2]; Steve Grubbs[3]; Pilar Hermida[2]; Erik Hofmeister[1]
[1]Auburn University; [2]Cummings School of Veterinary Medicine at Tufts University; [3]Boehringer-Ingelheim;

The purpose of the study was to evaluate adrenocorticotropin hormone (ACTH) responses following the administration of three different doses (1.0 mg, 0.5 mg and 0.25 mg) of thyrotropin releasing hormone (TRH) to Miniature horses with and without pituitary pars intermedia dysfunction (PPID). A total of 20 client-owned Miniature horses were enrolled in the study. Group 1 (non-PPID control horses) consisted of 10 horses less than 10 years of age with no evidence of clinical signs of PPID and/or a normal basal ACTH concentration. Group 2 consisted of 10 horses over 15 years of age with clinical evidence of PPID and/or an increased (>35 pg/ml) basal ACTH concentration. A complete physical examination, basal blood draw for measuring ACTH concentrations and a questionnaire regarding health issues were completed on the first visit. On days 0, 14 and 28, both groups of horses underwent TRH stimulation tests using three different doses of TRH given intravenously on different calendar days at least two weeks apart. Testing was performed between the months of February and June. Three doses of TRH were evaluated: 1.0 mg, 0.5 mg, and 0.25 mg. On the day of testing, approximately 5 ml of blood was drawn into each EDTA tube via jugular venipuncture and blood was collected before and 2, 5, 10 and 20 minutes after TRH administration. Blood and anticoagulant were mixed gently by inverting tubes several times and then the tubes were chilled immediately in either an ice bath or a refrigerator. Tubes were centrifuged within 4 hours of collection and plasma was separated and transferred to cryovials for storage at -20°C. Frozen plasma was shipped with ice packs via overnight courier service. Plasma ACTH concentrations were measured using a solid-phase, two site sequential, chemiluminescent immunoassay designed for Immulite at Cornell University Animal Health Diagnostic Center. Receiver operator characteristic (ROC) curves were plotted for each time point with each TRH dose and Younden's value was calculated and used to determine the best cutoff point for diagnosis. Analyses were performed using a commercially available statistical software package (GraphPad Prism v.8) and $p<0.05$ was considered significant. The ROC analysis results were examined to determine diagnostic performance, the 1.0 mg TRH dose and 10 minute time point had 100% sensitivity and 90% specificity with a cutoff value of 130 pg/ml, compared to the 0.5 mg TRH dose at 10 minutes which had a 100%sensitivity and 80% specificity with a cutoff value of 121 pg/ml. In contrast, the 0.25 mg TRH dosage had 90% sensitivity and 80% specificity with a cutoff value of 104.3 pg/ml at 20 minutes post TRH injection. All other combinations had lower sensitivity and specificity values. Thyrotropin releasing hormone (TRH) dosages of 0.5 mg and 1.0 mg have the best sensitivity and specificity at 10 minutes post injections for consistent identification and diagnosis of miniature horses with PPID. A larger group of horses should be examined in the future to determine whether a lower TRH dose can be used as a diagnostic test in miniature horses.

Abstract E57: Prerenal and Renal Failure in Horses: Retrospective Study in an Equine Hospital (1987–2019)

Daniel Jean[1]; Guillaume Bon[2]; Romane Warlop[3]
[1]University of Montreal; [2]Cliniques de Conques; [3]Private practice

Renal failure in horses is described in the literature but few objective information is available on the survival rate and the prognostic factors in a large number of equine patients. The purpose of this clinical study was to identify equine patients with renal failure, determine the survival rate as well as the prognostic factors, and establish correlation between clinical data and necropsy findings. The retrospective study was performed with medical files of horses presented with hypercreatinemia (>150 umol/l) combined with at least one urology examination during hospitalization between 1987 and 2019 at the Veterinary Teaching Hospital at the University of Montreal. Equine patients were divided into three groups: 1) hypercreatinemia with a normal urology, 2) acute renal failure, and 3) chronic renal failure considering the clinical context. Two hundred sixty-three horses met the criteria. One hundred thirty-eight horses were in acute renal failure, 94 in prerenal failure and 31 in chronic renal failure. The short-term survival rate was 75,5% (prerenal failure), 60% (renal failure) and 55% (chronic renal failure) respectively. The survival rate for the three groups increased throughout period and was correlated to the lowest average creatinine value observed in these patients in the last decade compared to the period between 1987 to 2006. The mortality rate was directly correlated to the highest creatinine value of equine patients during hospitalization. In conclusion, the survival rate of renal failure improved during this three decades period and could be explained by the improvement of therapeutic management throughout period.

Abstract E58: The Thyrotropin-Releasing Hormone Procedure Produces Repeatable ACTH Concentrations in PPID-Negative and PPID-Positive Horses

Rhonda M. Hoffman[1]; John Haffner[1]; Steve Grubbs[2]
[1]Middle Tennessee State University; [2]Boehringer Ingelheim

Pituitary Pars Intermedia Dysfunction (PPID) affects approximately 20% of horses aged 15 and older. Advanced cases of PPID in horses exhibiting generalized hypertrichosis are easily identified, but diagnosing horses with subtle clinical signs of early PPID without generalized hypertrichosis is more challenging. Basal ACTH has a lower diagnostic sensitivity in horses with early compared to advanced PPID. Compared to basal ACTH, the thyrotropin-releasing hormone (TRH) stimulation procedure has been shown to have a greater sensitivity of detecting horses with subtle signs of PPID, but it is unknown if the ACTH response to TRH is repeatable in individual horses. The purpose of this study was to conduct TRH stimulation tests at 4-week intervals, beginning in February and ending in June, in horses with and without PPID to determine the repeatability of ACTH response after TRH stimulation. Twelve horses (PPID+ and PPID–) with a mean age of 18.8 yrs (range 12 to 25 yrs) were used. Basal ACTH concentration from blood collected on Day -45 was used to identify PPID status of each horse. On Day 0, all horses had blood collected for basal ACTH followed by IV administration of 1 mg TRH. Blood was then collected exactly 10 min post-TRH administration (T10-ACTH). Blood samples were centrifuged and plasma was frozen at -80C until ACTH analysis using a chemiluminescent immunoassay (Immulite) at the Animal Health Diagnostic Laboratory, Cornell University, Ithaca, NY. The TRH stimulation procedure was repeated on Days 28, 56, 84 and 112. These subsequent samples were compared to the T10-ACTH samples collected on Day 0. Data were analyzed using a mixed model with repeated measures to compare T10-ACTH and the percent increase of ACTH after TRH stimulation, using horse as the subject and Day as the repeated effect. Pearson's correlation coefficients were used to examine relationships between T10-ACTH on Days 28, 56, 84 and 112 to the T10-ACTH on Day 0. Bland-Altman plots were constructed to compare T10-ACTH on Days 28, 56, 84 and 112 to the T10-ACTH on Day 0. Of the 12 horses, the TRH stimulation indicated 5 negative and 5 positive for PPID, with 2 horses equivocal. The average ACTH in PPID- horses was 17.6 ± 0.7 pg/mL (basal) and 61.4 ± 9.2 pg/mL (T10-ACTH), with a 349 ± 41% increase in ACTH after TRH stimulation. The average ACTH in PPID+ horses was 43.5 ± 3.6 pg/mL (basal) and 410 ± 58 pg/mL (T10-ACTH), with a 948 ± 194% increase in ACTH after TRH stimulation. Repeated measures analysis indicated no effect of Day on T10-ACTH ($P=0.40$) or the percent increase of ACTH after TRH stimulation ($P=0.12$). Pearson's correlation coefficients indicated strong relationships between T10-ACTH on Day 0 and all other days ($R>0.70$, $P<0.01$). Bland-Altman plots indicated an average Day bias of 27 pg/mL in all horses compared to Day 0, with a Day bias of 10 pg/mL in PPID– and 43 pg/mL in PPID+ horses. The Immulite inter-assay CV was 9.3%, which accounts for most of the observed Day bias. In conclusion, the TRH stimulation procedure produces repeatable ACTH concentrations in samples collected 10 min after administration of TRH in horses collected at 4-week intervals over 112 days beginning in February through June. **Acknowledgments**: The authors would like to acknowledge Boehringer Ingelheim Animal Health and the Middle Tennessee State University Horse Science program for supporting this work.

Abstract E59: Methodologic Comparison of Two Analyzers for Measurement of Glucose and Creatinine Concentrations in Sick Horses

Diego E. Gomez[1]; Sebastien Buczinski[2]; Megan Palmisano[3]; Shannon Darby[3]; Rob Mackay[3]
[1]University of Guelph; [2]University of Montreal; [3]University of Florida

Accurate and efficient assessment of blood glucose concentration is critical in the clinical management of many pathological conditions in equine medicine. Plasma creatinine concentration is the most widely used and commonly accepted criterion in the medical laboratory for the evaluation of renal function. Methodological interferences may significantly affect the results of these two analytes and might influence therapeutic decisions. The aim of this study was to determine the level of agreement between 2 different analyzers for determining concentrations of glucose and creatinine in sick adult horses. Blood samples from 124 sick horses were used in this study. Samples for both analyzers were collected simultaneously by venipuncture of the jugular vein. Glucose and creatinine were measured using a Beckman Coulter AU480 Chemistry analyser® (BCCA) and a Nova Biomedical Stat Profile® (NBSP), respectively. Passing-Bablok regression and Bland-Altman analysis were performed to determine the level of agreement between the 2 analyzers. The level of agreement between the two analyzers in detecting hyperglycemia (glucose >120 mg/dL) and hypercreatininemia (creatinine >2 mg/dL) was explored using the Kappa coefficient test using prespecified cut-offs. McNemar's test was used to evaluate whether the measurement of glucose and creatinine with different analyzers are equally likely to identify an hyperglycemia and hypercreatininemia. On admission, NBSP detected 64/119 (54%) horses with hyperglycemia whereas BCCA identified 73/119 (62%). NBSP detected 28/122 (23%) with hypercreatininemia while BCCA identified 36/122 (30%). The results of Passing-Bablok regression failed to identify significant methodologic differences between analyzers for measuring glucose and creatinine (the Intercept (I) 95% CI include 0 and the slope (S) was equal to 1 for both regressions). The Bland-Altman plot indicated that the mean bias (95% limits of agreement) for glucose was -4.7 mg/dL (-70 to 60 mmol/L) and for creatinine was 0 mg/dL (-3 to 3). Kappa coefficient analysis showed a fair agreement between analyzers for detection of hyperglycemia (K=0.43, 95%CI, 0.27 to 0.59) and hypercreatininemia (K= 0.58, 95%CI, 0.41 to 0.74) in sick horses. McNemar's test showed that both analyzers were equally likely to detect hyperglycemia (P=0.11) and hypercreatininemia (P=0.07) in sick horses. Methodologic differences in measuring glucose and creatinine by two different analyzers were not detected in this study.

Abstract E60: Investigation of Neurological and Nephrogenic Side Effects of Polymyxin B Administration in Healthy Horses

Julia N. van Spijk[1]; Katrin Beckmann[2]; Meret Wehrli Eser[1]; Martina Boxler[3]; Martina Stirn[4]; Thea Rhyner[5]; Lanja Saleh[6]; Angelika Schoster[1]
[1]Clinic for Equine Internal Medicine, Vetsuisse Faculty, University of Zurich; [2]Section of Neurology, Department of Small Animals, Vetsuisse Faculty, University of Zurich; [3]Institute of Forensic Medicine, University of Zurich; [4]Clinical Laboratory, Department of Clinical Services, Vetsuisse Faculty, University of Zurich; [5]Equine National Center; Dana Kaelin - Clinic for Equine Internal Medicine, Vetsuisse Faculty, University of Zurich; [6]Institute of Clinical Chemistry, University Hospital Zurich

Polymyxin B (PolyB) is used in horses due to its antiendotoxic effect. Neurologic and nephrogenic side effects limit its use in human medicine. The objective of this study was to investigate side effects of PolyB in healthy horses. In a prospective, randomized, blinded cross over trial six healthy horses received treatment A (PolyB 6000 IU/kg IV q 12 h for 84 h) followed by treatment B (PolyB as before + gentamicin 10 mg/kg IV q 24 h for 84 h) or treatment B followed by treatment A with a wash out period of 3 weeks. Ataxia was scored by three blinded observers on video recordings obtained daily before, during, and three days after treatment. Serum concentration of PolyB [PolyB] and gentamicin [gentamicin] were measured daily, serum creatinine and urine analysis every other day. An ordinal mixed multivariable model was used to evaluate the effect of treatment group, number of PolyB dose, time since last dose, and [PolyB] and [gentamicin] on ataxia score. All horses developed proprioceptive ataxia (median: 1–3/5) in both treatment groups. A statistically significant effect of an increasing number of PolyB doses ($p<0.001$) and treatment group ($p<0.01$) was seen. Median ataxia score increased from 0 to 2/5 (treatment A) and from 0.5 to 3/5 (treatment B). Three days after treatment, median ataxia score decreased to 0/5 and 1/5 in treatment group A and B, respectively, with a statistically significant effect of time since last dose ($p<0.001$). [PolyB] and [gentamicin] had no effect on ataxia score. Weighted kappa of intra-rater agreement of ataxia was 0.5, 0.7, and 0.9. Inter-rater agreement was 0.3–0.6. Clinical, laboratory, and the remaining parts of neurological exam were within normal limits during the study period. PolyB administration led to transient ataxia in healthy horses. Severity was dependent on duration of therapy and concurrent treatment with gentamicin. No nephrotoxic side effects were observed.

Abstract E61: Physiological Changes Associated with Estrous Cycle on Adrenocorticotropic Hormone and Insulin Concentration in Mares

François-René Bertin; Gemma Hicks; Natalie Fraser
The University of Queensland

Pituitary pars intermedia dysfunction (PPID) and insulin dysregulation (ID) are both associated with poor reproduction performance and failure to conceive can trigger endocrine testing. The physiological changes associated with a normal estrous cycle on adrenocorticotropic hormone (ACTH) and insulin are poorly described, and test interpretation can be difficult. We hypothesized that estrous cycle would alter ACTH and insulin concentrations. Eighteen mares with ages ranging from 3 to 18 years and body condition scores ranging from 4 to 6/9 with no clinical signs suggestive of or laboratory findings consistent with PPID and ID were followed during their reproduction season. ACTH and insulin concentrations were measured at the periovulatory period, the postovulatory period as well as 14 and 28 days after ovulation. Determination of the stage in the estrous cycle was determined by ultrasound and progesterone concentrations. All the mares were inseminated at ovulation and gestation was then confirmed by ultrasound. Mares were grouped by age (younger or older than 12 years of age) and by gestation status (pregnant or not). Eleven mares became pregnant and age was not associated with pregnancy. There was a significant effect of estrous cycle stage and gestation on ACTH concentrations with higher ACTH concentrations at 14 days after ovulation (p=.0023) and in pregnant mares (p=.0097). No significant effect of estrous cycle stage or gestation on insulin concentrations was detected. ACTH concentrations change with estrous cycle and pregnancy. Therefore, knowledge of a mare's reproductive status might be beneficial when interpreting an ACTH concentration.

Abstract E62: Omega-3 Fatty Acid Incorporation Into Phosphatidylcholine in Plasma, Synovial Fluid, and Surfactant From Supplemented Horses

Undine Christmann; Courtney Hancock; Audrey Emery; Jesse Poovey; Casey Hagg; Stacy Anderson; Seth Chapman; Eric Mattson; Paul Wood
College of Veterinary Medicine, Lincoln Memorial University

Omega-3 fatty acids are known for their ability to modulate inflammation. Horses with asthma or osteoarthritis frequently receive these supplements in conjunction with other treatments. Docosahexaenoic acid (DHA) and eicosapentaenoic acid (EPA) are integrated into lipids which can be used to measure their uptake into different body compartments. The purpose of this study was to evaluate DHA and EPA incorporation into phosphatidylcholine in plasma, synovial fluid, and surfactant from horses. Twenty healthy horses were included in this study and randomly assigned to the supplement or control group. Supplemented horses received 8.3 g EPA and 5.3 g DHA orally 1x/day for 90 days. Horses were evaluated on days 0 (baseline), 30, 60, and 90. Samples included blood, synovial fluid, and bronchoalveolar lavage fluid. Sample lipidomic profiles were determined using high-resolution mass spectrometry. Data were analyzed with a mixed-effects repeated measures ANOVA (Tukey-Kramer posthoc test; values of $p<0.05$ considered as significant). Significantly higher relative levels of DHA and EPA containing phosphatidylcholine species (PC 40:6 and PC 40:5) were noted in plasma, synovial fluid, and surfactant from supplemented horses compared to control horses and in supplemented horses on days 30, 60, and 90 of collection versus baseline. Levels of PC 40:6 and PC 40:5 were stable within the group of control horses with no significant differences between sampling times. In conclusion, incorporation of DHA and EPA into phosphatidylcholine reliably reflects their uptake in different body compartments. The biological effects and clinical relevance of these findings remain to be determined.

Abstract E63: Insulin Dysregulation Variables in Equine Metabolic Syndrome are Similar in Horses with Previous Non-Endocrinopathic Founder

Rhonda M. Hoffman; John Haffner
Middle Tennessee State University

The objective of this study was to compare measures of insulin dysregulation using a Frequently Sampled Intravenous Glucose Tolerance test (FSIGT) in two types of foundered horses: those with Equine Metabolic Syndrome (EMS) versus horses that had a single bout of founder caused by non-endocrinopathic stress or injury, with both founder types compared to normal, never foundered, controls. The hypothesis was that measures of insulin dysregulation would be greater in horses with EMS compared to control horses and those with a previous non-endocrinopathic founder. Eighteen horses, averaging 503 ± 53 kg BW, 6.5 ± 0.9 BCS, and aged 12 ± 5 yrs were used. Of these, 6 horses had chronic, repeated laminitis and founder associated with EMS; 6 previously had a single bout of founder not related to EMS, named Other Non-Endocrinopathic Founder (ONEF, including concussion, overheating, injury, or acute illness), and 6 were normal controls (CON), having no history of laminitis. All horses tested negative for Pituitary Pars Intermedia Dysfunction. Lateral radiographs of front feet indicated coffin bone rotation in all foundered horses and normal anatomy in CON horses. None exhibited acute phase laminitis, with the last incidence reported at 12 ± 5 mos for EMS and 60 ± 24 mos for ONEF horses ($P<0.0001$). Age, BW, and BCS were not different between groups ($P=0.12$). Prior to testing, horses were housed overnight in stalls and provided low-NSC mature grass hay and water. Jugular catheters were placed, and basal samples collected. For the FSIGT, 300 mg glucose/kg BW was infused IV, and 30 mU insulin/kg BW administered 20 min later. During the FSIGT, 31 venous samples were collected, at 1, 2, 3, 4, 5, 6, 7, 8, 10, 12, 14, 16, 19, 22, 23, 24, 25, 27, 30, 35, 40, 50, 60, 70, 80, 90, 100, 120, 150, 180 and 240 min after the initial glucose dose. The 240 min sample was included due to expected insulin dysregulation and slower glucose clearance in the EMS horses. Blood samples were collected into heparinized sample tubes and placed in ice water until plasma was removed 10 to 20 min later, then frozen at –4°C. A colorimetric assay was used to measure glucose, and RIA for insulin. The FSIGT partitions glucose disposal into insulin-independent glucose clearance, or glucose effectiveness (Sg), and insulin-dependent glucose disposal, or insulin sensitivity (Si). Both Sg and Si are described in minimal model analysis by two differential equations solved simultaneously using relative changes in blood glucose and insulin as influenced by the FSIGT. The minimal model estimates acute insulin response to glucose (AIRg) using the area under the curve for the first 20 min of the FSIGT, which represents pancreatic β-cell responsiveness. The disposition index (DI) is the multiplied product of Si x AIRg and accounts for the appropriateness of the AIRg relative to Si. The Si, Sg, AIRg and DI were calculated from the FSIGT data using MinMod Millennium and WinSAAM software. Data were tested for normality and analyzed using a mixed model with repeated measures using horse as the subject and sample time as the repeated effect, to compare the effects of group (EMS, ONEF, CON) on basal glucose and insulin, Si, Sg, AIRg and DI. Table 1 summarizes means ± SE of horse clinical characteristics and variables related to insulin dysregulation. Basal insulin was higher in EMS than ONEF ($P=0.033$) and CON horses ($P=0.013$), but CON and ONEF were not different ($P=0.41$). Insulin sensitivity was lower in EMS than CON horses ($P=0.015$), but ONEF were not different from either CON ($P=0.19$) or EMS

(P=0.46). The AIRg was greater in EMS than CON horses (P=0.034), but ONEF was not different from either CON (P=0.23) or EMS (P=0.37). There was no difference between groups in basal glucose (P=0.43), Sg (P=0.13) or DI (P=0.37). The lower Si and higher AIRg in EMS vs CON horses indicated expected insulin dysregulation similar to previous reports. The lack of difference in Sg of EMS and CON separates these EMS horses from previous reports of obese horses, which exhibit higher insulin-independent glucose clearance. Body condition scores of the EMS horses were not different from ONEF or CON, so perhaps obesity, but not EMS per se, affects peripheral tissue glucose transporters that act independent of insulin. Basal insulin was greater in EMS vs ONEF but similar in ONEF and CON, which suggests a difference between effects of founder type. Basal insulin in ONEF horses did not meet the cutoff for identification as laminitis-prone (defined as >20 mIU/L, ECEIM Consensus Statement, 2019). Insulin sensitivity in ONEF was not different from either EMS or CON, and was in between the cutoff criteria in horses previously defined as insulin dysregulated (<1.0 L x mIU-1 x min-1 x 10-4) or insulin sensitive (>1.5 L x mIU-1 x min-1 x 10-4). These results, combined with the lack of difference in AIRg between ONEF and EMS, suggest that ONEF horses may be insulin dysregulated at a degree below clinical detection. The endocrine status of the ONEF horses at the time of their single bout of founder is unknown. Thus, it is also unknown whether a sub-clinical insulin dysregulation existed that increased the risk of a single-bout of founder in the ONEF horses, or if the one-time non-endocrinopathic founder produced endocrine changes that resulted in persistent sub-clinical insulin dysregulation. The American Quarter Horse Foundation's support of this work is gratefully acknowledged.

Table 1. Clinical characteristics and variables associated with insulin dysregulation (mean ± SE), including insulin sensitivity (Si), insulin-independent glucose disposal (Sg), acute insulin response to glucose (AIRg), and disposition index (DI) in horses with Equine Metabolic Syndrome (EMS) or horses that had a single bout of Other Non-Endocrinopathic Founder (ONEF), compared to normal controls (CON).

	CON	ONEF	EMS	*P*
Body weight, kg	527 ± 8	485 ± 30	483 ± 22	0.36
Body Condition Score	6.3 ± 0.2	6.2 ± 0.5	7.2 ± 0.5	0.12
Age, yrs	11 ± 3	13 ± 3	12 ± 2	0.84
Last acute laminitis	Never	60 ± 24 mos[a]	12 ± 5 mos[b]	**< 0.0001**
Basal Glucose	94.9 ± 2.7	92.6 ± 6.4	99.3 ± 7.4	0.43
Basal Insulin, $mU \bullet L^{-1}$	3.87 ± 1.02[b]	10.8 ± 5.0[b]	33.0 ± 10.4[a]	**0.013**
Si, $L \bullet mU^{-1} \bullet min^{-1} \times 10^{-4}$	3.36 ± 1.03[a]	1.47 ± 0.86[ab]	0.60 ± 0.46[b]	**0.015**
Sg, $min^{-1} \times 10^{-2}$	2.13 ± 0.31	1.43 ± 0.40	1.36 ± 0.30	0.13
AIRg, $mU \bullet min \bullet L^{-1}$	260 ± 58[b]	477 ± 135[ab]	874 ± 247[a]	**0.034**
DI, $\times 10^{-2}$	9.72 ± 4.61	4.56 ± 1.95	5.46 ± 4.57	0.37

[ab] Means in rows with different superscripts differ by the P-value shown in each row.

Abstract E64: Outcome and Survival in 9 Horses Diagnosed with Hematopoietic Neoplasia and Treated with Cytotoxic Chemotherapy

Hannah Manning; Erin Pinnell; Adam Bercz; Teresa Burns
The Ohio State University

Hematopoietic neoplasia in the horse is rare; once diagnosed, it can be treated with various cytotoxic agents. Horses appear relatively tolerant of the adverse effects of cytotoxic chemotherapeutics; however, little information exists in the literature regarding the use and safety of these drugs in equine practice. The administration of systemic chemotherapy in horses diagnosed with hematopoietic malignancies is also limited based on negative perceptions regarding treatment efficacy and cost. The objective of this study was to evaluate outcomes, survival, adverse events, and client satisfaction in horses diagnosed with hematopoietic neoplasia and treated with systemic chemotherapeutics. Medical records of horses diagnosed with hematopoietic neoplasia and treated with systemic chemotherapeutics from 2005–2019 were evaluated. Nine horses were selected for inclusion. Diagnoses included lymphoma in 77.8% (7/9) and leukemia in 22.2% (2/9) of patients. Survival time ranged from 8–556 days, with 66.7% (6/9) of patients surviving 0–11 months and 33.3% (3/9) of patients surviving 12+ months post diagnosis and treatment with systemic chemotherapy. Of the patients in our study population, 66.7% (6/9) had no adverse effects reported within the medical record. Side effects reported included self-limiting tachycardia, tachypnea, pyrexia, alopecia, and abortion. Of the clients that responded to surveys, their satisfaction with treatment was high, with none of them noting side effects in their horses and a majority reporting they would treat another horse with chemotherapy. Treatment of hematopoietic malignancies with cytotoxic chemotherapeutics appears to be safe, reasonable, and well tolerated in horses, and investigation to optimize therapeutic protocols is warranted.

Abstract E65: Glucose Stimulates GLP-2 Secretion From Equine Small Intestine

Melody A. de Laat; Poppy Sibthorpe; Robert Spence
Queensland University of Technology

Glucagon-like peptide-2 (GLP-2) is secreted by enteroendocrine cells in the small intestine in response to dietary nutrients and increases epithelial cell density and glucose transporter (SGLT-1) expression, which increases glucose uptake. Equine insulin dysregulation is associated with increased glucose bioavailability and plasma GLP-2 concentrations are higher in insulin-dysregulated, compared to metabolically healthy, ponies. This study aimed to determine *in vitro* whether intestinal GLP-2 secretion increases in response to glucose and whether inhibition of SGLT-1 would reduce GLP-2 concentration. Equine jejunum (n=6; in triplicate) was collected from an abattoir, rinsed with ice-cold saline and placed in transport buffer prior to removal of serosa and sectioning (~100 mg explants). Tissues were incubated for 60 min at 37°C with 0 or 12 mM added glucose to determine glucose-dependent GLP-2 secretion. Subsequently, fresh explants were incubated with or without two inhibitors (separately), phlorizin (1 mM; SGLT-1 inhibitor) and phloretin (0.67 mM; SGLT-1 and GLUT-2 inhibitor), for 60 min with 12 mM glucose. An ELISA was used to measure GLP-2 secretion which was normalized to protein concentration. Stimulation with 12 mM glucose increased (P=0.03) GLP-2 secretion. This glucose-dependent GLP-2 secretion was inhibited by both phlorizin (P=0.02) and phloretin (P=0.03). This experiment confirmed that equine jejunum secretes GLP-2 which increases in response to glucose stimulation and can be reduced with SGLT-1 inhibitors. These data contribute to understanding of mechanisms potentially associated with increased glucose bioavailability in insulin-dysregulated horses and provide preliminary information for future investigations of the pathophysiology of insulin dysregulation.

Abstract E66: Blood Vitamin C, Vitamin B and Cortisol Concentrations in Healthy and Critically Ill Foals

Katarzyna Dembek[1]; David Wong[2]; Lauren Young[1]
[1]Iowa State University; [2]VA-MD College of Veterinary Medicine

Sepsis is a common disease in neonatal foals with a variety of therapeutics being used to facilitate recovery. Some evidence suggests that administration of hydrocortisone, vitamin C and vitamin B may improve outcome in people with sepsis, but information is not available in regard to these variables in healthy or ill neonatal foals. The goal of this study was to document and compare the concentrations of vitamin C, vitamin B and cortisol in healthy and critically ill neonatal foals. We hypothesized that vitamin C and B concentrations will be lower in septic foals compared to healthy controls and that hypovitaminosis C and B will be associated with severity of disease and outcome in critically ill foals. Blood samples were collected from healthy (n=15), septic (positive blood culture and/or sepsis score >12 [n=17]) and sick non-septic (SNS; n=15) foals <1 week of age at the time of admission and 72 and 120 hours after admission. Vitamin C and B concentrations were measured with high performance liquid chromatography mass spectrometry and cortisol concentration was determined by immunoassay. Hypovitaminosis C and B was defined based on the lower level of 95% confidence interval for vitamins concentrations in healthy foals. The prevalence of hypovitaminosis C on admission (defined as vitamin C <7 ug/dL) was 58% and 26% in septic and SNS foals, respectively. Hypovitaminosis B (vitamin B <2 ng/mL) was more frequent in septic foals (47%) compared to SNS (13%), (P<0.05). In septic foals, vitamin B and C concentrations were lower than in healthy foals at 72 and 120 hours after admission (P<0.05). Cortisol concentration was higher in septic and SNS foals compared to healthy foals on admission (P<0.05). Low vitamin C and B concentrations were associated with severity of disease in neonatal foals. Vitamin C and B hypovitaminosis may contribute to a pro-inflammatory state in equine neonatal sepsis.

Abstract E67: The Prevalence of Cyathostomin Anthelmintic Resistance on Horse Farms in Prince Edward Island, Canada

Jaimie Amanda Butler[1]; Haley Greenbank[1]; Rebecca Parrish[1]; Martin Nielsen[2]; William Stoughton[1]
[1]Atlantic Veterinary College; [2]Gluck Equine Research Center, University of Kentucky

The majority of adult equines carry a parasite burden consisting mainly of nematodes, which include 40 different species of small strongyles (cyathostomins) and to a lesser extent large strongyles. There are currently three classes of anthelmintics to treat strongyle infections. These anthelmintics were initially efficacious in controlling cyathostomins; however, due to the widespread overuse of anthelmintics, cyathostomins have developed increasing resistance to the classes benzimidazoles and pyrimidines, with macrocyclic lactones displaying early signs of resistance. Detection of early resistance can be investigated using egg reappearance periods (ERP). The ERP is defined as the week cyathostomin egg shedding occurs after an effective anthelmintic treatment. Shortening of this period overtime is an indicator of anthelmintic resistance. The objective of this study was to investigate the prevalence of cyathostomin anthelmintic resistance to pyrantel pamoate and ivermectin, and to determine egg reappearance periods on horse farms in Prince Edward Island (PEI), Canada. One hundred and one horses on 15 horse farms across PEI, Canada were enrolled in the study. The number of horses per farm ranged from 4–11 with means of 6.9 and 6.2 for pyrantel pamoate and ivermectin groups, respectively. Fecal egg counts (FEC) were performed initially on 270 horses. Horses shedding >200 eggs per gram (epg) were treated with 6.6 mg/kg PO of pyrantel pamoate (n=101). Fecal egg counts were conducted every 2 weeks for 8 weeks post treatment. Once FEC were above 200 epg horses were dewormed with 0.2 mg/kg PO of ivermectin (n=87), and FEC were performed every 2–3 weeks for 7 weeks. Fecal egg count reduction tests (FECRT) and egg reappearance periods were used to detect the efficacy of each anthelmintic. Fecal egg count reduction tests detected pyrantel pamoate resistance on 2/14 farms. No resistance to ivermectin was detected using the FECRT. Egg reappearance periods detected no early resistance to pyrantel pamoate on susceptible farms, while ivermectin had a reduction in the ERP of 5 weeks on 2/12 farms. The prevalence of pyrantel pamoate cyathostomin resistance was less on PEI than other studies, while the shortened ERP of 5 weeks for ivermectin was comparable to findings of early resistance in other parts of the world. These findings will allow us to educate owners and veterinarians on appropriate anthelmintic protocols in PEI, and can be used as a baseline for continued monitoring of ERP and anthelmintic resistance in this region.

Abstract E68: Evaluation of a Clinical Sign Scoring System For Pituitary Pars Intermedia Dysfunction in Horses

Steven Grubbs[1]; Dwana Neal[1]; Tom Keefe[2]
[1]Boehringer Ingelheim; [2]Colorado State University

Pituitary pars intermedia dysfunction (PPID) has been described as the most common endocrinologic disorder of aged horses. The diagnosis consists of information obtained from current history, a physical examination to document clinical signs associated with PPID, and then to confirm these findings, diagnostically evaluate adrenocorticotropic hormone (ACTH). In many horses with reported PPID associated signs, results of ACTH evaluation were within normal reference range. It is possible that veterinarians were testing horses without clinical signs or with very subtle PPID-associated signs. In addition, certain PPID-associated signs, by themselves, may not be predictive of disease. Based on this dilemma, the purpose of this study was to obtain epidemiological information from a large population of horses, and then develop a clinical sign score (CSS) based on odds ratios. In all, 2,989 horses exhibiting one or more of the typical signs of PPID were enrolled in the study. At initial visit, a physical examination was conducted and blood drawn for basal adrenocorticotropic hormone (ACTH), insulin, and glucose. All samples were centrifuged, plasma shipped overnight and analyzed for ACTH, insulin, and glucose by the Animal Health Diagnostic Center, Cornell University, Ithaca, NY. The association between PPID status, based on ACTH results, and each of the demographic variables and test results for insulin and glucose were statistically evaluated individually using the Pearson chi-square test and collectively using multiple logistic regression analysis of PPID to provide odds ratios for PPID associated with the clinical signs. The CSS is the product of the odds ratios for PPID vs. all clinical signs collectively. Based on statistical analyses of the CSS data, a value of 1.5 for the CSS was selected as a cut-off for predicting PPID status. The higher the CSS the more likely a horse was PPID+. Clinical sign scores ranged from 0.60 to 5.51, with an overall geometric mean of 1.59. Using this model in the non-fall months, the CSS score with an ACTH cut-off of 35 pg/mL (or 50 pg/dL) had a sensitivity of 68.3% (70.9%) and a specificity of 58.5% (52.6%). During the fall months using an ACTH cut-off of 100 pg/mL, the CSS had a 69.4% sensitivity and a 62.3% specificity. Based on this population of horses, delayed shedding and abnormal sweating were significant predictors of PPID+ status. Although pot-belly/weight gain, regional adiposity, excessive thirst and recurrent infections are considered to be associated with PPID, odds ratios for each were <1.0. The CSS may be a viable option for veterinary practitioners to utilize to determine if evaluation of ACTH in a particular horse is necessary. For example, based on the results of this study, if only one of the following clinical signs are present: potbelly/weight gain, regional adiposity, excessive thirst or recurrent infections; evaluation of ACTH for PPID may not be recommended. Additional field-based studies are essential to further evaluate the positive and negative predictive value of the CSS for PPID in horses.

Abstract E69: Efficacy of Single Active and Combination Anthelmintics Against Equine Strongyles in Adult Horses

Edwina Wilkes; Natalie Ford; Kristopher Hughes
Charles Sturt University

Cyathostomins are considered the most important strongylid parasites of the horse worldwide. Intensive interval treatment regimens have resulted in the development of widespread anthelmintic resistance in cyathostomins. It is imperative that existing anthelmintics are used judiciously and monitoring is undertaken for development of anthelmintic resistance (AR) and to ensure that effective drugs are utilised. The aim of this study was to evaluate the efficacy of abamectin (ABA), oxfendazole (OXF) and a commercially available combination anthelmintic (ABA+OXF) against strongyles in adult horses. Horses (>2 yo) resident at Charles Sturt University were screened for inclusion in the study. Faecal samples were collected from each horse and three faecal egg counts (FECs) were performed on each sample using the modified McMaster method. Thirty six horses with >50 strongyle epg were recruited into the study and randomly allocated into one of four groups on treatment day (day 0): ABA+OXF, ABA, OXF and an untreated control group. Post-treatment FECs were determined on days 14, 21, 28, 35, 42, 49 and 56. Results from a single count (C1) and the average of three counts (X1-3) were used to calculate the arithmetic means (AM) for FEC on day 0 and day 14. Faecal egg count reductions (FECRs) were calculated using the AM of individual FEC results, AM of arcsin-transformed individual FECRs and correction of the AM of the treatment group FECR by changes in the AM of the control group FECR. A FECR <90% on d14 was considered to represent AR. Egg reappearance periods (ERP) were defined as the timepoint where FECR (calculated using AM of individual counts) was <90%. Using the results of C1 for FECRT, strongyles were susceptible to ABA+OXF (100%) and ABA (99.19–99.95%) and resistant to OXF (55.05–80.63%). Similarly, using the results of X1-3, strongyles were susceptible to ABA+OXF (99.83–99.98%) and ABA (99.64–99.81%) and resistant to OXF (70.76–86.12%). The ERP for ABA+OXF, ABA and OXF was 49 days, 35 days and 14 days, respectively. The results of this study demonstrated that the ABA-OXF combination was highly effective against benzimidazole-resistant strongyles and there was also evidence of greater egg suppression compared to the single active anthelmintics. The shortened egg reappearance period following treatment with abamectin is considered to be an early indicator of macrocyclic lactone resistance development. The prudent use of anthelmintic combinations may contribute to more sustainable control of endoparasites in horses.

Abstract E70: Duration of Effectiveness of Frozen/Thawed Thyrotropin Releasing Hormone to Stimulate ACTH Release in Horses

Steven Grubbs[1]; Kelley Jones[2]; Rhonda Hoffman[3]; John Haffner[3]
[1]Boehringer Ingelheim; [2]Around the Bend Veterinary Services; [3]Middle Tennessee State University

The thyrotropin releasing hormone (TRH) stimulation of ACTH due to the increased sensitivity compared to measurement of basal ACTH has increasingly been used as a diagnostic test for pituitary pars intermedia dysfunction (PPID). Many equine practitioners will freeze doses of TRH, then thaw, and drive to the farm to examine the horse. If the horse owner declines testing, the veterinarian has TRH that has been frozen and thawed. Anecdotally, TRH can only be frozen and thawed once for optimum and consistent response. The potency and stability over time of TRH is unknown after one freeze thaw cycle. It would be useful for veterinarians to understand if thawed TRH stored at refrigerated temperature can produce accurate results when administered to a horse at a later date. This study was designed to determine the duration of effectiveness of TRH following one freeze/thaw cycle when stored at 5°C over time in horses. Ten horses (PPID+ and PPID-) were enrolled, mean age 18.8 yrs (range 12 to 25 yrs). Horses were first paired by PPID status and randomized into 2 groups of 5 horses each prior to –Day 14. Thirty, 1 mg/ml doses of constituted TRH were frozen at -20°C on –Day 28. On –Day 14, ten doses of TRH were thawed at 5°C followed by thawing of the remaining (20 doses) TRH on Day 0. Thawed TRH was stored at 5°C until administration. On Day 0, all horses had blood collected for baseline ACTH (T0-ACTH) followed by administration of 1 mg TRH IV. Blood was then collected exactly 10 minutes post-TRH administration (T10-ACTH). The TRH stimulation procedure was repeated with the thawed TRH on Days 14, 28, 42 and 56. In order to avoid potential carryover effects of multiple TRH stimulation procedures administered every two weeks, horses in Group 1 had the TRH stimulation repeated on Days 14 and 42 whereas Group 2 horses had the TRH stimulation repeated on Days 28 and 56 days TRH post-thaw. All samples were centrifuged following collection and plasma was frozen (-80C) until analysis at Cornell Animal Health Diagnostic Center. Data were analyzed using a mixed model with repeated measures to compare T10-ACTH and the percent increase of ACTH after TRH stimulation, using horse as the subject and day as the repeated effect. Pearson's correlation coefficients were used to examine relationships between T10-ACTH on Days 14, 28, 42, 56 to the T10-ACTH on Day 0. There was no effect of Group ($P>0.25$), so when appropriate, data were combined for analysis. There was no effect of Day post-thaw on T10-ACTH ($P=0.13$) or the percent increase of ACTH after TRH stimulation ($P=0.36$). Pearson's correlation coefficients indicated strong relationships between T10-ACTH on Day 0 and all other days ($R>0.98$, $P<0.001$). The thyrotropin releasing hormone stimulation test produced repeatable ACTH concentrations in samples collected 10 min after administration of TRH in horses when using TRH that has been thawed and stored at 5°C for up to 56 days.

Abstract E71: Oclacitinib Maleate (Apoquel®) Dose Determination in Horses with Naturally Occurring Allergic Dermatitis

Marike Visser; Dawn Cleaver; Blair Cundiff; Vicky King; Gordon Sture
Zoetis

Oclacitinib maleate (Apoquel®) is a Janus Kinase/Signal Transducer and Activator of Transcription (JAK/STAT) inhibitor approved for use in dogs with atopic and allergic dermatitis. The efficacy and safety of two different oral dose levels (0.1 or 0.25 mg/kg) of Apoquel for the reduction of clinical signs of equine allergic dermatitis were evaluated in a randomized, placebo-controlled, double-masked, multicenter study. Horses were evaluated for pruritus (Days 0, 1, 2, 3, 5, 7, 14, 21, and 28) and skin lesions were scored (Days 0, 14, and 28) to establish a minimum effective dose. Once-daily oral dosing occurred beginning on Day 0 through 28(+2). Fifty-eight animals were randomized at an intended ratio of 1:1:1 into one of three treatment groups: T01 (0.0 mg/kg, placebo, n=19), T02 (oclacitinib maleate, 0.1 mg/kg, n=18), or T03 (oclacitinib maleate, 0.25 mg/kg, n=21) at four veterinary practices in the United States. Horses were required to be >12 months and ≤20 years of age, weigh ≥200 kg, have at least "mild itching" (≥40 mm) according to the Owner Assessment of Pruritus Visual Analog Score (PVAS), have clinical signs of allergic dermatitis, and be otherwise healthy. The primary efficacy variable was the PVAS. A significant treatment difference (P≤0.0938) was found for 0.25 mg/kg Apoquel compared to placebo beginning at Day 5 (i.e., Days 5, 7, 14, 21, 28). In addition, there was a significantly (P=0.0844) greater reduction in the percent change from baseline in the PVAS compared to placebo for the 0.1 mg/kg Apoquel group at Day 28 (-43.4% vs -20.4%) and for the 0.25 mg/kg Apoquel group (P≤0.0829) starting at Day 5 (-28.7% vs -12.4%) and continuing through Day 28 (-59.6% vs -20.4%). A comparison of the Lesion Score between placebo and 0.1 mg/kg Apoquel, and placebo and 0.25 mg/kg Apoquel detected no significant differences (P≥0.1316). The Lesion Score mean percent change from Day 0 (baseline) was a decrease of 26.6%, 51.5%, and 63.0% at Day 28 for 0 mg/kg (placebo), 0.1 mg/kg, and 0.25 mg/kg, respectively, showing a dose-dependent clinical improvement in skin lesions that was not significantly different (P≥0.1495). Adverse events were considered typical for a population of horses with allergic dermatitis and no horses were removed from the study due to an adverse event. Clinical pathology summaries revealed no effects that appeared to be clinically significant or biologically important. The results of this study demonstrated the efficacy and safety of Apoquel for the treatment of pruritus in horses following oral administration of 0.25 mg/kg once daily for 28 days with onset of effect as early as 5 days after initiation of treatment.

Abstract E72: Effect of Oral Trazodone on Intraocular Pressure in Healthy Horses

Alexandra Moss; Rachel Hritz; Rachel Hector; Kathryn Wotman
Colorado State University

Trazodone has been used successfully in other veterinary species to reduce anxiety and detrimental behaviors. It has been previously shown to elicit light sedation in horses when administered orally, therefore this medication may be used in horses that show anxiety while on stall rest. Prior literature regarding the effect of trazodone in humans and rabbits have identified potential ocular side effects; specifically, altered intraocular pressure (IOP) and vertical pupil diameter (VPD). This study evaluated the effect of oral trazodone on behavioral, physical, and selected ocular parameters in horses. Eight clinically normal horses (4 mares, 4 geldings, 9±5) received a single 6 mg/kg oral dose of trazodone. Baseline heart rate, respiratory rate, rectal temperature, gastrointestinal borborygmi, level of sedation and IOP and VPD in both eyes were assessed prior to administration of the medication. All parameters were measured again at 0.5, 1, 2, 4, 6, 8, 12, and 24 hours post administration. Mild sedation was evident in 7/8 horses from 0.5 to 8 hours. Mean IOP was significantly decreased (7 to 4 mmHg) for the same period as sedation was apparent. Temperature was reduced from 1 hour to 8 hours. VPD was only significantly reduced (1.2 mm) at 0.5 hours after trazodone administration. All horses in this study reached plasma trazodone concentrations of >130 ng/mL, which is described for anti-depressive effects in humans, and produced sedation in these horses. Trazodone alters ocular exam parameters in healthy horses, hence further research is warranted prior to its use in horses with ophthalmic disease.

Abstract E73: Fecal Extract From Obese Horses Induces Inflammation *In Vitro*

Paige Roth[1]; Jone Stanley[2]; Ana Chamoun Emanuelli[2]; Canaan Whitfield-Cargile[2]; Michelle Coleman[2]
[1]College of Veterinary Medicine, Texas A&M; [2]Texas A&M University

The pathophysiology of obesity in horses is likely complex and multifactorial, though an association with an inflammatory state is postulated. The source of this inflammation, however, is unknown. Studies in human medicine suggest that the microbiome plays a role in mediating a chronic inflammatory response in obese people. The objective of this study was to evaluate whether the enteric microbiome from obese horses could induce an inflammatory response in murine macrophages. Feces obtained via rectal palpation from 14 obese (Body Condition Score (BCS) ≥7) and 14 non-obese (BCS ≤5) horses residing on a single farm, with identical management practices, were included in the study. There was a significant ($P<0.05$) increase in levels of lipopolysaccharide (LPS) and pro-inflammatory cytokines Interleukin-1β (IL-1β), Tumor Necrosis Factor-α (TNF-α), and Interleukin-6 (IL-6) from macrophages treated with microbiome from obese horses compared to non-obese horses. Overall, this study indicates that the microbiome likely plays a role in mediation of an inflammatory response. Further, the increased levels of inflammatory markers induced by microbiome of obese horses suggest important differences in the enteric microbial composition of these horses, compared to non-obese animals, warranting further study.

Abstract E74: Effect of the Proximal Gastrointestinal Tract on Viability of Equine Probiotics: An *In Vitro* Study

Ana Berreta; Trevor Alexander; Clark Kogan; Claire Miller; Jamie Kopper
Washington State University

Probiotics are defined as live microorganisms that confer a health benefit to the host when administered in adequate amounts, indicating that the presence of viable microorganisms is essential. Assessment of human probiotics indicates that without enteric coating few microorganisms survive the proximal gastrointestinal (GI) tract. The objectives of this study were to determine the effect of exposure to conditions consistent with the proximal GI tract on viability of commercially available equine probiotics, and to establish if increasing the gastric pH, consistent with use of a proton pump inhibitor, would improve viability of probiotics. We hypothesized that exposure to the proximal GI tract would decrease the viability of microorganisms in probiotics. A list of 53 commercially available equine probiotics was made, of which 11 were selected using a random number generator. The microbial contents were assessed using culture dependent techniques before and after exposure to a simulated equine proximal GI tract. Statistical analyses were performed using the Wilcoxon Rank Sum Test. Significance was set at $P<0.05$. None of the products evaluated advertised the use of enteric protection. There was a significant reduction in the viability of microorganisms after exposure to conditions consistent with the proximal GI tract which, although still notable, was tempered by conditions consistent with use of a proton pump inhibitor. Results from this study suggest that, without enteric protection, the proximal GI tract has a negative impact on viability of microorganisms in commercially available equine probiotics.

Abstract E75: Pharmacokinetics of Intravenous and Intramuscular Ceftazidime in Neonatal Foals

Christina D. Marino[1]; Clare Ryan[2]; Londa Berghaus[3]; Brenton Credille[2]; Virginia Fajt[4]
[1]Veterinary Teaching Hospital, University of Georgia; [2]College of Veterinary Medicine, University of Georgia; [3]University of Georgia; [4]College of Veterinary, Medicine, Texas A&M University

Cephalosporin antimicrobials are commonly used to treat septic foals when an aminoglycoside is contraindicated. Some cephalosporins have recently been unavailable because of manufacturing shortages. Ceftazidime is a third-generation cephalosporin that has been used in clinical equine practice because of its expected coverage of gram-negative bacteria, renal safety profile, and availability. However, the plasma disposition has not been evaluated in foals. The goal of this study was to estimate the pharmacokinetic parameters of single intravenous and intramuscular doses of ceftazidime in neonatal foals. Six healthy 2–3 day old foals were given single 25 mg/kg doses of ceftazidime intravenously (IV) and intramuscularly (IM) in a cross-over design. The order of route of administration was randomly assigned with a 48-hour washout period between doses. No adverse reactions were noted during treatment. Non-compartmental analysis was used to estimate plasma pharmacokinetic parameters. The median T1/2 for IV and IM administration was 2.28 hours and 2.0 hours, and median AUC0 - last was 317 (µg/ml)/hr and 350 (µg/ml)/hr, respectively. The median Cmax after IM administration was 100.6 µg/ml, with a median Tmax of 0.7 hours. There were no statistically significant differences between IV and IM pharmacokinetic parameters ($p<0.05$). The 25 mg/kg dose of ceftazidime resulted in serum concentrations above the human CLSI Enterobacteriaceae susceptible breakpoint for 8 hours following IV and IM administration. Based on results of this study, IV or IM ceftazidime can be considered as an alternate third-generation cephalosporin for use in neonatal foals.

Abstract E76: Pilot Field Study Comparing the Microbiome in Horses With and Without Equine Glandular Gastric Disease

Linda J. Paul[1]; Aaron Ericsson[2]; Frank Andrews[3]; Michael Keowan[3]; Michael St. Blanc[3]; Heidi Banse[3]
[1]School of Veterinary Medicine, Louisiana State University; [2]University of Missouri; [3]Louisiana State University

The pathophysiology of equine glandular gastric disease (EGGD), a common disorder, remains poorly understood. A previous study using horses kept under identical management conditions showed gastric microbiome changes in association with EGGD. The objective of this study was to characterize the gastric microbiome in horses across multiple barns and differing management practices, to determine if differences between horses with and without EGGD persisted in a more varied population. It was hypothesized that differences in the microbiome would be associated with the presence of EGGD. Thirty client owned horses from various barns in the southern Louisiana area were enrolled. Gastroscopy was performed, EGGD grade was recorded, and pinch biopsies of endoscopically healthy glandular mucosa were collected. In horses with EGGD ≥2, biopsies of ulcerative lesions were also collected. Microbiome analysis was performed using the 16S rRNA gene. Microbial populations were compared by EGGD grade and barn. Biopsies with <1000 reads were excluded from analysis. No differences were detected between horses with and without EGGD, based upon similarity or diversity indices ($p>0.05$). When horses were grouped by barn, differences were detected among groups using the Jaccard ($p=0.0001$) and Bray-Curtis ($p=0.0033$) similarity indices. This was further supported by principal coordinate analysis. The influence of barn may be due to differences in management practices. Although no differences in microbial populations were observed with EGGD, this may be because of the small sample size and variability in gastric microbial populations among barns. Inclusion of additional horses may allow for identification of differences associated with EGGD.

Abstract E77: Pharmacokinetics of Oclacitinib following Oral and Intravenous Administration to Horses

Wendy Collard; John Thorn; Sarah Smith; Kenneth Feenstra
Zoetis

Apoquel® (oclacitinib maleate) is a Janus kinase inhibitor (JAKi) approved for the control of pruritus associated with allergic dermatitis as well as the control of atopic dermatitis (AD) in dogs. Allergic skin conditions can occur in other species such as cats and horses, with pruritus as a hallmark clinical sign that can lead to scratching and additional, self-induced skin damage. The objective of this work was to characterize the single dose pharmacokinetics of oclacitinib in horses following intravenous (IV) and oral administration. Two studies were conducted. In one study four horses received intravenous 0.25 mg/kg oclacitinib. In the second study an average dose of 0.2 mg/kg was achieved in six horses dosed orally by administering multiple 16 mg tablets based on individual body weight. In both studies blood samples for plasma were serially collected for 72 hours post dose, and oclacitinib plasma concentrations were determined by LC/MS/MS. Following IV administration at 0.25 mg/kg, the clearance of oclacitinib was low (1.43 mL/min/kg), the volume of distribution was moderate (0.705 L/kg), and the half-life was 9–10 hours. Following oral administration, oclacitinib was well absorbed with a mean Cmax of 204 ng/mL at a mean tmax of 3.7 hours. The oral half-life of 9 hours was similar to the IV half-life. In comparison to the half-life observed in dogs (4 hours), the half-life in horses is longer suggesting that once a day dosing could maintain the plasma concentrations needed to control pruritus in the horse.

Abstract E78: Safety of Repeat Gabapentin Dosing (40 or 120 mg/kg) in Adult Horses

Jenifer R. Gold; Tammy Grubb; Lais Malavais; Nicolas Villarino
Washington State University

Background: Neuropathic pain, as in laminitis, is largely refractory to standard analgesia. Gabapentin controls neuropathic pain in other species and may be effective in horses. Significant gabapentin-mediated pain relief is unproven in horses, however, possibly due to insufficient dosing, as the effective dose is unknown. Our previous study showed that horses, like humans, have disproportional absorption of gabapentin. However, the absorption is more proportional than in humans, and the absorption disparities are likely due to inter-horse variation. The proportionality is important as it allows a dosing recommendation based on linear pharmacokinetics but, single-dose proportionality studies do not allow evaluation of safety using a valid therapy length for neuropathic pain. Prior to initiating pharmacodynamic studies in painful horses, safety of an appropriate dosing regimen for laminitis of at least two weeks is required. With repeat gabapentin dosing in other species, adverse effects include; sedation/ataxia in multiple species and rarely acute liver injury and exacerbation of chronic renal disease in humans. No adverse effects have been reported in horses. **Methods**: In a randomized cross-over study using 6 adult horses, each. horse received a dose of 40 mg/kg (Trial 1) and 120 mg/kg (Trial 2) of gabapentin orally twice daily prior to eating. The horses had a 3-week washout in between doses. The horses had physical examinations, ataxia and sedation scores performed 3 times daily for 15 days. Biochemistry profiles were analyzed prior to initiation of (pregabalin), 7 and 14 days after the first gabapentin dose. Data were analyzed using Chi-Square, student-t tests and ANOVA, as required by data type. Significance was set at ($p<0.05$). **Results**: Over the 2-week study period none of the horses became sedate or ataxic with either dose of gabapentin. Biochemistry variables remained within normal limits. There were no significant differences within or between horses nor within or between trials for any variable. **Conclusion**: Repeat dosing of gabapentin is safe when dosed at 40 or 120 mg/kg twice daily for two weeks.

Table 1: Biochemistry variables for horses receiving repeated doses of gabapentin (40 mg/kg: Trial 1; 120 mg/kg: Trial 2) for two weeks

Trial 1	Glucose mg/dl	Creatinine mg/dl	BUN mg/dl	BUN/Creat	Calcium mg/dl	Total protein g/dl	Albumin g/dl
Pregabalin	90±8	1.45±0.55	11.66±1.6	10.33±6.97	10.75±0.57	6.50±0.57	2.75±0.10
Day 7	88±7	1.53±0.30	11.0±1.6	7.17±1.07	10.63±0.47	6.42±0.28	2.62±0.15
Day 14	89±3	1.60±0.29	11.17±2.73	7.00±1.41	10.70±0.27	6.38±0.19	2.67±0.09
Trial 2							
Pregabalin	87±3	1.38±0.23	10.17±1.77	7.50±1.89	10.33±0.21	6.76±0.39	2.67±0.11
Day 7	88±3	1.48±0.41	12.67±3.09	9.00±3.05	10.30±0.38	6.78±0.20	2.65±0.17
Day 14	86±10	1.63±0.39	10.0±3.05	6.67±3.40	10.75±0.38	6.85±0.26	2.67±0.11

Trial 1	Globulins g.dl	AST U/L	ALKP U/L	GGT U/L	Tbili mg/dl	CK U/L	LDH U/L
Pregabalin	3.93±0.18	305±115	132±8	17.8±4.2	1.72±0.60	116±26	788±254
Day 7	3.78±0.20	362±173	130±7	19.8±4.1	1.55±0.35	121±44	1128±664
Day 14	3.75±0.15	384±214	128±14	19.0±3.7	1.63±0.57	157±83	787±273
Trial 2							
Pregabalin	4.12±0.38	298±77	123±13	20.0±3.7	1.25±0.32	127±43	943±492
Day 7	4.15±0.18	276±70	111±11	17.3±2.4	1.68±0.52	106±30	869±406
Day 14	4.18±0.27	267±41	112±7	16.7±2.9	1.32±0.50	111±27	849±365

Abstract E79: Pulmonary Disposition of Gallium Maltolate after Oral Administration to Foals

Natalia Rodriguez[1]; Noah D Cohen[1]; Robert J. Taylor[2]; Lawrence R. Bernstein[3]; Angela I. Bordin[4]
[1]Texas A&M University; [2]College of Veterinary Medicine and Biomedical Sciences, Texas A&M University; [3]Gallixa LLC.; [4]Equine Infectious Disease Laboratory, TAMU

Emergence of isolates resistant to macrolides and rifampin creates great need for alternative antimicrobials for treating foals with *Rhodococcus equi* pneumonia. We have demonstrated *in vitro, ex vivo,* and *in vivo* activity of gallium maltolate (GaM) against *R. equi*. Although serum pharmacokinetics of GaM in foals have been reported, its pulmonary disposition following oral administration is unknown. It is important to determine the pulmonary disposition of GaM because virulent *R. equi* infect and replicate in alveolar macrophages after inhalation. Our objective was to characterize the pulmonary disposition of GaM in foals following a single oral dose. We hypothesized that oral administration of GaM would achieve concentrations exceeding the minimum inhibitory concentration (MIC) against *R. equi* in broncho-alveolar lavage (BAL) cells. Eight university-owned Quarter-Horse foals 4 months of age were studied. BAL fluid and cells were collected at time 0, 4, 8, 24, and 48 hours after GaM administration. Concentrations of GaM in the serum, BAL fluid, and BAL cells were determined by inductively coupled plasma-optical emission spectrometer or mass spectrometer, depending upon concentration. Time to maximum concentration (Tmax) for BAL fluid was 24 hours, and concentrations were below the MIC at all times. The Tmax in BAL cells was 48 hours and concentrations were above the MIC by 4 hours. Results indicated that GaM achieves therapeutic concentrations in BAL fluid cells, warranting evaluation for treating clinical pneumonia caused by *R. equi* in foals. If accepted for presentation, serum pharmacokinetics from these foals also will be reported.

Abstract E80: The Value of Transabdominal Ultrasound in Horses Presenting with Medical Gastro-Intestinal Disease

Gayle D. Hallowell[1]; Carolyn Bates[2]; Emma Peal[3]; Caroline Bullard[3]; James Bailey[4]; Adam Redpath[4]; Mark Bowen[4]
[1]School of Veterinary Medicine and Science, University of Nottingham; [2]Defence Animal Training Regiment; [3]Royal Army Veterinary Corps, Household Cavalry Mounted Regiment; [4]SVMS, University of Nottingham

To evaluate the frequency with which abnormalities are detected using transabdominal ultrasound in horses that present with weight loss, chronic diarrhoea, pyrexia of unknown origin or recurrent bouts of abdominal pain. Clinical and ultrasound case records were searched for abdominal evaluations over the period of 2015–2020 in horses >1 year of age. Signalment, ultrasound findings and final diagnoses were recorded as was outcome. Abnormalities recorded from abdominal ultrasound included: increased small (SI) or large intestinal (LI) wall thickness defined as >4 mm, abnormal intestine, abnormal appearance or volume of peritoneal fluid or abnormal abdominal organ or other. Other information recorded included specific intestinal wall thickness, other tests performed, treatments administered and any further findings. Recurrent colic was classified as >3 episodes in 3 months and chronic diarrhoea of >7 days duration. Images were obtained using a Vivid Q ultrasound machine (GE Healthcare) using both a 3 MHz convex and 7.5 MHz microconvex probes and saved as cineloops that were evaluated offline. 224 evaluations were found that had both abdominal ultrasound images and clinical reports or records. Horses ages ranged from 3–28 (13.5±4.8 years), were predominantly TB/part-breds (29%) or ID/part-breds (29%) with 59% geldings and 41% mares. 61% of evaluations were for horses with weight loss, 24% with recurrent colic, 9% with diarrhoea and 6% with other signs. Abnormalities were identified in 80% of examinations. Of those with abnormalities, 19% had thickened SI, 40% thickened LI, 21% thickened SI and LI, 4% abnormal motility, 8% abnormal peritoneal fluid and 8% with other abnormalities. Transabdominal ultrasound can identify abnormalities in a large percentage of horses that present with these clinical signs, making the time spent performing this technique worthwhile.

Abstract E81: Effect of Neonatal Dysphagia on Subsequent Racing Performance in Standardbred Horses

Barbara Delvescovo[1]; Kathleen Mullen[2]; Steven Eicker[3]; Dorothy Ainsworth[1]
[1]Cornell University; [2]Littleton Equine Medical Center; [3]Valley Agricultural Software King Ferry

We previously reported on a clustering of dysphagic foal cases occurring over a 5-year period (2012–2016) on a Pennsylvania (PA) Standardbred broodmare farm located amongst fracking operations (Mullen *et al.*, JVIM 33:2392, 2019). Another Standardbred broodmare farm located 250 miles away in New York (NY), owned by the same individual and operated identically, failed to experience similar cases of foal dysphagia over that same period. That investigation uncovered high concentrations of polycyclic aromatic hydrocarbons (PAHs) in the well water. We attributed the foal dysphagia to PAH exposure of late pregnant mares as the dysphagia prevalence was eliminated after a water filtration/treatment system (WFTS) was installed on the farm in August 2015. We were unaware of studies examining the effects of neonatal dysphagia on athletic performance and undertook the current investigation. We hypothesized that racing performance of formerly dysphagic foals would be negatively impacted. This prospective study was conducted from 2014–2017 evaluating foals born on the PA and NY farms; additional data were obtained from detailed farm records of foals born in 2012 and 2013. Foals (N=8) with dysphagia attributed to causes other than PAH environmental exposure (e.g. prematurity, sepsis, neonatal encephalopathy, selenium deficiency, palatal congenital abnormalities) were excluded from our analysis. The proportion of foals from both farms that raced, their age of first race and two custom-designed performance indices, Earnings Per Start Index (EPS) and Speed Index, were compared between the dysphagic and normal foals using Chi-Square and Wilcoxon Rank Sum Tests with $p<0.05$. 126 foals were born alive during the study period: 74 on the NY farm and 52 on the PA farm. Ten foals died prior to race training from causes unrelated to dysphagia, leaving 116 horses in the study. On the PA farm, 54% of unaffected and 72% (13/18) of the dysphagic foals raced; 70% (47/67) of the normal NY foals raced. Comparative analysis between the two farms revealed no significant difference in these percentages ($p=0.27$). Installation of the WFTS on the PA farm did not affect the percentage of foals that eventually raced ($p=0.71$). Performance analysis revealed no difference in the EPS, Speed Indices or age of first race between dysphagic and normal foals when either compared between (NY vs PA) or within (PA) farms ($p>0.05$). Our results suggest that although treatment of neonatal dysphagia incurs a financial burden for the owners, the athleticism of formerly dysphagic foals does not appear to be negatively impacted as measured by EPS and Speed indices.

Abstracts E82: Effects of Supplements Containing Turmeric and Devil's Claw on Equine Gastric Ulcer Scores in Horses

Frank M. Andrews[1]; Heidi Banse[2]; Michael St. Blanc[2]; Mary Retif[2]; Nicole Arana-Valencia[2]; Michael Keowen[2]; Frank Garza[2]; Lydia Gray[3]; Chin-Chi Liu[2]
[1]Equine Health Studies Program, Department of Veterinary Clinical Sciences, LSU; [2]LSU School of Veterinary Medicine; [3]SmartPak Equine

Supplements commonly fed to horses to decrease inflammation and pain might cause or worsen gastric ulcers. The purpose of this study was to determine if a supplement containing turmeric (TUM; Curcuma longa) and devil's claw (DC; Harpagophytum species) caused worsening of gastric ulcer scores in stall confined horses. Twelve clinically healthy adult Thoroughbred horses with Equine Gastric Ulcer Syndrome (EGUS) scores >0 were included in a non-crossover study design. Horses were stratified by EGUS score and assigned to either the treatment (TUM/DC supplement) or control (supplement without TUM/DC) group. The supplements were fed daily for 28 days. Gastroscopy was performed on days 0, 14, and 28. The EGUS score, non-glandular ulcer number (NGN) and severity (NGS) scores and glandular number (GN) and severity (GS) scores were recorded. In addition, body weight, gastric juice pH, packed cell volume (PCV), total protein (TP), and results of a point-of-care blood analyzer were evaluated on Day 0 and Day 28. A mixed ANOVA model in SAS was used to analyze the variables measured with treatment, day and their interactions. Assumptions of these models (linearity, normality of residuals, and homoscedasticity of residuals) and influential data points were also assessed by examining standardized residual and quantile plots. When a fixed effect was detected, Tukey *post-hoc* comparisons were performed with least square means for the effect. Significance was set at $P<0.05$. Mean EGUS and NGS scores were significantly lower in both treatment and control groups by Day 14 and 28, when compared Day 0. Blood parameters, body weight, and gastric juice pH did not change significantly change during the study. In conclusion, a supplement containing turmeric and devil's claw did not result in worsening of gastric ulcers or alter health parameters after 28 days of feeding.

Abstract E83: Multidimensional Analysis of Bronchoalveolar Lavage Cytokines Reveals Interferon-Γ as a Key Biomarker in Equine Asthma

Jane S. Woodrow[1,2]; Melissa Hines[1]; Carla Sommardahl[1]; Bente Flatland[1]; Kaori Davis[3]; Yancy Lo[2]; Zhiping Wang[2]; Mary Katherine Sheats[3]; Elizabeth Lennon[2]
[1]University of Tennessee; [2]University of Pennsylvania; [3]North Carolina State University

Equine asthma syndrome (EAS) consists of two broad groups, defined as: mild to moderate (mEAS) and severe (sEAS). Each group remains a heterogenous population with continued need to clarify their pathogenesis. The role of pro-inflammatory cytokines, both type 1 and type 2 cytokines, in addition to anti-inflammatory cytokines, such as transforming growth factor (TGF)β-1 and bone morphogenetic proteins (BMP), need to be further elucidated in EAS groups in order to best mange and treat the patient. The objectives of this study were to: (1) analyze the cytokine profile present in BALF that defines endotypes of equine asthma, and (2) analyze anti-inflammatory molecule gene expression levels in immune cells of the BALF. Cytokine concentrations in BALF supernatant was evaluated with a commercially available multiplex immunoassay. Multidimensional scaling (MDS) analysis was performed in order to visualize on a two-dimensional figure the clustering of samples based on age, sex, cytokine concentrations, and neutrophil percentage. Gene expression results were normalized to β-actin and relative quantification was performed using 2-ΔΔCt method. BALF multiplex showed that horses with asthma had significantly higher TNFα (mEAS: 6.54 [4.79–7.69] pg/mL, p=0.013; sEAS: 11.31 [3.67–21.77] pg/mL, p=0.0009) concentrations compared to healthy horses (2 [0.8–3.63] pg/mL); as well as significantly elevated CXCL8 (mEAS: 14.41 [8.74–22.48] pg/mL, p=0.024; sEAS: 19.83 [16.86–27.82] pg/mL, p<0.0001), concentrations compared to healthy horses (4.54 [1.26–8.88] pg/mL). Additionally, IFNγ was significantly elevated in sEAS (35.28 [21.08–75.45] pg/mL, p=0.012) compared to healthy horses (21.08 [21.08–21.08] pg/mL). MDS analysis demonstrated that IFNγ differentiates severe from moderate asthma, and that TNFα and CXCL8 are key biomarkers of equine asthma endotype. Tgfβ1 was significantly increased in mEAS immune cells versus healthy (p=0.02), and although statistical differences were not seen in BMP7 expression, a subset of horses diagnosed with sEAS had markedly elevated BMP7 expression. Elevated BMP7 in a subset of sEAS horses may signify over exuberant response to counteract airway remodeling and fibrosis due to BMP7's ability to block Tgfβ1-driven fibroblast activation. These results will help further define EAS immunopathology, which could improve understanding of asthma endotypes, and potentially identify novel therapeutic strategies.

Abstract E84: A Retrospective Study of Exercise-Induced Pulmonary Hemorrhage and Asthma in Barrel Racing Horses in Texas

Laszlo Hunyadi[1]; Munashe Chigerwe[2]; Emily Sundman[3]
[1]Texas Tech University; [2]University of California Davis; [3]Equine Sports Medicine & Surgery

This study was to evaluate the bronchoalveolar lavage (BAL) fluid cytological results of barrel racing horses with exercise-induced pulmonary hemorrhage (EIPH), asthma, or both. A retrospective study was conducted of 95 horses diagnosed with non-infectious respiratory disease and BAL results at a private practice in Texas. EIPH only was diagnosed in 28/95 (30%) BAL samples, asthma only was diagnosed in 25/95 (26%) BAL samples, and EIPH and asthma was diagnosed in 42/95 (44%) BAL samples. A history of EIPH was not predictive of a diagnosis of EIPH, or asthma. Of the BAL results consistent with asthma only, the primary elevated cell type was mast cell; followed by samples with both elevated eosinophils and mast cells. No cytological differences in frequency of elevated inflammatory cell types were found between horses diagnosed with EIPH and those without a diagnosis of EIPH. The study showed a substantial percentage (44%) of the horses examined had evidence of both EIPH and asthma on BAL cytologic evaluation. Comorbidities of EIPH and asthma is a common diagnosis in barrel racing horses in this study.

Abstract E85: Clinical Response in Horses with Severe Equine Pasture Asthma to Allergen Specific Immunotherapy

Ann Chapman[1]; Michelle Woodward O'Gorman[2]; Mike Keowan[2]; Frank Garza[2]; Frank Andrews[2]
[1]Louisiana State University; [2]School of Veterinary Medicine, Louisiana State University

Severe equine pasture asthma (EPA) is a debilitating respiratory condition of grazing horses in the southeastern U.S. The condition is characterized by airway inflammation, bronchial hyper-reactivity, chronic cough and dyspnea. This disorder has a strong association with hot and humid climate and high environmental fungal spores and grass pollens. The goals of this study were to collect intradermal and serum allergy test results from horses with severe EPA, and to evaluate clinical responses to allergen specific immunotherapy, formulated based on both test results, over the course of one year while allowing continuous exposure to pasture aeroallergens. Seven university owned horses with severe equine pasture asthma were enrolled in the study based on BAL inflammation and clinical scoring system (presence of tachypnea, nostril flaring, abdominal lift, spontaneous or inducible cough, nasal discharge, abnormal tracheal sounds, and abnormal pulmonary auscultation during summer months). All horses underwent serum and intradermal allergy at the onset of expected clinical remission (November). Horses were randomly assigned to either treatment group (n=4) or controls (n=3) and were housed together on mature grass pastures for the duration of the study. Horses within the treatment group received subcutaneous injections of dilutions of selected allergens over the course of the study using a standard protocol (Table 1). Standing thoracic radiographs and bronchoalveolar lavage cytology were performed on all horses on prior to starting therapy, and then approximately at days 90, 180, 270, and 365. Horses were observed daily and evaluated at pre-determined intervals using the established clinical respiratory scoring system by blinded evaluators. Horses that exhibited severe exacerbation of clinical disease were administered short term rescue therapy. Total clinical scores improved significantly in both groups ($P<0.05$) during the remission period (Day 28) and worsened during the exacerbation season (Day 85–270). A time effect was noted in mean BAL neutrophilia (43.3%±30.6) in all horses ($P<0.05$). There was no significant difference in total clinical scores and BAL neutrophilia between treatment groups. In conclusion, allergen specific immunotherapy alone did not improve clinical score or BAL inflammation in horses with equine pasture asthma following a year of therapy. Future larger studies examining allergen specific immunotherapy, potentially over longer periods of time in younger animals, along with environmental management in horses with EPA are warranted.

Table 1. ASIT treatment protocol

Day	Vial 1	Vial 2	Vial 3
1	0.1 ml		
3	0.2 ml		
5	0.4 ml		
7	0.8 ml		

9	1 ml		
11		0.1 ml	
13		0.2 ml	
15		0.4 ml	
17		0.8 ml	
19		1 ml	
21			0.1 ml
23			0.2 ml
25			0.4 ml
27			0.8 ml
29			1 ml
39			1 ml
59			1 ml

Continue every 3 weeks until respiratory flare

Every 14 days when flaring begins

Equine 3D Electro-Anatomical Mapping in Sinus Rhythm Elucidates the Electrocardiogram and Cardiac Electrical Activation Pattern

Gunther van Loon[1]; Lisse Vera[1]; Annelies Decloedt[1]; Stijn Schauvliege[2]; Tim Boussy[3]; Glenn Van Steenkiste[1]
[1]Equine Cardioteam, Ghent University; [2]Ghent University; [3]AZ Groeninge

Understanding the depolarisation pattern of the equine heart under normal physiological conditions, and its relationship to the surface electrocardiogram (ECG), is of uppermost importance for research on pathophysiology of complex arrhythmias. The purpose was to use a 3D electro-anatomical mapping system under general anaesthesia to evaluate the qualitative and quantitative atrial and ventricular depolarisation patterns and correlation with the surface ECG in 7 healthy horses in sinus rhythm. Bipolar activation maps of the endocardium were analysed. The first atrial activation was located at the height of the crista terminalis. Only one interatrial conduction pathway was recognised. The first and second P wave deflections represented the right and left atrial depolarisation, respectively. His bundle electrograms could be recorded in 5 out of 7 horses. Left ventricular activation (Fig. 1 to 3) started at the mid septum, right ventricular activation apically from the supraventricular crest. This was followed by separate depolarisations at the height of the mid free wall. Further ventricular depolarisation occurred in an explosive pattern. Electrically active tissue could be found in all pulmonary veins. In contrast to findings of previous studies, all parts of the ventricular depolarisation contributed to the surface ECG QRS complex. This study provides a reference for the normal sinus impulse endocardial propagation pattern and for conduction velocities in equine atria and ventricles. This information allows to improve ECG reading and develop better recording techniques, such as new 12-lead ECG configurations, in horses.

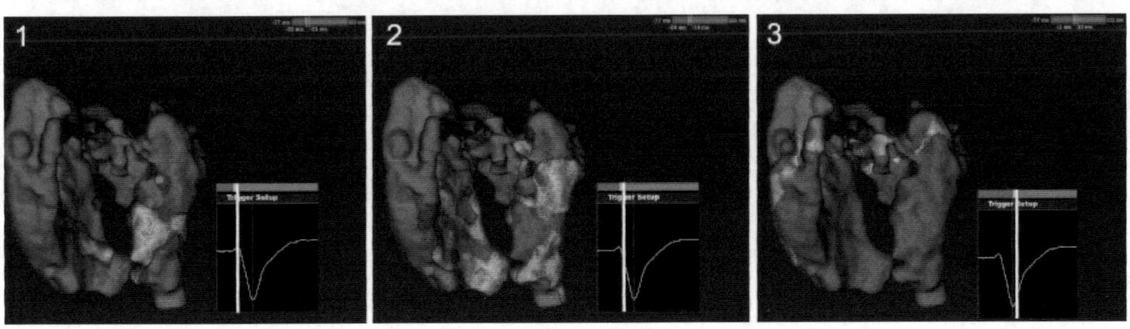

MARCKS Is a Novel Target in Equine Asthma Syndrome

Mary K. Sheats; Kaori Davis
College of Veterinary Medicine, North Carolina State University

The goal of this study was to investigate the MARCKS (Myristoylated Alanine Rich C Kinase Substrate) protein as a potential novel therapeutic target for Equine Asthma Syndrome (EAS). Our lab, and others, previously established a role for MARCKS in both neutrophil and macrophage inflammatory functions. In the current study, we hypothesized that MARCKS protein is upregulated in BAL cells in horses with asthma; and that inhibition of the MARCKS protein with a MARCKS-specific inhibitor peptide known as MANS, will diminish the zymosan-induced respiratory burst of equine alveolar macrophages, and peripheral blood neutrophils, *ex vivo*. Bronchoalveolar lavage (BAL) was performed as previously described using a 3 m cuffed Bivona ® tube and 300 mLs sterile saline. BAL cytology (400 cell count differential), physical, and rebreathing exam, were used to classify horses as normal, mild/moderate, or severe EAS. Total MARCKS protein expression in BAL cell lysates was quantified using an equine specific MARCKS ELISA (MyBioSource, San Diego, CA, USA) and normalized to total protein (BCA assay) in cell lysate. Equine alveolar macrophages were isolated by adhesion. Peripheral blood neutrophils were isolated by ficol density gradient centrifugation. Zymosan-induced respiratory burst was measured as luminol-enhanced chemiluminescence. Cells were pretreated with 0, 10, 25, 50, or 100 μM MANS (MARCKS-specific inhibitor peptide), or 100 μM RNS (control peptide). Normalized MARCKS protein expression was significantly higher in BAL cell lysates from horses with mild/moderate (n=14) and severe asthma (n=5), compared to BAL cell lysates from horses with normal lower airway cytology (n=6) (One-way ANOVA, $p<0.05$). MANS peptide pretreatment, but not RNS, significantly attenuated zymosan-induced respiratory burst of equine alveolar macrophages and peripheral blood neutrophils (One-way ANOVA, $p<0.05$). These findings determine that MARCKS protein expression is altered in BAL cells from horses with EAS, indicating a possible role for MARCKS in the pathophysiology of EAS, and support MARCKS inhibition as a potential therapeutic strategy to alter the inflammatory response of airway cells including alveolar macrophages and neutrophils.

The Development of a Therapeutic Antibody for Equine Endocrinopathic Laminitis

Samira Rahnama[1]; Niveditha Vathsangam[2]; Robert Spence[1]; Carlos Medina-Torres[3]; Christopher Pollitt[3]; Melody de Laat[1]; Simon Bailey[2]; **Martin N. Sillence**[1]
[1]Queensland University of Technology; [2]The University of Melbourne; [3]The University of Queensland

Although therapeutic monoclonal antibodies (mAb) are used in human medicine, few are under development for small animals and none are available for horses. Nor are there any registered medications to prevent or treat endocrinopathic laminitis, which is associated with hyperinsulinaemia. Because insulin receptors are scarce in the hoof, it was proposed that high concentrations of insulin damage the lamellae by overstimulating IGF-1 receptors (IGF-1R). Anti-IGF-1R mAb have been used successfully in human cancer patients, with few adverse effects. Thus, they could provide a useful therapeutic option for laminitis. A recombinant human anti-IGF-1R mAb was re-engineered to avoid an immune response in the horse, and expressed in CHO cells. *In vitro*, the equinized mAb was highly selective for binding to equine IGF-1R over insulin receptors (>60-fold), as measured by radioligand binding studies using lamellar and liver cell membranes. However, the mAb blocked the stimulatory effects of both peptides on cell proliferation in lamellar explants (100%, $p<0.001$), as measured by 3H-thymidine uptake. *In vivo*, when Australian Standardbred horses were subjected to a prolonged hyperinsulinaemic-euglycaemic clamp over 48 h (n=6), pre-treatment with a low dose of the mAb reduced the sinking of the distal phalanx (47%, $p<0.05$), measured radiographically, and markedly lessened the stretching of the secondary epidermal lamellae, as observed in histological sections ($p<0.05$). The results confirm that insulin can induce laminitis via IGF-1R, and demonstrate the potential of an anti-IGF-1R mAb for preventing or treating this disease.

Inflammatory Bowel Disease in Horses

Gayle D. Hallowell, MA, VetMB, PhD, DACVIM (LAIM), DACVECC, DECVSMR, PFHEA, FRCVS
Sutton Bonington, Leicestershire, UK

Introduction

Inflammatory bowel disease (IBD) in horses is associated with clinical signs of weight loss, diarrhoea and/or recurrent colic[1] likely due to changes in absorption of nutrients and water and effects on gastrointestinal motility[1]. Inflammatory bowel disease was proposed to be driven by an inappropriate immune response to bacteria, viruses, parasites and dietary allergens[1]. Diagnosis of inflammatory bowel disease in horses is achieved using a variety of techniques including abnormal oral glucose or xylose absorption tests, increased intestinal wall thickness on transabdominal or transrectal ultrasound, partial thickness rectal and transendoscopic duodenal biopsies, or full-thickness intestinal biopsies[2]. The sentinel publication in the early 2000s on inflammatory bowel disease[1] in horses made for grim reading regarding prognosis, although perhaps did not match clinical impressions of this disease. This retrospective case review was published at a time when better-quality ultrasound machines were available, making transabdominal ultrasound in horses feasible, and these machines could generate good-quality, diagnostic images allowing accurate assessment of intestinal wall thickness. This poor prognosis was recently challenged in another, larger prospective publication, where initial response to therapy was 75% with 3-year survival reported to be 65%.

Types of Inflammatory Bowel Disease

There are five types of IBD described in the horse. These include lymphocytic-plasmacytic enterocolitis, granulomatous enteritis, diffuse eosinophilic enterocolitis, idiopathic focal eosinophilic enterocolitis and multisystemic eosinophilic epitheliotrophic disease (MEED)[1]. All these disorders can affect both the small and large intestine. It has been suggested that a sixth subtype affecting the large intestine that resembles ulcerative colitis in humans has a high prevalence in normal horses, but the clinical significance of these findings are yet to be elucidated (Kerbyson and Knottenbelt, personal communication). MEED and GE are the only subtypes of IBD that are associated with disease of tissues beyond the intestinal tract. In MEED, changes in the skin and other organs are commonly reported, and in GE changes in the skin and lymph nodes can be seen[1]. Granulomatous enteritis and MEED have most commonly been reported in horses <4 years of age with a reported increased prevalence in Standardbreds[1]; the breed predilection may suggest that there is a genetic component[3]. The other three subtypes have been identified in any age of horse and variety of breeds[1,2,4]. It has been proposed that lymphocytic-plasmacytic enterocolitis may be a precursor to intestinal lymphoma as in other species[1,4-6]. Granulomatous enterocolitis has histopathological similarities to Crohn's disease in humans and is more likely to be associated with anaemia, likely secondary to reduced absorption of nutrients or associated with chronic disease[1]. Surgery to undertake resection and anastomosis may be curative for focal idiopathic eosinophilic enterocolitis[1,7].

Diagnostic Approach to the Suspected Case of IBD

There are a multitude of diseases that can result in weight loss, diarrhoea and recurrent colic that are more likely than IBD. It is therefore important to ensure that these more common diseases have been ruled out. This would include ruling out (as far as possible) dental, parasitic and other infectious diseases, sand accumulation, renal or hepatic failure, intestinal motility disorders and gastric disease[2]. Specific tests that can be considered valuable during evaluation of a suspected IBD case would include faecal worm egg count, blood biochemistry, oral glucose or xylose absorption test, transabdominal (Figures 1 and 2) and transrectal ultrasonography, partial-thickness transendoscopic duodenal and rectal mucosal biopsies and full-thickness intestinal biopsies obtained by laparoscopy or laparotomy[2-4].

Figure 1

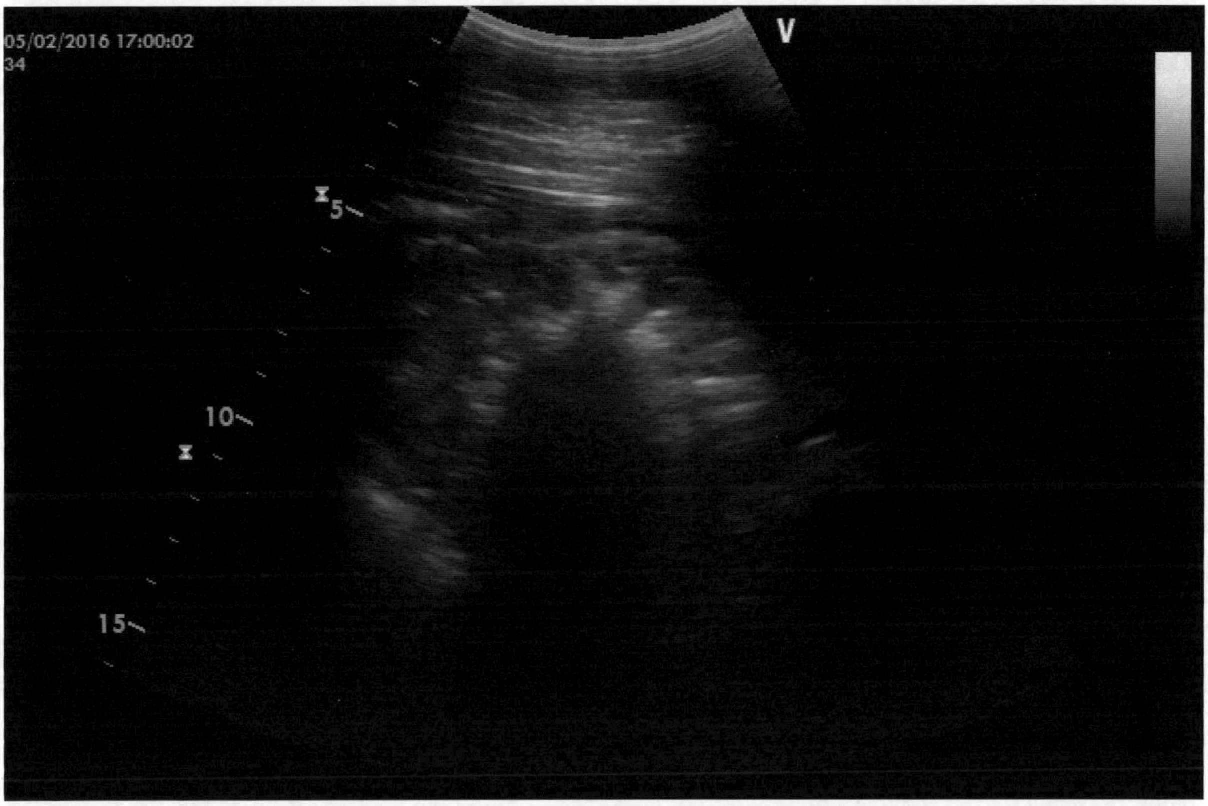

This is a transabdominal image of the left ventral colon imaged ventrally at the level of the left 14th intercostal space in an 8-year-old Irish draught gelding that presented with weight loss and ventral oedema and hypoalbuminaemia. The horse was treated with a tapering course of oral dexamethasone and oral azathioprine, which had minimal effect on the clinical, clinicopathological and ultrasonographic findings. Post-mortem examination revealed eosinophilic enterocolitis that affected much of the colon. No endoparasites were identified, nor was there evidence of encysted small redworm.

Figure 2

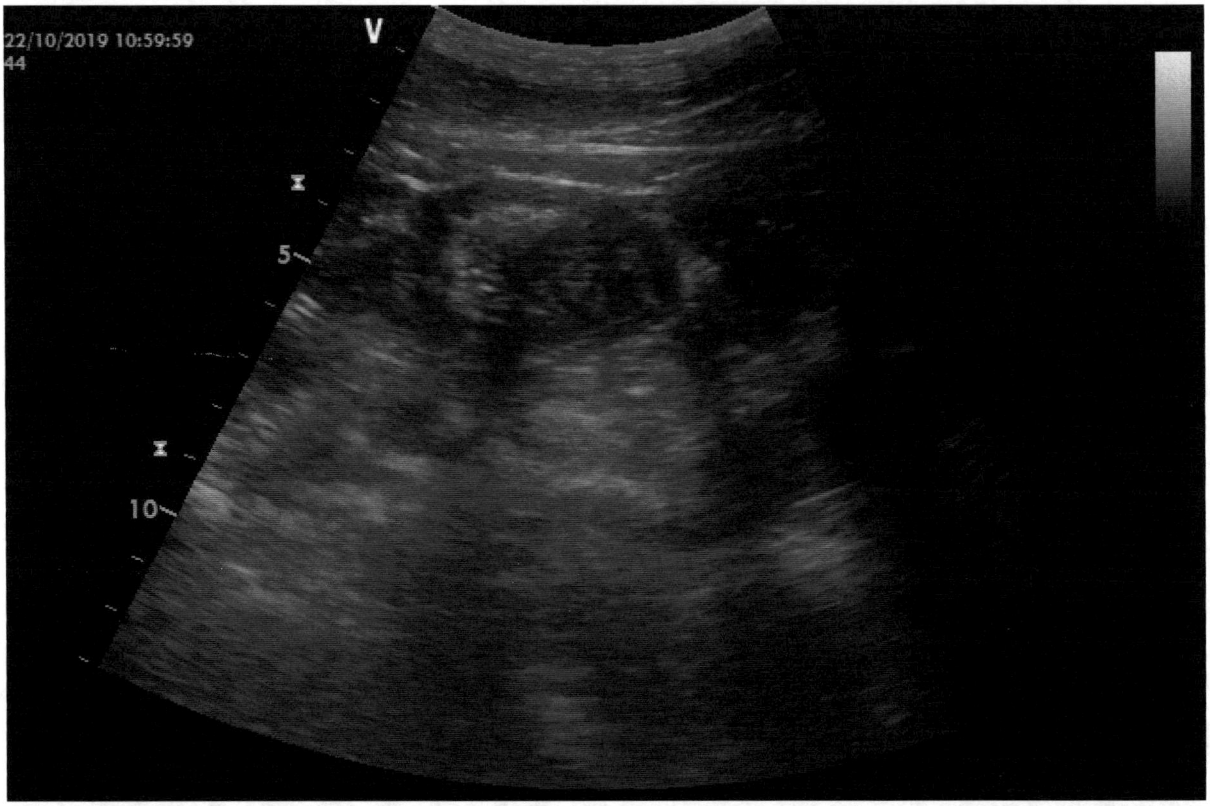

This is an image of the duodenum imaged transabdominally at the mid-to-dorsal region of the right 13th intercostal space in a 16-year-old Irish sport horse gelding that presented with weight loss and poor appetite with a body condition score of 2/9. Transabdominal ultrasound revealed generalised increase in wall thickness of both the small and large intestines. An exploratory laparotomy performed to obtain full-thickness intestinal biopsies recovered samples that revealed disseminated lymphocytic-plasmacytic enterocolitis. The horse was treated with a 3-month tapering course of oral dexamethasone and oral azathioprine and parenteral nandrolone laurate. After approximately 2.5 months, the horse began to gain weight slowly and reached body condition score 4/9 approximately 6 months after initial presentation.

Clinical, Clinicopathological Signs and Diagnostic Tests Valuable for Diagnosing IBD

The most common presenting signs with IBD include weight loss, combined weight loss and recurrent colic and recurrent colic alone[2], and in certain subtypes of IBD diarrhoea and inappetence[4]. Ventral oedema can be observed secondary to hypoalbuminemia[2,4]. Increased fecal worm egg counts have been reported to be negative or have a low prevalence[2,4].

Clinicopathological signs most commonly identified are hypoproteinemia, hypoalbuminemia and abnormal absorption of glucose or xylose[1-4]. Anemia was reported in one study[1], although breed-specific reference ranges were not stated to have been used. In addition, increased serum activity of gamma-glutamyltransferase was reported in horses with MEED[1].

Abnormalities on partial-thickness transendoscopic duodenal and rectal mucosal biopsies are variable and likely to relate to the type of IBD and extent of that disease. These abnormalities are reportedly between 0–82% for rectal mucosal biopsies[1-4], where abnormalities were most common in between 40–50% of cases. Transendoscopic duodenal biopsies did not yield a diagnosis in one study[3]. Identification of thickening of the intestinal wall using transabdominal ultrasonography was reported in 40–50% of cases[2,4].

INFLAMMATORY BOWEL DISEASE IN HUMAN PATIENTS

Inflammatory bowel disease in humans is largely grouped into two subtypes: Crohn's disease or ulcerative colitis, and both subtypes wax and wane between remission and clinical disease[8]. In contrast to horses, a great deal is known about the natural history and epidemiology of IBD in human patients. Overall, there is no gender difference in prevalence for Crohn's disease or ulcerative colitis[8]. Both of these disorders are initially recognised in people in their 20s–40s[9], although more cases are being recognised in children as early as 6 months of age[10]. Both subtypes of IBD require lifelong management with maintenance drugs and then additional pharmacological agents for flareups, and then when both of these strategies fail, surgery is required to remove affected sections of intestine[11]. The use of biological drugs such as infliximab and adalimumab has become a mainstay for IBD management in people[11]. These drugs are genetically engineered from living organisms and target specific proteins and enzymes that are responsible for initiation and perpetuation of inflammation. Targets include tumour necrosis factor alpha (infliximab, adalimumab, golimumab and certolizumab pegol) and two anti-integrin agents (natalizumab and vedolizumab) with direct and indirect actions also on interleukins, adhesion molecules and colony-stimulating factors[12]. These drugs are very expensive[11] and associated with significant side effects, although an improvement on high-dose corticosteroid therapy[12].

Epidemiology of IBD in Human Patients

It is known that there is a complex relationship between genetics, the environment, bacteria, viruses and parasites and cereal and meat proteins[8]. The relative roles of these factors are somewhat controversial, but it is accepted that the enteric microbiota play a central role in the pathogenesis of IBD[9,13].

Genetics plays a complex role in IBD, as 160 susceptibility loci/genes have been associated with the development of the disease[9,14]. It is postulated that the first year of life is a critical time for development of the individual's microbiome development and that this microbiome helps establish maintenance of homeostasis versus the development of inflammation[9,13,15]. It is proposed that early-life antibiotic exposure, which alters the intestinal microbiome, is associated with increased risk of development of Crohn's disease later in life[15,16]. In multiple studies, it has been shown that being raised in less hygienic environments plays a protective role[17]. Other protective factors include lower rates of antibiotic use, increased numbers of pets and livestock, larger family sizes and increased exposure to enteric pathogens, specifically *Helicobacter pylori*[17] and helminths[18,19], and fibre in the diet. Risk factors include NSAID administration[20,21], breastfeeding[22], gastrointestinal infections (particularly *Salmonella* and *Campylobacter*[23]), high carbohydrate/fat diet, living in an urban setting[24], air pollution and stress[25]. There is an association between IBD and vitamin D deficiency[26]; this may play an inciting role or be a consequence of reduced intestinal absorption and activity leading to less sunlight exposure.

Prevalence of IBD in Human Patients

The prevalence of IBD in the developed world is around 0.5–1% of the population, with the highest prevalence in Canada, U.S., Northern Europe and Australia suggesting that genetics, diet or both are likely to be significant contributing factors[8,11]. The lowest prevalence is in Asia and other developing countries, although IBD cases have been steadily increasing over the last 2–3 decades and correlate with industrialisation and westernisation, suggesting that diet, lifestyle and environmental factors are important in the etiology of IBD[8]. As IBD emerges in developing countries, ulcerative colitis is seen first, followed by a rise in the number of cases of Crohn's disease[27].

Inflammatory Bowel Disease in Dogs

Epidemiology of IBD in Dogs

Many of the epidemiological risk factors discussed in humans also apply to dogs. These include a likely genetic predisposition, multifactorial etiologies and role of commensal bacteria[28]. As discussed above in human patients, it has been shown that dogs also have dysbiosis with IBD. This dysbiosis shows overlapping yet distinct dysbiosis networks[29]. Evaluation of the microbiota is a currently very active area of research for evaluation of a variety of diseases. What is currently unclear is whether this change in microbiota is a cause or effect of the disease process. Even more frustrating is that we are currently unsure of how to make medium- to long-term changes to the microbiota in dogs, and this is also true in other species and applies to pre- and probiotics and fecal transplantation.

Diagnosis of IBD in Dogs

As in other species, a presumptive diagnosis is made having ruled out other causes of hematochezia, diarrhoea and weight loss, and diagnosis is confirmed with an intestinal biopsy. Clinicopathological indices of disease severity, as in people, can be assessed using biomarkers that include perinuclear anti-neutrophil antibody (pANCA), C-reactive protein (CRP) and calprotectin.

Management of IBD in Dogs

It is accepted that based on the multifactorial nature of IBD, not every case will respond to the same management; but, as in humans, both anti-inflammatories and antibiotics may play a role in treatment, and both species may require further pharmacological intervention in order to manage 'flareups'[28]. In addition to traditional steroids such as prednisolone and dexamethasone, other pharmacological agents that manipulate the inflammatory response are used and include sulphasalazine, budesonide and cyclosporine. Sulphasalazine has immunosuppressive, antibacterial and anti-inflammatory properties; as it is metabolised to its active component in the large intestine, it is probably most useful for large intestinal disease. Budesonide is regarded as a very potent glucocorticoid that may be associated with fewer side effects. Cyclosporine is an immunosuppressive agent. Antibiotics that are commonly implemented for IBD in dogs that have concurrent anti-inflammatory effects include metronidazole and tylosin. These antibiotics not only remove pathogenic organisms, but also change the microbiota and reduce the commensal bacterial loads that are not tolerated by dogs with IBD due to their dysfunctional GI immune system, ultimately suppressing cell-mediated immunity[28]. Other newer agents being used include chlorambucil, cyclophosphamide (which are immunosuppressants and immunomodulators) and 5-aminosalicyclates (which is an anti-inflammatory agent); these drugs can be used in conjunction with steroids or as an alternative[29]. There is currently limited evidence in the literature for consistent benefits of pre- and probiotics[28].

One of the mainstays for management of IBD in dogs, as in people, is manipulation of the diet. This usually begins by undertaking an exclusion diet using those with hydrolysed proteins or free of chicken and rice, the most common proteins found in canine diets. It is reported that 50–100% of dogs respond to dietary changes[28].

Conclusions and What Can We Learn From Other Species to Benefit the Horse

As equine clinicians, assuming the pathophysiology in the horse is similar to other species, it appears that we are trying to simplify treatment and assume that this is a short-term condition that will be resolved with steroids. In fact, we need to look at the inciting course and work on dietary management to simplify, and ideally the protein source that causes or incites the clinical signs. This involves considering the fibre source and changing if appropriate, reducing the carbohydrate content of the diet and simplifying where appropriate to one or two sources, adding in corn oil (which may have anti-inflammatory properties) and supplementing vitamin D. In the authors' experience, it can be challenging to simplify the diet, as horses are reluctant to eat simple carbohydrates. In addition, consideration of surgery if medical management is unsuccessful (especially if there may be a focal lesion that could be resected), particularly if the next option is euthanasia. In addition, we need to manage client expectations so that they understand that this is likely to be a chronic condition and may need further treatment for 'flareups'.

As well as management of the condition when it occurs, we also need to consider the impact of early therapy and diet in foals when we look at the risk factors in other species. These would include use of anthelmintics and antibiotics, exposure to gastrointestinal infections and rearing in very clean conditions. Ultimately what we lack are large, multi-centre studies with definitive diagnoses on these cases, thorough case histories and ultimately an understanding of the natural history of this condition in the horse.

References

1. Schumacher J, Edwards JF, Cohen ND. Chronic idiopathic inflammatory bowel diseases of the horse. *J Vet Intern Med*. 2000;14(3):258–265.
2. Kaikkonen R, *et al*. Diagnostic evaluation and short-term outcome as indicators of long-term prognosis in horses with findings suggestive of inflammatory bowel disease treated with corticosteroids and anthelmintics. *Acta Vet Scand*. 2014;56:35.
3. Metcalfe LV, *et al*. A retrospective study of horses investigated for weight loss despite a good appetite (2002–2011). *Equine Vet J*. 2013;45(3):340–345.
4. Kemper DL, *et al*. Equine lymphocytic-plasmacytic enterocolitis: a retrospective study of 14 cases. *Equine Vet J Suppl*. 2000;32:108–112.
5. Salem PA, *et al*. "Mediterranean abdominal lymphoma," or immunoproliferative small intestinal disease. Part I: clinical aspects. *Cancer*. 1977;40(6):2941–2947.
6. Jacobs G, *et al*. Lymphocytic-plasmacytic enteritis in 24 dogs. *J Vet Intern Med*. 1990;4(2):45–53.
7. Edwards GB, Kelly DF, Proudman CJ. Segmental eosinophilic colitis: a review of 22 cases. *Equine Vet J Suppl*. 2000;32:86–93.
8. Ponder A, Long MD. A clinical review of recent findings in the epidemiology of inflammatory bowel disease. *Clin Epidemiol*. 2013;5:237–247.
9. Denson LA, *et al*. Challenges in IBD research: update on progress and prioritization of the CCFA's research agenda. *Inflamm Bowel Dis*. 2013;19(4):677–682.
10. Benchimol EI, *et al*. Increasing incidence of paediatric inflammatory bowel disease in Ontario, Canada: evidence from health administrative data. *Gut*. 2009;58(11):1490–1497.
11. Rocchi A, *et al*. Inflammatory bowel disease: a Canadian burden of illness review. *Can J Gastroenterol*. 2012;26(11):811–817.
12. Danese S, Vuitton L, Peyrin-Biroulet L. Biologic agents for IBD: practical insights. *Nat Rev Gastroenterol Hepatol*. 2015;12(9):537–545.
13. Sartor RB. Microbial influences in inflammatory bowel diseases. *Gastroenterol*. 2008;134(2):577–594.
14. Cho JH, Brant SR. Recent insights into the genetics of inflammatory bowel disease. *Gastroenterol*. 2011;140(6):1704–1712.
15. Olszak T, *et al*. Microbial exposure during early life has persistent effects on natural killer T cell function. *Science*. 2012;336(6080):489–493.
16. Kronman MP, *et al*. Antibiotic exposure and IBD development among children: a population-based cohort study. *Pediatrics*. 2012;130(4):e794–803.
17. Molodecky NA, Kaplan GG. Environmental risk factors for inflammatory bowel disease. *Gastroenterol Hepatol (N Y)*. 2010;6(5):339–346.
18. Chu KM, *et al*. Childhood helminth exposure is protective against inflammatory bowel disease: a case control study in South Africa. *Inflamm Bowel Dis*. 2013;19(3):614–620.
19. Elliott DE, Weinstock JV. Where are we on worms? *Curr Opin Gastroenterol*. 2012;28(6):551–556.

20. Cipolla G, *et al.* Nonsteroidal anti-inflammatory drugs and inflammatory bowel disease: current perspectives. *Pharmacol Res.* 2002;46(1):1–6.
21. Ananthakrishnan AN, *et al.* Aspirin, nonsteroidal anti-inflammatory drug use, and risk for Crohn disease and ulcerative colitis: a cohort study. *Ann Intern Med.* 2012;156(5):350–359.
22. Baron S, *et al.* Environmental risk factors in paediatric inflammatory bowel diseases: a population based case control study. *Gut.* 2005;54(3):357–363.
23. Porter CK, *et al.* Infectious gastroenteritis and risk of developing inflammatory bowel disease. *Gastroenterol.* 2008;135(3):781–786.
24. Soon IS, *et al.* The relationship between urban environment and the inflammatory bowel diseases: a systematic review and meta-analysis. *BMC Gastroenterol.* 2012;12:51.
25. Lerebours E, *et al.* Stressful life events as a risk factor for inflammatory bowel disease onset: a population-based case-control study. *Am J Gastroenterol.* 2007;102(1):122–131.
26. De Silva P, Ananthakrishnan AN. Vitamin D and IBD: more pieces to the puzzle, still no complete picture. *Inflamm Bowel Dis.* 2012;18(7):1391–1393.
27. Bernstein CN, Shanahan F. Disorders of a modern lifestyle: reconciling the epidemiology of inflammatory bowel diseases. *Gut.* 2008;57(9):1185–1191.
28. Jergens AE, Simpson KW. Inflammatory bowel disease in veterinary medicine. *Front Biosci (Elite Ed).* 2012;4:1404–1419.
29. Vazquez-Baeza Y, *et al.* Dog and human inflammatory bowel disease rely on overlapping yet distinct dysbiosis networks. *Nat Microbiol.* 2016;1:16177.

Pain Pathophysiology, Assessment and Management in Large Domesticated Animals

Khursheed Mama, DVM, DACVAA
Fort Collins, CO, USA

INTRODUCTION

Assessment and management of pain in large animal species remain challenging and have significant implications for their wellbeing and, in many cases, survival. While some information from other species may be applicable, responses to medications and therapies may differ. In food-producing animals, the additional consideration of drug residues and withdrawal times may limit options. Newer drug formulations and alternative routes of administration in addition to increasing awareness and training of other modalities offer hope in providing comprehensive care.

PAIN PATHOPHYSIOLOGY

Pain is defined by the International Association for the Study of Pain as "an unpleasant sensory and emotional experience associated with actual or potential tissue damage or described in terms of such damage". Hence perception is needed to feel pain, and magnitude may be influenced by other factors such as anxiety, anticipation, etc. Nociception results from activation of receptors (nociceptors) located at the distal ends of both unmyelinated (C) and poorly myelinated (Aδ) nerve fibers. Impulses conducted by these nerve fibers are processed at the dorsal horn of the spinal cord and ascend through the two major pain pathways, spinothalamic and spino-reticular thalamic, to the brain. Withdrawal responses to a nociceptive input may be mediated at the level of the spinal cord, but perception requires nociceptive input to the brain and will generally result in some form of avoidance behavior or clinical manifestation (e.g., lameness) designed to protect the animal from further damage. Metabolic (e.g., change in glucose regulation), physiological (heart rate, blood pressure, etc.), and neuroendocrine (e.g., cortisol, catecholamines) responses are also associated with nociceptive input to the brain and may occur in conscious or unconscious (e.g., anesthetized) animals.

The acute pain state is typically resolved by removal of stimulus or medical intervention. The descending inhibitory pathways originating from the brain activated by the initial nociceptive input also play a role in terminating acute pain. In addition to analgesic medications, non-painful afferent inputs (mediated by Aβ fibers) such as massage or ice at the site of injury can help diffuse the magnitude of pain perceived by providing input at the wide dynamic range neurons in the dorsal horn of the spinal cord. If stimulus continues and interventions are not instituted or not successful, peripheral (mediated via chemicals such as prostaglandins, leukotrienes, neuropeptides, and nerve growth factors) and central sensitization (mediated via neuromodulates [such as substance P] that activate *N*-methyl-d-aspartate [NMDA] and tachykinin receptors) can occur, leading to chronic changes and sustained pain. Hypersensitivity (or a reduction in the intensity of the stimulus required to cause pain) and allodynia (in which a typically nonpainful stimulus is now perceived as pain) are manifestations of this. Exaggerated responses to a noxious stimulus (hyperalgesia) and persistence of pain after removal of a noxious stimulus (hyperpathia) may also occur.

Visceral pain (originating from the gastrointestinal tract or other organs) is transmitted by parasympathetic and sympathetic afferent nerve fibers. The autonomic fibers involved in transmitting noxious inputs from visceral organs are diffuse and overlap extensively, resulting in difficulty in localizing the site of pain.

Neuropathic pain, which is yet another form of pain, may result from injury to the peripheral or central nervous system. This may originate from surgical pain, trauma, and radiation injury, among other causes.

Pain Assessment

Due to the complexity of the pain pathways and animal responses, influence of disease medications and treatment modalities on physiological parameters, stress hormones, biomarkers, etc., a comprehensive assessment is advocated to try minimizing the impact to the animal and its quality of life. Lack of immediate availability of some measurements (e.g., joint biomarker, glucocorticoids) adds to the challenges. Assessment is further confounded by species differences, the most obvious being prey versus non-prey animals, but training (as, for example, to reduce escape behaviors) may also impact accurate evaluation.

In an attempt to overcome these challenges, there has been significant recent work on the development of multifaceted pain assessment tools that consider physiologic (e.g., heart rate, blood pressure) and behavioral (both spontaneous and in response to interaction or manipulation) indices developed specifically for a given species in different clinical scenarios (e.g., post-arthroscopy or colic in the horse). In addition to assessment of presence or absence of certain parameters (e.g., shifting weight), the frequency within a measurement period is also correlated with degree of pain or discomfort. Modeled after rodent scales, the grimace scale for horses correlates six specific facial expressions with the degree of pain; for example, ears stiffly backward can indicate a greater degree of pain versus none if ears face forward. In small ruminants, other factors (such as lack of herding behavior) are associated with lack of wellbeing. In young animals, signs may differ from the adult; for example, foals may grind teeth if painful or exhibit unwillingness to lie down but will stamp hind feet while buckling in the front, etc.

These differences in behavior and the diversity of available assessment tools highlight the importance of training individuals in how to use the scales and what to look for. This is especially important when multiple people are responsible for assessment and will increase chance of consistent categorization of the pain state and help dictate degree and type of intervention and its value. It is also noteworthy that most pain scales are developed to assess acute pain. Chronic pain and quality-of-life scales (which include activity, appetite, good days vs. bad, etc.) are also available for some species, but less work has been done in this area. Other tools such as force plates or gait analysis to assess lameness and subtle changes may also be used.

Therapeutic Options

While nociceptive pain is considered an adaptive or protective strategy, and the associated inflammatory response is considered necessary for the healing process, preventing progression to a maladaptive or chronic state, where the pain is disproportionate to the insult and no longer serves a purpose, should be a goal of treatment. Many drug classes with choices within them are available to the veterinarian to help manage acute pain. **Nonsteroidal anti-inflammatory drugs** (NSAIDs) are commonly used to help normalize the inflammatory response generated by tissue injury. In equids and ruminants, NSAIDs are often administered pre-procedurally or immediately post injury to help minimize inflammation. They may be continued in the postoperative period at the dose, dosing interval, and route appropriate to the species and circumstance. In addition to systemic administration, topical administration may be considered if toxicity is a concern. Caution or concurrent use of gastroprotectants may be warranted in foals, as gastrointestinal tract ulceration tends to be more prevalent. **Corticosteroids** that also have anti-inflammatory properties are less commonly administered systemically due to other physiological effects, but they are often used to decrease intraarticular inflammation.

Drugs used for sedation (e.g., **α_2-adrenergic agonists**) or induction of anesthesia (e.g., **ketamine**) often also have analgesic properties. For example, α_2-adrenergic agonist drugs mediate their analgesia through descending inhibition and in addition to being given systemically have been evaluated for epidural, perineural, and intraarticular use. These alternate routes can decrease cardiovascular and gastrointestinal side effects often associated with this drug class. Oral preparations (of detomidine and dexmedetomidine) are available, and while their efficacy as analgesics by this route has not been evaluated, they can reduce stress by calming the animal. Ketamine, an NMDA receptor antagonist, is believed to play a role in mitigating spinal facilitation of pain and may help in management of pain where the nociceptive input is ongoing. In addition to intraoperative use, low doses may be administered postoperatively.

Opioids are a broad drug class and include μ-agonists (e.g., morphine, methadone, fentanyl), partial μ-agonists (e.g., buprenorphine), and κ-agonists/μ-antagonists (e.g., butorphanol, nalbuphine). They are frequently used in combination with sedatives or tranquilizers to enhance sedation in horses, but studies evaluating their analgesic efficacy when given systemically continue to provide mixed results. Additionally, they can cause excitement when given alone, and high doses negatively influence gastrointestinal motility. Differential effects for some drugs are noted between foals and adults, and more recent studies have shown that efficacy and side effect profiles between horses may also be due to genetic polymorphisms. Interestingly, alternative routes of administration (e.g., intraarticular, perineural, epidural) show efficacy with reduced side effects and should be considered for management of pain.

Local anesthetic drugs including lidocaine, mepivacaine, and bupivacaine have been commonly used for diagnostic purposes (e.g., perineural or intraarticular administration in horses) and to facilitate interventional procedures (e.g., paravertebral or inverted L in cattle). Perioperative nerve blocks, epidural administration, and long-term administration using perineural catheters for pain relief are also described. Systemic use of lidocaine for pain management (potentially beneficial as a result of anti-inflammatory and free radical scavenging properties) is a more recent development, as is use of intravenous regional anesthesia for distal limb pain. Most recently, a sustained-release liposomal formulation of bupivacaine (approved for use in dogs and cats with an approximately 72-hour period of efficacy) has been advocated for use in specific circumstances.

Many other medications including the short-acting anticholinergic butylscopolamine, COX-3 inhibitor acetaminophen, bisphosphonates, tramadol, gabapentin, etc., have been considered in specific circumstances with newer drugs and immune-modulating compounds constantly emerging.

Other Techniques and Modalities

The author has less direct familiarity with many of the following treatments or options, and for some the supporting evidence is limited. However, many are unlikely to cause harm and may provide some relief to animals who are not helped by traditional therapies, and hence their inclusion. When evaluated, most of the work has been done in horses.

As alluded to previously, **epidural** (and in some cases spinal) administration of drug may be warranted. This can be done using a needle or catheter pending the level of the lesion and anticipated duration of treatment. The site (lumbosacral versus sacrococcygeal or intercoccygeal) varies with species and desired effect. Drugs similarly are selected based on desired effects. Sacroiliac injections, ultrasound-guided facet blocks, etc., are increasingly used to diagnose and treat the source of pain. Autologous treatment with **interleukin-1 receptor antagonist protein** (IRAP), an anti-inflammatory cytokine, has been shown when injected intraarticularly to improve lameness in horses with osteoarthritis (as compared to placebo).

Cold or low-level lasers, thought to modulate distribution of local inflammatory mediators, are anecdotally reported to have beneficial effects for musculoskeletal pain without causing tissue trauma. **Extracorporeal shockwave therapy** provides focused, pulsed, high-energy waves to an injured area with a goal of causing neovascularization and tissue remodeling in horses with a variety of musculoskeletal injuries. There are mixed reports of success in managing pain. **Acupuncture and electro-acupuncture** modulate nociceptive input at the dorsal horn of the spinal cord to provide analgesia, but again the efficacy has been mixed.

The application of manual low-amplitude, high-velocity thrusts or the practice of **chiropractic manipulation** is intended to improve tissue function and restore normal joint motion and has been shown to benefit the kinematics and response to mechanical nociception in horses with back pain.

Summary

While challenges remain, the availability of drug and non-drug-based modalities to manage pain is increasing, as are individuals with focused effort in this area.

Case Examples

The following are sample cases that illustrate potential ways of managing primarily acute pain. Not all drugs or techniques described are approved; scientific evidence for their safety or efficacy may be limited. While additional techniques (e.g., acupuncture, massage, physical therapy) may also be viable, these examples focus primarily on pharmacological intervention. The analgesic plan for each case example represents one **possible** multimodal approach for analgesia and should not be considered exclusive. As pain is individual, what might work in one animal may not be adequate or result in a greater degree of side effects in another.

An approximately 750-kg, 4-year-old bucking rodeo bull kicks at the chute while it awaits its turn and has a degloving laceration from mid metatarsus to the level of the fetlock. The plan is to try to attend to the wound on site (at least bandage if possible) and then transport the bull to a nearby veterinary facility where additional diagnostics can be performed and a repair attempted. While a fierce bucking bull, he has a reasonable demeanor during husbandry-related activities.

- **For transport, consider:**
    - Keeping him in the chute/squeeze if available
    - Non-pharmacologic support: bandage, stabilization of the limb
    - If standing sedation is needed, consider an intramuscular combination of xylazine (0.02 mg/kg), ketamine (0.04 mg/kg), and butorphanol (0.01 mg/kg). This is colloquially known as 'ket stun.' For this bull, the doses would be approximately 15 mg xylazine, 7.5 mg butorphanol, and 60 mg ketamine. These doses can be increased or given IV if needed (but brief recumbency may result).
    - Nonsteroidal anti-inflammatory: e.g., phenylbutazone 6 mg/kg intravenously or 6–10 mg/kg by mouth. (Note this is a once-daily treatment in cattle, as phenylbutazone has a longer half-life than in horses. Dose may need to be decreased on subsequent days.)

- **For in-clinic evaluation, consider:**
    - Additional medications for radiographs or other diagnostics
    - Sacrococcygeal epidural with morphine and saline
    - Intravenous fluids and laboratory evaluation if need to anesthetize; decide if this should wait since bull likely not fasted
- **During anesthesia:**
    - Regional limb perfusion with lidocaine if vessels are intact and/or perineural blocks with longer-acting local anesthetic
    - Can consider low-dose ketamine CRI during anesthesia maintenance

A 22-year-old pony mare with chronic bilateral forelimb laminitis which flares up periodically. She receives oral firocoxib currently.

- In addition to supportive care (shoes/support, footing adjustments, ice, etc.), consider:
- Increasing dose of firocoxib transiently (possibly re-load to target higher concentration)
- Perineural injections with local anesthetic; if effective, could consider liposomal bupivacaine or addition of morphine to the injection, which anecdotally is thought to help with pain management
- Addition of tramadol (10 mg/kg PO twice daily) provides up to three days of pain relief
- Injectable ketamine (0.6 mg/kg/hour) in addition to tramadol (5 mg/kg PO twice daily) can potentiate duration of analgesia

REFERENCES

References are available upon request.

The Stewart Approach to Acid-Base Can Change Your Life: Strategies for Clinical Cases

Rose D. Nolen-Walston, DVM, DACVIM (LAIM)
Kennett Square, PA, USA

The term acid is derived from Latin "acidus," meaning "sour taste." As our understanding of chemistry progressed in the 19th century, two mechanistic approaches to acidity arose. The first, from Arrhenius in 1884, who won a Nobel Prize for identifying that salt in solution created an acidic solution of sodium and chloride ions. About two decades later, Naunyn expanded that thought to conclude that the acid-base balance of the blood could be determined by sodium cations that were base-forming, and chloride anions that were acid-forming. However, by 1923 the Bronsted and Lowry definition of acids as H^+ donors and bases as H^+ acceptors gained primacy, and laid the groundwork for the first attempts to apply physical chemistry principles to biological acid-base derangements. Henderson, and later Hasselbalch focused on the alkali buffer effects of bicarbonate, and popularized their eponymous equation.

The "Boston" Carbon Dioxide/Bicarbonate Approach

In 1963, Schwartz and Relma developed an approach that used the Henderson-Hasselbalch equation to describe acid-base imbalances in body. It introduces the equilibrium equation showing a relationship between water, carbon dioxide, carbonic acid, and bicarbonate which was determined by a dissociation constant:

$CO_2 + H_2O \leftrightarrow H_2CO_3 \leftrightarrow HCO_3^- + H^+$

Their paper opened by stating that "disorders of acid-base equilibrium can be caused by primary disturbances in the pulmonary regulation of the concentration of carbonic acid in body fluids, by changes in the concentration of bicarbonate and other buffer anions induced by metabolic disturbances or by some combination of the two." Known as the "Boston Approach," acid-base balance and correction was taught for fifty years using the concept that there were four main derangements of pH in the blood: respiratory acidosis and alkalosis, and metabolic acidosis and alkalosis. The principle is that the body "compensated" for a primary derangement of, say, respiratory acidosis from hypoventilation (high PCO_2) by inducing a compensatory metabolic alkalosis (the kidneys retaining bicarbonate). This approach was effective for simple acid-base disorders but showed limitations when mixed, complex derangements were being evaluated, especially those involving non carbon-dioxide-based moieties.

The Siggard-Andersen "Copenhagen" Base Excess Approach

During the seventies, a pair of Danish researchers focused on the "metabolic aspect" of acid-base balance. They noted that as carbon dioxide could affect bicarbonate levels, and hemoglobin was also a potent buffer, bicarbonate was not ideal as a stand-alone value to assess non-respiratory acid-base derangements. Siggard and Andersen introduced the concept of "base excess," the amount of hydrochloric acid or sodium hydroxide required to return a liter of blood with a standardized PCO_2 and Hb to a pH of 7.40. Since then, BE has been developed into a calculated, rather than a measured value.

The Anion Gap Approach

Around the same period, Emmett and Nairns developed an alternative method to assess unmeasured contributors to metabolic acid-base derangements: anion gap = ([Na^+] + [K^+]) −

($[Cl^-] + [HCO_3^-]$). Increases in the AG could be attributed to unmeasured anions such as ketones, sulfates and lactate. In the days before lactate assays were commonly available, this was an extremely useful calculation for veterinarians as it was one of the most efficient methods for identifying lactic acidosis. However, in large animal species where aspirin poisoning is rare and diabetic ketoacidosis is not a common diagnosis, the advent of stall-side lactate monitoring has made the calculation of the anion gap less valuable.

The Stewart Approach

While these approaches were being popularized in hospitals throughout the world, a young physiologist called Peter Stewart took a completely different approach, and reverted to the Arrhenius definition of an acid, that was founded on dissolved cations and anions, rather than simply Bronsted-Lowry definition using the concentration of H^+. He posited that the independent determinants of pH in the plasma were:

1. **PCO_2**: the pressure of dissolved CO_2 in the plasma, as measured by blood gas analysis, and controlled by alveolar ventilation. This means that HCO_3^- and other carbon dioxide species are **dependent** variables, and follow changes in pH rather than create them. This rejects the Henderson-Hasselbalch model.
2. **Strong ion difference**: The difference in total charge between strong cations (Na^+, K^+, Ca^{++}, Mg^{++}) and strong anions (Cl^-, lactate, sulfate, ketones, etc.). Strong ions are fully dissociated at physiological pH.
3. **A_{TOT}**: The nonvolatile weak acids. These include albumin, globulins, and inorganic phosphate, and are not fully dissociated at physiological pH.

The Stewart approach results in six possible acid-base derangements that can occur independently and in combination: respiratory acidosis or alkalosis; low SID acidosis or high SID alkalosis; and high A_{TOT} acidosis and low A_{TOT} alkalosis. The SID can be defined as the apparent SID (SID_a), calculated as the sum of all non-chloride anions, or the effective SID (SID_e), which does not include the unmeasured ions. The Strong Ion Gap (SIG) is the difference between SID_a and SID_e, and thus provides an accurate estimation of unmeasured ions such as ketones, sulfates, etc.

The Gamblegram

Stewart maintained that any explanation of acid-base would have to follow the law of electroneutrality (i.e., that the sum of the anionic charges equals the sum of the cationic charges). Additionally, Stewart reminded us that the body also follows the law of Conservation of Mass, where no new material can be "created." Therefore, the total amount of sodium and chloride in the body remains the same—the height of those columns can change only with dilution, excretion or ingestion, thus making them independent variables. Bicarbonate, however, can be formed from the dissociation of carbonic acid, as seen previously:

$CO_2 + H_2O \leftrightarrow H_2CO_3 \leftrightarrow HCO_3^- + H^+$

Based on this, James Gamble created the Gamblegram to show the relation between concentrations of cations and anions in the blood (see figure below). One important feature of the Gamblegram is that it shows how, based on the dissociation constants of water and carbonic acids, more (or less) HCO_3^- can be formed within the body as levels of strong ions (predominantly Na and Cl) change relative to each other, based on the disease process. The bicarbonate, therefore, can "fill in the gap" in the anion column to maintain electroneutrality, making its concentration **dependent** on the magnitude of the SID, not an **independent** cause of pH change.

Figure 4 from Muir (2017): Strong ion gap (SIG) is the difference between SID_a and SID_e

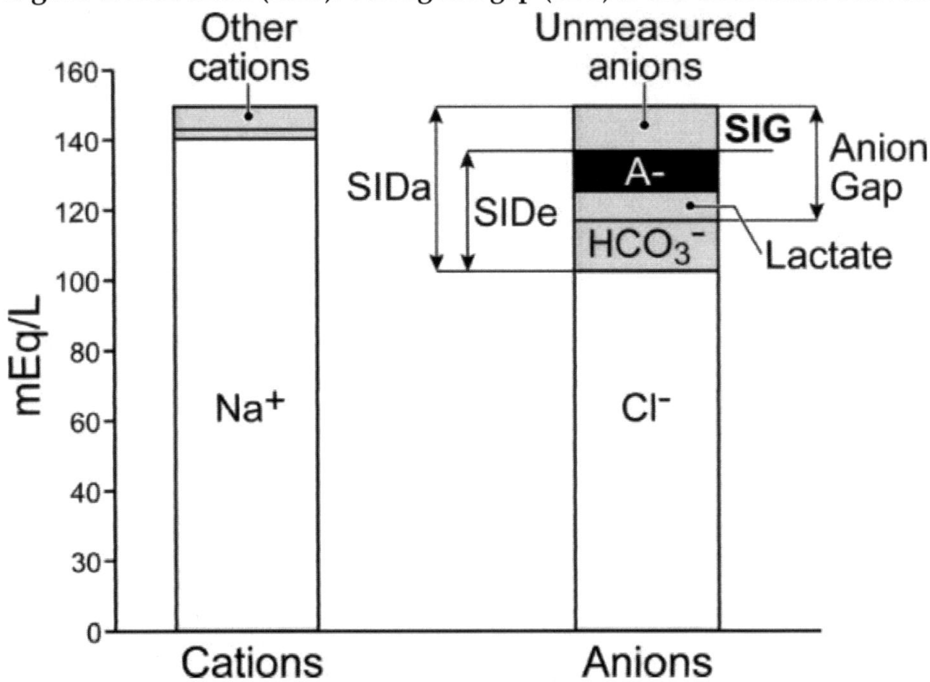

The SIG is an accurate measure of the unmeasured anions present in plasma. [In this figure, A^- == A_{TOT}].

MODIFYING THE STEWART APPROACH FOR CLINICAL PRACTICE

Unfortunately, Stewart's original papers used extremely complex fourth order polynomial equations to use his independent variables to solve for pH, and they did not catch on widely at the time. However, in the early years of the 21st century his theories gained a resurgence of interest among nephrologists, intensivists and clinicians with an interest in acid-base physiology. Since then, several attempts have been made to simplify his work and make it more applicable for practicing clinicians. The veterinarian, Peter Constable, constructed the foremost of those simplifications and his approach has gained considerable traction both in human and veterinary medicine. All of these systems rely on the application of the three independent variables, pCO_2, SID, and A_{TOT}, and therein lay the first problem. SID and pCO_2 can be easily measured using standard blood gas and clinical chemistry analyzers, but assessment of total nonvolatile acids is technically difficult.

Additionally, Constable showed that there were significant species differences in the dissociation constant and concentration of total nonvolatile weak acids (A_{TOT}) were quite different in horses versus humans: $[A_{TOT}]_{equine}$ had a mean of 15.0 mEq/L versus 20.0 in man, and K_a 2.22 x 10^{-7} Eq/L versus 3.0 in man. As direct measurement of A_{TOT} is not practical in clinical settings, he demonstrated that in the adult horses it could be approximated from total protein via refractometry using the formula: $[A_{TOT}]$ = TPP x 2.24. Subsequent work by Viu (2016) provides normal values for SID and A_{TOT} in healthy foals of various ages and with various analyzers. In foals under 1–2 days of age, passive transfer may not be complete, leading to lower values for the globulin portion of A_{TOT}. Even after passive transfer is complete, though, TPP for foals is lower than it is in adults. Although higher values for phosphorus in juveniles make up for this to a degree, although A_{TOT} is lower in foals <6 months of age than it is in older foals and adults. Viu concludes, "the normal interval of SID_a in neonatal foals is 34–46 mEq/L for reference laboratory … Regarding A_{TOT}, the reference interval in neonatal foals is 10.3–16.6 mEq/L obtained in a reference laboratory, and is 11.0–17.5 mEq/L [… for] refractometry." A_{TOT}

is calculated using the following formula: A_{TOT} = 2.24 x [Alb] + 1.4 x [Gb] + 0.59 x [Pi], but was demonstrated to be accurately approximated in foals 1–21 days using the formula: total plasma protein [TPP] x 2.5.

Since then several apps and programs have been introduced which allow practitioners to enter in the relevant parameters and obtain a mathematical assessment of the components of complex acid-base disorders. Story, in 2016, published an excellent back-of-the-envelope approach for humans that is easily converted to foals. He states that assuming that normal base excess (BE) approximates zero, simple calculations can be used to assess the effect of strong ion difference (SID) and changes in the concentrations of albumin, lactate, and "other" (unmeasured) ions on BE. For adult humans, he uses the formula:

$BE_{adult\ human}$ = [Na - Cl - 35] + [1- lactate] + [0.25 * (42 - Alb)] + [other ions]

In this equation the other ions represent the unmeasured ions of the SIG. These are mostly commonly the anions that cause a high anion-gap acidosis, and are often recalled using the acronym "MUDPILES." Since L-lactate is accounted for separately in this formula, in humans these would include ketones, aspirin and other salicylates like Pepto-Bismol, ethylene glycol, uremic toxins, and acetaminophen.

Using data from Constable and Viu, this can be adjusted for adult horses and neonatal foals using the following approach:

$BE_{adult\ horse}$ = [Na + K - Cl - **40**] + [1 - lactate] + [**15** - (**2.2*TPP**)] + [other ions]

$BE_{foal\ 1\text{-}21}$ = [Na + K - Cl - **36**] + [**2.6** - lactate] + [**14** - (**2.5*TPP**)] + [other ions]

In this formula, lactate and the electrolytes are entered in mEq/L, and TPP in mg/dL. Negative values indicate a base deficit, or acidosis, and positive values show an alkalosis. This can be further broken down into the specific components of the SID effect, the lactate effect, the TPP effect, and the "unmeasured ions" effect, to localize the specific problem. In horses and ruminants, the list of [other ions] could also include D-lactate.

Using the Stewart Approach to Make Custom Fluids

Using the Stewart Approach to analyze acid-base imbalances has several advantages. The first is that if your clinical assessment of the respiratory and neurologic system is within normal limits, a lot of information can be gleaned from basic chemistry analysis without the need for frequent blood gases. The most common acid-base derangements in horses with colitis are hyponatremia, leading to a low SID acidosis; hyperlactatemia, exacerbating the low SID academia; and hypoproteinemia, which results in a moderate low A_{TOT} alkalosis. When selecting fluids for a severely compromised patient with diarrhea, I will often make up "custom solutions" rather than using stock fluids.

So for example, consider a 4-day-old foal, 50 kg presenting for severe diarrhea of 2 days' duration. The foal has a sodium of 121 mEq/L and a chloride of 95 mEq/L. The potassium is 2.1 mEq/L, and the TPP is 4.1 g/dL. The bicarbonate is low, but as a Stewart-ista, I set it on the backburner for now. The lactate reads at 10.1 mmol/L, and the blood glucose is 53 mg/dL. Using a semi-quantitative Stewart Approach, I determine that the SID3 (Na + K – Cl) is 28, demonstrating a low SID acidosis (SID3 in a neonatal foal should be ~36). The effect on BE of this SID acidosis is therefore 36 - 28, which gives an SID base deficit of [-8 mEq/L]. The hyperlactatemia also contributes a further -7.5 mEq/L effect on BE. However, the TPP is 4.1 g/dL, giving us a mildly low A_{TOT} of 10.25 mEq/L, leading to an effect on BE of +4 mEq/L. Without even measuring $PaCO_2$ or pH, we can already do a great deal to describe this foal's acid-base abnormalities and create a fluid plan for him. For those who enjoy looking at the bicarbonate, Muir (2017) reminds us of the intuitive effect of crystalloid SID on HCO_3-: "SID >

[HCO_3-] increases plasma pH (alkalosis); SID < [HCO_3-] decreases plasma pH (alkalosis); and SID = [HCO_3-] yields no change in plasma pH."

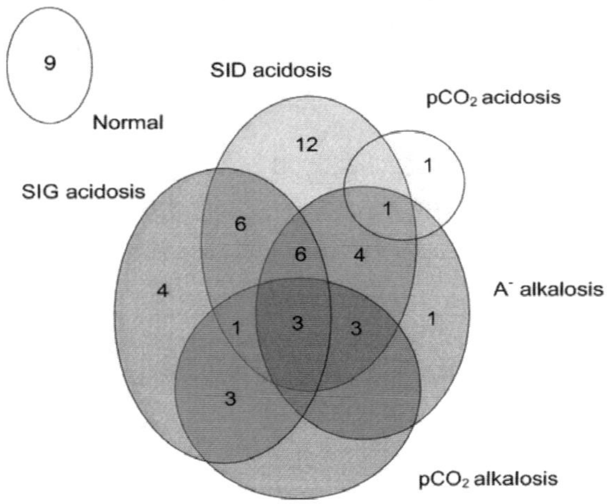

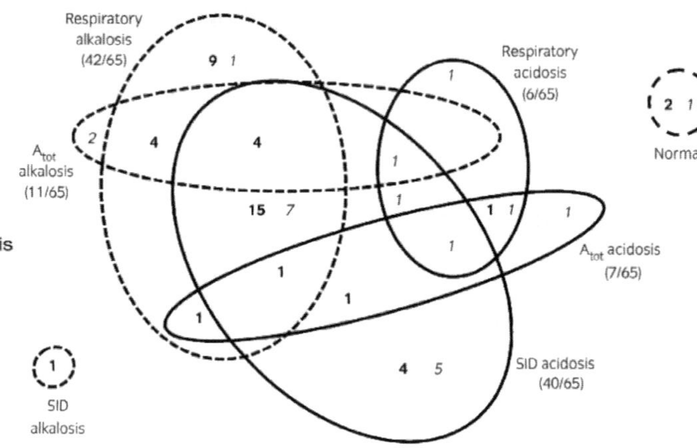

Figure 2 from Gomez (2013) (left): Interpretation of acid-base disorders on admission for each horse [n=54, with diarrhea/colitis] by the physicochemical approach. SIG strong ion gap; pCO_2 partial carbon dioxide pressure; SID strong ion difference; A− total net negative charge of blood proteins.
Figure 1 from Viu (2015) (right): Interpretation of acid-base imbalances on admission by the physicochemical approach, in surviving (bold) and nonsurviving neonatal foals (italics). Disorders resulting in acidosis are shown as solid lines and disorders resulting in alkalosis as dashed lines. The number of foals with each acid-base disorder (simple and mixed) is shown within parentheses.

Table 1. Pertinent values for commonly used intravenous fluids

Fluid	Na (mEq/L)	Cl (mEq/L)	K (mEq/L)	SID_3 Na+K-Cl)	Dextrose (mg/L)	mOsm
Plasma	140	100	4	36		~300
0.9 % saline	154	154	0	0		308
Lactated Ringer's	130	109	4	25*		310
Plasma-Lyte 148 A	140	98	5	43		295
Plasma-Lyte 56 M in 5% dextrose	40	40	13	13	5000	112 (363)
Isotonic bicarb (1.4% $NaHCO_3$)	150	0	0	150		300
8.4% bicarb (1 mEq/ml)	1000	0	0	1000		2000
Hypertonic saline (7.2%)	1232	1232	0	0		2464
5% dextrose	0	0	0		5000	0 (314)

*LRS also contains a racemic mixture of 28 mEq/L of D- and L-lactate, which, if not metabolized, can add to the SID and cause lactic acidosis effects. Both the gluconate and lactate in the balanced electrolyte solutions must be metabolized for the solutions to contain an effective SID of >0.

After bolus resuscitation, the construction of "custom fluids" then requires goals to be set for SID, K, dextrose, and osmolality, and then using the above table to create a solution that meets those goals. So, for the foal in question, one might decide that in order to increase the SID to closer to 40 in the face of ongoing electrolyte losses, a fluid with SID of ~60 would be effective. Although it is not complicated to use the delta of volume of distribution for sodium replacement (which, although it should seem that it would be restricted to ECF, actually functionally has a volume of distribution equal to TBW due to fluid shifts from changes in ECF osmolarity) plus expected losses, in reality this math has not been productive in my hands. The variation in diarrheic fluid and solute losses is impossible to predict, and so I tend to ballpark a starting replacement rate at ~10–20 mEq/L of SID above my goal plasma value. The potassium is then added in in the form of KCl at 2 mEq/L, dextrose in the form of 50% dextrose, and then SID tweaked using 8.4% $NaHCO_3$ at 1 mEq/L, with an effective SID of 1000 mEq/L. It is simple enough to then calculate the exact values per liter, but if the added volumes are <200 ml, I often do not correct for total volume slightly higher than 1L. Below is an example of a calculation chart I might use to create specific fluids for this foal. It should be noted that the dextrose level here is significantly below 8 mg/kg/min, and would only be used if the foal were also getting enteral nutrition. For foals being held on GI rest, I institute TPN through a dedicated line after the first 12 hours.

Table 2. Sample custom fluid calculations for a horse with diarrhea and low SID acidosis

Fluid	Volume (ml)	Na (mEq)	Cl (mEq)	K (mEq)	SID_3	Dextrose (mg)	mOsm
Plasmalyte 148	1000	140	98	5	43		295
KCl (2 mEq/ml)	20		40	40	0		80
50% dextrose (500 mg/ml)	100					5000	(314)
8.4% $NaHCO_3$ (1 mEq/ml)	13	13	0		1000		13
TOTAL - per bag	1133	153	138	45	60	5000	388
TOTAL - per liter		135	122	40	53	4413	342

Before ending this discussion, though, I think it is imperative to be clear that for all of the flaws of the original ("Boston") bicarbonate/carbon dioxide analysis of acid-base, the Stewart approach is not without its detractors. Many of the tenets held dear by Stewart acolytes have been challenged in detailed, evidence-based ways. While I have not found a clinical approach to acid-base that I believe offers more clinical utility than Stewart, the critique by Kurtz (2009) challenges many of the mechanistic assumptions on which the strong ion approach relies. I strongly recommend it to those who wish to evangelize the "Canadian" method to their colleagues and trainees.

References

1. Constable PD. A simplified strong ion model for acid-base equilibria: application to horse plasma. *J Appl Physiol.* 1997;83(1):297.
2. Friesen JH. An app to calculate bedside Stewart acid-base equations. *Anesth Analg.* 2016;123(6):1638.
3. Gomez DE, *et al.* Physicochemical approach to determine the mechanism for acid–base disorders in 793 hospitalized foals. *J Vet Intern Med.* 2015;29(5):1395.
4. Gomez DE, *et al.* Physicochemical interpretation of acid-base abnormalities in 54 adult horses with acute severe colitis and diarrhea. *J Vet Intern Med.* 2013;27(3):548.
5. Harvey AM. Classics in clinical science: James L. Gamble and "Gamblegrams". *Am J Med.* 1979;66(6):904–6.
6. Kishen R, *et al.* Facing acid-base disorders in the third millennium—the Stewart approach revisited. *J Nephrol Renovasc Dis.* 2014;7:209.
7. Kurtz I, *et al.* Acid-base analysis: a critique of the Stewart and bicarbonate-centered approaches. *Am J Physiol Renal Physiol.* 2008;294(5):F1009.
8. Magder S, Emami A. Practical approach to physical-chemical acid-base management. Stewart at the bedside. *Ann Am Thorac Soc.* 2015;12(1):111.
9. Muir W. Effect of intravenously administered crystalloid solutions on acid-base balance in domestic animals. *J Vet Intern Med.* 2017;31(5):1371.
10. Spital A, Sterns RD. The paradox of sodium's volume of distribution: why an extracellular solute appears to distribute over total body water. *Arch Intern Med.* 1989;149(6):1255.
11. Stämpfli HR, *et al.* Clinical utility of serum biochemical variables for predicting acid-base balance in critically ill horses. *Vet Clin Pathol.* 2014;43(4):547.

12. Stämpfli HR, *et al*. Weak acid-concentration Atot and dissociation constant Ka of plasma proteins in racehorses. *Equine Vet J Suppl*. 1999;(30):438.
13. Story DA. Stewart acid-base: a simplified bedside approach. *Anesth Analg*. 2016;123(2):511.
14. Viu J, *et al*. Simplified strong ion difference approach to acid-base balance in healthy foals. *Vet Emerg Crit Care*. (San Antonio). 2016;26(4):549.
15. Viu J, *et al*. Acid base imbalances in ill neonatal foals and their association with survival. *Equine Vet J*. 2017;49(1):51.

Respiratory Gene Technology® as a Potential Treatment for Equine Asthma

Sanni Hansen, DVM, PhD; Louise Laustsen, DVM; Charlotte Hopster-Iversen, DVM, PhD, DECEIM; Nina D. Otten, DVM, PhD; Rune Bech, DVM; Simon Byrgesen, DVM; Julie Fjeldborg, DVM, PhD
University of Copenhagen, Copenhagen, Denmark

Introduction

Corticosteroids are considered the mainstay treatment option for equine asthma. Alternative treatment options to glucocorticoids for equine asthma are desirable due to withdrawal time and risk of laminitis when using corticosteroids. Challenges are especially the obese horse with metabolic disease and thereby increased risk of laminitis when treated with corticosteroids. But also the high performance horses where the withdrawal time due to corticosteroid treatment postpones the participation in competitions.

Respiratory gene technology® (RGT) is a form of autologous conditioned serum (ACS). The principle of ACS was first introduced to the market in 1980 for treatment of joint disease. The research behind the product showed an increase in interleukin-1 receptor antagonist (IL-1RA), which is a naturally anti-inflammatory cytokine that inhibits the action of IL-1.[1,2] IL-1 pathways affect the inflammatory response, by granulocyte recruitment, non-specific airway reactivity and mucin production. The incubation of the serum within specially designed containers increased the amount of IL-1RA several fold.[3,4]

Research Project

The aim of this study was to evaluate if serum and RGT, a commercial kit to produce autologous conditioned serum, were effective in reducing neutrophils and mast cells in bronchoalveolar lavage (BAL) of racehorses with cytological evidence of mild equine asthma (MEA).

Thirty-six thoroughbred racehorses in active training were enrolled in the study: a control group (n=11) with normal BAL values, an RGT treatment group (n=12), and a serum treatment group (n=13). An endoscopy including BAL was performed before (T0), after a six-week treatment period including 12 intramuscular injections of RGT or serum (T6), and as a follow-up 10 weeks after last treatment (T16).

Study Results

For the control group, no difference in BAL cytology between any of the time points was found. A significant decrease in BAL neutrophils (RGT $p<0.01$; serum $p<0.01$) and BAL mast cells (RGT $p=0.02$; serum $p<0.01$) was found for both the RGT and the serum treatment groups between T0 and T6. A significant decrease between T0 and T16 was found for BAL neutrophils in the serum treatment group ($p=0.05$) and for BAL mast cells in both the serum and the RGT treatment groups (RGT $p<0.01$; serum $p=0.03$). No difference was found between T0 and T16 for BAL neutrophils for the RGT group ($p=0.49$).

Study Summary

Preliminary results from this study showed that intramuscular treatment with either RGT or serum was effective in reducing BAL neutrophils and mast cells in horses with MEA. Further studies are necessary to evaluate the long-term effects of RGT treatment as well as the effect of treatment in horses diagnosed with severe equine asthma.

References

1. Hraha TH, *et al.* Autologous conditioned serum: the comparative cytokine profiles of two commercial methods (IRAP and IRAP II) using equine blood. *Equine Vet J.* 2011;43(5):516–521.
2. Arend WP, *et al.* Interleukin-1 receptor antagonist: role in biology. *Annu Rev Immunol.* 1998;16:27–55.
3. Dinarello CA, *et al.* Immunological and inflammatory functions of the interleukin-1 family. *Annu Rev Immunol.* 2009;27:519–550.
4. Hernandez ML, *et al.* IL-1 receptor antagonist reduces endotoxin-induced airway inflammation in healthy volunteers. *J Allergy Clin Immunol.* 2015;2:379–385.

Abstract F01: The Effect of Potassium Levels on Electrocardiographic Data in Calves with Neonatal Diarrhea

Osman Safa Terzi[1]; Erdal Kara[2]; Ece Irmak Alpsoy[2]; Demet Ayhan[2]; Bensu Merve Arikan[2]; Hasan Albasan[3]
[1]Department of Internal Medicine, Faculty of Veterinary Medicine, University of Ankara, Ankara, Turkey; [2]Ankara University; [3]Pet Depot Veterinary Group

Hyperkalemia in newborn calves with diarrhea can result in severe cardiac rhythm abnormalities. The purpose of present study was to evaluate the potassium levels along with electrocardiographic (ECG) data in 36 calves (age <30 days) with neonatal diarrhea from 3 different dairy farms during December 2019 in Ankara, Turkey. Of 36 calves, 14 had hyperkalemia and 22 had normokalaemia. Hyperkalemic calves (cK+: 5.84–8.33) had significantly (p=.0001) deeper T wave in lead II than calves with cK+ within the reference range. There was no significant difference in P, R, S amplitudes, P, QRS, T durations, PR intervals, or any other ECG findings between hyperkalemia and normokalaemia of calves with neonatal diarrhea. Acidemia and metabolic acidosis frequently evident in the calves with hyperkalemic diarrhea that were reported with cardiac rhythm abnormalities in the previous studies. On the other hand, in the present study, hyperkalemic neonatal calves with diarrhea had only deeper T wave but no other abnormalities; however, acidemia and metabolic acidosis were not severe. Holter monitoring should be warranted for hyperkalemic calves with diarrhea.

Abstract F02: Obstipation in Pet Pigs Presented to a University Teaching Hospital: 20 Cases

Kallie J. Hobbs; Diego Gomez; Sally DeNotta; Aitor Gallastegui
University of Florida

The aim of this retrospective study was to describe the signalment, history, clinicopathologic and diagnostic imaging findings, and post-mortem lesions in 20 pet pigs diagnosed with obstipation at a university teaching hospital. Treatments and clinical outcomes are also discussed. Medical records of pet pigs admitted to the Large Animal Hospital at the University of Florida between 2007 and 2019 were reviewed for a diagnosis of obstipation. Obstipation was defined as the absence of feces for 48 hours or greater, with evidence of intestinal distention from gas or fecal accumulation in the rectum and/or distal colon on either computed tomography (CT) or radiographs (RD). Data was collected regarding clinical signs, hematologic and biochemical profiles, blood gas analysis (BG), and diagnostic imaging findings from the date of admission. Information on treatments, if surgery was pursued, duration of hospitalization, and clinical outcome was recorded. Range and median were used to describe population characteristics. Twenty cases met the inclusion criteria. At the time of diagnosis of obstipation, the most frequently observed clinical signs were lethargy (n=20, 100%), anorexia (n=19, 95%), tachypnea (n=16, 80%), vomiting (n=12, 60%), tachycardia (n=10, 50%) and decreased appetite (n=2, 10%). Hematologic abnormalities included lymphopenia (<5.30 k/uL, n=17, 85%), leukopenia (<11 k/uL, n= 8, 40%), and toxic changes in neutrophils (n=8, 40%). Biochemical alterations included hypokalemia (<4 mEq/L, n=12, 60%) and hypoglycemia (<65 mg/dL, n=6, 30%). The most common diagnostic imaging findings included moderate gastric distention (n=13, 65%), gas-distended small intestinal loops (n=10, 50%), and moderate amounts of ingesta throughout the gastrointestinal tract (n=17, 85%). All 20 pigs had a combination of two or more of these findings. Medical therapy consisted of fluid therapy via intravenous/rectal/oral, combination intravenous/oral or combination rectal/oral in all cases. Additional therapies included antimicrobial therapy (n=17, 85%), non-steroidal anti-inflammatory drugs (n=11, 55%), carbonated soda beverage (n=6, 30%), exploratory laparotomy (n=6, 30%), and anti-emetics (n=3, 15%). Of the 20 pigs, 14 (70%) were discharged from the hospital and 6 (30%) were euthanaized. The median time of hospitalization for surviving pigs was 3 days (range: 2–7 days). The median time of hospitalization for pigs that underwent surgery and survived was 5 days (range: 3–6 days) whereas the median time of hospitalization for pigs that were managed medically was 3 days (range: 0.8–7 days).The median time of hospitalization for non-surviving pigs was 24 hours (range 0.5–3 days). Based on the observations in this study, obstipation in pet pigs carries a moderate to good prognosis. Diagnostic imaging findings such as moderate gastric distention and moderate amounts of ingesta throughout the gastrointestinal tract are commonly identified on CT or radiographs of obstipated pigs. Leukopenia, lymphopenia and hypokalemia were the most common blood work abnormalities present in the pigs included in this study.

Abstract F03: Influence of Sample Volume and Time on Rumen Juice Analysis in Cattle

Suzanne A. C. Clergue[1]; Sarah Depenbrock[2]; Munashe Chigerwe[2]
[1]Large Animal Hospital, VMTH, UC Davis; [2]School of Veterinary Medicine; UC Davis

Literature describing rumen juice (RJ) analysis as a diagnostic tool is based on analysis of a 10 mL sample directly after sampling. However, it may be challenging to obtain 10 mL of RJ from some patients, and clinical circumstances may delay the RJ analysis. This study quantified the effect of sample volumes, 2, 5, 10, 50 and 100 mL; and time-to-analysis on RJ analysis after 30 and 60 minutes (min). Two liters of RJ were obtained from a cannulated donor cow 21 times. The samples were subdivided into 1 set per time point. A set was composed of 2 duplicates of each sample volume. Samples were analyzed at 0, 30, and 60 min after sampling. Analysis of RJ consisted of pH, methylene blue reduction time (MBRT), and protozoal motility scoring. At 30 and 60 min, pH of the 2 mL and 5 mL samples were significantly ($p<0.05$) higher than pH of the 10 mL, 50 mL, 100 mL samples. The MBRT was significantly prolonged for 2 mL samples compared to 10, 50 and 100 mL samples at all time points. The pH and MBRT at 60 min were significantly higher than at 0 min for all sample volumes. There was no significant difference in protozoal motility for all sample volumes at all times. This study indicates that the interpretation of RJ analysis may be altered by sample volumes less than 10 mL or 30 to 60 min delays in analysis.

Abstract F04: An Alternative Approach to Evaluating Brix Refractometer Performance for Assessment of Beef Colostrum Quality

Lisa Gamsjaeger[1]; Ibrahim Elsohaby[2]; Jennifer Pearson[1]; Michel Levy Ibrahim Elsohaby[1]; Deborah Haines[2]; Edmond Pajor[1]; Claire Windeyer[1]
[1]University of Calgary; [2]Western College of Veterinary Medicine

The Brix refractometer is a practical tool that can be used to assess colostral Immunoglobulin G (IgG) concentration, but studies evaluating its performance in beef colostrum and identifying appropriate thresholds to detect low-IgG colostrum are sparse. The objective of this study was to assess Brix refractometer performance for estimating beef colostrum IgG concentrations, focusing on determination of the threshold for identification of colostrum containing <100 g/L IgG. A total of 416 beef colostrum samples from 11 cow-calf operations in Alberta, Canada, were available for this retrospective study. The IgG concentrations were measured by radial immunodiffusion (RID, Lab A=416) and estimated by Brix refractometry in three different laboratories to allow assessment of interlaboratory agreement (Lab A=364, Lab B=271, Lab C=220). A Spearman correlation coefficient was assessed between RID and Brix (Lab A) results. Interval likelihood ratios were calculated to determine the best Brix threshold for detecting colostrum containing <100 g/L of IgG. Optimal thresholds were also evaluated using the misclassification cost-term term approach with different possible prevalences of low-IgG colostrum and different plausible ratios of false-negative to false-positive costs. Test characteristics were calculated for the selected threshold. Concordance correlation coefficients (CCC) among Brix results of the three laboratories were determined and Bland-Altman analysis was performed to assess interlaboratory agreement. The mean colostrum IgG concentration was 149.6±38.7 g/L. Brix refractometry and RID results were positively correlated (r=0.72). A threshold of <24% Brix was best for detecting colostral IgG concentrations of <100g/L. For most considered scenarios, using a threshold of 24% Brix was also economically beneficial. Sensitivity, specificity and accuracy at this threshold were 88.9%, 91.3% and 91.2%, respectively. The CCC among Brix values measured at the three laboratories was high (Lab A/B: CCC=0.89; Lab A/C: CCC=0.96; Lab B/C: CCC=0.96) and Bland Altman analysis showed small mean differences (-1.2 to 0.09% Brix) and narrow limits of agreements (-4.8 to 2.4 % Brix) among laboratories. In conclusion, Brix refractometer performance for assessment of colostral IgG concentrations in beef colostrum was good. The use of Brix refractometry can therefore be recommended to aid colostrum management decisions on farms. Beef calves whose dams have colostrum Brix values of <24% may require larger volumes of colostrum and/or benefit from supplementation with colostrum from a different dam or a commercially available colostrum supplement.

Abstract F05: Bacterial Culture and Susceptibility of Samples Taken from Septic Foot Lesions of Adult Beef Cattle

Kelsey E. Walker[1]; Tamara Gull[2]; John Middleton[1]; Pamela Adkins[1]
[1]Veterinary Health Center, University of Missouri; [2]Veterinary Medical Diagnostic Laboratory, University of Missouri

Lameness in cattle can have a major impact on animal welfare and producer profits. The purpose of this study was to determine the most common bacterial species associated with sole abscess and deeper septic processes of the foot in beef cattle, and to determine antimicrobial susceptibility patterns of the cultured microorganisms. Adult beef cattle that presented to the University of Missouri Food Animal Clinic and were diagnosed with a sole abscess or deep digital sepsis were enrolled. Exudate was sampled using a bacterial culture swab (ESwab, Copan Diagnostics, Murrieta, CA) during assessment of the lesion. The exudate was aerobically and anaerobically cultured, isolates were identified using matrix-assisted laser desorption/ionization time of flight mass spectrometry, and antimicrobial susceptibility was performed using Sensititre plates (Sensititre BOPO6F Vet AST Plate, Thermo Fisher Scientific, Waltham, MA). To date, 31 samples from 31 cattle have been collected, and 134 bacterial isolates have been grown in culture. Most samples (28/31) had polymicrobial growth, with a median number of five isolates per sample (range: 1–13). *Trueperella pyogenes*, *Escherichia coli*, and *Streptococcus uberis* were the most commonly isolated bacteria, but over forty species have been identified. Susceptibility profiles (n=15) of the most commonly isolated bacterial species revealed that most (12/15) were pansusceptible to tested antimicrobials. Resistance to oxytetracycline and tulathromycin was noted in 3 of 9 *Trueperella* pyogenes isolates. Results suggest that bacteria isolated from septic foot lesions are primarily opportunistic pathogens that are generally susceptible to commonly used antimicrobials in food animal practice.

Abstract F06: Sudden Death in Weaned Lambs and Calves Due to *Strongyloides papillosus* in the United States

Toby L. Pinn-Woodcock[1]; Gerald Duhamel[1]; Manigandan Lejeune[2]; Belinda Thompson[2]
[1]College of Veterinary Medicine, Cornell University; [2]Cornell Animal Health Diagnostic Center

The Cornell Animal Health Diagnostic Center (AHDC) has confirmed two outbreaks of sudden death in apparently healthy weaned dairy calves in New York associated with the nematode, *Strongyloides papillosus*. While *S. papillosus* is known to cause symptoms of ill thrift and diarrhea in young ruminants, these are the first United States reports describing an association with sudden death of weaned calves raised in confinement housing. Researchers previously linked *S. papillosus* hyperinfection to sudden death in weaned calves and lambs in Japan, and suggested sudden death caused by *S. papillosus* is secondary to fatal arrhythmias, often without significant findings on post-mortem exam. To further investigate the prior occurrence of bovine and ovine sudden death associated with *S. papillosus* in the United States, anamnestic animal data, post-mortem and ancillary diagnostic results from ovine and bovine cases submitted to the AHDC with a history of sudden death and *S. papillosus* ova detection on fecal examination were evaluated from 2007 through 2019. A total of 10 cases were identified meeting the inclusion criteria, consisting of 2 bovine and 8 ovine cases. Five of these cases (50%) had open post-mortem diagnoses and clinical presentations similar to those previously described in Japanese *S. papillosus* outbreaks, including weaned juveniles (3 lambs; 2 calves) during the late summer and fall. These cases originated from New York state (n=4) and Massachusetts (n=1) and had *S. papillosus* fecal egg counts ranging from 3,100 to 29,000 eggs per gram. The remaining 5 cases identified had definitive post-mortem diagnoses, including bronchopneumonia, *Clostridium novyii* hepatitis, listeriosis, haemonchosus, urolithiasis, abomasitis, meningitis and enteritis, with *S. papillosus* fecal egg counts ranging from 2 to 1,250 eggs per gram. These findings suggest that cases of sudden death associated with the presence of *S. papillosus* may have been previously given an open post-mortem diagnosis at the AHDC. *S. papillosus* should be considered a novel differential diagnosis for sudden death in weaned lambs and calves in the northeastern United States.

Abstract F07: Urine Acidification Using Oral Methionine in Miniature Pigs

Rachel E. Oman; Pamela Adkins; Corrinn Bromfield
University of Missouri

Obstructive urolithiasis is a life-threatening condition affecting miniature pigs yet there is limited information regarding medical management of the condition in this species. The objective of this study was to determine the effect of oral methionine on urine pH in miniature pigs. A prospective crossover study design was used. Methionine (VET Pharmaceuticals, Miramar, FL) was administered orally every 24 hours for six days at 100, 200 and 300 mg/kg to twenty male, juvenile, intact miniature pigs. All pigs were individually housed and fed Mazuri mini-pig active adult diet (Mazuri Exotic Animal Nutrition, St. Louis, MO) at 2% of body weight. Urine pH was measured twice daily on free-catch urine samples. Serum electrolytes and renal enzymes were monitored for evidence of systemic acidification or electrolyte abnormalities. The urine pH decreased significantly when administered at 100 mg/kg ($p<0.03$) and 300 mg/kg ($p<0.02$) with a mean decrease of 0.52 and 0.50, respectively. All pigs remained systemically healthy throughout the study. Oral methionine appears to be a safe and effective means of acidifying urine in miniature pigs. The degree of acidification achieved in this study may not be sufficient to dissolve uroliths in clinical cases of obstructive urolithiasis. However, oral methionine has potential to serve as a component of medical management of obstructive urolithiasis in miniature pigs.

Abstract F08: One-Year Cross-Sectional Study of Dermatological Lesions in 433 Dairy Cattle in a Veterinary Teaching Hospital

Eloi Guarnieri; Frédéric Sauvé; David Francoz
Faculté de Médecine Vétérinaire, Université de Montréal

Little literature is available on the prevalence of bovine skin diseases. The objective of this study was to describe the dermatological lesions (DL) in dairy cattle admitted at the Faculty of Veterinary Medicine (FVM), Université de Montréal, from July 1, 2018, to June 30, 2019. A study-dedicated dermatological lexicon was first developed by the authors. Over a year, all dairy cattle admitted at the FVM were included in the study. Dairy cattle readmitted or without integumentary examination within 48 hours of admission were excluded. DL involving feet and ear canals were also excluded. The morphological and location of the skin and the oral mucosal DL were recorded by a trained observer. On the 610 cattle admitted at the FVM, 433 dairy cattle were included. Most of them were Holstein breed (90%) and females (98%). The mean age was 3.3 years old (+/- 2.8 years). Of these 433 cattle, skin and oral mucosal lesions were observed in 91% and 9% of the cases, respectively. Most cattle had at least one hock (55%), one carpus (47%) or one stifle (23%) affected. Crusts (55%), callus (54%) and alopecia (51%) were the most common skin DL. Erosions (43%) and ulcers (20%) were the most common oral mucosal lesions. This cross-sectional study highlighted the high prevalence of the DL in this specific cattle population. Considering these results, further studies would be important to determine etiologies, economic and clinical impacts of these DL.

Abstract F09: Arrival Risk Factors Associated with Morbidity in Milk and Grain Fed Veal Calves in Québec

Abdelmonem Mohamed; Julie Berman; Sébastien Buczinski; Simoun Dufour; David Francoz
Faculty of Veterinary Medicine, Montreal University

Inadequate transfer of passive immunity (TPI), arrival weight, seasons, purchase price and calf suppliers are important determinants of morbidity and mortality in veal calves. The objective of this prospective cohort study was to identify the risk factors, already present at arrival, and associated with subsequent individual treatment in Québec's veal calves. Thirty calves were randomly enrolled in each of 59 milk and grain fed veal lots. Blood samples were collected during the 1st week after arrival and evaluated using the brix refractometer (adequate TPI if ≥8.4%). Data regarding arrival weights, purchase price (CAD/pound), origin, arrival season and feed (milk-fed or grain-fed veal) were recorded upon arrival. Mortality, individual treatments and slaughter weights were obtained at the end of each production cycle. Generalized linear mixed models were used to analyze the association between morbidity (1 or more individual antimicrobial treatment) as a dependent variable and the different risk factors as predictors while controlling putative confounders. A total of 1,668 calves were included, and 432 calves were individually treated. In the multivariable model, treatment odds were higher in calves with failure of immunity transfer with odds ratio [OR] 1.5 (95% confidence interval [CI] 1.15–2.15, p=0.004). An interaction between season and arrival weight was found. Interestingly, calves weighing between 42.5 and 45.5 kg arriving in winter had lower odds of treatment than heavier calves arriving in the same season (p=0.01). We will evaluate the impact of those different risk factors on odds of mortality and on average daily gain during the feeding period.

Abstract F10: Clinical and Necropsy Findings Associated with Increased Risk of Natural Death in Critically-Ill Neonatal Crias

Erica Dorsey[1]; Daniela Bedenice[2]
[1]Becker College; [2]Cummings School of Veterinary Medicine, Tufts

The study purpose is to identify conditions associated with natural death of critically-ill neonatal llamas and alpacas, for which a decision of euthanasia was not reached during hospitalization. The specific aim was to improve targeted diagnostic evaluation and treatment recommendations in neonatal crias, where clinical signs of illness can remain insidious or vague. A retrospective cohort analysis of necropsy and clinical data was performed, including 61 critically-ill crias admitted to a university hospital for intensive care management under 2 weeks of age. All data were presented descriptively and compared between patients that were euthanized and those which died naturally, using univariate and multivariate analyses (SPSS-22). In the cohort of 61 non-surviving crias (57 alpacas, 4 llamas; 32/61 females), none of the measured clinical or laboratory admission parameters differed significantly between crias that were ultimately euthanized (31/61, 50.8%) or died naturally (30/61, 49.2%). Similarly, median duration of hospitalization did not differ between mortality groups. Pulmonary (46/61, 75.4%), hepatic (26/61, 42.6%) and cardiac pathology (20/61, 32.8%) was most commonly observed on necropsy, with a site of infection identified in 39/61 (63.9%) animals. Multivariate analyses determined that suppurative pneumonia (OR: 4.6, 95% CI: 1.2–17; P=0.023), prematurity (OR: 6.7, 95% CI: 1.4–32.6; p=0.019) and older admission age (OR: 1.15 per day increase in admission age, 95% CI: 1.002–1.33; p=0.046) significantly increased the risk for natural death. These data suggest that advanced diagnostic modalities to facilitate early, etiological identification of infection and pulmonary disease should be explored to enable improved, targeted treatment and illness outcome.

Abstract F11: A Survey of Coccidiosis Shedding Among Small Ruminants on St. Kitts and the Appalachian Area

Jerry Roberson[1]; Ninian Cameron-Blake[2]; Phippa Gibbons[1]; Jennifer Ketzis[2]
[1]College of Veterinary Medicine, Lincoln-Memorial University; [2]School of Veterinary Medicine, Ross University

The purpose of these two similar studies was to determine the presence of coccidiosis and contrast the shedding of coccidia in sheep and goats by age group on the island of St. Kitts and in an area of the Appalachia. Small ruminant producers in both locations were contacted in person, by telephone or email until 50 farms in both locations agreed to participate. Convenience fecal samples were collected across age groups on each farm. A modified McMasters test was performed to determine the coccidia per gram. Simple descriptive statistics were applied across four age groups: <six months, six months to <1 year, yearlings and adult (two years and older). The coccidia shedding was highest among those <6 months of age in both regions (Table 1). There was a clear linear trend in decreased shedding as age increased in the Appalachia region. This linear trend was not apparent for small ruminants of St. Kitts. Although some farms utilized coccidiostats, the same linear trend can still be noted. Overall, coccidia shedding on St. Kitts was lower across all species and age groups when compared to the Appalachia region. Because shedding of coccidia decreases substantially from young to old, it appears that a natural immunity develops by adulthood. Coccidiosis control via coccidiostats was not an option on St. Kitts but may be of value in the Appalachia region in young stock but of little value in adults.

Table 1. Average coccidia shedding (coccidia per gram) by small ruminant species, location, coccidiostat usage and age group

Category	<6 months	6 months–<1 year	Yearling	Adult (>2 years)
Appalachia (AP)	5319	2610	1278	281
St. Kitts (SK)	1722	547	303	458
AP ovine	4128	2581	585	188
SK ovine	1433	676	253	428
AP caprine	8847	2708	2292	404
SK caprine	2525	380	378	516
AP coccidiostats	4639	1525	498	299
AP no coccidiostat	8492	4632	2660	280

Abstract F12: Preliminary Investigation of Xenotransfusion as a Therapeutic Modality for Anemia in Goats

Joe Smith[1]; Austin Viall[2]; Ryan Breuer[3]; Paul Plummer[2]; Amanda Kreuder[2]; Rebecca Walton[2]
[1]College of Veterinary Medicine, Iowa State University; [2]ISU; [3]University of Wisconsin

Treatment of anemia is a common and challenging clinical condition to treat in goats. Despite being a common sequelae of gastrointestinal parasite infection, some veterinary clinics do not have access to caprine whole blood or packed red blood cells. Xenotransfusion is the administration of blood or blood products from one species to another, and is used successfully in small animal emergency situations. Since cattle are more readily available in most teaching hospitals, and can donate larger volumes of blood, the objective of this preliminary study was to evaluate the suitability of donor bovine blood for caprine recipients. This study was divided into 2 phases. For the first phase blood samples were collected under informed consent from 15 bovine and 15 caprine patients. For compatibility evaluation of each combination, the plasma in the centrifuged tube was for major and minor cross matching. For the second phase two goats were then transfused with bovine whole blood and monitored for reactions. Transfused blood cell kinetics were subsequently evaluated. Both goats were euthanized 14 days post transfusion and necropsied. The majority (11/15) of cross matches were compatible between caprine recipients and bovine donors. The live goat transfusions were well tolerated, although an anaphylactic event was observed in 1 goat. No abnormal changes were noted on postmortem examination. Transfused cells were easily recognizable on blood film analysis 5 days post transfusion. Xenotransfusion with bovine whole blood may provide a therapeutic modality for emergency treatment of goats with anemia.

Abstract F13: Validation of a Predictive Model for Downer Cows Presented to a Hospital

Alexandra Gariépy[1]; Maria Puerto Parada[1]; Juan Carlos Arango-Sabogal[1]; Marie-Ève Bilodeau[2]; André Desrochers[1]; David Francoz[3]; Sylvain Nichols[1]; Gilles Fecteau[1]
[1]Faculté de Médecine Vétérinaire, Université de Montréal; [2]Hôpital Vétérinaire Victoria; [3]Université de Montréal

Establishing an early prognosis would help reducing unnecessary treatments on downer cows with a low chance of survival. In a previous study, a multivariable logistic regression model used data from 1471 down cows presented to the Centre Hospitalier Universitaire Vétérinaire to predict the outcome. Variables included were: age, days of recumbency, PCV (%), heart rate, endotoxemia, AST (U/L), creatinine (umol/L), phosphate (mmol/L) and total CO_2 (mmol/L). This study objective was to assess the predictive ability of the model on a new series of downer cows treated at the same referral hospital. The study population included all recumbent dairy cattle older than 15 months presented between 2017 and 2019 (n=137 cases). Information regarding lactation status, hospitalization (outcome, diagnosis) and blood analysis results (CBC, biochemistry) were obtained. The sensitivity and specificity of the model were computed at various probability threshold. For all predictors, the mean and 95%CI (exact or Poisson when required) were estimated. The study population was similar to the previous population except for the following variables that had a statistical difference of the mean: days of recumbency, PCV, AST, creatinine and total CO_2. The overall survival was 61%: survival (n=83) and non-survival (n=54). Model sensitivity was 89%, specificity 46% and positive and negative predictive values 73% and 71% respectively, using the optimal probability cut-off point for the model (50%). The model appears to be an interesting tool, if combined with a complete physical examination to rule out fatal condition. Study limitations include possible selection bias.

Abstract F14: Clinical Safety Data for the Use of Pantoprazole in Hospitalized Cattle, a Retrospective Study

Joe Smith[1]; Amanda Kreuder[2]; Austin Kosusnik[2]; Jonathan Mochel[2]
[1]College of Veterinary Medicine, Iowa State University; [2]Iowa State University

In human patients, multiple adverse effects have been reported from the use of pantoprazole including hematologic and electrolyte abnormalities, as well as anaphylaxis and edema. However, currently no data exists regarding these adverse effects after administration of pantoprazole in cattle. To evaluate the clinical safety of pantoprazole in cattle, medical records of all cattle administered pantoprazole at the Food Animal and Camelid Hospital of Iowa State University over an approximately five year period were retrospectively analyzed for adverse effects. Records were scrutinized for animals with CBC and/or blood chemistry data available immediately before and after administration of pantoprazole and analyzed for percent change in observed values between the testing timepoints. Paired data were then evaluated by paired T-tests or Wilcoxon testing. Forty-three eligible patients were identified. Pantoprazole was administered intravenously (1 mg/kg) or subcutaneously (2 mg/kg). Mild hypomagnesemia was observed post-pantoprazole administration in ten cattle (<2.1 mg/dL); however, no significant changes were noted when compared to baseline (p=0.7241). Significant changes were noted in markers of hepatic (p=0.0059) and renal function (p=0.0293), however, these were not clinically significant (decreased BUN, AST). Anaphylaxis post-pantoprazole administration was not observed, however seven cattle displayed edema after dosing. Clinicians should be aware of the potential for adverse events in hospitalized cattle being administered pantoprazole and monitor patients accordingly. While these preliminary retrospective results indicate that pantoprazole may be a safe adjunctive therapy in cattle, additional studies are warranted to further determine the safety and toxicity for cattle.

Abstract F15: Change of Antimicrobial Prescriptions in Calves in Switzerland after the Launch of Antimicrobial Use Guidelines

Alina Hubbuch[1]; Ruth Peter[1]; Barbara Willi[2]; Sonja Hartnack[3]; Cedric Muentener[1]; Hanspeter Naegeli[1]; Christian Gerspach[5]
[1]Institute of Veterinary Pharmacology and Toxicology, Vetsuisse Faculty, University of Zurich; [2]Clinic for Small Animal Internal Medicine, Vetsuisse Faculty, University of Zurich; [3]Section of Epidemiology, Vetsuisse Faculty, University of Zurich; [5]Clinic for Ruminants, Department for Farm Animals, Vetsuisse Faculty, University of Zurich

The increasing threat of antimicrobial resistance promotes the need for antibiotic stewardship programs to foster responsible antimicrobial use. Therefore, guidelines for prudent antimicrobial use supported by an online antimicrobial stewardship tool (AntibioticScout.ch) were introduced in Switzerland in December 2016. The objective of the present study was to evaluate antimicrobial prescriptions for Swiss calves before (2016) and after (2018) the launch of these guidelines. To that end, cases of calves with pneumonia, diarrhea and otitis from a university hospital and eight private practices in Switzerland were included. Data on anamnesis, clinical findings, diagnostic work-up and antimicrobial treatment were collected. Type and percentages [95% CI] of antimicrobial prescriptions were compared between 2016 and 2018. Of the total number of calves, 88.2% [85.4–90.6] in 2016 (n=625) and 88.4% [85.7–90.7] in 2018 (n=655) were treated with antibiotics. The use of highest priority critically important antimicrobials (HPCIAs) decreased from 52.7% [48.6–56.9] in 2016 to 38.0% [34.2–41.9] in 2018; this decrease was found at the university hospital and in private practice and in cases with pneumonia and diarrhea. Particularly the use of fluoroquinolones decreased (2016: 43.1% [39.2–47.2]; 2018: 31.1% [27.6–34.8]). Overall, the number of first line treatments increased from 12.8% [10.4–15.6] in 2016 to 20.2% [17.3–23.4] in 2018. In cases with pneumonia, first line treatments increased (2016: 15.3% [11.6–19.9]; 2018: 26.5% [21.8–31.9]) and third line treatments decreased (2016: 43.5% [38.0–49.3]; 2018: 27.9% [23.1–33.3]); this was seen in cases treated at the university hospital, whereas in private practice only a decrease of third line treatments was observed. In cases with diarrhea, more second line at the expense of unjustified antimicrobials were prescribed at the university hospital in 2018. Antimicrobial treatment of calves with otitis did not change from 2016 to 2018. After the introduction of AntibioticScout.ch, more prudent use was observed in the treatment of calves with pneumonia and diarrhea as less HPCIAs, especially fluoroquinolones, and more first line treatments were prescribed. However, the overall frequency of antimicrobial treatment did not change and the use of HPCIAs was still common in 2018. Therefore, further antimicrobial stewardship activities are necessary.

Abstract F16: Effect of Age and Pregnancy Status on Pharmacokinetics of Flunixin in Dairy Does

Joe Smith[1]; Tara Marmulak[2]; Joan Rowe[3]; Lisa Tell[3]
[1]College of Veterinary Medicine, Iowa State University; [2]CSU; [3]UC Davis

The objective of this study was to investigate the effect of age and reproductive status on the pharmacokinetics of flunixin meglumine when administered by intravenous and subcutaneous routes to dairy goats The animals were 8 pregnant dairy does aged 3.75±3.33 years (range: 1–9 years) and 20 nulliparous dairy does aged 0.71±0.04 years (range: 0.66–0.79 years). Does were given flunixin meglumine via intravenous (1.1 mg/kg) and subcutaneous (1.1 mg/kg) routes. Blood samples for quantifying flunixin and 5-hydroxy flunixin were collected prior to dosing and at predetermined time points. Plasma flunixin and 5-hydroxy flunixin concentrations were quantified using liquid chromatography and mass spectroscopy. Pharmacokinetic parameters were estimated using noncompartmental methods. When comparing young does to older pregnant does there was an effect of age and reproductive status on pharmacokinetic parameters after intravenous administration for clearance (increased, p=0.0019); area under the curve (decreased, p=0.0001); mean residence time (decreased, p=0.0001); and volume of distribution (increased, p=0.0238). There were significant differences in time to maximum concentration (decreased, p=0.0179); clearance (Cl/F) (increased, p=0.0071); area under the curve (decreased, p=0.0002); and mean residence time (decreased, p=0.0012) when comparing the younger to older doe groups after subcutaneous dosing. No significant differences were noted for bioavailability between the groups (p=0.6780) Age and reproductive status influence the pharmacokinetics of flunixin after intravenous and subcutaneous administration to dairy does.

Abstract N01: Functional Evaluation of Microactuators to Prevent Hematoma Shunt Obstruction in an *In Vitro* Hydrocephalus Model

Dillon Devathasan[1]; R. Timothy Bentley[2]; Ángel Enríquez[1]; Hyowon Lee[1]; Stephanie Thomovsky[2]; Craig Thompson[2]; Qi Yang[1]
[1]Purdue University; [2]Veterinary Teaching Hospital, Purdue University

Mechanical obstruction accounts for a high failure rate in ventriculoperitoneal shunting. We evaluated the ability of a magnetic microactuator to micronize obstructive hematoma fragments within ventriculoperitoneal shunts. An *in vitro* hydrocephalus model was used. Phosphate-buffered saline was continuously driven into a 100 ml beaker representing the cranial vault. In five controls, outflow from the beaker was via a ventricular catheter (VC), valve and distal catheter. Five treatment shunts were identical except the VC contained a microactuator. Porcine blood (85 ml) was added to the beaker and entered the shunts. Hydrodynamic resistance within the beaker was measured. Over a 7-hour experiment, time over threshold pressure of 40 mm Hg (TOT), peak obstructive pressure and end-experiment outflow volume were measured. The site of all occlusions was recorded. Four of five control shunts developed 6 obstructions (3 VC; 3 valve and distal catheter) compared to 1 treatment shunt (1 VC). Control shunts demonstrated persistently high pressure (TOT, 293.48 minutes) compared to treatment shunts (TOT, 0.62 minutes) ($p=0.006$). Mean peak obstructive pressure was 213.0 mm Hg and 41.1 mm Hg in the control and treatment groups, respectively ($p=0.001$). At end-experiment, significantly larger volumes were collected from treatment than control shunts ($p=0.047$), further supporting prolonged patency of treatment shunts. These data suggest that a microactuator is an effective, non-invasive method for preventing ventriculoperitoneal shunt obstruction. Microactuator-embedded shunts conferred protection along the entire length of the shunt (VC and VDC), maintaining a pressure <40 mm Hg for the entire experiment duration in 3/5 shunts.

Abstract N02: Pharmacokinetics of a Novel Cytosine Arabinoside Subcutaneous Protocol in Dogs with Meningoencephalomyelitis of Unknown Etiology

Hilary A. Levitin; Kari Foss; Devon Hague; Jennifer Reinhart; Timothy Fan
University of Illinois

Meningoencephalomyelitis of unknown etiology (MUE) is a common cause of non-infectious inflammatory disease of the central nervous system in dogs. It is an important disease in the field of veterinary medicine given the severity of neurological signs that can be seen at the time of diagnosis, its guarded prognosis, and the large financial investment required by owners for diagnostic testing and treatment. Cytosine arabinoside (CA) has become an attractive treatment to be given with corticosteroids as CA is associated with favorable survival times and low incidence of systemic side effects in dogs. Various dosing recommendations exist, most of which indicating inpatient treatment, with no consensus of a standard of care. The purpose of this study was to evaluate the pharmacokinetics of a novel one-day SC CA protocol in dogs with (MUE). Eight client-owned dogs that had MRI and CSF findings compatible with antemortem diagnosis of MUE, along with pertinent negative infectious disease testing, were prospectively enrolled in the study. All eight dogs received a standard CA constant rate infusion (CRI) protocol (200 mg/m^2 IV over 24 h) as treatment for MUE. Four weeks later, the SC protocol (50mg/m2 every 2 h for 4 treatments) was administered. Prior to initiating each protocol, a CBC and chemistry were performed to assess for adverse effects. Blood samples were collected 0, 1, 2, 3, 4, 5, 6, 7, 8, 10, 12, 14, 16, 20, and 24 h after initiating each protocol. Plasma CA concentrations were measured using a validated high-pressure liquid chromatography assay. Steady state was defined as plasma CA concentrations exceeding 1μg/mL at 1 and 8 h following initiation of treatment, consistent with previous studies. The Wilcoxon signed rank test was used to compare pharmacokinetic parameters of interest between protocols. No adverse effects were observed in CBC and chemistry panels performed prior to administering each CA protocol. Peak CA concentration (Cmax) for the SC protocol was significantly higher than for the CRI protocol, measuring 3.4μg/mL (1.6–9.7μg/mL) and 1.09μg/mL (0.77–1.67μg/mL), respectively (p=0.0156). CA concentration at 1 and 8 h following initiation of treatment was significantly higher for the SC protocol compared to the CRI protocol (p=0.0078). CA concentrations were above 1μg/mL for a significantly longer period of time for the SC protocol versus the CRI protocol, measuring 9.25 h (4.5–10.5 h) and 3.125 h (0–9.75 h), respectively (p=0.039). The investigated SC protocol administered over 8 h provided superior CA plasma levels for a longer duration of time than a standard CRI protocol administered over 24 h. Furthermore, the proposed SC protocol is amenable to outpatient treatment, reducing stress associated with hospitalization and limiting repeated IV catheterization.

Abstract N03: The Bloody Study: Effect of Hemodilution on Cerebrospinal Fluid Analysis in Dogs with Neurological Disease

Rachel Lampe; Anne Barger; Kari Foss; Devon Hague
University of Illinois

Elevations of total nucleated cell count (TNCC) or total protein (TP) concentration in the cerebrospinal fluid (CSF) of dogs with neurologic disease can help categorize lesions in the central nervous system. Iatrogenic blood contamination at the time of CSF collection is a common occurrence, which can make diagnostic interpretation difficult. The purpose of this study was to identify the best method of correction of TNCC in CSF, based on the amount of hemodilution present. Paired sequential samples of CSF were prospectively obtained from a single collection site and patient in which one sample had visibly more hemodilution than the other. The TP concentration, red blood cell (RBC) count, and TNCC were analyzed within 30 minutes of collection. A complete blood count (CBC) was collected within 24 hours of the CSF collection. Two linear regression models were fitted, and the results were compared with 3 previously suggested correction formulas. Pearson's correlations coefficients were calculated for each correction formula with the difference between the TNCC in the hemodiluted and clear sample. Sum squared errors were compared amongst the 5 correction formulas. Nineteen paired samples were collected. The differences between the hemodiluted and clear samples were significant for TNCC ($p<0.00042$), RBC count ($p<0.0001$) and TP ($p=0.0013$). Using a ratio of 500:1 and 1000:1, RBC:WBC yielded a correlation of $r=0.78$. A previously suggested correction formula that included the WBC and RBC from the CBC yielded a correlation of $r=0.98$. Two linear regression models fitted using the CSF TNCC, RBC, and the peripheral blood RBC, with one additionally including the peripheral WBC, both had a correlation of $r>0.99$. The linear regression model including both the peripheral RBC and WBC had the lowest sum squared errors. The results of the study indicate that hemodilution has an effect on the TNCC, RBC, and TP in CSF. All correction methods including CBC parameters were more accurate than evaluating the hemodiluted CSF parameters in isolation. A linear regression model including the RBC and WBC from the CBC provided the most accurate correction of TNCC in hemodiluted CSF.

Abstract N04: The Role of Prophylactic Omeprazole in Dogs Treated Surgically for Thoracolumbar Intervertebral Disc Herniation

Jaya M. Mehra[1]; Melissa Lewis[1]; Katie Tolbert[2]
[1]Purdue University; [2]Texas A&M University

Gastrointestinal (GI) signs and bleeding are commonly reported in dogs with thoracolumbar intervertebral disc herniation (TL-IVDH). Dogs are frequently treated with proton pump inhibitors (e.g., omeprazole) during hospitalization, however, their efficacy in reducing GI signs in these dogs is unknown. We hypothesized that in dogs undergoing surgery for acute TL-IVDH, omeprazole would not decrease the frequency of GI signs compared to placebo. Client-owned dogs undergoing hemilaminectomy for acute TL-IVDH. Randomized double-blinded placebo-controlled prospective clinical trial. Placebo or omeprazole (1 mg/kg) was orally administered q 12 h for 5 days during hospitalization. Development of clinical GI signs were recorded daily and fecal occult blood (FOB) testing, PCV, and blood urea nitrogen (BUN)/creatinine ratios were performed at specific time points. A fisher's exact test ($p<0.05$) was used to compare the frequency of GI signs or clinicopathologic abnormalities between groups and to investigate an association between pre-operative anti-inflammatory drug use and development of clinical GI signs. Twenty-one dogs (12 group A, 9 group B) were enrolled with mean SD age of 6.5±2.6 years. GI signs developed in 12/21 dogs (57%; 7/12 group A, 5/9 group B) which was not different between groups. Diarrhea was common (38%; 5/12 group A, 3/9 group B), but clinical GI bleeding was rare (9%; 1/12 group A, 1/9 group B). FOB was positive in 7/21 dogs (33%; 3/12 group A, 4/9 group B) and postoperative BUN/creatinine ratio was elevated in 5/21 dogs (24%; 4/12 group A, 1/9 in group B); these differences were not significant between groups. Anti-inflammatory drugs were used prior to enrollment in 15/21 dogs (71%; 8/12 group A, 7/9 group B) and were not significantly associated with development of GI signs. GI signs were common in dogs undergoing surgery for acute TL-IVDH. Short-term, prophylactic omeprazole treatment did not decrease development of GI signs compared to placebo.

Abstract N05: What's Wrong with Rongeuring? Comparing Pneumatic Burr and Rongeuring Techniques for Thoracolumbar Hemilaminectomies

Barbara Lindsay
Center for Veterinary Specialty and Emergency Care

This retrospective study compares the neurological outcomes of small dogs undergoing a thoracolumbar hemilaminectomy for intervertebral disc herniation (IVDH), performed via a standard technique using a pneumatic burr or a rongeuring technique using a combination of Lempert and Kerrison bone rongeurs. The secondary aim was to determine if the surgical time was significantly different between the two techniques. Exclusion criteria included: dogs weighing more than 8 kilograms, neurological pathology other than IVDH, multiple site non-continuous hemilaminectomies, and incomplete medical records. An adjusted Modified Frankel Score (MFS) was assigned to each dog prior to surgery, day after surgery, at discharge, and at recheck. Anesthetic records were evaluated to accurately record the surgical time of each procedure (cut to closure of incision). Records of 297 separate surgical procedures were analyzed (2012–2019); 173 were performed with a pneumatic burr and 124 performed with rongeurs. Mean age of dogs who underwent the rongeuring technique were older (6.85 y) compared to the pneumatic burr group (5.96 y). No significant difference was present in the pre-surgical adjusted MFS between the two groups nor was the length of hospitalization. Neurological scores at each time point for the two surgical groups were analyzed using Mann-Whitney tests, and no difference was found (all p values <0.05). Surgical time was found to be significantly shorter in the rongeur group (83 minutes), compared to the pneumatic burr group (113 minutes). In conclusion, rongeuring provides a safe surgical alternative in small dogs with IVDH, which may also allow for a shorter anesthetic time.

Abstract N06: Pharmacokinetics of Mebendazole in Canine Plasma and Cerebrospinal Fluid: A Pilot Study

Amy B. Yanke[1]; Kendall Day[2]; Amanda Taylor3; Crisanta Cruz-Espindola[1]; Christina Hargis[1];; Dawn Boothe[1]
[1]College of Veterinary Medicine, Auburn University; [2]College of Veterinary Medicine, University of Florida; [3]MedVet Columbus

Novel therapies are needed for treatment of gliomas. Fenbendazole and mebendazole demonstrated anti-neoplastic effects on high grade canine glioma cells at *in vitro* mean inhibitory concentrations (IC50) of 10 ng/ml for mebendazole versus approximately 150 ng/ml for fenbendazole. Our study aimed to titrate the dose of mebendazole necessary to achieve potentially effective concentrations (10 ng/ml) of mebendazole in cerebrospinal fluid (CSF) when administered orally to dogs. We hypothesized that an oral dose of 100 to 200 mg/kg would be necessary. Beagle cross dogs received 100 mg/kg (n=4) or 200 mg/kg (n=5) and blood samples were collected intermittently for 60 hrs. Mebendazole was quantitated in canine plasma and CSF using high performance liquid chromatography (lower limits of quantitation 10 and 5 ng/ml, respectively). Data was subjected to noncompartmental analysis. The mean peak plasma concentrations (Cmax; ng/ml) were, for 100 and 200 mg/kg, respectively, 201+88 occurring at 7+2 hr and 181+80 at 15+4.5 hr. The respective areas under the curve (AUC: ng/ml/hr) were 2350 640 vs 3072+1502. Average concentrations (ng/ml) were 45+8.4 vs 67+29. For CSF, Cmax was 11+11 vs 20+6.1 and AUC 88+56 vs 287+114. Relative bioavailability in CSF vs plasma was 5 to 10%. No adverse events were noted. This study demonstrates a dose concentration relationship for mebendazole in dogs and that the *in vitro* IC50 for gliomas can be achieved in CSF at 100 mg/kg, although 200 mg/kg might be more prudent. Optimal target concentrations in CSF for treatment of canine gliomas remains to be determined.

Abstract N07: Evaluation of Serum High-Mobility Group Box 1 Concentration in Dogs with Epilepsy

Yoonhoi Koo; Taesik Yun; Hakhyun Kim; Ji-Houn Kang; Mhan-Pyo Yang; Byeong-Teck Kang
Laboratory of Veterinary Internal Medicine, Veterinary Teaching Hospital, College of Veterinary Medicine, Chungbuk National University

Elevation of inflammatory mediators' concentrations in the brain decreases the seizure threshold which contributes to epileptogenesis. High-mobility group box 1 (HMGB1) is one of the key mediators of neuroinflammation. It has been shown that serum HMGB1 levels increased in laboratory animal models and human patients with epilepsy. However, studies of serum HMGB1 concentrations in epileptic dogs have not been reported. We hypothesized that there would be elevated levels of serum HMGB1 in dogs with epilepsy and that HMGB1 levels may vary according to the epilepsy etiology, epilepsy status, and treatment. Blood samples were collected from 28 epileptic dogs, 12 dogs with non-epileptic brain disease, and 26 healthy dogs. Meningoencephalitis of unknown etiology and brain tumors without seizure were referred to as non-epileptic brain diseases. Epilepsy dogs were divided into idiopathic epilepsy (14 dogs) and structural epilepsy (14 dogs). Serum HMGB1 concentrations were estimated using the canine HMGB1 enzyme-link immunosorbent assay kit. Dogs with epilepsy regardless of underlying causes had significantly higher serum HMGB1 levels than healthy dogs ($p=0.001$). Dogs with structural epilepsy had significantly higher serum HMGB1 levels than healthy dogs ($p=0.003$), however, dogs with non-epileptic brain disease did not. Moreover, serum HMGB1 concentrations of idiopathic epilepsy with epilepsy course >3 months were increased compared to with epilepsy course ≤3 months ($p=0.019$). However, serum HMGB1 concentration was not correlated with seizure frequency, brain lesion volume, seizure control status and duration of treatment Serum HMGB1 concentrations of epilepsy dogs were significantly higher than healthy dogs, however dogs with non-epilepsy did not, indicating that serum HMGB1 could be a biomarker of epileptic dogs. Furthermore, the elevation of serum HMGB1 concentration in epileptic dogs was found to be independent of seizure frequency, brain lesion volume, seizure control status, and duration of treatment. These results suggest that serum HMGB1 could be used as an early diagnostic biomarker for epilepsy in dogs.

Abstract N08: Cannabidiol Disposition After Single Oral Dosing in Fasted and Fed Dogs

Dawn M. Boothe[1]; C. Cruz-Espindola[1]; R. Gillette[2]; R. Strunk; C. Warner[2]
[1]Auburn University; [2]Travco Products Inc

Cannabidiol (CBD), one of two major phytocannabinoids, is largely void of psychotropic effects. Increasingly it is being used for a variety of canine conditions. Orally administered CBD undergoes first pass metabolism, leading to approaches that increase oral bioavailability (F). The purpose of this study was to determine if oral bioavailability of a full spectrum hemp oil (NMXCB1220™) can be improved in dogs when administered as a soft chew, or, as in humans, with food. Normal, apparently healthy beagle dogs (n=8) were studied three times using a non-randomized triple cross-over design with a 7-day washout. CBD (2 mg/kg) was administered orally as hemp-oil in fasted, followed by fed dogs and finally as a soft chew. Blood was collected intermittently for 24 hrs. Plasma CBD and THC were quantitated using LC-MS/MS with MRM (validated in dogs). Following noncompartmental analysis, key parameters were (mean+sd, fasted oil followed by fed oil and fasted soft chew): maximum plasma concentration (Cmax; ng/ml) 110+61 vs 289+127 vs 272+130 at time to maximum concentration (hr) of 3.5+1.4 vs 2.3+0.7 vs 3.8+0.7; area under the curve (ng/ml/hr) 1672+2543 vs 1292+592 vs 1100+379; and elimination half-life (hr) of 6.4+3.3 vs 6.3+1.7 vs 5.0+2. The relative bioavailability (F) of CBD in fed vs fasted animals was 1.78+0.89 and fed hemp oil vs fasted soft chews 1+0.3. THC was detectable in all preparations (Cmax <30 mcg/ml). Feeding increased hemp oil CBD bioavailability in dogs in this study. Fasted dogs receiving a soft chew had similar concentrations to that achieved with industrial hemp oil in fed dogs.

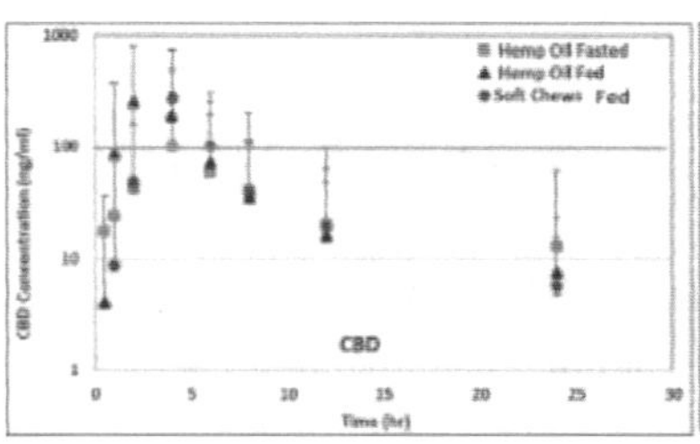

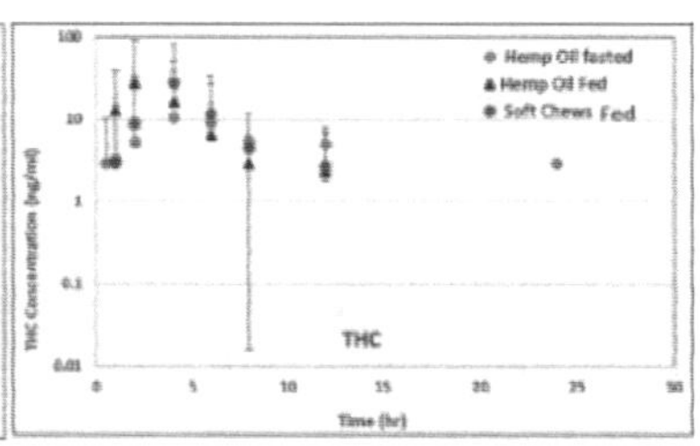

Abstract N09: The Use of Magnetic Resonance Spectroscopy to Differentiate Canine Brain Masses

Sarah B. Deluty[1]; Lynn Griffin[2]; Rebecca Packer[2]
[1]CVM, Colorado State University; [2]Colorado State University

Magnetic Resonance Imaging (MRI) is a common modality used to aid in the diagnosis of canine brain tumors; however, limitations remain for accurate determination of mass lesion type. MR spectroscopy (MRS) can be obtained with the diagnostic MRI procedure, and is a quantitative measure of brain metabolism that may eventually provide information to help predict histologic type, grade, outcome, and individualized treatment targets. The aim of this study was to evaluate the use of MRS in the context of canine brain masses, and evaluate whether or not MRS could be used to differentiate neoplastic from non-neoplastic masses, as well as differentiate histologic type of tumor. Data from 144 clinical cases of naturally-occurring canine brain tumors and inflammatory lesions that presented to the Colorado State University Veterinary Teaching Hospital from 2006–2019 were available for analyses. 44 of these cases provided met the inclusion criteria for paired MR spectroscopy and histological data for correlative study. Choline (Ch), creatine (Cr), lactate (LL), myoinositol (mI) and NAA brain metabolites, and calculated values of Cr+Ch, Ch/Cr, Cr/Ch, NAA/Cr, NAA/Ch, Cr/NAA and Ch/NAA, were explored for their correlation to the final etiology of brain dysfunction. NAA/Ch differed significantly between normal brain versus glioma ($p=0.0082$), and glioma versus inflammatory lesions ($p=0.0118$). Based on our data, MRS may be used as additional evidence to differentiate glioma from inflammatory brain lesions, providing and alternative non-invasive diagnostic tool to clinicians.

Abstract N10: Magnetic Micro-Actuator Enabled Catheters for Ventriculoperitoneal Shunting: Magnetic Resonance Safety and Artifacts

R. Timothy Bentley[1]; Angel Enriquez[2]; Hyowon Lee[2]; Tiffany Lyle[2]
[1]Veterinary Clinical Sciences, Purdue University; [2]Purdue University

Our previous micro-actuator studies have shown reduction of ventriculo-peritoneal shunt obstruction *in vitro* and *in vivo*. Magnetic resonance imaging (MRI) safety and artifacts still need characterization, including gradient echo sequences which are extreme conditions for magnetic implants. Ventricular catheters were implanted into the lateral ventricles of four pigs (2 control, 2 micro-actuator). MRI was performed post-operatively and one month later. After neurological observation for one month, pigs were sacrificed for gross and histological evaluation. All pigs returned to neurologically normal post-operatively, except contralateral menace response deficits. MRI revealed no evidence of micro-actuators induced mechanical or thermal injury, with no signal changes in adjacent brain parenchyma. Minor post-operative MRI findings, including contrast-enhancement limited to the surgical site, were the same in both groups and resolved at one month. Ventriculomegaly was mild in both groups. Micro-actuator associated artifacts unchanged at one month included striking large, 4-lobed signal voids on gradient echo sequences. Small, mainly hypointense artifacts were present on T1-weighted, T2-weighted, and fluid attenuation inversion recovery sequences. Grossly, maximal ventricular dimension was 1–1.5 cm in both groups. Flocculent material remained in the ventricle of 2 treatment and 1 control pig. The implant tract was associated with gliosis or inflammatory infiltrates in all 4 pigs. There is apparent safety upon MRI of micro-actuator enabled ventricular catheters, with no clinical, imaging or pathological differences from the control group. MRI micro-actuator artifacts overlapped in appearance with surgical hemorrhage, except a characteristic, 4-lobed gradient echo appearance.

Abstract N11: Evaluation of "Underreporting of Seizures" with Electroencephalography (EEG) in Canine Epilepsy

Masayasu Ukai[1]; Thomas Parmentier[1]; Miguel Cortez[2]; Andrea Fischer[3]; Luis Gaitero[1]; Hannes Lohi[4]; Stephanie Nykamp[1]; Tarja Pääkkönen[4]; Veronique Sammut[5]; Fiona James[1]

[1]Ontario Veterinary College, University of Guelph; [2]The Hospital for Sick Children (SickKids); [3]LMU Munich; [4]University of Helsinki; [5]VCA West Los Angeles Animal Hospital

Background—In people, objective data from electroencephalography (EEG) are mainly used to diagnose epilepsy, measure seizure frequency and evaluate efficacy of anti-seizure drugs. Conversely, many epilepsy studies in veterinary neurology use subjective data, e.g., owner-based questionnaires or histories. The possibility of underreporting of seizures using only subjective data, similar to the phenomenon known in human medicine, is unknown in veterinary epileptology. **Hypothesis/Objective**—This study examined the correlation between reported seizure frequency and EEG frequency of ictal paroxysmal discharges (PDs) to determine whether the seizure underreporting phenomenon exists in veterinary epileptology. **Animals**—Thirty-three ambulatory EEG recordings in epileptic dogs that showed more than one ictal PDs. Dogs with status epilepticus were excluded. **Methods**—Retrospective observational study. Ictal PDs were manually counted over the length of recording to obtain the frequency of EEG seizures. Reported seizure frequency from owners was categorized into four groups indicating weekly, daily, hourly and minutely seizure. Spearman rank test was used for correlation analysis. **Results**—The coefficient value (rs) comparing reported seizure and ictal PD frequency is 0.368. Other rs comparing history against various seizure types are: 0.296 for all types without myoclonic epilepsy, 0.206 for myoclonic epilepsy, and 0.374 for absence seizures. **Conclusions**—Only a weak correlation exists between frequency of reported seizure from caretakers (subjective data) and ictal PDs on EEG (objective data). Subjective data may not be reliable enough to determine true seizure frequency given the discrepancy with EEG-confirmed seizure frequency. Confirmation by prospective study would be ideal.

Abstract N12: Juvenile-Onset Motor Polyneuropathy in Siberian Cats

Melissa Lewis[1]; Kelly Crawford[2]; Dayna Dreger[1]; Kari Ekenstedt[1]
[1]Purdue University; [2]Med Vet

Polyneuropathies are infrequently described in cats, with a subset exhibiting genetic predisposition as reported in several specific breeds. Our objective was to characterize a novel motor polyneuropathy in a family of related Siberian cats. Clinical data and pedigree information were obtained from the medical records and breeder of eight closely related Siberian cats including four clinically affected and four clinically unaffected individuals. Electrodiagnostic and muscle/nerve biopsy samples were obtained from one affected cat. Follow-up information was obtained for all affected cats including any relapses and treatments administered. Onset of signs ranged from four to ten months in affected cats. Clinical signs were characterized by progressive or waxing/waning neuromuscular weakness (4/4), normal sensory function (4/4), and variably decreased withdrawal reflexes (3/4) and plantigrade stance (2/4). All cats returned to normal neurologic function within several weeks; two cats had a recurrence of weakness within a month of initial signs from which they recovered fully. In one cat, electromyography and motor nerve conduction studies showed multicentric, distally distributed spontaneous activity, normal conduction velocity, reduced amplitude, and polyphasia. Histologic evaluation of muscle and nerve biopsies in that cat showed recent denervation and intraneural edema. All cats were treated with robenacoxib with two subsequently switched to tapering courses of prednisolone. Pedigree analysis of the affected family supports an autosomal recessive, single-gene mode of inheritance, although a genetically complex/polygenic condition cannot be ruled-out. Further genetic investigations are underway. We describe a novel motor polyneuropathy in juvenile Siberian cats characterized by self-limiting weakness with potential relapse.

Abstract N13: Relationship between Admission Variables in Dogs with Brain Herniation: A Retrospective Study in 54 Dogs

Jiwoong Her[1]; Katherine Gerken[2]; Erik Hofmeister[2]; Amy Yanke[2]; Ashley Peters[3]; Jin Yoon[4]; Lenore Bacek[5]; Kendon Kuo[2]
[1]Department of Clinical Sciences, College of Veterinary Medicine, Auburn University; [2]College of Veterinary Medicine, Auburn University; [3]Veterinary Emergency Group; [4]Veterinary Surgical Centers; [5]BluePearl Veterinary Partners

Brain herniation is one of the most frequent life-threatening neurological emergencies. This study aimed to document the admission systolic blood pressure (SBP), heart rate (HR), and Modified Glasgow Coma Scale (MGCS) in dogs with or without brain herniation and to determine the correlation with brain herniation, to provide a specific indicator for brain herniation in the emergency setting. Medical records from one hospital were compiled to identify dogs presenting for neurological signs with a brain MRI. Based on MRI findings, dogs were divided into two groups: dogs with evidence of herniation or with a normal brain MRI (control group). The two groups were compared for HR, SBP, MGCS, SBP-HR, age, and weight. A total of 54 dogs with brain herniation were included. The control population consisted of 40 dogs. The herniation group had significantly higher SBP (p=0.0078), greater SBP-HR difference (p=0.0006), and lower MGCS (p<0.0001) compared to control group. A cut-off value of SBP >178 mm Hg, SBP-HR >60, and MGCS <14 provide a specificity of 90 to 98 percent to identify brain herniation. A combination of SBP >140 and HR <80 provided 24% sensitivity and 100% specificity to diagnose dogs with brain herniation (p<0.0001). High SBP, a greater SBP-HR, a combination of higher SBP and lower HR, and low MGCS upon admission were associated with brain herniation in dogs with neurological signs. Early recognition of these abnormalities may help veterinarians to suspect brain herniation and determine timely treatment.

Table 1. Admission variables between Case (herniated) and Control dogs. Data are presented as mean ± SD [95% CI] or median ± interquartile range (range)

	Case (n = 54)	Control (n = 40)	*P* value
Age, years	7.8 ± 3.5 [6.8-8.7]	7.1 ± 4.2 [5.8-8.4]	0.3915[*]
Weight, kg	17.3 ± 13 [13.7-20.8]	20.7 ± 14.9 [15.9-25.5]	0.2374[*]
Lactate, mmol/L	2.0 ± 1.2 [1.6-2.4]	2.6 ± 2.1 [1.8-3.3]	0.1677[*]
HR, beats/minute	102 ± 32 [93-111]	114 ± 31 [104-124]	0.0625[*]
BP, mm Hg	151 ± 32 [142-159]	133 ± 28 [124-142]	**0.0078**[*]
SBP-HR	49 ± 47 [36-62]	19 ± 34 [8-30]	**0.0006**[‡]
MGCS score	16 ± 2.5 (3.0-18.0)	17 ± 2 (13-18)	**< 0.0001**[¶]
MGCS: motor activity	5 ± 1.0 (1.0-6.0)	5 ± 1 (4-6)	0.3331[¶]
MGCS: brainstem reflexes	5 ± 1.0 (1.0-6.0)	6 ± 0.5 (4-6)	**0.0005**[¶]
MGCS: level of consciousness	5 ± 1.0 (1.0-6.0)	6 ± 1 (4-6)	**< 0.0001**[¶]

Note: [*]Student's *t* test, [‡]Welch-Satterthwaite *t* test, [¶]Mann-Whitney *U* test. $P < 0.05$ is considered significant. Statistically significant *P* values highlighted in bold.

Table 2. Data and Multivariable (SBP+HR), Sensitivity and Specificities

SBP+HR	PPV*	Sensitivity	Specificity	*P* value
HR < 60, SBP > 120	3/3	6%	100%	0.2588[§]
HR < 60, SBP > 140	3/3	6%	100%	0.2588[§]
HR < 60, SBP > 160	1/1	2%	100%	1.0000[§]
HR < 80, SBP > 120	13/13	24%	100%	< 0.0001[ǁ]
HR < 80, SBP > 140	13/13	24%	100%	< 0.0001[ǁ]
HR < 80, SBP > 160	6/6	11%	100%	0.0364[§]
HR < 100, SBP > 120	18/19	33%	98%	0.0002[ǁ]
HR < 100, SBP > 140	17/17	31%	100%	< 0.0001[ǁ]
HR < 100, SBP > 160	7/7	13%	100%	0.0191[§]
HR < 120, SBP > 120	29/37	54%	80%	0.0002[ǁ]
HR < 120, SBP > 140	26/31	48%	88%	0.0002[ǁ]
HR < 120, SBP > 160	12/15	22%	93%	0.0453[ǁ]

Note: *PPV = positive predictive value, [§]Fisher's exact test, [ǁ]Log-likelihood ratio test.

Abstract: N14 Serum and Cerebrospinal Fluid GFAP and pNF-H Concentrations in Dogs with Meningoencephalomyelitis of Unknown Etiology

Lauren Green; Christopher Mariani; Laura Ruterbories; Natasha Olby; Peter Early; Karen Munana
North Carolina State University Veterinary Hospital

Meningoencephalomyelitis of unknown etiology (MUE) is a common cause of neurologic dysfunction in dogs. Despite treatment, approximately 33% of dogs don't survive longer than three days. No consistent factors have been identified that predict prognosis. Glial fibrillary acidic protein (GFAP) and phosphorylated neurofilament heavy chain (pNF-H) are structural proteins found within astrocytes and neurons, respectively, and are released into the cerebrospinal fluid (CSF) and blood when these cells are damaged. We hypothesized that these proteins might be useful biomarkers of prognosis in dogs with MUE. The objectives of this study were to compare CSF and serum concentrations of GFAP and pNF-H from dogs with MUE to samples from clinically normal dogs and to investigate the ability of these biomarkers to predict survival to hospital discharge. Serum and CSF samples from dogs diagnosed with MUE (n=13 and 40) and clinically normal dogs (n=5 and 10) were retrieved from a biorespository. Concentrations of GFAP and pNF-H were determined using commercially available ELISA kits. There were no clinically relevant differences in serum concentrations of either analyte between groups. CSF concentrations of GFAP and pNF-H were increased in MUE dogs compared to control dogs ($p=0.0071$ and $p<0.0001$, respectively). No differences were noted in GFAP ($p=0.183$) or pNF-H ($p=0.773$) when comparing dogs that failed to survive to discharge versus dogs that survived. CSF, GFAP and pNF-H may be useful biomarkers of MUE, although this study did not support a role in predicting survival.

Abstract N15: Comparing Baseline T-cell Activation, by IL-2 Expression, between Dogs with Immune-meditated Diseases and Healthy Dogs

Alison Little; Caitlin Riggs; Michaela Beasley; Santosh Katarukonda; Andy Shores; Andrew Mackin; Robert Wills; Henrique Lupiano; Todd Archer
Mississippi State College of Veterinary Medicine

Cyclosporine is commonly utilized for the treatment of immune-mediated diseases. Although pharmacokinetic monitoring of this drug is widely available, cyclosporine blood concentrations may poorly reflect clinical response in dogs. Human studies have found that pharmacodynamic testing of patients receiving cyclosporine may be preferable. Based on advances in human cyclosporine monitoring, a PCR-based assay measuring RNA expression of IL-2 in activated T cells has been developed and validated in dogs by the Mississippi State Pharmacodynamic Laboratory. We hypothesize that dogs with immune-mediated diseases will have a higher baseline level of IL-2 expression compared to healthy control dogs, due to an overactive immune system in the disease state. The purpose of this study was to compare the degree of baseline (pre-treatment) T-cell activation, as calculated by the difference between unactivated and activated delta Ct IL-2 expression, between healthy dogs and dogs with meningoencephalitis of unknown origin (MUO), inflammatory bowel disease (IBD), immune-mediated hemolytic anemia (IMHA). Three mLs of blood were collected from healthy dogs (9), as well as dogs definitively diagnosed with MUO (8), IBD (9) and IMHA (9) prior to treatment with cyclosporine. Blood was submitted to the Mississippi State Pharmacodynamic Laboratory for baseline RT-qPCR analysis of IL-2 mRNA expression. A linear model determined the least square mean for the difference in unactivated and activated IL-2 dCt was higher in the diseased states (7.122 for IMHA, 7.573 for IBD, 9.419 for MUO) as compared to the healthy dogs (5.279), with a higher number signifying more activation. Differences in least squares means with Tukey adjustment for multiple comparisons determined that dogs with MUO and IBD had significantly higher values than healthy dogs. Dogs with IMHA had higher values than healthy dogs but a significant difference was not detected. No significant differences in values were detected between the three disease states. The baseline degree of activation of T cells was higher in disease states as compared to healthy dogs, likely due to the overactive immune system in these immune-mediated diseases. While it has recently been shown that healthy dogs demonstrate significant suppression of IL-2 expression within 24 hours of oral dosing with cyclosporine, in our experience dogs with immune-mediated disorders often do not respond that quickly. An increased degree of T-cell activation in disease states, as compared to healthy dogs, may be one possible explanation for the additional time it appears to take for a clinically relevant response to occur in patients with immune-mediated disorders.

Abstract N16: Potential Use of Dry Surface Electrodes for Electroencephalography (EEG) in Dogs

Michal Hazenfratz; Shane Bateman; Luis Gaitero; Fiona James; Julia Luca; Thomas Parmentier
University of Guelph

Electroencephalography (EEG) records the cerebral cortical electrical activity via scalp electrodes. A common type of electrode is the subdermal wire electrode (SWE) for which sedation is frequently required for placement due to its mild invasiveness. Surface electrodes such as spring-loaded (SLE) and PressOn (POE) electrodes are less invasive but have never been evaluated in dogs. We hypothesized that the dry surface electrodes would remain functional and perform equivalently to the SWEs. EEG was recorded on 6 awake dogs over 6 hours. Surface electrodes were randomized to one of 8 locations above an SWE, creating 4 POE-SWE and 4 SLW-SWE comparison pairs. Epochs were selected throughout the recordings and loss of electrode signal, electrode impedance, as well as concordance with spectral analysis, were compared between electrode types. After the first hour, SLE vs. SWE retention was 15/24 vs. 24/24 (p-value=0.003) and POE to SWE was 21/24 to 24/24 (p-value=0.233). After 6 hours, SLE vs. SWE retention was 12/24 vs. 23/24 (p-value=0.001162) and POE to SWE was 15/24 vs. 20/24 (p-value=0.1939). Impedance was significantly higher for both SLE and POE compared with SWE (p-value<0.001). Preliminary concordance for spectral analysis showed insufficient agreement between the two types of surface electrodes and the SWEs. SWE consistently outperformed the novel dry surface electrodes, but the POE may have potential for clinical use.

Abstract N17: Evaluation of Three-Dimensional Printing and Virtual Rendering as Teaching Tools for Cerebrospinal Fluid Collection

Megan Lin[1]; Leontine Benedicenti[2]; Amy Thibault[2]; Kenneth Drobatz[2]
[1]Red Bank Veterinary Hospital; [2]School of Veterinary Medicine, University of Pennsylvania

The use of three-dimensional (3D) models and virtual rendering are becoming more common in veterinary medical education. In this prospective study, a novel 3D-printed model of a canine skull with its C1 and C2 vertebrae, and a 3D virtual rendered video were created to teach students about the anatomy and procedure of a cerebrospinal fluid (CSF) collection. The goals of this study were to: 1) compare the effects of three teaching methods (text reading, video of clinical demonstration, 3D virtual rendered video) on successful outcomes of students performing a CSF collection on a physical 3D-printed model, and 2) evaluate students' learning satisfaction from using the 3D-printed model. This study was conducted on 56 fourth-year veterinary students enrolled in the Penn Vet Neurology service clinical rotation from 2019–2020. Students were randomly divided into three groups: 1) reading a text description with a 2D image (19 students), 2) watching a clinical video of a neurologist performing CSF collection on a dog with voice-over of the text description (19 students), 3) watching a 3D virtual rendered video with voice-over of the text description (18 students). After studying their teaching method for 5 minutes, students were given 3 attempts to perform CSF collection on the 3D-printed model. Students were evaluated on palpation of correct anatomical landmarks on the model and a CSF collection outcome was recorded as successful if fluid was obtained from the model. The Fisher's exact test with pairwise comparisons was used to determine significant differences in outcomes between the three groups. When comparing students' success rates of correct anatomical landmark palpation on the 3D-printed model, there was no significant difference between the 3D virtual rendered video and text groups ($p=0.180$), or between the text and clinical video groups ($p=1.000$). However, there was significant difference between the clinical video and 3D virtual rendered video groups ($p<0.046$), demonstrating that the clinical video method was more effective for students to achieve correct anatomical landmark palpation. When comparing students' success in CSF collection outcomes on the 3D-printed model, there was no significant difference between all three groups in success of obtaining fluid on the first attempt, second attempt, third attempt, or overall success of all three attempts combined ($p>0.05$ for all comparisons). Following the study, students ranked their level of satisfaction regarding if using the 3D-printed model was helpful for learning about the CSF collection procedure (scale from 1–5, 1 being not helpful, 5 being very helpful). Results were collected from 54 of the 56 total students (96%), with a mean rating of 4.037. Students in the 3D virtual rendered video group were also asked to rank if having 3D virtual rendering shown in lectures would be helpful (same scale as previous). Results were collected from 17 of the 18 students (94%), with a mean rating of 4.58. These results suggest that while students have high learning satisfaction with the use of 3D models and virtual rendering, further research is required to evaluate the true benefits and correlation of 3D tools on students' learning outcomes when compared to traditional teaching methods.

Abstract N18: Aging Dogs with Spontaneous Brain Microhemorrhages Have Diminished Interthalamic Adhesion Size: A Comparative MRI Study

Curtis W. Dewey[1]; Emma Davies[1]; Philippa Johnson[1]; Marissa O'Donnell[2]; Simon Platt[3]; Mark Rishniw[1]; Kelsey Robinson[3]; Joseph Sackman[2]
[1]College of Veterinary Medicine, Cornell University; [2]Long Island Veterinary Specialists; [3]College of Veterinary Medicine, University of Georgia

Spontaneous brain microhemorrhages (SBM) in elderly people occur with Alzheimer's disease but also occur in the absence of dementia. A common cause of SBMs in people is cerebral amyloid angiopathy (CAA), which is associated with brain atrophy. CAA patients often present for transient neurologic dysfunction. Brain microhemorrhages have been described in older dogs; it is unclear if these are associated with brain atrophy. Small interthalamic adhesion (IA) size on MRI is a reliable indicator of brain atrophy in canine cognitive dysfunction (CCD). We hypothesized that aging dogs with SBM (based on T2*) presenting for neurologic dysfunction but without cognitive decline would have small IA sizes compared with successfully aging dogs. The objective of this study was to compare IA size between 3 groups of aging (>9 yrs) dogs: 1) neurologically impaired dogs with SBM but without cognitive dysfunction (17) 2) dogs with CCD (16) and 3) successfully aging dogs (controls-26). Common clinical complaints for SBM dogs was recent onset of central vestibular dysfunction, seizures, or both. MR images from 59 dogs were reviewed. IA (height and mid-sagittal area) and total brain volume (TBV) measurements were acquired. IA measurements, normalized to TBV were compared between groups. Controls had significantly larger IA measurements and normalized IA area than SBM and CCD dogs, but these did not differ between SBM and CCD. CCD patients had fewer microhemorrhages than SBM dogs ($p<0.05$). SBMs in aging dogs is associated with brain atrophy like CCD, but may represent a distinct disease category.

Abstract N19: Canine Cognitive Dysfunction Patients Have Reduced Hippocampal Volumes Compared with Aging Controls: An MRI Study

Curtis W. Dewey[1]; Marissa O'Donnell[2]; Simon Platt[3]; Mark Rishniw[1]; Kelsey Robinson[3]; Joseph Sackman[2]
[1]College of Veterinary Medicine, Cornell University; [2]Long Island Veterinary Specialists; [3]College of Veterinary Medicine, University of Georgia

Hippocampal atrophy is a key pathologic and MRI feature of human Alzheimer's disease (AD). Hippocampal atrophy has not been documented via MRI in canine cognitive dysfunction (CCD), which is considered the dog model of human AD. The purpose of this retrospective comparative volumetric MRI study was to compare total hippocampal volumes between successfully aging (control) dogs and dogs diagnosed with CCD. Mimics® software was used to derive total hippocampal volumes and total brain volumes from the MRI studies of 42 aging dogs (>9 yrs): 16 dogs diagnosed with CCD and 26 successfully aging controls. Total hippocampal volume normalized to total brain volume was significantly less for CCD patients compared with control dogs ($p<0.05$). The results of this study suggest that-similar to human AD-hippocampal atrophy is a pathological feature of CCD. This finding has potential importance for both investigating disease mechanisms related to dementia as well as future hippocampal-targeted therapies.

Abstract N20: Molecular Insights into Protein Aggregation of Mutant Superoxide Dismutase 1 in Degenerative Myelopathy

Shintaro Kimura[1]; Ryo Honda[1]; Yuji O. Kamatari[1]; Zenichiro Kato[1]; Hideaki Hara[2]; Sadatoshi Maeda[1]; Hiroaki Kamishina[1]
[1]Gifu University; [2]Gifu Pharmaceutical University

In this study, we focused on a canine neurodegenerative disease (degenerative myelopathy: DM) associated with mutant superoxide dismutase1 (SOD1). Several studies have reported that both T18S and E40K mutations in SOD1 promoted the formation of insoluble aggregates in neurons and glial cells. However, the molecular mechanism of the protein aggregation remains unknown. The objective of this study was to clarify the aggregation mechanism of mutant SOD1 proteins (T18S, E40K). Wild-type (WT) and mutant (T18S, E40K) recombinant canine SOD1 proteins were expressed in *Escherichia coli* and inactive (apo) and active (holo) types were purified. The aggregate formation was measured over time using Thioflavin-T (Th-T) assay. Transmission electron microscopy (TEM) was performed using JEM-2100F. Further, the denaturation midpoint (Tm) was calculated from the denaturation curve obtained by circular dichroism spectrum measurement under heating. Th-T fluorescence intensity was significantly increased in apo-T18S and apo-E40K SOD1 compared to WT and corresponding holo-SOD1, respectively. Curvy fibrous aggregates were observed both in apo-T18S and apo-E40K SOD1 by negatively stained TEM. There were no aggregates in WT or holo-SOD1. The Tm values of apo-SOD1 were $\approx 30°C$ lower than those of holo-SOD1, whereas no significant difference was found between WT and E40K. The T18S mutation significantly reduced the Tm value by $\approx 5°C$ both in the apo- and holo-SOD1. The apo-SOD1, but not holo-SOD1, had low Tm values and a high propensity to aggregate, indicating that the unstable apo-form might be the initial intermediate in SOD1 aggregation. The T18S mutation reduced the structural stability of SOD1. However, the E40K mutation did not alter the thermal stability of SOD1. The E40K mutation decreased the negative net charge which enhances intermolecular electrostatic forces between unfolded polypeptides compared to WT and T18S. In conclusion, we clarified that the T18S and E40K mutation promoted SOD1 aggregation through distinct mechanisms, involving destabilization of the native structure or reduction of the repulsive negative charge, respectively.

Abstract N21: Prion-Like Propagation of Mutant Superoxide Dismutase-1 in Canine Degenerative Myelopathy

Nana Tanaka; Hiroaki Kamishina; Sadatoshi Maeda; Masatoshi Inden; Yuji Kamatari; Shintaro Kimura
Gifu University

Canine degenerative myelopathy (DM) is a chronic, progressive and fatal neurodegenerative disease. DM-affected dogs have homozygous mutations in the superoxide dismutase 1 (SOD1) gene. The accumulation of misfolded protein aggregates in motor neurons is implicated as an important pathological process of DM. However, the mechanism of SOD1 protein aggregation and accumulation is largely unknown. In recent studies, it has been reported that cell-to-cell transmission of misfolded protein aggregates in SOD1-mediated amyotrophic lateral sclerosis (ALS) in humans. In the present study, we investigated the propagation of SOD1 protein aggregates in DM using canine SOD1 expressing cells. Wild-type (WT) and mutant (E40K) canine SOD1 genes tagged with GFP or DsRed were co-transfected into mouse neuroblastoma cells. The intracellular SOD1 aggregate formation rate was calculated under a fluorescence microscope. The surfactant soluble and insoluble fractions were extracted from transfected cells, and the relative amount of protein in each fraction was quantified by Western blotting. In addition, cells transfected with WT and E40K plasmids labeled with different fluorophores were co-cultured and the localization and morphology of each protein were evaluated. Further, E40K-SOD1 protein aggregates prepared using dithiothreitol was added to the culture medium of WT transfected cells. The aggregate formation rate in the cells was calculated as above. The percentage of cells containing aggregates was markedly higher in WT which were co-transfected with E40K than that in WT co-transfected with WT. The relative amount of WT insoluble fraction protein was increased by co-transfection with E40K. In co-culture of WT and E40K transfected cells, WT and E40K proteins were co-localized within the same cells and formed aggregates. Further, the aggregate formation rate of WT was increased by addition of E40K aggregates. Canine mutant SOD1 protein E40K not only undergoes aggregate formation but also induces aggregation of native SOD1 protein. Additionally, it was suggested that E40K aggregates could be taken up into cells and induce aggregate formation of native-fold proteins.

Abstract N22: 3D-Printed Drill Guides for Canine Thoracic Spinal Surgery

Christopher L. Mariani[1]; Ola Harrysson[2]; Josh Zlotnick[2]
[1]College of Veterinary Medicine, North Carolina State University; [2]North Carolina State University

Various conditions require surgical stabilization of the canine thoracic vertebral column. However, placement of implants into these vertebrae, particularly those cranial to T12, is challenging due to narrow implantation corridors and proximity to critical structures including the spinal cord, aorta and thoracic cavity. We hypothesized that 3D-printed guides could successfully constrain drill trajectories in these vertebrae to safe, predetermined implantation corridors. Five canine cadavers were obtained and the thoracic vertebral columns were disarticulated, leaving the paraspinal musculature intact. The vertebral segments were imaged with computed tomography and the resulting studies were imported into Mimics (Materialise). The thoracic vertebrae from T8-13 were individually segmented and 3D objects were created and exported into 3Matic (Materialise). Drill guides were printed on a Projet MJP 2500 Plus (3D Systems) using Visijet M2 Rigid White material. The cadaveric vertebral segments were positioned to approximate a dog in ventral recumbency and a simulated surgical approach was made. The guides were placed and drill tracts were created using a surgical drill with a 2.7mm drill bit. The vertebral columns were then re-imaged and imported into Mimics. Accuracy of the intervention was assessed by measuring deviation from the intended entry point as a linear measurement and angular deviation in 3 planes. A total of 58 drill tracts were created in 30 vertebrae. The overall mean entry point deviation was 1.4 mm (range 0.4–3.4mm) and the overall mean angular deviation was 5.1° (range 1.5–10.8°). This additive manufacturing technique results in drill tracts with accuracy acceptable for clinical use.

Abstract N23: Prevalence of Radiculopathies Associated with Type 1 Intervertebral Disc Disease on MRI in Dogs

Solene Diop[1]; Isabelle Masseau[2]; Joane Parent[2]
[1]Highcroft Veterinary Referrals; [2]Faculté de Médecine Vétérinaire, Université de Montréal, Saint-Hyacinthe, QC, Canada

Hansen type I intervertebral disc disease (IVDD) is common in dogs, affecting predominantly chondrodystrophic breeds. Extruded nucleus pulposus in the vertebral canal or intervertebral foramen can induce the onset of radiculopathies by direct compression of the nerve roots. Interestingly, radiculopathies in humans have been reported even in the absence of a compressive component, hypothetically from local inflammation induced by the presence of nearby extruded disc material. The aims of this study were (1) to determine the prevalence of radiculopathies, with or without nerve root compression, using magnetic resonance imaging (MRI) in dogs with type I IVDD, and (2) to investigate a possible association between nerve root anomaly detected on MRI and signalment or clinical data. Medical record database was searched for any dog diagnosed with type 1 IVDD on MRI between January 2009 and March 2019. Studies were included upon availability of T1 weighted post-gadolinium images. Each imaging study was reviewed for evidence of radiculopathy based on the identification of one or more of the following imaging features: (1) visible compression of the nerve root by extruded material, (2) thickening of the nerve root (in comparison to the contralateral nerve root) or (3) nerve root enhancement on post contrast T1 weighted sequences. Dogs without MRI features of radiculopathy were assigned to the control group. Thirty-one dogs met the inclusion criteria. Nerve root lesions were identified in 17 dogs (radiculopathy group). Fourteen dogs were included in the control group. Twelve dogs with radiculopathies had a compressive component. For a majority of dogs in the radiculopathy group, the extrusion was lumbar or lumbo-sacral (L4-L6: 6/17, L7-S1:3/17); the extrusion was thoracolumbar (T3-L3) for only 3 cases, cervical (C1-C5) for 2 cases and centered on the cervical intumescence (C6-T2) for 3 cases. Of the 5 dogs without compressive component, radiculopathies were diagnosed based on unilateral thickening of the nerve root (n=2), nerve root enhancement (n=1), and for 2 cases, both nerve root enhancement and thickening (n=2). There was no association between the presence of radiculopathy and the degree of spinal cord compression ($p=0.18$) or with any of the demographic data). Dogs with radiculopathies were presented later after the onset of clinical signs than dogs without nerve root lesions ($p=0.03$). Radiculopathies with or without nerve root compression can be identified on MRI in dogs with type I IVDD. Their diagnosis relies heavily on acquisition of post contrast images. Standardization of MRI protocols with sequences suited for detection of radiculopathy may contribute to a better recognition and thereafter, a more thorough understanding of their potential contribution to the clinical signs, for example pain.

Abstract N24: Impact of von Willebrand Factor on Dogs with Hansen Type 1 Disc Extrusions

Danny Sack; Sarah Stephan; Curtis Dewey; Emma Davies
Cornell University

This study describes the clinical impacts of von Willebrand factor (VWF) deficiency and outcome of two dogs with Hansen Type 1 disc extrusion. Two otherwise healthy Doberman Pinschers with disc extrusions had extensive bleeding within the vertebral canal secondary to previously undiagnosed VWF deficiency. Hematology, biochemistry, VWF and MRI were performed in both dogs. One dog underwent surgical decompression and both outcomes are described. Both dogs had progressive T3-S3 myelopathies (one non-ambulatory paraparetic, one paraplegic with no pain sensation). T2* sequences identified hemorrhage extending the length of 8 to 11 vertebral bodies in both dogs, leading to extensive compression of the spinal cord. Both were VWF deficient (AG of 25 and 40), resulting in a more extensive and compressive spinal cord lesion than typical even in hemorrhagic disc extrusions. One dog received surgical decompression with preoperative cryoprecipitate and desmopressin acetate and improved post operatively recovering normal ambulation after 6 weeks. The deep pain negative dog was euthanized. VWF deficiency resulted in greater extent and degree of compression with rapid ongoing progression. Prompt surgical intervention and a larger surgical site than typical in other hemorrhagic disc extrusions appear critical factors in outcome. Surgical decompression may not be feasible in some dogs due to extension of lesion and concerns for stability. Neither dog had a previous VWF test, with one dog previously spayed. After hour testing for VWF is not available and mucosal bleeding time is not specific. Standardised testing of young dogs from predisposed breeds is recommended prior to emergency situations.

Abstract N25: Development of a Non-invasive Diagnostic Technique to Assess Neuromuscular Disease

Rell L. Parker[1]; Chelsea Crowe[1]; Marguerite Knipe[1]; Adrian Harrison[2]; Peter Dickinson[1]
[1]University of California–Davis; [2]University of Copenhagen

Needle recorded and stimulated compound muscle action potentials acquired under general anesthesia are standard practice for the assessment of motor nerve conduction and neuromuscular transmission in dogs. Requirement for general anesthesia is limiting for initial diagnosis and serial assessment of disease progression and therapeutic monitoring. We hypothesized that acoustic myography (AMG), with surface recording of muscle pressure waves following surface stimulation of a motor nerve, is feasible for the assessment of motor nerve latency values and repetitive nerve stimulation in anesthetized and non-anesthetized dogs. Standard needle electrode electrophysiological assessment of motor nerve conduction velocity and repetitive nerve stimulation was done in 4 anesthetized laboratory-bred beagle dogs (aged 13–16 months) with stimulation of the fibular nerve at the level of the hock, stifle and hip and recording over the fibularis brevis or tibialis cranialis muscles. AMG recordings using a piezoelectric AMG device (CURO-Diagnostics ApS, Bagsværd, Denmark) were then acquired using needle electrode or skin surface stimulation at the same locations. Latency and repetitive stimulation recordings were generated in 8 pelvic limbs from 4 beagles. Stimulation amplitude was between 2.5–25 mA for surface stimulation, compared to 0.8–5.3 mA for needle stimulation. Stimulation was well tolerated in non-anesthetized subjects. Repetitive stimulation was performed at frequencies from 1–50 Hz. Correlation of standard needle electrode derived latencies and repetitive stimulation potentials was influenced by temporal sampling limitations of the CURO unit (2 kHz). AMG recording from the fibularis brevis muscle was limited by muscle volume and vascular pulse artifacts. Surface stimulated AMG recordings in non-anesthetized and anesthetized animals is feasible to measure latency and repetitive nerve stimulation. Modification of the recording device to increase the sampling rate (10 kHz) and establishment of reference range data will be necessary for clinical application. Pilot data from clinical patients has demonstrated the potential utility of AMG in defining neuromuscular junction fatigue in non-anesthetized dogs with myasthenia gravis.

Abstract N26: Correlation between Canine Cognitive Dysfunction Clinical Metrology Instruments, Cognitive Testing and Plasma Neurofilament Light Concentrations

Wojciech K. Panek[1]; Margaret Gruen[1]; Freya Mowat[2]; David Murdoch[3]; Natasha Olby[1]
[1]College of Veterinary Medicine, North Carolina State University; [2]University of Wisconsin-Madison; [3]Duke University Medical Center

Aging dogs suffer from Cognitive Dysfunction Syndrome (CDS), a condition in which cognitive decline is associated with amyloid pathology and cortical atrophy. Diagnosis is made through physical examination, elimination of systemic/metabolic conditions and completion of clinical metrology instruments (CMI) by owners. The purpose of this study was to evaluate the correlation between two frequently used CMIs, testing of different cognitive domains and plasma neurofilament light chain concentration (pNfL) in aging dogs. Senior and geriatric dogs were recruited. Owners completed the Canine Dementia Scale (CADES), and Canine Cognitive Dysfunction Rating scale (CCDR); dogs underwent physical and neurological examinations, routine blood work and urinalysis. Testing of executive control, working memory, social cues, and sustained attention was performed and pNfL was measured. Comparisons between CADES, CCDR scores, cognitive testing scores, and pNfL were made using logistic regression. Fourteen dogs from eight breeds were recruited (9.3 to 15.6 years). CADES scoring classified four dogs as severe CDS, two as moderate, two as mild and six as normal. CCDR identified four dogs at risk of CDS and ten as normal. A significant positive relationship was found between CADES score and pNfL ($p=0.015$). CADES score was associated with poor performance on sustained attention ($p=0.004$) and executive control ($p=0.0005$). The CCDR scale did not correlate with cognitive testing or pNfL. Our findings suggest that a multi-dimensional approach utilizing a combination of CADES, cognitive testing (sustained attention, executive control) and pNfL concentration can differentiate dogs affected with age-related cognitive dysfunction from cognitively healthy aged dogs.

Integrative Management of the Neurologic Patient: Surgery, Rehabilitation, and Alternative Medicine

Theresa E. Pancotto, MS, DVM, DACVIM (Neurology), CCRP
Blacksburg, VA, USA

INTRODUCTION

Neurologic patients can be some of the most challenging, yet most rewarding, to work with from the perspective of a rehabilitation practitioner. While many (>90%) of the average "back" or acute IVDD (Intervertebral disc disease) dogs with intact pain perception will improve significantly with surgical intervention alone regardless of the postoperative care, nonsurgical cases may require more therapeutic assistance. Additionally, many other neurological conditions, both surgical and not, may be dramatically helped by physical medicine.

The bulk of physical rehabilitation consists of active therapeutic exercises and passive manual therapies (massage, joint mobilization), but also includes application of various modalities (electrical stimulation, photobiomodulation), as well as incorporation of alternative treatment strategies (acupuncture, electromagnetic therapy) for some patients.

There is a lack of published literature supporting specific protocols after neurologic injury; however, the field of research in this area is growing. One of the difficulties studying rehabilitation is that the nature of the subject is highly tailored to the individual. Blanket protocols may fail due to variabilities in patient motivation, demeanor or ability. We also lack a sophisticated understanding of the nature of injury and recovery of some severe types of neurologic conditions such as acute, deep pain negative dogs.

The best treatment plan for any patient will come from taking a multimodal approach, although not necessarily all at once. Pain must be addressed first and foremost. If conventional analgesic therapy fails, alternative options should be implemented. Therapeutic exercises may be selected based on trial and error and modalities can be added based on the underlying condition and comorbidities. The effective practitioner is patient and creative; what works for one dog, may not work for another. Dogs, like people, can get bored; so you will not want to use the same exercises and/or the same modalities each visit.

SURGERY

Though surgical intervention is expensive and invasive, there are some conditions that are best treated surgically such as spinal fracture/luxations, including atlantoaxial luxations. Although conservative therapy is technically an option, the results are often not as good or as rapid as with surgical stabilization. Because these injuries are unstable, active rehabilitation must be delayed until proper healing has occurred to avoid worsening injury. This could be weeks to months depending on the age and health of the animal.

Although as a neurologist, I consider many other conditions for surgical candidacy, this is not always the best for the patient or client. For surgical spinal conditions, the pros and cons need to be discussed and considered for each individual family.

The impact surgery has on spinal stability should be considered prior to initiating a rehabilitation plan. While most surgical procedures are not inherently destabilizing, they can leave areas more susceptible to injury from minor trauma. High impact activities (climbing, running, jumping) should be avoided in the acute postoperative period, usually for a minimum of 4–6 weeks.

There are data to show that IVDD patients, who stay in the hospital with or without rehabilitation during the immediate postoperative period, have a better outcome than those who were treated at home. The implication, though speculative, that the nature of the caretaker and the rest provided at a veterinary surgical center offers some advantage to home care that may or may not be performed as directed. Similar results were found in dogs undergoing routine TPLO, where results were better for dogs who underwent postoperative rehabilitation with a certified therapist versus at home therapy only.

Pharmaceutical Therapy

An uncomfortable patient cannot willingly participate in therapeutic exercise and there is no place for "tough love" when it comes to managing pain. Inciting further discomfort through manipulation or exercise will only lead to deterioration of the human-animal bond and create a scenario that leads to an unhappy and potentially fractious patient.

If conventional pharmaceutical therapy fails, either due to ineffectiveness or intolerance of side effects, non-pharmaceuticals should be considered (see below). Photobiomodulation, acupuncture, electrical stimulation, and pulsed electromagnetic field therapy all have analgesic properties and may be used independently or in conjunction. Different types of physical activity may be more accepted based on the analgesic advantage of the treatment. For example, underwater treadmill walking may offer some pain relief by reducing the mass supported by the patient. It also offers heat, compression via hydrostatic pressure, and massage if jets are used. These physical methods of pain reduction may allow more activity to be performed in the water than what might be possible on land.

There are numerous tools available to help objectively assess pain in veterinary patients: Colorado Pain Score, Glasgow Pain Score, Canine Brief Pain Index, and Feline Musculoskeletal Pain Index, to name a few.

Physical Medicine/Rehabilitation

Therapeutic Exercise

This is the most important aspect of physical rehabilitation but it is also the most difficult! While the list of prescribed and ideal exercises for a particular down dog may be long, getting said dog **do** those things is another story. Initially figuring out what activity creates the movement you want can take time as well as lots of treats. The first few sessions need to be very rewarding, otherwise future sessions will not likely be productive. It's also important to remember not only how to teach dogs to get up, but to teach them how to get down. We have a tendency to focus on the 'up and walking' aspect but restoration of complete mobility and full independence requires an animal to be able to flex their limbs while under a load.

Manual Therapy

Passive manual therapies, such as massage and passive range of motion (PROM), are particularly helpful in recumbent dogs. These patients are susceptible to atrophy, decubitus, joint stiffness, and decreased lymphatic flow. PROM helps maintain joint health by increasing production and distributing joint fluid throughout the joint. Massage can relieve tension, particularly in areas of compensation, and improve lymphatic return. Massage techniques can be augmented with electrical stimulation for further pain relief. Various massage tools are available to improve depth of tissue work or increase the working surface area. I like 'flossing' to help reduce edema. A flossing band is a tight therapeutic band that can be wrapped from distal to proximal along an edematous limb. It can be left in place for 3–5 minutes and/or while gentle PROM is performed.

Joint mobilizations are often not indicated for neurologic injury; however, concurrent conditions may dictate their implementation. Tui-na is a branch of Chinese medicine that involves manual manipulation of tissues to improve function and harmony of the body.

Assistive Devices

Many assistive devices are used to restore patient mobility. However, they can also be used to support or alter a patient's movement patterns.

Modalities

Electrical Stimulation/TENS/NMES

Electrical stimulation includes electrical stimulation for pain control as well as for muscle contraction. The latter, which causes visible muscle contraction, can slow disuse atrophy but causes a fair amount of discomfort and is not always well tolerated by patients. TENS units are portable units for delivering electrical stimulation for pain control.

Photobiomodulation/Laser Therapy

Laser therapy has many potential benefits but it is a complementary therapy and is not a substitute for appropriate therapeutic exercise and manual intervention, when indicated.

Therapeutic Ultrasound

Therapeutic ultrasound provides a way to focally heat deeper tissue layers to improve elasticity. With respect to neurologic recovery, this can be helpful in addressing tightness in areas of compensation such as the shoulders and iliopsoas.

Extracorporeal Shockwave Therapy

Much like therapeutic ultrasound, ESWT is most useful for managing areas of over-compensation or comorbidities (tendinitis, fasciitis). Older units are very loud and sedation is usually needed.

Complementary/Alternative Medicine

Pulsed Electromagnetic Field Therapy

Two recent studies demonstrated improved pain control in the postoperative period using pulsed electromagnetic field therapy in dogs undergoing hemilaminectomy for IVDD. There were no significant differences in long-term neurologic recovery, however. PEMF is currently being studied for treatment of osteoarthritic pain in cats as a potential alternative to NSAIDs.

Acupuncture

An ancient art, acupuncture is over 2000 years old. A Cochrane review concluded that there is evidence that acupuncture can be helpful in managing pain. Unfortunately, much of the veterinary literature evaluating acupuncture is anecdotal.

Chinese Medicine

There may be other benefits to a full Chinese medical approach that includes food therapy and herbal medications. These are beyond my scope and experience.

Cats

The neurologic cat is a special creature who will often defy statistical odds, thus making her very difficult to discuss. Fear-free management strategies are a must when trying to rehabilitate cats in hospital. I have also found that cats tend to do better with less frequent visits, so we aim for an every-other-week schedule.

References

1. Zidan N, *et al*. The effect of electromagnetic fields on post-operative pain and locomotor recovery in dogs with acute, severe, thoracolumbar intervertebral disc extrusion: a randomized placebo-controlled, prospective clinical trial. *J Neuro Trauma*. 2018;35(15):1726–36.
2. Alvarez LX, *et al*. Effect of targeted pulsed electromagnetic field therapy on canine postoperative hemilaminectomy: a double-blind, randomized, placebo-controlled clinical trial. *J Am Anim Hosp Assoc*. 2019;55(2):83–91.
3. Monk ML, *et al*. Effects of early intensive postoperative physiotherapy on limb function after tibial plateau leveling osteotomy in dogs with deficiency of the cranial cruciate ligament. *Am J Vet Res*. 2006;67:529–536.
4. Zidan N, *et al*. A randomized, blinded, prospective clinical trial of postoperative rehabilitation in dogs after surgical decompression of acute thoracolumbar intervertebral disc herniation. *J Vet Intern Med*. 2018;32(3):1133–1144.
5. Hodgson MM, *et al*. Influence of in-house rehabilitation on the postoperative outcome of dogs with intervertebral disc herniation. *Vet Surg*. 2017;46:566–573.
6. Kathmann I, *et al*. Daily controlled physiotherapy increases survival time in dogs with suspected degenerative myelopathy. *J Vet Intern Med*. 2006;20(4):927–932.

A Novel Ultrasound Guided Cerebral Biopsy Device

Andy Shores[1]; Michaela Beasley[1]; Alison Lee[2]; Talisha Moore[1]; Ryan Taylor[1]; Alsion Little[1]
[1]College of Veterinary Medicine, Mississippi State University; [2]Mississippi State University

The objective of this study was the development and validation of a simple, low cost, disposable cerebral biopsy device visible with intraoperative ultrasound (IOUS). The designed biopsy device was first tested on non-animal tissue, then on a cadaver. Clinical trials were conducted on eight patients with intracranial lesions to obtain biopsy specimens for microscopic examination. The biopsy obtained from a non-animal specimen was used to evaluate size of the sample. Cadaver samples were evaluated by microscopic examination. In the final phase of the project, samples were obtained and examined from eight clinical patients after a standard craniectomy/craniotomy using IOUS guidance. Biopsy samples were successfully obtained from the cadaver and the clinical patients using the device. Microscopic examination of the eight clinical samples by pathologists resulted in a histopathologic diagnosis representative of the intracranial lesions seen on advanced imaging and intraoperatively. Quality of the biopsy was measured by visual and ultrasound imaging evidence and the ability of pathologists to evaluate the sample and render a diagnosis based on the samples. Microscopic evaluation of formalin-fixed samples revealed representative tissues and were of diagnostic quality in all patients. The novel biopsy device provided high-quality, diagnostic samples with no observed negative effect on the patients either clinically or as determined by IOUS of the site post-biopsy. The device is disposable and offers a good alternative to other cerebral biopsy techniques. The novel cerebral biopsy device coupled with intraoperative ultrasound offers a simple, reliable method of obtaining quality biopsy in patients with intracranial masses.

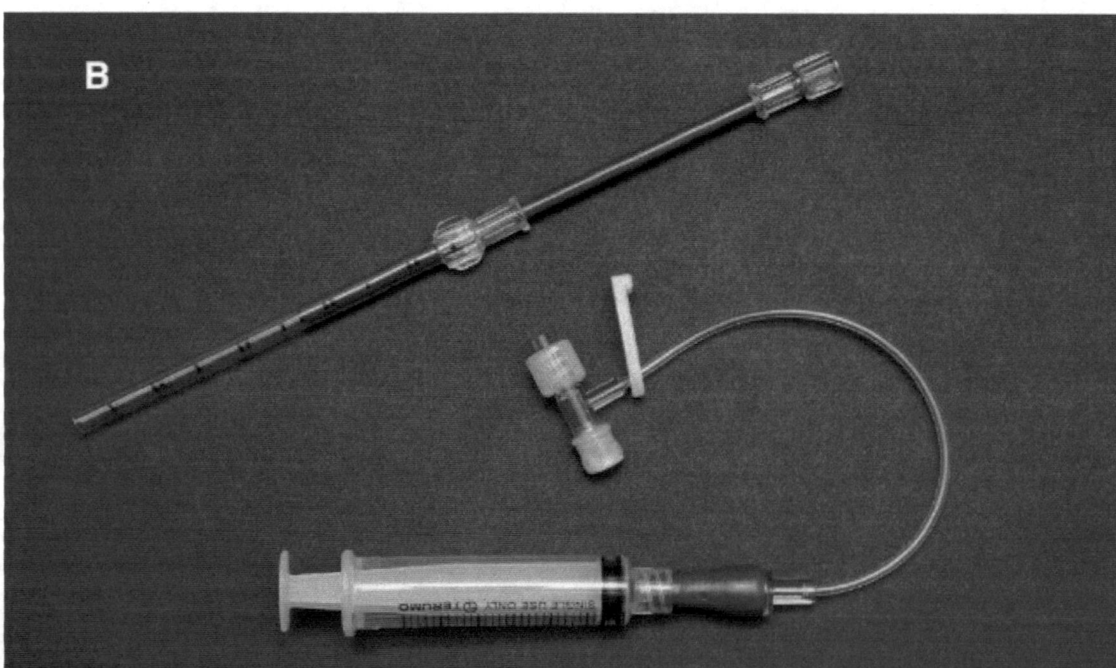

Activity and Sleep Quality in Idiopathic Epileptic Dogs Compared to Age- and Breed-Matched Controls

Starr Cameron[1]; Megan Barry[1]; Heidi Barnes-Heller[2]; Kylie Grady[1]
[1]University of Wisconsin-Madison; [2]Barnes Veterinary Services

The purpose of this study was to compare activity levels and sleep quality in dogs with idiopathic epilepsy to age and breed matched controls. The study population included 67 dogs diagnosed with idiopathic epilepsy between one and nine years old that were receiving at least one anti-epileptic drug (AED) and a seizure frequency of at least one seizure every three months. The control population consisted of age and breed matched control dogs with no diagnosed medical conditions that were not receiving any medications. Activity was measured using a canine-specific activity monitor (FitBarkTM), which was attached to the dogs' collars at all times. A sleep score was also calculated (as a percentage) by measuring the amount of activity (restlessness) during a specified four-hour window of sleeping time (decided by the owner). Dogs with idiopathic epilepsy receiving AEDs had, on average, a 20% decreased in activity compared to the control population, which was statistically significant (p value=1.24e-11). When breed was considered, epileptic Border Collies and German Shepherd Dogs on AEDs had significantly lower activity levels compared to controls (p values of 0.007 and 0.011, respectively). Sleep scores were also significantly different between the epileptic and control groups, with the epileptic dogs having better sleep quality (less restlessness) compared to the control population (p value=0.033). Breed was not statistically significant for sleep score. In conclusion, epileptic dogs receiving AEDs are statistically less active and have better sleep quality compared to age and breed matched controls.

Architectonic-Based Cortical Brain Atlas for the Canine

Philippa Johnson[1]; Kathlene Graham[2]; John Loftus[1]; Wen-Ming Luh[3]; Benjamin Rivard[1]; Andrew White[2]; Erica Barry[1]
[1]College of Veterinary Medicine, Cornell University; [2]University of Sydney; [3]National Institute of Aging

Rapid advancement of advanced neuroimaging techniques, data processing and analysis in neuroimaging has allowed researchers to investigate the brain in novel ways. Stereotaxic cortical brain atlases divide the cortex into architectonically distinct regions which are thought to correlate to function. They have become a vital tool for tissue segmentation and cortical localization and allow for a standardization of processing and analysis in functional, diffusion and structural MRI techniques. Although architectonically based stereotaxic cortical brain atlases have been created for humans, rodents, primates and felines, to date, none have been made available for the canine. In this study we created a canine brain atlas from high resolution T1-weighted MRI data generated *in vivo* from 30 neurologically healthy, mixed breed, mesaticephalic dogs. This atlas includes a population template and a cortical atlas. The effect of normalization of brain data from canine populations with differing skull shapes to the population template was tested. The cortical atlas was created by manual segmentation of the cerebral cortex to delineate regions with distinct myeloarchitecture. When dogs with differing skull shapes where normalized to the final population template, brachycephalics demonstrated the highest warping of neural structures. The myeloarchitectonic based cortical parcellation resulted in generation of 108 cortical regions per hemisphere grouped into frontal, sensory-motor, perisylvian, parietal, cingulate and occipital lobar areas. This study introduces a foundational tool to the field of veterinary neuroimaging, opening doors for novel advanced MRI techniques to be used to improve our understanding of neuroanatomy and neuropathology in the dog.

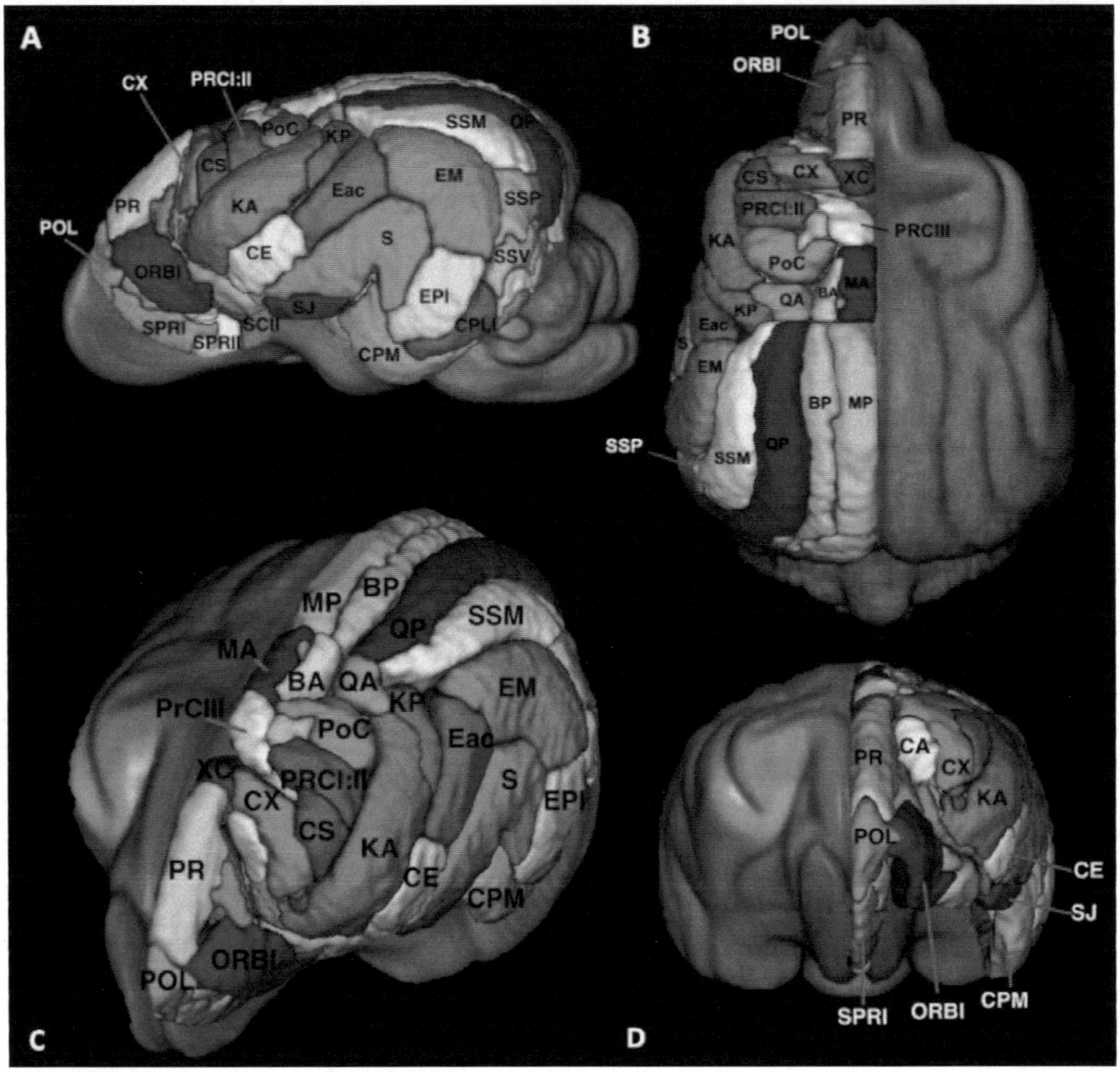
A
CX
PRCI:II
PoC
KP
SSM
QP
CS
EM
KA
Eac
SSP
PR
POL
S
CE
SSV
ORBI
SJ
EPI
SCII
CPLI
SPRI
SPRII
CPM
B
POL
ORBI
PR
CS
CX
XC
PRCI:II
PRCIII
KA
PoC
MA
QA
BA
KP
Eac
S
EM
BP
MP
QP
SSP
SSM
C
MP
BP
SSM
QP
MA
BA
QA
EM
PrCIII
KP
PoC
Eac
XC
PRCI:II
S
CX
EPI
CS
KA
PR
CE
CPM
ORBI
POL
D
PR
CA
CX
KA
POL
CE
SJ
SPRI
ORBI
CPM

Corpus Callosum Diffusivity Alteration is a Component of Healthy Aging in the Canine

Philippa Johnson[1]; John Loftus[1]; Wen-Ming Luh[2]; Erica Barry[1]
[1]College of Veterinary Medicine, Cornell University; [2]National Institute of Aging

The aging brain is known to alter in structure, connectivity and diffusivity in the human. Most notably the corpus callosum is observed to exhibit alterations in diffusivity values and, associated with this, right-left cerebral connectivity is reduced. The aging canine brain has been observed to reduce in volume however the effect of aging on corpus callosum diffusivity has not been evaluated. In this study we applied the novel technique of voxel-based-analysis to identify the extent and location of significant diffusivity abnormalities in the brain of healthy aged canines. Diffusion tensor imaging data were acquired from 20 neurologically and behaviorally normal mesaticephalic canines grouped into young (aged between 2–6 years) and old (aged between 10–11 years) cohorts. The data were skull stripped, corrected for noise, phase distortion, motion and eddy-current distortion. Tensor metrics FA and MD were calculated and voxel-based statistics applied a two-sample t-test to identify significant differences between age groups. Family-wise error corrected was used to correct for multiple comparisons. There were significant decreases in FA within the body and dorsal fibers for the corpus callosum in FA in the aged canine group when compared to the younger group. Similarly, there were significant decreases in MD in the corpus callosum in the aged group though these decreases were less widespread. Healthy aging is associated with decreases in FA and MD within the corpus callosum. This is similar to the diffusivity changes present within humans and may reflect alterations in structural connectivity between hemispheres.

From Inception to Realization: How 3D Printing Can Improve Your Patient Care

Fred Wininger, VMD, MS, DACVIM (Neurology)
Charlotte, NE, USA

3D printing (3DP) or additive manufacturing (AM) is becoming commonplace in the world of medicine. For some time, medical technologists have taken advantage of frequently used tomographic studies to create 3D digital models for diagnostic assessment and therapeutic planning. The conversion of these digital models into tangible objects furthers their value and can be done with surprising ease given new technological advantages. The potential value of 3DP extends well beyond scaled models and is being explored for customizable surgical tools, biomechanical testing, prosthetic implantation, and the creation of live tissue. The function of this talk is to introduce basic principles of 3DP including the state of print technology, best practices to create anatomic models from medical imaging, and an introduction to some of the more advanced applications.

Though any tomographic/cross-sectional imaging modality may be used for 3D printing, computerized tomography is used most commonly with the best results because of its spatial and contrast resolution. CT creates precise, small, isotropic voxels which promote the highest resolution prints. Further, the tissue of traditional interest has excellent contrast resolution, particularly with the administration of contrast agents. The most highly resolved 3D prints are when the tissue of interest is easily segmented from adjacent tissues based on proximity and thresholded contrast. Because of the range of densities the modalities can discern, it is the most preferable.

Once the tissue of interest is identified most imaging software programs can create a surface rendering. The surface rendering creates a "manifold" of single triangles meshed together in a complex polygonal pattern. These manifolds generate a variety of file formats, the most common of which being .stl or .obj. While these file formats can be readily printed, most models are further refined to optimize their appearance and integrity. Common changes include "remeshing" to improve manifold integrity, "smoothing" to eliminate imperfections in the image acquisition, supports to keep unconnected bones in their proper orientation, and "hollowing" to use less printing medium and keep models light.

Once complete, model files are exported to a "G Slicer" program. Most modern printers come with their own proprietary printing software that utilizes a slicer format. Though many different forms of printers are available, the two most commonly used outside of high end laboratories are fused deposition modelling(FDM) and stereolithography (SLA).

Because of their availability and cost, extrusion techniques such as FDM are the most commonly encountered in non-commercial use. A filament is melted and then layered on a platform through an extruder. FDM anatomic models have been reported to aid in surgical planning and pre-operative molding of metallic implants. However, FDM's lower resolution and thermoplastics materials may limit its utility in implantable prosthetics.

Light polymerized techniques such as STL solidify liquid resin into complex polymers with light of specific wavelengths. These techniques have higher resolution, are recently cost effective, and have the potential for biocompatibility, as the polymers are methacrylate based. However, their biomechanical properties are unclear, they have more complex post-processing, and are prone to printing error. In addition to anatomic modeling they are useful in custom jig manufacturing for implant guidance as well as casting molds for metallic fabrication.

Binding techniques such as selective laser sintering are attractive because of their high resolution and biocompatible/biomechanical materials such as polyether ether ketone (PEEK) and titanium. Though too expensive for on-site printing, private companies will custom print pieces from validated STL files. These techniques may have the greatest potential for custom implantable prosthetics. Printers that can create additively modeled live tissue are unavailable outside cutting edge laboratories.

The software capabilities and availability of 3DP are advancing as quickly as the hardware. Despite the fact that much of the medically branded programs are cost-limiting to most veterinarians, many of the general software packages are free to users and are even cloud based. In general the programs can be crudely divided into computer aided design (CAD) or artistic formats. CAD formats are generally geometrically based and ideal for orthopedic applications as they have a variety of pre-planned implants that can be scaled and altered to match currently available tools. Commonly used CAD programs range from the most capable **SolidWorks** to the freeware **Tinkercad**, a program created for children but often used by engineers. The artistic hardware is more often used for freeform creation and can sculpt or form new shapes around the anatomic meshes provided. Examples of these programs are **Meshmixer**, **Rhino**, and **Invesalius**. The combination of these programs facilitates the creation of tools/jigs and implants that can integrate with complex anatomy but fit our traditional hardware.

Perhaps the greatest value of 3DP is that any tool that can be imaged can be prototyped given the expertise of the computer engineer. As material options advance in SLA and FDM technology, the utility for 3DP grows, perhaps more for veterinary medicine than other medical fields because of the anatomic variety of our patients and the cost limitations of our clientele. There is a potential that a 3D printer will be as essential of a surgical tool as an autoclave in the foreseeable future. It is an exciting time for the manufacturing of parts by the hands of the clinicians that will use them.

Progress on Tricky Issues in Spinal Cord Injury: Deep Pain Negative Paralysis—Not an All-or-Nothing Phenomenon

Natasha J. Olby, VetMB, PhD, DACVIM (Neurology)
Raleigh, NC, USA

Introduction

Acute paralysis is one of the most common presenting problems to neurologists, and in these cases, prognosis is an extremely important question for owners and veterinarians. For animals with acute injuries, the most overwhelming predictor of outcome is initial injury severity. As such, animals that are paraplegic with no pain perception are assessed as having a clinically complete spinal cord lesion and drop into a lower category in terms of outcome; the prognosis depends on the initiating cause. For example, dogs with acute intervertebral disc extrusions (IVDE) due to Hansen type 1 disc degeneration have an approximately 50–60% chance of recovery with decompressive surgery, while dogs with vertebral fractures and luxations have a poor chance of recovery (<10%).[1-4] The difference in outcome between these two different diagnoses speaks to the differences in underlying pathology and highlights the rather limiting floor effect of our assessment of injury severity. In other words, this group of paraplegic deep pain negative (or clinically complete) dogs encompasses a wide range of injury severities. To add to this, the recovery of ambulation without the recovery of pain perception is a well-recognized phenomenon, and these animals are frequently referred to as 'spinal walkers.' A rigorous definition of a spinal walker is an animal that can walk independently with no descending connections across the lesion, and so effectively the assumption being made here is that lack of pain perception means that these dogs have a complete spinal cord injury.[5] If indeed these animals are true spinal walkers, this raises the very pertinent question about why some dogs develop this functionality and others do not, leading of course to how we can enhance that recovery. This presentation will examine the evidence for the presence of descending connections and describe the diversity of injury that is present in dogs that are deep pain negative, with a focus on factors that are associated with recovery of ambulation.

Frequency of Deep Pain Negative Walking in Dogs

Following acute, surgically decompressed IVDE, 7/18 (39%) dogs that remained deep pain negative at 1 month after injury recovered the ability to walk independently over the subsequent 18 months.[1] In this same population of dogs, a recovery rate of 59% (48/81 dogs) has been reported following intensive physical therapy.[6] In dogs that suffer trauma, there is a report of 2 of 9 dogs recovering walking, but in both dogs, there was no displaced fracture visible on radiographs.[1] In an experimental setting, there is a report of spontaneous recovery of locomotion in 6/6 adult dogs undergoing spinal cord transection (and re-transection in some cases) by around 3 months after injury.[7] This is in contrast to early experimental work, in which limb-stepping movements were observed, but functional walking was not recovered.[8] This difference in frequency of occurrence between dogs in a clinical and experimental setting is interesting and difficult to explain. Is it related to the level of the injury or the focal, clean nature of a surgical transection? We know that there are clear differences between species and between injury types that influence the recovery of walking. For example, cats can be trained to spinal walk effectively and reliably, while mice, rats, and people can be trained to produce stepping movements but cannot be trained to generate fully weight-bearing locomotion.[5]

Neural Circuitry

Locomotion is the result of interplay between the lower motor neuron, the local spinal cord circuitry, and ascending and descending impulses. The spinal cord circuitry is organized into central pattern generators (CPGs) that receive extensive afferent input and are driven by descending commands. The CPG result in the delivery of coordinated impulses to different groups of muscle in such a fashion that locomotion ensues. These pattern generators are extremely complex and extend over multiple levels of the spinal cord. There is disagreement about whether a CPG is modulated by afferent input but driven by descending input, or vice versa, and the ability to manipulate these CPGs in clinical patients has only recently been explored using epidural and transcutaneous stimulation.

Do Deep Pain Negative Dogs Have Complete Spinal Cord Lesions?

The most obvious explanation for recovery of walking in a subset of dogs with spinal cord injuries is that they have less severe injuries than others. The classic pattern of pathology following an acute contusive injury to the spinal cord is loss of the center of the spinal cord—radiating out from the grey matter and frequently preserving a peripheral ring of intact subpial axons. It is telling to learn that people with chronic, apparently complete lesions can recover some voluntary control of movement (of a toe, for example) following epidural spinal stimulation. It has also been shown that both sensory and motor-evoked potentials can be conveyed across clinically complete lesions in humans. In a comparable manner, two different groups have now reported intact translesional conduction in dogs that are persistently deep pain negative following a spinal cord injury, and a detailed DTI study demonstrated intact connections. Finally, it has been observed that some of these dogs will wag their tails voluntarily when they see their owners, again suggesting a voluntary, descending input. Taking this further, looking simply at the MRI lesion extent, there is enormous variability in both cross-sectional area and longitudinal extent of the lesion. The take-home message from these observations is firstly, that many dogs that persistently lack deep pain do still have functional connections across their lesions, and secondly, there is a wide range of pathology type and extent in dogs that lack deep pain perception.

Recovery Without Pain Perception

Following a severe spinal cord injury, there is a short period of spinal shock with gradual recovery of spinal reflexes over a period of 48–72 hours.[9,10] The recovery of hindlimb muscle tone, with subsequent development of spasticity and the development of crossed extensor reflex and mass response are variable, and, in my experience, the timing and extent of this recovery can be used to predict recovery of locomotion even in the absence of pain perception. In people, there is strong evidence that the spinal cord segments below the level of a complete lesion lose excitability, making them less sensitive to residual descending input.[11] This excitability can be improved using epidural or transcutaneous stimulation.[12] The presence of strong flexor spasticity in dogs is associated with recovery, as is a low threshold to elicit the H reflex; both findings likely reflect the excitation state of the spinal cord circuitry.[13,14] Dogs with more focal lesions are more likely to develop this flexor spasticity and to recover than dogs with extensive lesions that affect multiple levels of spinal cord circuitry. Similarly, dogs with intact motor-evoked potentials and evidence of intact axons on DTI MRI are more likely to recover.[13,15,16] The state of excitation can be improved by administration of 4-aminopyridine in dogs, a drug shown to enhance stepping in chronically paralyzed dogs.[17,18] More recently, we have found that transcutaneous spinal stimulation will abruptly change the excitation state of the spinal cord circuitry in these dogs.

What Does This Mean?

It is important to understand that the spectrum of injury severity does not end with a clinical finding of no pain perception and that a finding of no deep pain does not mean that there are no functional descending and ascending axons crossing the lesion. It is important (indeed a central understanding in veterinary neurology) to consider deep pain presence or absence as an important prognostic indicator at the time of an acute lesion, but it is also important to understand that the absence of pain perception is less central to prognosis once the patient gets beyond the 1-month timepoint. Finally, development of fully weight-bearing locomotion in chronically deep pain negative dogs is more likely in dogs with focal lesions and evidence of connections across their lesion, and these dogs can be identified by their strong flexor withdrawal reflex, early development of spasticity and crossed extensor reflexes, and, in some cases, by a voluntary tail wag. Interventions that increase the excitability of the local spinal cord circuitry can aid in recovery of locomotion.

References

1. Olby N, Levine J, Harris T, Muñana K, Skeen T, Sharp N. Long-term functional outcome of dogs with severe injuries of the thoracolumbar spinal cord: 87 cases (1996–2001). *J Am Vet Med Assoc.* 2003;222(6):762–769.
2. Jeffery ND, Barker AK, Hu HZ, Alcott CJ, Kraus KH, Scanlin EM, *et al.* Factors associated with recovery from paraplegia in dogs with loss of pain perception in the pelvic limbs following intervertebral disk herniation. *J Am Vet Med Assoc.* 2016;248(4):386–394.
3. Aikawa T, Fujita H, Kanazono S, Shibata M, Yoshigae Y. Long-term neurologic outcome of hemilaminectomy and disk fenestration for treatment of dogs with thoracolumbar intervertebral disk herniation: 831 cases (2000–2007). *J Am Vet Med Assoc.* 2012;241(12):1617–1626.
4. Aikawa T, Shibata M, Asano M, Hara Y, Tagawa M, Orima H. A comparison of thoracolumbar intervertebral disc extrusion in French bulldogs and dachshunds and association with congenital vertebral anomalies. *Vet Surg.* 2014;43(3):301–307.
5. Edgerton VR, Courtine G, Gerasimenko YP, Lavrov I, Ichiyama RM, Fong AJ, *et al.* Training locomotor networks. *Brain Res Rev.* 2008;57(1):241–254.
6. Gallucci A, Dragone L, Menchetti M, Gagliardo T, Pietra M, Cardinali M, *et al.* Acquisition of involuntary spinal locomotion (spinal walking) in dogs with irreversible thoracolumbar spinal cord lesion: 81 dogs. *J Vet Intern Med.* 2017;222:762–766.
7. Handa Y, Naito A, Watanabe S, Komatsu S, Shimizu Y. Functional recovery of locomotive behavior in the adult spinal dog. *Tohoku J Exp Med.* 1986;148(4):373–384.
8. Sherrington CS. Flexion-reflex of the limb, crossed extension-reflex, and reflex stepping and standing. *J Physiol (Lond).* 1910;40(1–2):28–121.
9. Blauch B. Spinal reflex walking in the dog. *Vet Med Small Anim Clin.* 1977;72(2):169–173.
10. Smith PM, Jeffery ND. Spinal shock—comparative aspects and clinical relevance. *J Vet Intern Med.* 2005;19(6):788–793.
11. Harkema S, Gerasimenko Y, Hodes J, Burdick J. Effect of epidural stimulation of the lumbosacral spinal cord on voluntary movement, standing, and assisted stepping after motor complete paraplegia: a case study. *The Lancet.* 2011;377(9781):1938–1947.
12. Angeli CA, Edgerton VR, Gerasimenko YP, Harkema SJ. Altering spinal cord excitability enables voluntary movements after chronic complete paralysis in humans. *Brain.* 2014;137(Pt 5):1394–1409.
13. Lewis MJ, Howard JF Jr, Olby NJ. The relationship between trans-lesional conduction, motor neuron pool excitability, and motor function in dogs with incomplete recovery from severe spinal cord injury. *J Neurotrauma.* 2017;34(21):2994–3002.
14. Lewis MJ, Olby NJ. Development of a clinical spasticity scale for evaluation of dogs with chronic thoracolumbar spinal cord injury. *Am J Vet Res.* 2017;78(7):854–861.
15. Lewis MJ, Cohen EB, Olby NJ. Magnetic resonance imaging features of dogs with incomplete recovery after acute, severe spinal cord injury. *Spinal Cord.* 2017;55:182–189.

16. Lewis MJ, Yap P-T, Mccullough S, Olby NJ. The relationship between lesion severity characterized by diffusion tensor imaging and motor function in chronic canine spinal cord injury. *J Neurotrauma*. 2018;35(3):500–507.
17. Lim JH, Muguet-Chanoit AC, Smith DT, Laber E, Olby NJ. Potassium channel antagonists 4-aminopyridine and the T-butyl carbamate derivative of 4-aminopyridine improve hind limb function in chronically non-ambulatory dogs; a blinded, placebo-controlled trial. *PLoS ONE*. 2014;9(12):e116139.
18. Lewis MJ, Laber E, Olby NJ. Predictors of response to 4-aminopyridine in chronic canine spinal cord injury. *J Neurotrauma*. 2019;36(9):1428–1434.

Progress on Tricky Issues in Canine Spinal Cord Injury: Role of Advanced Imaging in Determining Injury Severity and Functional Outcome

Melissa Lewis, VMD, PhD, DACVIM (Neurology)

INTRODUCTION

Magnetic resonance imaging (MRI) has become widespread in the diagnosis of canine spinal cord injury (SCI). While this imaging modality routinely guides treatment decisions in the acute setting, there is also a growing role for MRI in quantifying severity of injury and predicting prognosis and functional outcome, especially among severely injured dogs. Additionally, studies of specialized applications of MRI including diffusion tensor imaging (DTI) are becoming more prevalent in dogs and are enhancing our understanding of how structural lesion characteristics relate to short and long-term functional outcomes following injury.

CONVENTIONAL MRI IN ACUTE AND CHRONIC SCI

Both qualitative and quantitative features of acute myelopathies on MRI have been extensively described in dogs. While the degree of extradural compression on MRI has not been associated with injury severity or prognosis, the presence and dimensions of intramedullary hyperintensity on T2-weighted (T2W) images and length of spinal cord compression have been associated with outcome following acute SCI in dogs. MRI features suggestive of progressive myelomalacia (PMM) have also been described. The presence and length of a T2-weighted intramedullary hyperintensity and the length of cerebrospinal fluid attenuation on HASTE (single-shot turbo spin echo) sequence have been associated with development of PMM, supporting a role for MRI in identification of this devastating condition.

Among dogs with severe SCI left with permanent neurologic impairment, the MRI appearance of the lesion in the subacute to chronic setting is less well-documented. Intramedullary areas of malacia and cavitation as well as spinal cord atrophy have been shown to be common at the lesion epicenter while the frequency of post-traumatic syringomyelia was variable and fibrosis was uncommon. Additionally, apparently complete compromise of spinal cord tissue on transverse images (i.e., no identifiable normal intramedullary areas at the epicenter) was noted in approximately 30% to more than 80% of dogs between two studies. Interestingly, longer length of the compromised region was associated with less recovery of motor function and independent ambulation, supporting a relationship between specific imaging features and functional status in the chronic setting.

DIFFUSION TENSOR IMAGING IN ACUTE AND CHRONIC SCI

Diffusion tensor imaging (DTI), a variation on diffusion weighted imaging, relies on cellular diffusion of water molecules to construct images. DTI allows quantitative microstructural assessment of spinal cord tissue following injury and can detect changes such as axonal damage, demyelination, cavity formation and loss of tissue architecture. Additionally, quantitative information can be leveraged to create a representation of spinal cord white matter tracts, known as tractography.

DTI in dogs has been shown capable of differentiating injured from healthy controls and acute from chronic injury. Additionally, restricted diffusion in the acute setting might be a useful marker of injury severity and predictor of functional outcome among severely affected dogs. For dogs evaluated in the chronic setting after severe injury, DTI has been associated with motor function and tractography detected loss of spinal cord integrity. The role of DTI as a noninvasive imaging biomarker and prognostic indicator requires further investigation.

Importantly, optimization and standardization of imaging protocols and processing techniques are needed in order for DTI to be incorporated into routine clinical use in SCI patients.

REFERENCES

1. Alisauskaite N, Spitzbarth I, Baumgartner W, *et al*. Chronic post-traumatic intramedullary lesions in dogs, a translational model. *PLoS One*. 2017. doi.org/10.1371/journal/pone.0187746.
2. Balducci F, Canal S, Contiero B, Bernardini M. Prevalence and risk factors for presumptive ascending/descending myelomalacia in dogs after thoracolumbar intervertebral disk herniation. *J Vet Intern Med*. 2017;31:498–504.
3. Boekhoff TM, Flieshardt C, Ensinger EM, Fork M, Kramer S, Tipold A. Quantitative magnetic resonance imaging characteristics: evaluation of prognostic value in the dog as a translational model for spinal cord injury. *J Spinal Disord Tech*. 2012;25:E81–7.
4. De Risio L, Adams V, Dennis R, McConnel FJ, Platt SR. Association of clinical and magnetic resonance imaging findings with outcome in dogs suspected to have ischemic myelopathy: 50 cases (2000–2006). *J Am Vet Med Assoc*. 2008;233:129–135.
5. De Risio L, Adams V, Dennis R, McConnell FJ. Association of clinical and magnetic resonance imaging findings with outcome in dogs with presumptive acute noncompressive nucleus pulposus extrusion: 42 cases (2000–2007). *J Am Vet Med Assoc*. 2009;234:495–504.
6. Gilmour LJ, Jeffery ND, Miles K, Riedesel E. Single-shot turbo spin echo pulse sequence findings in dogs with and without progressive myelomalacia. *Vet Radiol Ultrasound*. 2017;58:197–205.
7. Griffin JF, Davis MC, Cohen ND, Young BD, Levine JM. Quantitative magnetic resonance imaging in a naturally occurring canine model of spinal cord injury. *Spinal Cord*. 2015;53:278–284.
8. Ito D, Matsunaga S, Jeffery ND, Sasaki N, Nishimura R, Mochizuki M, Kasahara M, Fujiwara R, Ogawa H. Prognostic value of magnetic resonance imaging in dogs with paraplegia caused by thoracolumbar intervertebral disk extrusion: 77 cases (2000–2003). *J Am Vet Med Assoc*. 2005;227:1454–1460.
9. Lewis MJ, Yap PT, McCullough S, Olby NJ. The relationship between lesion severity characterized by diffusion tensor imaging and motor function in chronic canine spinal cord injury. *J Neurotrauma*. 2017;34:1–8.
10. Lewis MJ, Cohen EB, Olby NJ. Magnetic resonance imaging features of dogs with incomplete recovery after acute spinal cord injury. *Spinal Cord*. 2018;56:133–141.
11. Okada M, Kitagawa, Ito D, *et al*. Magnetic resonance imaging features and clinical signs associated with presumptive and confirmed progressive myelomalacia in dogs: 12 cases (1997–2008). *J Am Vet Med Assoc*. 2010;237:1160–1165.
12. Penning V, Platt SR, Dennis R, Cappello R, Adams V. Association of spinal cord compression seen on magnetic resonance imaging with clinical outcome in 67 dogs with thoracolumbar intervertebral disc extrusion. *J Small Anim Pract*. 2006;47:644–650.
13. Wang-Leandro A, Siedenburg JS, Hobert MK, Dziallas P, Rohn K, Stein VM, Tipold A. Comparison of preoperative quantitative magnetic resonance imaging and clinical assessment of deep pain perception as prognostic tools for early recovery of motor function in paraplegic dogs with intervertebral disc herniations. *J Vet Intern Med*. 2017;31:842–848.
14. Wang-Leandro A, Hobert MK, Alisauskaite N, Dziallas P, Rohn K, Stein VM, *et al*. Spontaneous acute and chronic spinal cord injuries in paraplegic dogs: a comparative study of *in vivo* diffusion tensor imaging. *Spinal Cord*. 2017;1-9. doi:10.1038/sc.2017.83.
15. Wang-Leandro A, Hobert MK, Kramer S, *et al*. The role of diffusion tensor imaging as an objective tool for the assessment of motor function recovery after paraplegia in a naturally-occurring large animal model of spinal cord injury. *J Transl Med*. 2018;16:258.
16. Yoon H, Kim J, Moon WJ, *et al*. Characterization of chronic axonal degeneration using diffusion tensor imaging in canine spinal cord injury: a quantitative analysis of diffusion tensor imaging parameters according to histopathological differences. *J Neurotrauma*. 2017;34:3041-3050.
17. Yoon H, Moon WJ, Nahm SS, *et al*. Diffusion tensor imaging of scarring, necrosis, and cavitation based on histopathological findings in dogs with chronic spinal cord injury: evaluation of multiple diffusion parameters and their correlations with histopathological findings. *J Neurotrauma*. 2018. doi:10.1089/neu.2017.5409.

Progress on Tricky Issues in Spinal Cord Injury: Facilitating Clinical Trials on Spinal Cord Injury in Private Practice

Nick Jeffery, BVSc, PhD, MSc, DECVS, DECVN, FRCVS
College Station, TX, USA

INTRODUCTION

Based on laboratory experiments in rodents there is a multitude of potential treatments for spinal cord injury in dogs that could readily be introduced into everyday clinical veterinary neurology. Before such putative therapies become widely adopted it is important that they are properly tested in clinical trials, otherwise there is a risk that many animals will receive unnecessary or ineffective treatment.

THE PROBLEM

There are many practical obstacles to running clinical trials in canine spinal cord injury, especially when focusing on cases presenting in the acute phase. For instance, there is controversy regarding selection from pre-clinical data of the most promising interventions to go forward, but probably the most difficult obstacle of all is the need to integrate clinical trial methods into private referral hospitals: how can research become a built-in component of day-to-day clinical work?

The need for more general buy-in to participation in trials amongst neurologists arises because, realistically, the benefits of new interventions, if any, are not going to be readily apparent in small patient cohorts. The huge variability in injury severity and location and the variation in genetics, size and functionality between individuals with SCI implies that the effects of new therapies will need to be evaluated in a large number—probably many hundreds—of patients. There is also a need for long-term follow-up: it is well-recognized that full recovery following spinal cord injury may not be attained for at least 3–6 months after injury. Unfortunately, while academic institutions have many individuals trained in design, implementation and analysis of clinical studies, the majority of dogs with spinal cord injury are treated in referral hospitals. So, there is a need to maintain and develop robust collaborations between academic institutions and private hospitals.

This issue might also be re-phrased as **'What are the obstacles to carrying out research in referral practice?'** In general, this question can be reduced down to time and money and, in business, time usually **is** money!

POTENTIAL SOLUTIONS

The burden of taking part in clinical research must be minimized; participation must become easy, quick and inexpensive. How can this be done?

1. **Reduce the burden of data collection**. This takes two forms: i) using a minimal number of simple outcome measures, perhaps even so simple that owners are able to collect them (as an analogy there is a move toward patient-centered outcomes in human medicine); ii) methods that are easy to use - such as electronic data collection forms, perhaps also available on a smartphone.
2. **Simplifying randomization techniques**. Not only is there is a need for randomization, but the procedure also must prevent blinded personnel from knowing what the next allocation will be. Fortunately, appropriate methods need not be complicated; cases can be randomized simply and easily in most veterinary clinical trials by concealing treatment allocation in numbered opaque envelopes. In corporate practices it might be possible to randomize automatically according to accession number (but avoiding allocation by odd versus even, which is not random).

3. **Responsibility for design and statistical analysis can be handled by a central organizer**—usually at an academic center. Whilst it is important to consult on trial design with participants in various types of clinic, especially in view of the need to streamline process, the formal aspects of design and statistical analysis can be handled by coordinating academic centers.
4. **Promotion of clinical research as a benefit to the owner and the referring veterinarian** that will aid bonding to the referral clinic. The frequent and prolonged follow-up that is required for many studies on spinal cord injury need not only be an unwelcome task but can also be viewed as providing reassurance to both owner and referring veterinarian of the commitment of the referral center to high standards and optimal outcome.
5. **Provision of support for the techs, interns and residents** who may often be required to take part but for whom the benefits of research will not be so readily apparent. Many staff and junior clinicians have the perception that research is pie in the sky and will never make a difference to care of their patients; however, this attitude can often be overcome, or at least challenged, by helping them to realize the critical roles they play in the research team. Frequent reporting of progress and interesting outcomes back to the team as a whole and inclusion of techs and junior clinicians in the final documents aids in improving research morale. There may also be a role for more tangible incentivization such as food and drink vouchers and thank-you events.

Finally, facilitating clinical trial participation is a tough nut to crack and optimal paths to the final destination are likely to require much discussion amongst veterinary neurologists, residents, interns and support staff—but it should be possible to make it happen!

Abstract O01: Feline T-Cell Low-Grade Intestinal Lymphoma: A Novel Model of Lymphomagenesis According to the One-Health Concept

Valérie G. Freiche[1]; Nathalie Cordonnier[1]; Maria Elena Turba[2]; Thierry Molina[3]; Olivier Hermine[3]; Julie Bruneau[3]; Lucile Couronné[3]
[1]Ecole Nationale Vétérinaire d'Alfort; [2]Laboratorio Genefast; [3]Hôpital Universitaire Necker

Feline T-cell low-grade intestinal lymphoma (T-LGIL) is an indolent disease, recognized as the first digestive neoplasm in cats. The condition is characterised by the infiltration of the gastrointestinal tract by neoplastic T-lymphocytes. T-LGIL shares common clinical, paraclinical and ultrasonographic features with lymphoplasmacytic enteritis (LPE). Establishing a final diagnosis is challenging. Moreover, the disease shares similarities with digestive indolent T-cell lymphoproliferative disorder in human. As a result, translational research in feline T-LGIL was considered relevant to improve understanding of the human condition, which is rare and still poorly defined. We prospectively analysed clinical, paraclinical data and full-thickness small intestinal biopsies from 22 domestic cats diagnosed with T-LGIL and 22 cases diagnosed with LPE. A novel extensive histopathological and molecular study including T-cell receptor clonality analysis was performed on all samples. Separate assessment of the epithelium and the lamina propria was achieved. Differentiation criteria between feline T-LGIL and LPE included villous atrophy, lymphocytic cryptitis, depth of infiltration, apical-to-basal gradient of cellularity, nest and plaques identification and fibrosis extent within the lamina propria revealed by Trichrome de Masson's staining. CD3 and Ki67 expression levels in lamina propria and intra epithelial lymphocytes (IEL) were significantly increased in T-LGIL cases as compared with LPE cases. All T-LGIL were PhosphoStat3- and PhosphoStat5+ in contrast to LPE. Clonality assessment revealed monoclonal TCR□rearrangement in 82% of the T-LGIL cases and in 40% of the IBD cases. This prospective study specifically focuses on the pathogenesis and diagnosis of T-LGIL: a new model of lymphomagenesis has emerged from this comparative research and suggests a continuum between inflammatory enteropathy towards digestive lymphoma. Finally, strong similarities between feline T-LGIL and human indolent T-cell gastrointestinal lymphoproliferative disorders are highlighted.

Abstract O02: External Beam Radiation Therapy for the Treatment of Canine Appendicular Osteosarcoma: 77 Cases

Carissa J. Norquest[1]; Brian Flesner[1]; Charles Maitz[1]; Jeffrey Bryan[1]; Melanie Moore[2]; Tara Ehling[3]; Jimmy Lattimer[1]
[1]University of Missouri—Veterinary Health Center; [2]Care Center—Cincinnati; [3]University of Missouri—Veterinary Health Center (Wentzville)

Radiation therapy (RT) is used for local pain alleviation in dogs with appendicular osteosarcoma, especially among dogs that are poor surgical candidates for amputation. Recently published fracture rates for stereotactic body RT are >60%. However, historical fractionated protocols lack time to fracture and fracture rates. The objectives of this retrospective study were to determine fracture rate, tolerability, progression-free interval and survival time of dogs receiving RT (coarse or fine fractionated) for appendicular osteosarcoma. Seventy-seven dogs that received RT between 2006–2018 as part of treatment for appendicular osteosarcoma were available for evaluation. Seventy-one received coarse fractionation while the remaining six received fine fractionation. The overall pathologic fracture rate was 31%. Pathologic fracture rate was significantly higher for dogs that received fine fractionation RT (83%) compared to dogs that received coarse fractionation RT (26%), p=0.0096. The overall progression free interval (PFI) and overall survival time (OST) were 77 days and 109 days, respectively. Factors that increased both PFI and OST included improvement within 30 days of starting RT, completion of RT, and administration of chemotherapy. Increased serum ALP negatively affected both PFI and OST. Survival was significantly shorter in dogs that received coarse RT without bisphosphonates than those that received coarse RT with zoledronate (p=0.020, log-rank). In conclusion, coarse fractionated RT is recommended over fine fractionated RT due to lower risk of pathologic fracture and similar PFI. Prospective evaluation of combined coarse fractionation RT and zoledronate, especially for dogs with poor surgical candidacy, is warranted for the treatment of canine appendicular osteosarcoma.

Abstract O03: Evaluation of the Association between Interleukin-6 and Thrombopoietin Concentrations with Thrombocytosis in Dogs with Carcinoma

Adrienne B. Cheney[1]; Deborah Knapp[2]; George Moore[2]; Andrew Woolcock[2]
[1]College of Veterinary Medicine; Purdue University; [2]Purdue University

Carcinoma-associated thrombocytosis is well documented in people, and pre-treatment thrombocytosis has been shown to predict prognosis and response to therapy in certain epithelial tumors. In carcinoma-associated thrombocytosis, the tumor has been shown to produce interleukin-6 (IL-6) and thrombopoietin (TPO) causing increased thrombopoiesis. Thrombocytosis has been evaluated in dogs, and neoplasia is a common diagnosis in 25–55% of these patients. Carcinoma is the most common tumor diagnosed in dogs with concurrent thrombocytosis and neoplasia, however the association between carcinoma and thrombocytosis in dogs has only been investigated retrospectively. The objective of this proof-of-concept study was to evaluate the concentrations of IL-6 and TPO in dogs diagnosed with carcinoma with or without a thrombocytosis. We hypothesized that IL-6 and TPO concentrations would be greater in dogs with carcinoma compared to controls, regardless of platelet count. We further hypothesized that IL-6 and TPO concentrations would be greater in dogs with carcinoma and thrombocytosis when compared to dogs with carcinoma and normal platelet counts. Dogs with a histologic diagnosis of carcinoma were included in the study, and pre-treatment EDTA blood was collected for complete blood count including platelet, and serum was collected for measurement of IL-6 and TPO by enzyme linked immunosorbent assay (ELISA). A second group of healthy control dogs was recruited for comparison of blood platelet counts and serum concentrations of IL-6 and TPO. Data were non-parametrically distributed, and summary descriptive statistics presented as median [range]. Nonparametric numerical data for groups was assessed with the Wilcoxon rank sum test and Spearman rank correlation coefficient. One-hundred and sixteen dogs were included in the study; 61 of these were diagnosed with carcinoma, and 55 dogs were healthy controls. The median platelet counts of the two groups were 349×10^3/mL [161–660] and 263.5×10^3/mL [146–459], respectively. Of the dogs with carcinoma, 12/61 (19.6%) had a thrombocytosis with the median platelet count in this sub-group being 548.5×10^3/mL [533-660]. The median concentration of IL-6 was not different between the carcinoma and control dogs (9.70 pg/ml [0–181.53] vs. 3.03 pg/ml [0–280.77], p=0.155). Median TPO concentration was significantly higher in the dogs with carcinoma when compared to controls (87.42 pg/ml [0–>600] vs. 15.99 pg/ml [0–>600], $p<0.001$). There was no difference in IL-6 concentrations when the dogs with carcinoma were stratified by platelet count. Median TPO concentration was higher in the dogs with carcinoma and thrombocytosis when compared to those with normal platelet counts, but this difference did not reach significance (80.16 pg/ml [0–302.435] vs. 44.84 pg/ml [0–>600], p=0.11). There was no significant association between TPO concentration and platelet count (r=-0.13, p=0.32). These findings confirm that TPO concentrations are significantly increased in dogs with carcinoma, regardless of platelet count. While we did not establish TPO as a predictor of carcinoma-associated thrombocytosis, this may be due to the low number of patients with a concurrent thrombocytosis in this study. TPO is likely to be one of multiple factors which can impact platelet number, production, and consumption in dogs with carcinoma. Future research is necessary to determine if TPO is a valuable biomarker in the diagnosis, staging, or treatment of canine carcinoma.

Abstract 004: Prognostic Utility of Computed Tomography Radiomic Features for Canine Lung Tumors: An Analytical Study

Hannah Able; Amber Wolf-Ringwall; Aaron Rendahl; Christopher Ober; Davis Seelig; Chris Wilke; Jessica Lawrence
University of Minnesota

Tumor heterogeneity is a well-established marker of tumor behavior and it has been associated with prognosis in human lung tumors. Quantitative analysis of computed tomography (CT) radiomic features is an indirect measure of tumor heterogeneity. The purpose of this study was to extract CT radiomic features from canine primary pulmonary tumors and correlate features to histopathologic diagnosis or survival. First-order statistical-based CT texture features were extracted from segmented tumor volumes. Time to tumor progression (TTP) and survival were calculated as days (d) from the date of CT scan. Sixty-nine tumors from 67 dogs were evaluated. Fifty-seven tumors were classified as carcinomas and 12 as non-carcinomas. Fifteen dogs had metastasis. All dogs were treated with surgical resection; 15 dogs received postoperative chemotherapy. Median tumor volume was 34 cm^3 (0.1–1196 cm^3). There was wide variation in first-order statistics. Mean Hounsfield units (HU) ratio ($p=0.022$) and median mean HU ratio ($p=0.021$) were significantly higher in carcinomas than non-carcinomas. Tumor sphericity was strongly correlated to volume ($rs=1.0$) and mean HU was strongly correlated to median HU ($rs=0.9$); other parameters were not correlated with each other. Median TTP and overall median survival time (MST) were 229 d and 322 d, respectively. MST was significantly longer ($p=0.0092$) for carcinomas (357 d) compared to non-carcinomas (56 d). When carcinomas were considered separately, volume was significantly associated with TTP ($p<0.0001$) and MST ($p<0.0001$). Metastasis at diagnosis significantly decreased MST (78 d versus 407 d; $p=0.008$). Further study of radiomic features in canine lung tumors is warranted, particularly given that it non-invasively provides additional tumor data.

Abstract O05: High Ki67 Expression is Associated with Poor Prognosis in Canine B-cell Chronic Lymphocytic Leukemia

Emily D. Rout[1]; Julia Labadie[2]; Janna Yoshimoto[1]; Kaitlin Curran[3]; Anne Avery[1]
[1]Colorado State University; [2]Fred Hutchinson Cancer Research Center; [3]Oregon State University

Human B-cell chronic lymphocytic leukemia (BCLL) has a highly variable clinical course. Canine BCLL is generally considered an indolent disease, but previous studies demonstrate a wide range in survival times. Our objective was to evaluate clinical outcome in a larger population of BCLL patients and identify clinical or flow cytometric prognostic factors. A retrospective study was performed evaluating clinical presentation, flow cytometry features and overall survival in 121 BCLL patients. BCLL was defined as >5,000 lymphocytes/µL in the blood with an expansion of small-sized CD21+ B cells (>60% of the lymphocytes) by flow cytometry. Three breed groups were represented: small breed dogs (n=55) due to increased risk of BCLL; Boxers (n=33) due to preferential use of unmutated immunoglobulin genes, which is associated with poor prognosis in human BCLL; other breeds (n=33) to compare outcome to small breed cases. Proliferation, determined by the percent of Ki67-expressing B cells by flow cytometry, was evaluated in 39/121 cases. The median overall survival time (MST) for all cases was 10.5 months (range, 1 day - 55 months). Boxers had significantly shorter survival (MST, 5.9 months) than non-Boxers (MST, 14.1 months; p=0.0003), and there was no significant difference in survival between small breeds and other non-Boxer breeds. Cases with high Ki67 (>40% Ki67-expressing B cells) had significantly shorter survival (MST, 5.8 months) than low Ki67 cases (MST undetermined) among all cases evaluated (p=0.023), and non-Boxer cases (p=0.028). In conclusion, BCLL demonstrated a variable clinical course and Boxer dogs and cases with high Ki67 had more aggressive disease.

Abstract O06: Clinical Differences in Aberrant Feline T-cell Leukemia Phenotypes

Klaudia Z. Polak[1]; Emily Rout[1]; Julia Labadie[2]; Paul Avery[1]; Anne Avery[1]
[1]Colorado State University; [2]Fred Hutchinson Cancer Research Center

The three major neoplastic phenotypes seen in cats with lymphocytosis include CD4+CD5+ T-cell, CD4-CD8-CD5+ (double negative [DN]) T-cell, and CD5-low-expressing T-cell. CD4+CD5+ T-cell leukemia is the most common and has a prolonged clinical course, while the two aberrant T-cell phenotypes, DN T-cell and CD5 low T-cell, are rarer and have significantly shorter survival times. We previously described the poorer prognosis of the aberrant T-cell leukemia phenotypes, but small numbers of cases were evaluated. All CD5 low cases were grouped together regardless of CD4 and CD8 expression. Our objective was to examine a larger cohort of cases with aberrant T-cell phenotypes, particularly among CD5 low cases, and investigate clinically important differences within the DN, CD4+, or CD8+ CD5 low subtypes. We performed a retrospective study evaluating clinical data and survival information in 54 cats with a flow cytometry diagnosis of an aberrant T-cell phenotype (CD4+ with low CD5 expression, CD8+ with low CD5 expression, DN with low CD5 expression, and DN with normal CD5 expression) in peripheral blood. The aberrant T-cell phenotypes, DN CD5+, CD4+ CD5 low, CD8+ CD5 low, and DN CD5 low, had statistically significant ($p<0.0001$) differences in median survival times (MST) (613 days [n=18], 146 days [n=13], 139 days [n=5], and 24 days [n=18], respectively). Although the DN CD5+ phenotype has the longest MST, survival was significantly shorter in cases with splenomegaly and/or splenic mass (223 days; $p=0.0081$) compared to cats without splenic abnormalities (MST, 774 days). Presence of anemia was significantly different between phenotypes ($p=0.021$), and was more common in the DN CD5 low cats (67%) compared to DN CD5+ cats (18%). Peripheral lymphadenopathy across all phenotypes was rare. Blood smear interpretation of DN CD5+ cases reveals predominantly small mature lymphocytes, whereas DN CD5 low lymphocytes were variable and ranged from small to large and immature in appearance. In conclusion, there were marked differences in overall survival between aberrant T-cell leukemia phenotypes. Cats with DN CD5 low T-cell leukemia have a very poor prognosis and tend to be anemic, whereas cats with DN CD5+ T-cell leukemia have more indolent disease, and splenic abnormalities are associated with reduced survival.

Abstract O07: Characterization of Monoclonal Gammopathies in Patients with Normal Total Proteins

Christina Jeffries; R. Adam Harris; Paul Avery; A. Russell Moore
Colorado State University

Unexplained hyperproteinemia or hyperglobulinemia have been considered indications to perform serum protein electrophoresis (SPE) to rule out a myeloma associated serum paraproteinemia (M-protein). We sought to characterize M-proteins in samples with normal total proteins (≤7.5 g/dl for canine) submitted to a veterinary diagnostic laboratory.

The archives at Colorado State University's Clinical Pathology Laboratory were searched for samples from canine patients that had SPE performed in 2019. The final interpretation and clinical data was recorded and evaluated. Additionally, cases with a M-protein that were confirmed by SPE and immunofixation (IF) performed between 2014 and 2019 and had a TP ≤ 7.5 g/dL were characterized for albumin and globulin concentration, albumin globulin ratio (A:G), age, M-protein isotype and concentration.

In 2019, 127 canine samples had SPE performed, of which 68 had IF. 56 samples had a diagnosed M-protein. 31 cases of M-protein were confirmed by IF. M-protein containing samples were not more likely to have hyperproteinemia (p = 0.11, OR 1.79, 95% CI 0.97 - 3.7), hyperglobulinemia (p = 0.83, OR 0.89, 95%CI 0.30 - 2.7), or hypoalbuminemia (p = 0.37, OR 1.57, 95% CI 0.60 - 4.5). IF evaluation was available at M-protein diagnosis for 11/13 (84.6%) samples with a normal total protein, and 7/8 (87.5%) samples with a normal globulin. To confirm that IF was an integral part of the initial diagnosis, 7 electrophoretograms with an IF confirmed M-protein, normal total protein and normal globulin were reviewed by 2 pathologists blinded to the IF results; the M-protein could not be identified in any of the cases, though a suspicious restricted band was noted in 5 samples. Due to lack of IF on all samples, the overall incidence of M-proteins could not be determined. Twenty-seven cases with TP ≤ 7.5 g/dl and a SPE and IF confirmed diagnosis of M-protein were identified between 2014 and 2019. Records were available for 11/27 cases which indicated SPE and IF were performed due to concern for an underlying paraproteinemia secondary to multiple myeloma or secretory plasma cell tumor or B-CLL. The mean age at diagnosis was 9.5 years (Min 3, Max 14). M-protein was identified in samples with a total protein as low as 5.2 g/dl. The mean albumin concentration was slightly decreased at 2.7 g/dl (Min 1 g/dl, Max 3.6 g/dl). The mean globulin concentration was slightly elevated at 3.9 g/dl (Min 2.4 g/dl, Max 6.3 g/dl) yet was within normal limits in 10/27 cases. The mean A/G was slightly decreased at 0.7 (Min 0.2, Max 1.2). The M-protein concentration mean was 1.3 g/dl (Min 0.4 g/dl, Max 3.6 g/dl) and could not be quantified in three cases. The incidence of M-protein isotypes was 70.4% IgA (19/27), 14.8% IgM (4/27), 11.1% IgG (3/27), and 3.7% (1/27) light chain only.

A normal total protein or normal globulin concentration in dogs does not rule out the possibility of an M-protein, especially if there is a decreased A:G. IF may be helpful in these circumstances to identify low amplitude M-proteins that may be missed with SPE alone. Additional studies are needed to determine if the A:G, age, and protein isotype distributions seen in this study are similar in dogs with the more typical M-protein and hyperproteinemia.

Abstract O08: Exosomes Derived from Canine Lymphoma Cells Induce M1 Polarization of Monocytes.

Akiyoshi Tani; Hirotaka Tomiyasu; Hajime Asada; Yuko Goto-Koshino; Koichi Ohno; Hajime Tsujimoto
The University of Tokyo

Exosomes are small extracellular vesicles released from various types of cells. Exosomes play pivotal roles in intercellular communications and act as shuttles by transmitting signals and transferring their contents. In human medicine, it has been shown that exosomes derived from tumor cells exert immunomodulating effects on immune cells such as monocytes and lymphocytes in tumor microenvironment. Meanwhile, there have been few studies that examined these functions of exosomes in canine tumors. Therefore, the purpose of our study was set to investigate the immunomodulating effects of exosomes derived from a canine cell line (CLBL-1) of diffuse large B cell lymphoma (DLBCL), which is one of the most common tumors in dogs, on monocytes. Canine monocytes were isolated from peripheral blood samples of healthy beagle dogs. The experimental animal care procedures were approved by the Animal Use and Care Committee of the University of Tokyo (P16-172). At first, peripheral blood mononuclear cells (PBMCs) were isolated by density gradient centrifugation, and CD14 positive monocytes were isolated from PBMCs using magnetic-activated cell sorting (MACS) method. Exosomes were extracted from FBS-free RPMI medium after culture of CLBL-1 for 24 hours using commercially available exosome isolation reagent. Two-hundred thousands of monocytes were incubated in RPMI medium containing 10% exosome-depleted FBS with or without 80 μg/ml of CLBL-1-derived exosomes for 4 hours. After incubation, total RNAs were extracted from adherent monocytes, and the relative quantities of IL-1β, IL-6, IL-10, TNFα and CD163 mRNAs were examined by RT-qPCR. In addition, changes in gene expression profiles were comprehensively examined by RNA-seq analysis using NextSeq 500. Read bases were aligned to the canine reference genome (CanFam 3.1), and differentially expressed genes (DEGs) were extracted as those with >1.5-fold changes in expression levels and raw p-value < 0.1. An online resource, DAVID v6.8, was used to identify biological functions and pathways associated with extracted DEGs, and a value of $p<0.05$ was considered significant in this analysis. In RT-qPCR analysis, it was shown that the expression levels of IL-1β, IL-6 and TNFα genes were significantly increased in exosome-treated monocytes compared with control monocytes. On the other hand, there was no significant difference in expression levels of IL-10 or CD163 gene. In RNA-seq analysis, 214 genes were extracted as DEGs between exosome-treated and control monocytes. Among these genes, 118 genes were upregulated and 96 genes were downregulated in exosome-treated monocytes compared with control monocytes. Analysis using DAVID with these DEGs showed that these genes were significantly associated with intracellular signaling pathways including NF-kappa β, TNF, and NOD-like receptor signaling pathways. Results of RT-qPCR indicated that treatment with exosomes derived from CLBL-1 increased the expressions of cytokines associated with the phenotype of M1-macrophage, which is classically considered to be pro-inflammatory macrophages, in monocytes. RNA-seq analysis revealed that extracted DEGs between exosome-treated and control monocytes were significantly associated with NF-kappa β signaling pathway, which was also reported to be activated in human monocytes or macrophages by treatments with exosomes derived from ovarian, breast and gastric cancers. It has been reported that the activation of NF-kappa β signaling pathway leads to activate monocytes and inhibition of NF-kappa β signaling in monocytes resulted in the reduction of tumor formation in a mouse model of human hepatocellular carcinoma. The results of this study indicated that exosomes derived from CLBL-1 activated monocytes through intracellular signaling including NF-kappa β pathway and increased the expressions of pro-inflammatory cytokines associated with M1-macrophage phenotype. The intercellular communications through exosomes might be potential targets of therapeutics for canine lymphoma, and further studies are warranted to investigate the roles of exosomes in pathophysiology of canine lymphoma.

Abstract O09: The VIGOR Clinical Trial: Anti-Tumor Immunity Induced by Neoadjuvant Oncolytic Virotherapy in Spontaneous Osteosarcoma

Kelly M. Makielski[1]; Aish Sathyanarayan[2]; Michael Henson[3]; Kathleen Stuebner[3]; Alexandru-Flaviu Tabaran[3]; Ingrid Cornax[3]; M. Gerard O'Sullivan[3]; Andrea Chehadeh[3]; Donna Groschen3; Kelly Bergsrud[3]; Lauren Mills[3]; Milcah Scott[3]; Aaron Sarver[3]; Michael Farrar[3]; Stephen Russell[2]; Shruthi Naik[2]; Jaime Modiano[3]
[1]Department of Veterinary Clinical Sciences, University of Minnesota; [2]Mayo Clinic; [3]University of Minnesota

Oncolytic virotherapy is an emerging therapeutic modality resulting in tumor destruction through viral mediated tumor lysis and induction of antitumor immunity. Vesicular stomatitis virus (VSV) is an immunogenic oncolytic virus with tropism for osteosarcoma that is safe and showed preliminary signals of efficacy in dogs with spontaneous cancer. The aims of this placebo controlled clinical study were to evaluate neoadjuvant VSV therapy and characterize anti-tumor immune responses in dogs with spontaneous appendicular osteosarcoma. Twenty-eight dogs were randomized to neoadjuvant VSV or placebo arms, followed by amputation and carboplatin chemotherapy. Tumor tissue and peripheral blood mononuclear cells (PBMCs) were collected pre- and post-treatment to characterize local and systemic antitumor immune responses. Osteosarcoma cell lines were established from each dog to characterize *in vitro* VSV susceptibility. VSV was well-tolerated with mild, transient fever and acute, clinically inapparent cytokine responses. Tumor histopathology showed focal necrosis and inflammation in tumors from VSV-treated dogs; RNA sequencing confirmed heterogeneity of immune cell infiltrates and cell cycle activation in the tumors. Massive parallel sequencing identified lymphocyte clones in tumors, lymph nodes, and PBMCs. Interim analyses suggest survival outcomes exceed those of historic control populations. Tumor pathology, *in vitro* correlates, and molecular immune responses are being correlated to survival outcomes. In conclusion, neoadjuvant VSV has an excellent safety profile, with preliminary evidence of clinical efficacy. These studies may shed light on the roles of viral oncolysis in initiating antitumor immunity and on biomarkers that are predictive of clinical efficacy in this heterogeneous canine cancer.

Abstract O10: Successful Treatment of Cutaneous Neoplasias with Electrochemotherapy in Horses

M. Carolina Duran[1]; Valentina Soto[1]; Javier Ojeda[2]
[1]Universidad Austral de Chile; [2]Universidad Austral

Electrochemotherapy (ECT) with bleomycin was used to treat different cutaneous neoplasias in horses. ECT-treatment sessions consisted in infiltration of tumoral masses with bleomycin (0,5 mg/cm^3) using a 21 G needle followed by electroporation with electric pulses from an eight-needle-electrode. All treatments were performed under general anesthesia and repeated as needed every 3–4 weeks. A total of 6 sarcoid, 5 melanoma and 4 squamous cell carcinoma (SCC) masses were diagnosed and treated. Surgical excision of all sarcoids was performed followed by ECT-treatment. The remained margins of 3 sarcoids were treated with 2 ECT-treatment sessions while three with a total of 3 sessions. Surgical excision of melanomas (large perianal masses) and SCC (upper and lower eyelid masses) was not performed. Two smaller melanoma masses required 2 ECT-treatment sessions, three bigger melanomas a total of 3 sessions. Two lower eyelid SCC masses were treated with 1 treatment session, two upper eyelid SCCs required 2 ECT-treatment sessions. Response to treatment was considered as over 70% size reduction of the treated masses. Complete response in all 15 tumour masses treated with ECT with bleomycin was obtained, no reoccurrence was noted 4 months after the last treatment. Results show that ECT with bleomycin is an effective, safe, and simple local treatment of cutaneous neoplasias in horses. Nevertheless, general anesthesia is required for each treatment session and depending on the mass, several session and surgical excision is needed.

Perianal melanoma masses **before** (a, b) and **after** (c, d) three ECT-treatment sessions

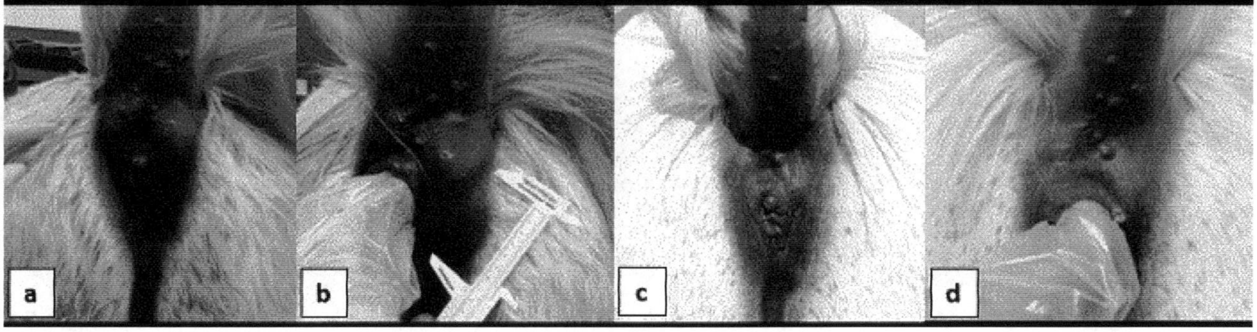

Abstract O11: Inflammatory Response and Body Composition of Bitches with Breast Tumor Supplemented with Omega-3 and Glutamine

Fabio Alves Teixeira[1]; Brana Bonder[1]; Mariana Porsani[1]; Lucas Gonçales[1]; Julio Nagashima[1]; Clair de Oliveira[1]; Júlio César Balieiro[1]; Karina Pfrimer[2]; Gabriel Santos[1]; Denise Fantoni[1]; Marcio Brunetto[1]; Cristiana Pontieri[3]
[1]School of Veterinary Medicine and Animal Science, University of São Paulo, São Paulo, Brazil; [2]Ribeirao Preto Medical School, University of São Paulo; [3]Nutritional Development Center, Grandfood Industry and Commerce LTD (Premier Pet)

There is hypothesis that immunonutrients as omega-3 fatty acids and glutamine can help treatment of oncologic patients due to their role in the immune and inflammatory response. This study aimed to verify inflammatory response and body composition of bitches with mammary tumor after mastectomy, during this three immunonutrients consumption. Under the approval of the Ethics Committee on Animal Use, twelve bitches with mammary neoplasia were divided into two groups: A—food without glutamine, EPA and DHA (91.4 g of protein; 57.1 g of fat and 5.0 g of fiber/1000 kcal); B—high protein food (60.4; 31.5 and 7.8) enriched with glutamine (3.5 g/1000 kcal) and fish oil (2.3 g of EPA+DHA/1000 kcal). They receive diets since 21 day before until 30 days after surgery. Serum measurements of TNF-α, IL-6, IL-10, IGF-1, C-reactive protein (CRP) (Milliplex Map panel) and determination of body composition (deuterium method) were performed pre- and until post-surgical moments. Statistical tests were used to compare the immunonutrients effects. There were no differences in the concentrations of different cytokines ($p>0.05$) and CRP ($p=0.51$) between groups. Group B had a higher concentration of IGF-1 ($p=0.04$), a higher percentage of muscle mass ($p<0.01$) and less body fat ($p<0.01$). The inclusion of glutamine and omega-3, in the amounts evaluated in this study, were not sufficient to modulate the inflammation of bitches with breast tumors submitted to unilateral mastectomy, but they demonstrated a potential beneficial effect on the maintenance of body muscle mass.

Abstract O13: Targeted Therapy Pevonedistat Promotes Canine Melanoma Cell Death Through DNA Replication and Senescence (VCS Award Winner)

Elizabeth A. Wood; Zhanping Lu; Shuai Jia; Anna Assumpção; Matthew Van; Mike Huelsmeyer; David Vail; Xuan Pan
University of Wisconsin-Madison

MLN4924 (pevonedistat) is a potent and selective NEDD8-activating enzyme (NAE) inhibitor. The NEDD8-regulated neddylation system is responsible for the regulated degradation of intracellular proteins with important cellular functions in cancer cell growth, apoptosis, angiogenesis and metastasis. In human melanoma, inhibition of NAE results in induction of DNA re-replication, S phase cell cycle arrest, DNA damage and apoptosis. The study aimed to assess the anti-cancer effect of MLN4924 on canine malignant melanoma cell lines and patient samples and to elucidate the underlying mechanisms. Canine melanoma cell lines and primary patient samples were evaluated for cell viability after incubation with varying concentrations of MLN4924 or dimethyl sulfoxide. Apoptosis, cell proliferation and senescence assays were performed to address underlying mechanisms of MLN4924-mediated anti-tumour effects. Gene expression of seven previously identified deregulated genes in human melanoma was compared in sensitive vs resistant samples. MLN4924 treatment significantly reduced the viability of canine melanoma cell lines and primary samples in a dose- and time-dependent manners. MLN4924 promoted cell apoptosis and inhibited cell growth through induction of DNA re-replication and cell senescence. While the majority of canine melanoma samples demonstrated sensitivity at nanomolar ranges, some samples were resistant to the treatment. Modulation of P21 levels correlated with canine melanoma cell sensitivity. These results provided justification for further exploration of MLN4924 as a treatment of canine melanoma.

Abstract O14: Receptor Tyrosine Kinase Dysregulation and Biological Activity of Toceranib against Canine Urothelial Carcinoma Cell Lines

Daniela Korec[1]; Darian Louke[2]; Joelle Fenger[1]
[1]College of Veterinary Medicine, The Ohio State University; [2]The Ohio State University

Transitional cell carcinoma (TCC) accounts for >90% of canine malignant tumors occurring in the urinary bladder. TCC tumors are generally inoperable and unresponsive to traditional chemotherapy, indicating a need for more effective therapies. Toceranib phosphate (Palladia) is a multi-target receptor tyrosine kinase (RTK) inhibitor that exhibits potent activity against members of the split kinase family of RTKs, including vascular endothelial growth factor receptor, platelet-derived growth factor receptor, Kit, and Flt-3, resulting in both direct antitumor and antiangiogenic activity. Toceranib (TOC) demonstrated single agent activity against a variety of tumor types in a phase 1 study in dogs with cancer, including several carcinomas. In this clinical trial, 3 of 4 dogs with bladder TCC treated with TOC alone had stable disease for 10 weeks or greater. Preliminary retrospective studies suggested approximately 86.7% of dogs with TCC experienced clinical benefit (partial response or clinically meaningful disease stabilization) following toceranib (TOC) treatment; however, the basis for the observed responses to TOC is not known. The purpose of this study was to evaluate normal canine bladder tissues, primary bladder TCC tumors, and established TCC cell lines for the expression and activation of VEGFR1, VEGFR2, PDGFRα, PDGFRβ, and KIT to assess whether dysregulation of these RTKs may contribute to the biological activity of TOC. To provide an initial assessment of RTK expression, Real Time PCR was performed on primary TCC tissue samples (N=10) and TCC cell lines (N=5) to detect VEGFR2, PDGFRα, PDGFRβ, and KIT mRNA. Transcript for VEGFR2, PDGFRα, and PDGFRβ was detected in all TCC tissue samples and TCC cell lines; however, mRNA for KIT was not detectable in any samples. The Proteome Profiler™ Human Phospho-RTK Array Kit (R & D Systems) provided a platform to assess phosphorylation of 42 different RTKs in primary TCC tissue specimens using the available flash frozen tumor specimens and TCC cell lines. In concordance with these data, PDGFRα, and PDGFRβ were found to be phosphorylated in all tumor samples and cell lines and KIT activation was not observed on the arrays. While message for VEGFR2 was identified in all tumor samples and cell lines, all samples exhibited low basal phosphorylation levels of this RTK. Core samples from all tumor samples and normal bladder tissues were available for evaluation and a tissue microarray was constructed to evaluate expression of receptors of interest and determine VEGFR, PDGFRα, and PDGFRβ immunoreactivity and localization in tumor cells and supporting stroma. Studies are ongoing to evaluate the *in vitro* activity of TOC on cell viability, apoptosis, and VEGFR, PDGFRα, and PDGFRβ phosphorylation in TCC cell lines. Taken together, our findings demonstrate that known targets of TOC are expressed/activated in primary TCC tumors and TCC cell lines. Given the observed phosphorylation of PDGFRα and PDGFRβ, these RTKs merit further investigation as to their role in mediating the biology of TCC and their contribution to TOC's activity.

Abstract O15: Plasma Cytokeratin 18 and Fecal Alpha-1 Antitrypsin Concentrations in Dogs with Osteosarcoma Receiving Carboplatin Chemotherapy

Kate Taikowski[1]; Adam Rudinsky[1]; Darian Louke[1]; Emma Warry[2]; Joelle Fenger[1]
[1]College of Veterinary Medicine, The Ohio State University; [2]College of Veterinary Medicine Texas A&M University

Gastrointestinal (GI) toxicity is a frequently encountered adverse side effect in both humans and dogs receiving cytotoxic chemotherapy. Chemotherapy-associated GI mucositis can manifest as pain, diarrhea, and weight loss and represents a major dose-limiting side effect of chemotherapy that has a significant impact on patient quality of life. Cytokeratin 18 (CK18), one of the major components of intermediate filaments in simple epithelial cells, undergoes caspase-mediated cleavage upon epithelial cell necrosis and apoptosis. Cellular apoptosis, a hallmark of GI mucositis, results in the release of CK18 into the circulation and data generated in humans suggest that circulating CK18 levels correlate with severity of GI mucosal cell injury induced by graft-versus-host disease and drug-associated GI mucositis. More recently, enteric markers such as fecal alpha1-antitrypsin (A1-AT) that reflect increased intestinal permeability have been shown to correlate with the severity of lacteal dilation and GI protein loss in dogs. To date, no specific mechanistic biomarkers exist to aid in the screening/monitoring of pre-clinical GI toxicity in dogs induced by cytotoxic chemotherapy. As such, the goal of this study was to evaluate the clinical utility of plasma CK18 and fecal A1-AT levels as non-invasive biomarkers of GI toxicity induced by cytotoxic chemotherapy. To address this, we conducted a prospective cohort study in client-owned dogs (n=10) diagnosed with appendicular osteosarcoma that previously underwent standard limb amputation. Dogs were treated with carboplatin chemotherapy (300 mg/m^2 IV) on day 0. Plasma was collected on days 0, 7, and 21 and CK18 concentration was measured by ELISA (ABClonal). Feces was collected on days 0 and 21 and A1-AT concentration was measured by ELISA (Creative Diagnostics). We hypothesized that plasma CK18 and fecal A1-AT levels would increase following carboplatin administration due to drug-induced GI epithelial cell damage/apoptosis, and that plasma CK18 and fecal A1-AT levels would correlate with the severity of GI toxicity. Mean baseline plasma CK18 concentration was variable amongst patients; however, mean plasma CK18 concentration prior to carboplatin chemotherapy treatment was not significantly different from CK18 levels at day 7 or day 21 (day 0/baseline =11.66±6.52 ng/ml, day 7 =13.58±8.33 ng/ml, day 21 =12.03±6.57 ng/ml, mean±SD, p=0.23). There was significant intra and inter-patient variability in mean fecal A1-AT levels at baseline. Mean fecal A1-AT concentration did not change significantly from day 0 to day 21 (day 0 =18.28±7.38 ng/ml, day 21 =15.03±7.76 ng/ml, p=0.20). In this study population, only 1 patient developed a grade 1 diarrhea, other gastrointestinal toxicity following chemotherapy was limited to grade 1 inappetence; therefore, we were unable to determine the association of plasma CK18 and fecal A1-AT concentrations with the development of signs of GI toxicosis. In this study population, plasma CK18 and fecal A1-AT concentration were not clinically useful biomarkers for the detection of GI toxicosis secondary to carboplatin administration. Further prospective evaluation of CK18 and A1-AT as biomarkers of drug-induced GI toxicity is warranted in a larger cohort of dogs receiving cytotoxic chemotherapy.

Abstract O16: A Virome Sequencing Approach to Feline Oral Squamous Cell Carcinoma to Evaluate Causative Factors (VCS Award Winner)

Shirley Chu[1]; Jeffrey Bryan[1]; Obi Griffith[2]; Gayle Johnson[1]; Zachary Skidmore[2]; Kristine Wylie[2]; Todd Wylie[2]
[1]University of Missouri; [2]Washington University

Feline oral squamous cell carcinoma (FOSCC) may be the best naturally-occurring model of human head and neck squamous cell carcinoma (HNSCC). HNSCC can be broadly divided into human papillomavirus (HPV)-negative cancers and HPV-positive cancers where HPV is the causative agent. Previous studies in FOSCC have used both species-specific and species-nonspecific PCR primers that may be insensitive to the detection of PVs and other viruses that may be divergent from known sequences. ViroCap is a targeted capture and next generation sequencing tool that was designed to identify all known vertebrate DNA and RNA viruses. In this study we used a metagenomic approach using ViroCap for DNA viruses in 20 FOSCC, 9 normal feline oral mucosal, and 8 suspected PV positive control samples. We tested the hypothesis that viruses would be enriched in FOSCC compared to normal oral mucosa. The virome of the FOSCC and normal feline oral mucosa consisted of feline foamy virus in 7/20 and 2/9 (35% and 22%), feline torque teno virus in 2/20 and 0/9 (10% and 0%), alphaherpesvirus in 2/10 and 0/9 (10% and 0%), FIV (0% and 22%), Epstein-Barr virus in 1/20 and 0/9 (5% and 0%) and feline papillomavirus in 1/20 and 0/9 samples (5% and 0% respectively). *Felis catus* papillomavirus-3 was found in 1 of 20 FOSCC samples. A virus was not associated consistently with FOSCC. If PVs have a role in FOSCC it is at most a supplementary or uncommon role. FOSCC appears most closely related to HPV-negative HNSCC. Future research on FOSCC should focus on identifying genetic and environmental causes.

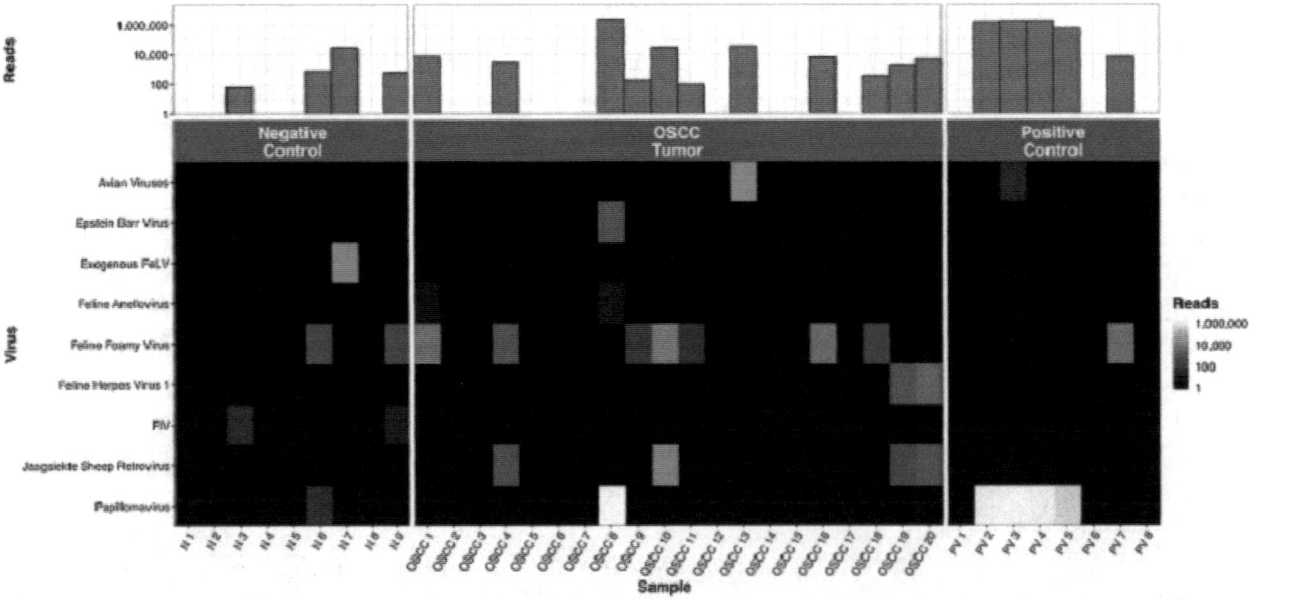

Abstract O17: Concurrent Treatment of Multiple Canine Mast Cell Tumors with Intratumoral Tigilanol Tiglate

Pamela D. Jones; Graham Brown; Justine Campbell
Qbiotics

Aims: To determine the safety and efficacy of tigilanol tiglate for the concurrent treatment of up to 3 mast cell tumors (MCT). Methods: Eighteen dogs diagnosed with forty-eight MCT were treated with tigilanol tiglate (1 mg/ml) by intratumoral injection. Patients received corticosteroids, H_1 and H_2 antagonists as part of treatment regime to minimise potential effects of degranulation reaction. On the day of treatment, the volume for each individual MCT was measured using the modified ellipsoidal method and the dose of tigilanol tiglate determined based on calculated volume of 0.5 mg per 1 cm^3 of tumor (50% v/v). Dogs were excluded if total tigilanol tiglate dose exceeded 0.25 mg/kg bodyweight or was greater than 5 mg. Patients were clinically assessed at days 1, 7, 14 and 28, and all adverse events recorded using VCOG classification. Efficacy was assessed for each treated tumor at day 28 as wither complete response (CR) or not-CR using RECIST criteria. Results: Forty-five tumors (93%) had achieved a CR at day 28 following a single injection of tigilanol tiglate. Only 3 tumors on 3 separate dogs did not achieve CR. All adverse events were graded mild to moderate with no or minimal intervention required. Conclusions: This small pilot study suggests that tigilanol tiglate was well tolerated and efficacious for concurrent local treatment of multiple MCTs. Clinical significance of the results: Tigilanol tiglate can be used for concurrent treatment of multiple canine MCTs.

Abstract O18: Arterial Thromboembolism of Metastatic Malignant Neoplasm Origin in Two Cats

Fernanda Vieira Amorim da Costa; Gabriela Schaefer; Isabella da Silva; Tayná Veronezi; Bárbara Rivas; Paula Ribeiro; Cristiano Gomes; Saulo Pavarini
Universidade Federal do Rio Grande do Sul

In cats, arterial thromboembolism (ATE) is an acute and severe clinical condition often related to hypertrophic cardiomyopathy, but there are few reports of its association with neoplasms. The aim of this study is to describe two cases of ATE of neoplastic origin. The first case report is a 14-year-old crossbreed female cat presented with sudden pelvic limb paralysis during 24 hours. Clinical and laboratory alterations included absence of a bilateral femoral pulse, hypothermia, hypotension, bradycardia, neutrophilic leukocytosis, lymphopenia, hyperproteinemia, azotemia, metabolic acidosis, hyperlactatemia, increased serum activity of ALT and hyperkalemia. The electrocardiogram indicated sinoventricular rhythm, and echocardiogram evaluation showed no alterations. Thorax radiographic evaluation revealed small areas of radiopacity in the pulmonary fields. Doppler ultrasonography evidenced a cranial aortic thrombus at the bifurcation of iliac arteries. The patient was hospitalized for clinical stabilization, which included fluid therapy, tramadol, dipyrone, acepromazine, heparin and clopidogrel. Abdominal aorta arteriotomy for thrombus removal was performed 24 hours after patient's admission. Blood gas analyses, performed sequentially, showed persistent metabolic acidosis and hyperkalemia, despite treatment with sodium bicarbonate, regular insulin associated with glucose, and calcium gluconate. The cat died in the postoperative period. Necropsy examination was compatible with pulmonary adenocarcinoma with lymph node metastasis. The second case report is a 12-year-old crossbreed female cat with a history of mastectomy two days earlier due to mammary carcinoma. Previous thorax radiographic evaluation revealed a discrete opacification of lung fields, and echocardiogram showed no alterations. The cat was admitted presenting acute pain and paralysis of the pelvic limbs. Physical examination revealed hypothermia, cold and painful pelvic limbs and, cyanotic foot pads. Laboratory findings demonstrate neutrophilia, thrombocytopenia, azotemia, increased serum activity of ALT and creatine phosphokinase (CPK). Clinical stabilization included fluid therapy, methadone, ampicillin, heparin and clopidogrel. During hospitalization, serum ALT and CPK levels decreased, but azotemia worsened significantly (creatinine 15.7 mg/dL). In addition, the cat developed oliguria (urine output <0.4 ml/kg/h) which was compatible with severe acute kidney injury (AKI). The patient presented hypotension and died few hours later. Necropsy showed several areas of nodular metastases in the lungs, compatible with metastatic mammary carcinoma by histopathological examination. There was a fibrin thrombus obstructing the aortic trifurcation and a firm whitish mass dorsal to that, which on microscopic examination showed neoplastic proliferation identical to those found in the lungs. Pelvic limb muscles showed areas of muscle necrosis. Infarction was observed on the left kidney, as well as several thrombosis areas. Both cats in the present study had major complications that were difficult to manage, such as ischemia and reperfusion syndrome associated with persistent hyperkalemia and AKI, culminating in death. The prognosis of this condition may be poor and neoplastic thromboembolism should be included as a differential diagnosis of cardiogenic thromboembolism.

Abstract O19: Anticancer Drug Treatment Increases Cancer Stem Like Cells in Canine Lymphoma Cells

Tomoko Okusa; Kenji Baba; Masaya Igase; Satoshi Kambayashi; Takuya Mizuno; Masaru Okuda
Yamaguchi University

Canine lymphoma is a type of malignant hematological tumors that develops frequently. The multidrug combination therapy used for the treatment is known to have a good outcome with great sensitivity. However, even in those good outcome cases, recurrent lymphoma occurs, resulting in their deaths. This is caused by the anticancer drug resistance. In human, a sub-population of lymphoma cells showing the cancer stem cell (CSC) like characteristics has been reported to increase by exposure to anticancer drugs. The purpose of this study is to clarify the relationship between the CSC characteristics and the anticancer drug resistance in canine lymphoma cells. In this study, the drug resistant cells were prepared using the canine lymphoma cell lines by continuous exposure to ACNU, doxorubicin or L-asparaginase, or by exposure to a HDAC inhibitor, phenyl butyrate (PB), for 48 hours. A colony formation ability in the soft agar, that was one of the CSC like characteristics, was high in the resistant cells, and it is thought that this CSC like characteristics participates in the resistant acquisition of the cancer cells. In particular, in the group exposed to PB, the colony formation ability was nearly twice higher than those in the unexposed group. It is necessary to identify a factor related with the drug resistance and the CSC like characteristics in the near future.

Chemotherapy-Induced Sepsis

Danielle DeCormier, LVT, VTS (Oncology), CCFP
Virginia Beach, VA, USA

INTRODUCTION

Chemotherapy is a common treatment used within veterinary oncology. There are several chemotherapeutic agents used to treat a variety of different types of cancer. Despite their differing mechanisms of action, the goal of each is the same: kill cancer cells. While doing their job, there are some expected, although not common, side effects. These include GI disturbances, as well as hematologic changes like neutropenia. While cancer is not often thought of as an emergency, chemotherapy-induced sepsis is one of the rare occurrences. Sepsis is generally thought of within the realm of critical care. While fundamentally the same, there are some very specific differences in approach to the cancer patient. These differences extend to the treatment and expectations for recovery as well.

CHEMOTHERAPY

Chemotherapy Basics

Chemotherapeutics are not widely used in veterinary medicine. They are often reserved for use within specialty veterinary practices by trained oncology teams. Most veterinary technology programs spend about one lecture period on their use. Due to this, not many technicians understand how these agents work. In very simple terms, chemotherapy acts against rapidly dividing cells and kills them. The different drugs do this in different ways, and not all of them are understood. Cancers grow and spread by shutting down or bypassing the body's normal cellular patterns and the cells reproduce unchecked. This is how chemotherapy works at the most basic level. Most chemotherapy, however, is not target or cancer specific. Other cells within the body are also rapidly dividing at the same time as cancer cells. The most common are those within the GI tract and white blood cells.

Side Effects

Some of the most commonly known facts about chemotherapy are the side effects. In humans, as well as cats and dogs, anorexia, nausea, and immune suppression can be expected. In humans, these are more pronounced because the doses are higher and the outcome is remission. In veterinary medicine, the goals center around quality of life, so the doses are much lower and the incidences of side effects are also lower. It is important to note that not all drugs affect animals the same way as human beings. A prime example is hair loss. This happens to most people undergoing chemotherapy treatment, but rarely in dogs, and almost never in cats, except their whiskers. This is due to the fact that human hair is most frequently in the growing phase, when it can be affected by chemotherapy, leading to balding. This is also why it grows back after therapy is finished or discontinued. Animal fur is usually in the resting phase, and only affected when it is actively growing. This is why it can appear thin and why it is slow to grow back after shaving for diagnostics or catheter placement. The exceptions to this are those breeds that need grooming due to constantly growing fur like bichons, sheepdogs, poodles, etc. The previous side effects mentioned are also due to chemotherapy causing cell death among the rapidly dividing cells of the GI tract and bone marrow. When it comes to the GI tract, inflammation and damage generally occur 3–5 days after the administration of chemotherapy. This is why clients are often instructed to watch for anorexia, vomiting, or diarrhea during this time. It is often self-limiting and resolves by about 6–9 days post-administration. The anorexia is usually a sign of nausea and the precursor to vomiting. The sooner these can be addressed with antiemetics and antidiarrheals, the better the experience will be for the pet and client.

Neutropenia

Bone marrow suppression caused by chemotherapy affects more than just the white blood cells, but the specific cell cycle of each cell type matters. Anemia is rarely caused by chemotherapy because red blood cells have a life span of 7–120 days. Chemotherapy affects the cells during reproduction, so only a small percentage of the red blood cell population is reproducing at one time. This is generally mild and only significant in the already anemic patient. Platelets are similar with their 5–10-day lifespan and are generally only affected with chronic chemotherapy use. Granulocytes have the shortest lifespan, with neutrophils having a lifespan of 4–8 hours. This gives chemotherapy time to disrupt a majority of cell division and replication. Each chemotherapy drug has what is referred to as the nadir period. This is when the neutrophil count is expected to be at its lowest point post-administration. For many of them, this is 7–10 days after administration, but a few have longer nadirs of 10–14 days, and some even have delayed nadirs for 14–21 days. This is the ideal time for a CBC to be checked, but also when getting an accurate history from a client is key. When the neutrophil count is below 1,000/µL, there is an increased chance for infection and sepsis.

Sepsis

Definitions

Sepsis is defined as the systemic inflammatory response to infection. Common sites for the bacteria to originate from in most septic cases are the GI tract, respiratory tract, severe dental disease, chronic urinary tract disease, and contaminated wounds. Severe sepsis includes the above but also associated with organ dysfunction and hypoperfusion or hypotension. Septic shock adds the layer of hypotension that is refractory to volume expansion. All of these are due to the prolonged presence of bacteria within the bloodstream. All types of sepsis are medical emergencies. Treatment for the septic patient includes hospitalization and immediate corrective therapies and antibiotics. Extensive diagnostics are often needed to find the cause and obtain the full extent of the condition. Cancer patients differ, as the cause is often known.

Predisposing Factors

Cancer offers multiple predisposing factors to sepsis. These include immune dysfunction caused by the cancer itself or treatment with corticosteroids, prevention of antibody production by splenectomy, malnutrition, prolonged hospitalization, and chemotherapy-induced. The most commonly studied and discussed is the chemotherapy-induced. How chemotherapy causes neutropenia and the resulting susceptibility to infection has already been discussed.

Presentation

Chemotherapy-induced sepsis can look like many other issues in cancer patients. The timeline and history obtained by clients is extremely important. Signs in the patient can be vague and nonspecific: lethargy, weakness, inappetance, and febrile. It is important to note that the lack of a fever does not impact the severity of the disease, and the patient's overall health must be taken into account. The nadir period of the drug must be considered and a CBC always performed. The absolute count of neutrophils is the most important number, not the percentage or total white blood cell count.

In general, septic patients will present with tachycardia, brick-red mucous membranes (increased CRT), hypotension, altered mentation, and GI signs. At first they can also have hyperglycemia followed by hypoglycemia. As it progresses to shock, the symptoms move to the other end of the spectrum with hypothermia, mm pallor, depression, and multi-organ failure. This can be avoided in the chemotherapy patient by strict adherence to follow-up exams and CBCs after any new chemotherapy treatment.

Treatment and Prevention

As previously discussed, most septic patients require extensive and aggressive treatment. Oncologists have a chart they use as a guideline for the treatment of neutropenia following chemotherapy and the possible sepsis:

	Prophylaxis	Grade 2/Mild toxicity	Grade 3/Mod toxicity	Grade 4/Severe toxicity
Neutropenia		**1000/μL**	**500–999/μL**	**<500/μL**
Broad-spectrum aerobic antibiotics	Not recommended	No	Oral Repeat CBC in 2–3 days Do not hospitalize	Oral IV if fever CBC in 24 hours
Broad-spectrum anaerobic antibiotics	No	No	No	Not routine unless refractory to aerobic antibiotics
Parenteral fluids (SQ or IV) and supportive care	No	No	Not routine unless febrile	Hospitalize if febrile

The chemotherapy patient is at a higher risk for infection and sepsis if they are hospitalized. So if they do not have a fever and are continuing to eat, it is best to send them home on oral antibiotics. The neutrophils have a rapid rebound rate and are often increasing 1–3 days after the nadir period. With this, the risk of infection is declining. Once a patient has experienced bone marrow suppression leading to neutropenia, the dose is altered to avoid it a second time. Prophylactic antibiotics are also often prescribed to be started along with the nadir period, usually 3–5 days after chemotherapy administration, through their next recheck and CBC appointment.

Conclusion

While chemotherapy-induced sepsis is a form of sepsis at the basic level, it is handled very differently. This side effect is prepared for and not out of the blue, as other emergent conditions can be. There are prescribed protocols to follow and the client can be prepared for this worst-case scenario, as well as what the plan will be ahead of time.

References

1. Biller B, Berg J, Garrett L, Ruslander D, Wearing R, Abbott B, *et al.* (2016). 2016 AAHA oncology guidelines for dogs and cats. *Journal of the American Animal Hospital Association*. 2016;52(4):181–204. doi: 10.5326/jaaha-ms-6570
2. Crump K, Thamm DH. *Cancer Chemotherapy for the Veterinary Health Team*. Chichester, West Sussex, UK: John Wiley & Sons; 2011.
3. Fullerton E. (n.d.). Chemotherapy-induced side effects: prevention and treatment. Retrieved from https://todaysveterinarynurse.com/articles/featureoncologychemotherapy-induced-side-effects-prevention-and-treatment/; Fall 2018.
4. MacDonald V. Chemotherapy: managing side effects and safe handling. *The Canadian Veterinary Journal = La Revue Vétérinaire Canadienne*. 2009;50(6):665–668.

5. Sorenmo KU, Harwood LP, King LG, Drobatz KJ. Case-control study to evaluate risk factors for the development of sepsis (neutropenia and fever) in dogs receiving chemotherapy. *Journal of the American Veterinary Medical Association*. 2010;236(6):650–656. doi: 10.2460/javma.236.6.650
6. Manuals Staff (n.d.). Infections caused by bacteria - Special Pet Topics. Retrieved from www.merckvetmanual.com/special-pet-topics/infections/infections-caused-by-bacteria.
7. Withrow SJ, Vail DM, Page RL. *Withrow and MacEwen's Small Animal Clinical Oncology*. St. Louis, MO: Elsevier; 2013.

Application of the International Myeloma Working Group Response Criteria in Dogs with Multiple Myeloma

A. Russell Moore; R. Harris; Christina Jeffries; Kate Vickery; Paul Avery
Colorado State University

Current recommendations for monitoring disease progression and response to treatment of immunoglobulin producing tumors in humans includes evaluation of serum or urine paraprotein (M-protein) concentration, involved free light chains and tumor size. Densitometry, the preferred method for qualifying M-proteins in human medicine, has recently been validated for use in the dog and was shown to outperform Ig-class specific radial immunodiffusion (RID) and ELISA based methods. A retrospective pilot study was performed to evaluate the applicability of the human based International Myeloma Working Group (IMWG) response criteria in dogs with multiple myeloma. Archives from 2016 through 2019 were reviewed for canine cases with a diagnosis of multiple myeloma, M-protein documented by serum protein electrophoresis (SPE) and immunofixation (IF) initial sample, and subsequent electrophoretic evaluation electrophoretic evaluation of serial samples. M proteins were quantified densitometrically from the electrophoretic study. When M-protein was not discernable on serial samples by SPE alone, repeat IF was performed, according to IMWG recommendations. Change in M-protein concentration was calculated and case progression was characterized using the IMWG response criteria. Available clinical history was reviewed. Survival was calculated as the time from initial electrophoretic evaluation to death or last known contact. Fifteen cases met inclusion criteria. This included a total of 69 SPE evaluations. A pre-treatment sample was evaluated in ten cases. All cases received some form of chemotherapy but this was not standardized. One case had a presumed solitary lesion with attempted surgical excision, but histopathology documented incomplete excision and chemotherapy was instituted. Total protein was within normal limits in 39 of 69 samples. Biochemically derived globulins were measured in 27 samples and were within normal limits in 13 samples. Complete response (CR, lack of M-protein by SPE and IF), was documented in one case. Six cases achieved very good partial response (VGPR, M-protein decreased by >90% but still detectable by IF) and an additional four cases met the criteria for a partial response (PR, 50–90% M-protein reduction). The cases that attained CR, VGPR or PR survived longer (median 630 days) than those that did not meet partial response criteria (M-protein change did not attain 50% decrease, median 333 days), log rank p=0.013, Figure 1. During treatment, four cases had M-protein defined progressive disease (>25% and minimal 0.5 g/dL M-protein increase) which correlated with concurrent or subsequent clinical deterioration. The majority of canine cases met IMWG criteria for a VGPR or PR although it is notable that only one dog achieved a CR. Preliminary data indicates that evaluation of electrophoresis-based M-protein detection and quantification using the human response criteria is possible in dogs with multiple myeloma and suggests that response measured in this manner correlates with survival. Evaluation of a larger cohort of cases in a prospective study is warranted.

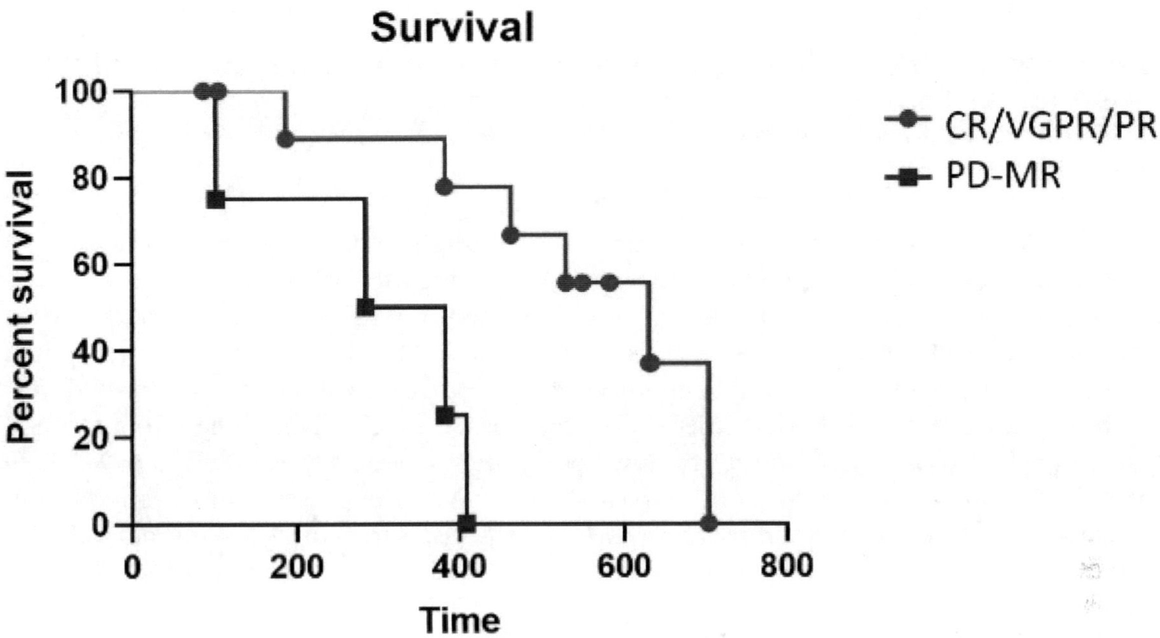
Survival
Percent survival
100
80
60
40
20
0
0
200
400
600
800
Time
CR/VGPR/PR
PD-MR

The Selective Inhibitor of Nuclear Export Verdinexor Exhibits Biological Activity Against Canine Osteosarcoma Cell Lines

Joelle M. Fenger; Justin Breitbach; Darian Louke; Mauria Watts; Alex Davies
The Ohio State University

Osteosarcoma (OS) is the most common malignant bone tumor in dogs. Despite aggressive surgical management and systemic chemotherapy, 90% of dogs still die due to chemotherapeutic-resistant metastatic disease. As chemotherapy remains the backbone for treatment of OS metastases, the development of combinational treatments with novel targeted molecular therapeutics and conventional chemotherapy represents a potential strategy to enhance therapeutic response. Exportin 1 (XPO1, also known as CRM1) is a chaperone protein responsible for the export of >200 target proteins out of the nucleus. Dysregulation of XPO1 activity is documented in a number of human cancers and XPO1 expression has been associated with the development of chemotherapy resistance. Prior studies in canine melanoma and mammary carcinoma cell lines have demonstrated that XPO1 is a relevant target for therapeutic intervention and recent phase I and II clinical trials evaluating a novel, orally bioavailable Selective Inhibitor of Nuclear Export (SINE) Verdinexor in dogs with spontaneous cancer demonstrate anti-tumor activity against non-Hodgkin lymphoma. In the present study, we sought to characterize the expression of XPO1 in primary canine OS tumor samples, OS cell lines and normal osteoblasts and evaluate the *in vitro* efficacy of Verdinexor alone or in combination with doxorubicin in canine OS cell lines. Real time PCR and Western blotting was performed to assess XPO1 transcript and protein expression in normal canine osteoblast cells, canine OS cell lines and primary OS tumor tissues. Canine OS cell lines and a subset of primary OS tumors had increased expression of XPO1 mRNA and protein compared to normal canine osteoblasts. All canine OS cell lines exhibited dose-dependent growth inhibition and increased caspase 3,7 activity in response to low nanomolar concentrations of Verdinexor (IC50 concentrations ranging from 21–74 nM). Notably, growth inhibition of normal canine osteoblast cell lines treated with Verdinexor was only observed at high micromolar concentrations (IC50 21 μM). The combination of Verdinexor and doxorubicin resulted in potent inhibition of cell viability and demonstrated synergetic activity in three canine OS cell lines. Concordantly, OS cell lines showed increased DNA damage as assessed by γH2AX immunofluorescence following combination treatment with Verdinexor and doxorubicin compared to either single-agent. These findings demonstrate that Verdinexor has biologic activity against canine OS cell lines at physiologically relevant doses and suggest that XPO1 inhibition in combination with standard doxorubicin treatment offers promising potential for chemotherapeutic intervention in canine OS.

Adoptive Natural Killer Cell Immunotherapy for Canine Cancers

William C. Kisseberth, DVM, PhD, DACVIM (Oncology)
Columbus, OH, USA

Introduction

Natural killer (NK) cells play an important role in the elimination of virally infected and malignant cells. NK cells are characterized by their ability to kill tumor targets without prior sensitization, and, unlike T cells, do not cause graft-vs.-host disease (GVHD) or cytokine release syndrome (CRS). NK cell numbers are decreased and function is impaired in many cancer patients; chemotherapy, radiation and anesthesia further deplete this essential subpopulation of immune effector cells. Further, immunosuppressive cytokines, such as TGFβ, in the tumor microenvironment cause further suppression of cellular immune responses. Thus, the core hypothesis of adoptive NK cell immunotherapy is that there exists a natural defect in innate immunity (a combination of cancer-induced reduction in NK cell numbers and immunosuppressive mechanisms resulting in suppressed function) that can be restored by adoptive transfer of NK cells.[10] In this presentation, we will review NK cell biology, the basis of NK cell immunotherapy, clinical trials of NK cells in human cancers, identification of canine NK cells and clinical studies of NK cells in dogs. Finally, an ongoing phase 1 study of adoptive NK cell immunotherapy in dogs with appendicular osteosarcoma will be discussed.

NK Cell Biology

NK cells are innate lymphoid cells that represent 5–15% of peripheral blood lymphocytes in humans.[1] NK cells have both cytotoxic and regulatory activity (reviewed[2]). They recognize malignant or virus-infected cells expressing danger signals (e.g., viral proteins, stress ligands, antibodies).[2] Their recognition of stressed target cells does not require antigen presentation and is not antigen specific; this is in contrast to T cells. In humans, NK cells are identified as CD3(-) lymphocytes that express CD56. NKp46 is an activating receptor broadly expressed on mature NK cells in most mammalian species, including dogs.[3,4]

NK cell activation is not antigen specific and represents a balance between activating and inhibitory signals. This is achieved through a balance between activating receptors for recognizing "danger" and inhibitory receptors that identify self. The net balance between these signals determines activation of NK effector function. NK cells express three main receptor families: natural cytotoxicity receptors (NCRs), C-type lectin receptors (CD94/NKG2), and killer cell immunoglobulin like receptors (KIRs) (reviewed[5]). NCRs are all activating receptors, while C-type lectin and KIR families include activating and inhibitory receptors. Activating receptors recognize ligands found commonly on tumor cells and virus-infected cells. Inhibitory receptors protect healthy cells from NK cell killing. MHC class I, expressed by healthy cells, provides NK cells with a "self" signal that is recognized by NK cell inhibitory receptors, preventing NK cell self-reactivity. Thus, NK cells use inhibitory receptors to recognize MHC class I as evidence of self and receive negative signals preventing self-attack, whereas activating receptors identify activating ligands associated with danger on target cells. NK cell activation thus is determined by a balance between danger and self-signals.[5]

Upon recognition of MHC class I, inhibitory KIRs send signals shifting the balance toward inhibition and protecting (normal) self. There are 15 functional KIR genes in the human genome; however, the number in the dog has yet to be determined (reviewed[6]). The absence of

MHC class I recognition is not in itself sufficient to induce NK cell activation—the presence of danger signals on target cells must be recognized by activating receptors. C-type lectin receptors can have activating or inhibitory functions. Structurally, they can be homodimers or heterodimers, with the heterodimeric forms consisting of CD94 and a Natural Killer Group 2 (NKG2) family member and the homodimeric form consisting of two NKG2D molecules (humans, mice).[7] Many other activating receptors can be found on NK cells, which rely on adaptor proteins to transduce activating signals. Activating receptors also include natural cytotoxicity receptor (NCR) family (NKp30, NKp44, NKp46), 2B4, DNAM-1 and others (reviewed[5]). NK cells also recognize antibody-coated cells and mediate target cell lysis through antibody-dependent cell-mediated cytotoxicity (ADCC).[8] NK cells bind the Fc end of IgG_1 and IgG_3 antibodies coating tumor cells, overriding any inhibitory signals. For many monoclonal antibody therapies, efficacy is highly dependent on this activity of NK cells.[8]

Upon activation, NK cells promote tumor cell death through multiple actions, including release of cytotoxic granules, death receptor pathway activation, and release of cytokines such as IFN-γ. NK cell degranulation results in release of perforin and granzyme into the immune synapse, which then induces target cell apoptosis through activation of caspase 3. Death receptor pathway activation results in increased expression of FASL and TRAIL, which interacts with death receptors FAS and TRAILR (DR4, DR5), inducing caspase 8 activation and apoptosis. Secreted IFN-γ stimulates antigen presentation to T cells and has anti-angiogenic effects (reviewed[5]).

NK Cell Immunotherapy

Adoptive NK cell therapy to increase the number and improve the function of NK cells has been tested in different settings and with different methods in people.[5,9] In many of these studies, NK cells were harvested by apheresis from healthy donors followed by T cell depletion to avoid GVHD and cytokine activation. This approach typically yields 2–5x10^6 NK cells/kg recipient weight delivered as a single dose. To circumvent many of the problems with this approach (cost, inconvenience, risks of apheresis), *ex vivo* NK cell propagation methods have been developed.[5,9] Historically, NK cells have been difficult to grow *in vitro* using cytokines with or without feeder cells. Interleukin 2 and 15 have been the most common cytokines used to support *in vitro* NK cell expansion in humans. More recently, the addition of IL-21 was found to further support NK cell proliferation and enhance cytokine production.[10] It is believed that different culture methods give rise to NK cells with different immunophenotypes.

Human Trials

In solid tumors, autologous lymphokine-activated killer (LAK) cell therapy with concomitant administration of IL-2 was investigated in the 1980s for the treatment of solid tumors nonresponsive to standard therapy.[11] Although a phase III trial for primary lung carcinoma patients using this approach was promising (5-year survival rate of 54.5%, compared to 33.4% in the control group, $p<0.001$), the approach had significant limitations including toxicity, cost, and logistics.[12]

A clinically relevant role for NK cells in the treatment of human hematopoietic malignancies was first demonstrated in the setting of HLA-mismatched hematopoietic stem cell transplant.[13] It was observed that when haploidentical stem cells from a donor matched for only one set of HLA alleles were used as an alternative stem cell source for acute myeloid leukemia (AML) patients lacking a HLA-matched donor, better outcomes (increased survival and decreased relapse) were obtained. This was attributed to NK cells from haploidentical donors finding that recipient leukemia cells had HLA differences and thus more likely to trigger NK cell activation. This results in a graft versus leukemia effect from direct killing of leukemia cells. Similarly, in the setting of HLA identical stem cell transplants for patients with AML or myelodysplastic syndrome (MDS), patients who were missing inhibitory KIR ligands had decreased relapse compared to those having all three ligands.[14]

Over 300 infusions in over 100 patients have been delivered to date in human clinical trials using NK cells expanded with K562-based feeder cell (mbIL15 or mbIL21) alone or with other agents.[10] These studies have demonstrated early evidence of efficacy with no dose limiting toxicities. Importantly, in the combined experience of a variety of different NK cell therapeutic products delivered to over 500 patients, there is only one report of GVHD related to NK cell infusions.[10]

Identification of Canine NK Cells

The surface markers and phenotypes at different stages of NK cell differentiation have not been defined in dogs, thus phenotypic identification of canine NK cells with surface markers has been problematic. In humans, NK cells can be defined most broadly as a CD3(-) CD56(+) lymphoid population with the $CD56^{bright}$ subpopulation generally expressing less CD16 and exhibiting greater cytokine production and $CD56^{dim}$ cells expressing high levels of CD16 and exhibiting potent cytokine production.[15] Canine NK cells are CD4(-) CD20(-) lymphocytes; however, as much as 30% of the population may express CD8. $CD5^{dim}$ cells, comprising about 15% of isolated canine PBMCs have been proposed as a candidate phenotypic marker for canine NK cells, as they fulfill many of the functional characteristics of NK cells, especially in the setting of IL-2 expansion and activation.[16] Additional NK cell phenotypes have been proposed based on *in vitro* expansion of cytotoxic large granular lymphocytes.[17] NCR1/NKp46 is widely considered to be a pan-species NK cell marker for mammals and has been identified on human, primate, rat, mouse, ovine and bovine NK cells.[3] Using a canine NKp46 antibody, CD3(-) NKp46(+) and CD3(-) NKp46(-) cell populations can be identified, with the CD3(-) NKp46(+) population comprising approximately 2–3% of the PBMCs.[18] These cells were more cytotoxic than the NKp46 (-) population. CD94 has recently been investigated as a candidate marker for canine NK cells using a canine-specific antibody. This antibody labels approximately 7.7% of isolated PBMCs. These cells were noted to have higher cytotoxicity than resting PBMCs.[19] Neither CD16 nor NKG2D, critical surface markers for human NK cells, have been conclusively demonstrated to be expressed on canine NK cells.

Canine Clinical Studies

Depending on the phenotypic definition used, NK cells represent between 2.5–15% of PBMCs under resting conditions.[6] As in people, different methods of expanding canine NK cell populations have been studied, including exposure of purified NK cells to different cytokines (IL-2, IL-12, IL-15, IL-21) and co-culture with feeder cell lines (EL08-1D2, K562).[6] Clinical studies of NK cells in dogs have relied on *ex vivo* expansion techniques. In two studies, canine large granular lymphocytes isolated from controls and dogs with cancer were stimulated with recombinant human IL-2 and cytotoxicity measured against the CTAC cell line.[20,21] Cytotoxicity was decreased in the cancer patients. Only a single study has been reported to date describing adoptive NK cell immunotherapy in the dog.[22] In this study, canine CD5-depleted NK cells were expanded with recombinant human IL-2 and genetically modified K562 feeder cells transduced with 4-1BBL (CD137L) and membrane-bound recombinant human IL-21. Cytotoxicity of the expanded NK cells was demonstrated in a soft tissue sarcoma PDX tumor *in vitro* and in a mouse model. Dogs with primary osteosarcoma tumors received palliative radiation therapy and two intratumoral injections of autologous expanded NK cells and recombinant human IL-2. Overall, there was limited systemic toxicity with one case of fever, chills, excessive salvation, and dehydration consistent with IL-2 toxicity. There were 3 cases of infection/tissue breakdown at the NK cell injection site. Although only a small number (10) of dogs were treated, dogs in the trial were found to have improved progression-free survival and decreased metastases when compared to historical controls.[22]

We have initiated a phase I clinical trial of adoptive allogeneic NK cell therapy in dogs with appendicular osteosarcoma and receiving standard of care limb amputation and adjuvant carboplatin chemotherapy. In this clinical trial, canine allogeneic *ex vivo* expanded NK cells are administered pre- and peri-operatively and then weekly post-operatively on non-chemotherapy weeks for a total of four cycles of chemoimmunotherapy. The *ex vivo* expanded cell product is manufactured using the buffy coats from healthy blood bank donors. T-cell depleted cells are stimulated using IL-2, co-cultured with K562 feeder cells expressing IL-21 and cultured with TGFβ to enhance resistance to TGFβ-mediated immunosuppression.[18,23] Further details of the clinical trial design, correlative assays and preliminary results will be presented.

References

1. Almeida-Oliveira A, Smith-Carvalho M, Porto LC, *et al*. Age-related changes in natural killer cell receptors from childhood through old age. *Hum Immunol*. 2011 Apr;72(4):319–29.
2. Vivier E, Artis D, Colonna M, Diefenbach A, Di Santo JP, Eberl G, Koyasu S, Locksley RM, McKenzie ANJ, Mebius RE, Powrie F, Spits H. Innate lymphoid cells: 10 years on. *Cell*. 2018 Aug 23;174(5):1054–1066.
3. Walzer T, Bléry M, Chaix J, *et al*. Identification, activation, and selective *in vivo* ablation of mouse NK cells via NKp46. *Proc Natl Acad Sci U S A*. 2007 Feb 27;104(9):3384–9.
4. Grøndahl-Rosado C, Bønsdorff TB, Brun-Hansen HC, Storset AK. NCR1+ cells in dogs show phenotypic characteristics of natural killer cells. *Vet Res Commun*. 2015 Mar;39(1):19–30.
5. Kannan GS, Aquino-Lopez A, Lee DA. Natural killer cells in malignant hematology: a primer for the non-immunologist. *Blood Rev*. 2017 Mar;31(2):1–10.
6. Gingrich AA, Modiano JF, Canter RJ. Characterization and potential applications of dog natural killer cells in cancer immunotherapy. *J Clin Med*. 2019 Oct 27;8(11).
7. Gunturi A, Berg RE, Forman J. The role of CD94/NKG2 in innate and adaptive immunity. *Immunol Res*. 2004;30(1):29–34.
8. Weiner LM, Surana R, Wang S. Monoclonal antibodies: versatile platforms for cancer immunotherapy. *Nat Rev Immunol*. 2010 May;10(5):317–27.
9. Baggio L, Laureano ÁM, Silla LMDR, Lee DA. Natural killer cell adoptive immunotherapy: coming of age. *Clin Immunol*. 2017 Apr;177:3–11.
10. Lee DA. Cellular therapy: adoptive immunotherapy with expanded natural killer cells. *Immunol Rev*. 2019 Jul;290(1):85–99.
11. Rosenberg SA, Lotze MT, Muul LM, *et al*. A progress report on the treatment of 157 patients with advanced cancer using lymphokine-activated killer cells and interleukin-2 or high-dose interleukin-2 alone. *N Engl J Med*. 1987 Apr 9;316(15):889–97.
12. Kimura H, Yamaguchi Y. A phase III randomized study of interleukin-2 lymphokine-activated killer cell immunotherapy combined with chemotherapy or radiotherapy after curative or noncurative resection of primary lung carcinoma. *Cancer*. 1997 Jul 1;80(1):42–9.
13. Ruggeri L, Capanni M, Casucci M, *et al*. Role of natural killer cell alloreactivity in HLA-mismatched hematopoietic stem cell transplantation. *Blood*. 1999 Jul 1;94(1):333–9.
14. Hsu KC, Keever-Taylor CA, Wilton A, *et al*. Improved outcome in HLA-identical sibling hematopoietic stem-cell transplantation for acute myelogenous leukemia predicted by KIR and HLA genotypes. *Blood*. 2005 Jun 15;105(12):4878–84.
15. Walzer T, Jaeger S, Chaix J, Vivier E. Natural killer cells: from CD3(-)NKp46(+) to post-genomics meta-analyses. *Curr Opin Immunol*. 2007 Jun;19(3):365–72.
16. Huang YC, Hung SW, Jan TR, *et al*. CD5-low expression lymphocytes in canine peripheral blood show characteristics of natural killer cells. *J Leukoc Biol*. 2008 Dec;84(6):1501–10.

17. Shin DJ, Park JY, Jang YY, *et al*. *Ex vivo* expansion of canine cytotoxic large granular lymphocytes exhibiting characteristics of natural killer cells. *Vet Immunol Immunopathol*. 2013 Jun 15;153(3–4):249–59.
18. Foltz JA, Somanchi SS, Yang Y, *et al*. NCR1 expression identifies canine natural killer cell subsets with phenotypic similarity to human natural killer cells. *Front Immunol*. 2016 Nov 23;7:521. eCollection 2016.
19. Graves SS, Gyurkocza B, Stone DM, *et al*. Development and characterization of a canine-specific anti-CD94 (KLRD-1) monoclonal antibody. *Vet Immunol Immunopathol*. 2019 May;211:10–18.
20. Funk J, Schmitz G, Failing K, Burkhardt E. Natural killer (NK) and lymphokine-activated killer (LAK) cell functions from healthy dogs and 29 dogs with a variety of spontaneous neoplasms. *Cancer Immunol Immunother*. 2005 Jan;54(1):87–92.
21. Funk J, Schmitz G, Bach U, *et al*. Influence of different tumour types on natural cytotoxicity (NK cell activity) and mitogen-induced lymphocyte proliferation in isolated blood lymphocytes from 110 dogs with tumours. *Res Vet Sci*. 2003 Apr;74(2):129–35. Erratum in: *Res Vet Sci*. 2003 Aug;75(1):87.
22. Canter RJ, Grossenbacher SK, Foltz JA, *et al*. Radiotherapy enhances natural killer cell cytotoxicity and localization in pre-clinical canine sarcomas and first-in-dog clinical trial. *J Immunother Cancer*. 2017 Dec 19;5(1):98.
23. Foltz JA, Moseman JE, Thakkar A, *et al*. TGFβ imprinting during activation promotes natural killer cell cytokine hypersecretion. *Cancers* (Basel). 2018 Nov 5;10(11).

Precision Medicine: Canine Cancer Genome Landscapes and Actionability Horizons

William P.D. Hendricks, PhD
Phoenix, AZ, USA

INTRODUCTION

Cancer leads to more dog deaths than any other disease. Millions of dogs, including more than half of geriatric dogs, are diagnosed with cancer each year.[1] A pressing need exists for new clinical tools (diagnostics and drugs) to empower veterinarians and pet owners to make decisions for canine patients with the hope of extending quality of life and survival. The field of "precision medicine" has matured over the past decade in human oncology at the intersection of cancer biology, genetics, genomics, therapeutics, and clinical diagnostics. While the promise of this field for bringing innovation to veterinary oncology may have often outstripped what it has delivered, we are now approaching a tipping point driven by new data, reagents, and drugs. Precision medicine is based on our understanding of cancer as a genetic disease that varies considerably between individuals and even between the same histologic tumor types. The genomics revolution of the past two decades has provided detailed maps of the underpinnings of many human cancers with the generation of hundreds of thousands of cancer genome sequences. These data are increasingly leveraged to guide development of new diagnostics, prognostics, and targeted treatments that are improving outcomes for human cancer patients. This revolution has now also crossed into veterinary oncology, particularly in pet dogs where more than one thousand cancer genomes have been published in the past five years. Although the path from mapping cancer genome landscapes to clinical translation is still under construction, the oncoming flood of genomic data stands poised to transform veterinary oncology. We and others are charting the genomic landscapes of naturally occurring canine cancers to guide development of new tools for veterinary oncology. Here we review understanding of cancer's genetic basis in humans and dogs, the genomic technology powering this understanding, and recent developments and opportunities in canine cancer genomics viewed through the lens of precision medicine.

HUMAN CANCER GENOME LANDSCAPES

Most human cancers (~80–90%) arise sporadically without a heritable genetic component, but cancer is nonetheless universally a genetic disease. Cancer's hallmark phenotypes (excessive cell proliferation, tumor formation, invasion, and metastasis) result from progressive cellular accumulation of pathogenic mutations in genes that regulate cell birth (oncogenes) and death (tumor suppressor genes).[2] Cancer gene mutations may arise through germline variation (predisposing pathogenic germline mutations) or may be acquired (somatic mutations) through environmental insult (e.g., cigarette smoke) and/or through simple replication error. Such mutations drive waves of clonal expansion in cell populations that become increasingly malignant. This model of stepwise cancer progression through mutation was established over 30 years ago in human colorectal cancer in which accumulation of mutations in cancer genes was shown to correlate with development of increasingly malignant colorectal tumors.[3]

Prototypical cancer genes include *BRAF* (one of the most commonly mutated oncogenes across cancers), *TP53* (one of the most commonly mutated tumor suppressor genes across cancers), and the *BCR-ABL* fusion formed by a translocation between chromosomes 9 and 22 (the Philadelphia chromosome) in chronic myelogenous leukemia (CML). Technology advances over the past four decades have enabled increasingly comprehensive mapping of cancer's genetic underpinnings. By the turn of the millennium, automated Sanger sequencing[4] had enabled sequencing of entire genomes, culminating in the completion of the Human Genome Project (HGP) draft by 2001.[5] The HGP's completion made possible larger-scale screens for mutations in gene families across dozens of cancers.[6] However, the invention of massively parallel sequencing technology (also known as next-generation sequencing or NGS) is ultimately what enabled the dramatic cost and throughput savings to fuel massive cancer genome sequencing projects across thousands of genes and dozens to thousands of samples. Large-scale efforts to map human cancer genomes within large teams and multi-institutional consortia began in earnest in 2005 as the National Institutes of Health launched The Cancer Genome Atlas (TCGA; www.cancer.gov/about-nci/organization/ccg/research/structural-genomics/tcga). The TCGA provided genomic data on 11,000 cases across 34 cancer types and 18 pan-cancer studies.[7] Similarly, the International Cancer Genome Consortium (ICGC; https://icgc.org)[8] included 76 projects across 21 cancer types with over 17,000 donors. A useful view of the wealth of genomic data available today is provided by the Catalogue of Somatic Mutations in Cancer (COSMIC; https://cancer.sanger.ac.uk/cosmic)[9], containing cancer mutation data curated from 26,829 papers. The September 2019 build includes 29.5 million mutations from 1.4 million samples. Additional powerful data portals include the cBioPortal for Cancer Genomics (data from 278 cancer studies) and the Genomic Data Commons (83,709 cases across 64 projects).

As the scope of cancer genome projects has grown, so has the complexity of the view of cancer's genetic basis. Not only does total mutation burden vary wildly—from pediatric cancers with very few mutations to cancers arising from mismatch-repair-deficiency or from mutagen-exposed tissues such as bladder, lung, and melanoma with hundreds to thousands of mutations—but so does the spectrum of mutations that can occur across patients with the same cancer type. As catalogued in COSMIC, 723 genes have been causally implicated in cancer. However, many of these genes converge on shared cellular pathways and phenotypic hallmarks.[2,6]

CANINE CANCER GENOME LANDSCAPES

Clear patterns of breed-specific cancer predisposition exist,[1] supporting that canine cancer is a genetic disease, likely with a higher proportion of heritable to sporadic cases relative to human cancer. A growing wealth of data has begun to elucidate canine cancer's genetic basis. The first canine cancer gene was discovered in 1999 when *KIT* mutations were found to be common in mast cell tumors.[10] At the time, the canine genome structure was still being characterized. The full high-quality sequence draft first emerged in 2005.[11] Notably, canine and human genomes share a high degree of similarity in their complement of genes as well as genomic structure, and gene and amino acid sequence. Cancer genes across the two species are particularly highly conserved with a mean shared protein identity of 91%. In the ensuing decade, numerous targeted sequencing, genome-wide association (GWAS), and cytogenomic studies identified early signs of additional canine cancer gene alterations. Many early canine cancer genetic studies have focused on identifying predisposing genetic variants from targeted sequencing or GWAS. Roles have been found in cancer for *MET* in Rottweilers,[12] *CDKN2A* in histiocytic sarcoma[13] and osteosarcoma,[14] *KITLG* in squamous cell carcinoma of the digit,[15] *BRCA1/2* and other loci in mammary tumors,[16,17] as well as *GNAI2*, hyaluronidase genes, and *DSCAM* in mast cell tumors[18]. Early studies of somatic mutations in sporadic canine cancers utilizing comparative genomic hybridization approaches to identify acquired karyotypic changes were conducted in lymphoma,[19] leukemia,[20] osteosarcoma,[21] canine transmissible venereal tumors,[22] and hemangiosarcoma[23]. This work began to uncover recurrent tumor-type- and sometimes breed-specific structural genetic alterations comparable to those seen in human cancers such as the *BCR-ABL* fusion in canine and human CML. Although stepwise accumulation of mutations in concert with pathologic tumor progression has not been as rigorously documented in dog as it has in human, nonetheless key studies have shown associations between genome-wide DNA copy number aberrations and tumor grade, such as seen in melanocytic lesions[24] and mast cell tumors[25].

In the absence of coordinated large-scale efforts, NGS-based canine cancer genomic data has emerged more slowly than human cancer data. Yet, growing investment from government, foundation, and academic institutions has begun to accelerate the pace of discovery. Today, ~1,200 canine cancer genome sequences have been published (compared to hundreds of thousands of human cancer genome sequences) across ~35 publications. Much of this work has occurred in the past few years, supported by NCI programs: the intramural Comparative Oncology Program and the Comparative Oncology Trials Consortium, extramural funding of canine and comparative cancer center supplements and U01s, the CTSA One Health Alliance, and the newly established Integrated Canine Data Commons.

Each genome landscaping study has revealed critical genomic characteristics of canine cancers that hold potential to reshape diagnosis, prognosis, and treatment in their respective tumor types. Often, overall genomic features and recurrently mutated cancer genes are shared between similar canine and human cancer types, but some unique genomic features of canine cancers are also consistently uncovered. Examples include: in B-cell lymphoma, frequent *TRAF3* mutations in 20–30% of canine (not common in human despite other shared mutations and gene expression profiles)[26,27]; in mast cell tumors, *KIT* in 20% of canine (80% of human)[25]; in osteosarcoma, shared structural complexity and *TP53* mutations (though *SETD2* mutations are relatively enriched in canine and *RB1* enriched in human)[28]; in angiosarcoma, shared *TP53* mutations (but enrichment for *PIK3CA* in canine and *PLCG1* and *PTPRB* in human)[29,30]; in human pediatric and canine glioma, shared receptor tyrosine kinase (RTK) and cell cycle genes, *TP53*, and *IDH1*[31]; in sun-shielded melanoma, shared absence of *BRAF* and enrichment for *RAS*, *MDM2*, *TP53*, and enrichment for structural changes[32,33]; in mammary carcinoma, high frequency of *PIK3CA*[34]; in both bladder cancer and prostate cancer in dogs, an enrichment for activating *BRAF* mutations relative to the human tumor types; in lung cancer, a low mutation burden in never-smokers, but an enrichment for *HER2* mutations in canine lung cancer[35]. Some notable shared features across canine cancer include frequent gains impacting *MYC* and *KIT*, as well as inactivating mutations impacting *SETD2*. Additionally, amidst an absence of canine cancer mutagenic etiologies (e.g., mismatch repair syndromes, cigarette smoking, sun damage, etc.), few if any hypermutated canine cancers have yet been found. Although much work remains to be done to comprehensively map canine cancer genome landscapes, each of these studies and others are not only uncovering new biological hypotheses, but they are also highlighting new diagnostic, prognostic, and predictive biomarkers that hold near-term translational value.

Clinical Utility of Cancer Genome Landscapes

Human cancer genomic data has consistently accelerated diagnostic and treatment innovation since the first recurrent chromosomal abnormality was discovered—the *BCR-ABL* fusion in CML. Not only is this fusion a diagnostic and prognostic biomarker, but imatinib—the first successful targeted therapy—was also developed to selectively target this gene product while preserving normal cells.[36] This mutation-targeting paradigm can sometimes vastly improve survival and therapeutic indices relative to chemotherapy. The majority of human cancer drugs approved and under development today are either immunotherapies or targeted small molecules developed against the products of genetic abnormalities (the mutant gene products themselves or pathway-level dependencies resulting from genetic changes). In fact, at least 20 FDA-approved human oncology treatment regimens (single drugs or combinations) are used in the first-line setting with required companion diagnostics for specific genetic aberrations. Response rates in these settings are significantly higher than those for non-genomically targeted therapies.[37] Hundreds of additional genomic biomarkers have also been shown via clinical and preclinical evidence to be associated with responses to approved and investigational drugs. Thus, more broadly in human cancer, genomic data now routinely supports oncology decision-making for individual patients through precision medicine tests that assay hundreds to thousands of biomarkers at once. Most academic centers and dozens of commercial laboratories now offer these human cancer precision medicine diagnostics.

Similar opportunities exist in canine cancer for the development and repurposing of genomically directed therapies and predictive biomarkers. To date, the single example of an FDA-approved drug for canine cancer remains the KIT inhibitor, toceranib (Palladia), for treatment of canine mast cell tumors that frequently bear *KIT* mutations. Unfortunately, data to support the role of *KIT* mutation as a predictive biomarker has been mixed, likely due in part to toceranib's broad activity against multiple RTKs alongside open questions about genomic background in mast cell tumors and the presence of other potential modifiers/targets of toceranib activity.[38] Nonetheless, the horizon for new canine cancer diagnostic biomarkers as well as new drug development and drug repurposing will continue to broaden as genomic landscapes are mapped with increasing resolution. For example, based on what is known of canine cancer genome landscapes today and inference from the 20 FDA-approved human treatment regimens with required predictive biomarker assessment, an estimated >20% of canine cancers may bear these biomarkers. As we continue to fill gaps in the canine cancer genome landscapes, these actionability horizons will continue to broaden.

REFERENCES

1. Dobson JM. Breed-predispositions to cancer in pedigree dogs. *ISRN Vet Sci.* 2013:1–23.
2. Hanahan D, Weinberg RA. Hallmarks of cancer: the next generation. *Cell.* 2011;144:646–674.
3. Fearon ER, Vogelstein B. A genetic model for colorectal tumorigenesis. *Cell.* 1990;61:759–767.
4. Sanger F, *et al.* Nucleotide sequence of bacteriophage phi X174 DNA. *Nature.* 1977;265:687–695.
5. Lander ES, *et al.* Initial sequencing and analysis of the human genome. *Nature.* 2001;409:860–921.
6. Vogelstein B, *et al.* Cancer genome landscapes. *Science.* 2013;339:1546–1558.
7. Weinstein JN, *et al.* The cancer genome atlas pan-cancer analysis project. *Nat Genet.* 2013;45:1113–1120.
8. Hudson TJ, *et al.* International network of cancer genome projects. *Nature.* 2010;464:993–998.
9. Forbes SA, *et al.* COSMIC: exploring the world's knowledge of somatic mutations in human cancer. *Nucleic Acids Res.* 2015;43:D805–D811.
10. London CA, *et al.* Spontaneous canine mast cell tumors express tandem duplications in the proto-oncogene c-kit. *Exp Hematol.* 1999;27:689–697.
11. Lindblad-Toh K, *et al.* Genome sequence, comparative analysis and haplotype structure of the domestic dog. *Nature.* 2005;438:803–819.
12. Liao A, McMahon M, London CA. Identification of a novel germline MET mutation in dogs. *Anim Genet.* 2006;37:248–252.
13. Shearin AL, *et al.* The MTAP-CDKN2A locus confers susceptibility to a naturally occurring canine cancer. *Cancer Epidemiol Biomarkers Prev.* 2012;21:1019–1027.
14. Karlsson EK, *et al.* Genome-wide analyses implicate 33 loci in heritable dog osteosarcoma, including regulatory variants near CDKN2A/B. *Genome Biol.* 2013;14:R132.

15. Karyadi DM, *et al.* A copy number variant at the KITLG locus likely confers risk for canine squamous cell carcinoma of the digit. *PLoS Genet.* 2013;9:e1003409.
16. Rivera P, *et al.* Mammary tumor development in dogs is associated with BRCA1 and BRCA2. *Cancer Res.* 2009;69:8770–8774.
17. Melin M, *et al.* Genome-wide analysis identifies germ-line risk factors associated with canine mammary tumours. *PLoS Genet.* 2016;12.
18. Arendt ML, *et al.* Genome-wide association study of golden retrievers identifies germ-line risk factors predisposing to mast cell tumours. *PLoS Genet.* 2015;11.
19. Thomas R, Smith K, Ostrander E, Galibert F, Breen M. Chromosome aberrations in canine multicentric lymphomas detected with comparative genomic hybridisation and a panel of single locus probes. *Br J Cancer.* 2003;89:1530–1537.
20. Breen M, Modiano JF. Evolutionarily conserved cytogenetic changes in hematological malignancies of dogs and humans—man and his best friend share more than companionship. *Chromosome Res.* 2008;16:145–154.
21. Thomas R, *et al.* Influence of genetic background on tumor karyotypes: evidence for breed-associated cytogenetic aberrations in canine appendicular osteosarcoma. *Chromosome Res.* 2009;17:365–377.
22. Thomas R, Rebbeck C, Leroi AM, Burt A, Breen M. Extensive conservation of genomic imbalances in canine transmissible venereal tumors (CTVT) detected by microarray-based CGH analysis. *Chromosome Res.* 2009;17:927.
23. Thomas R, *et al.* Genomic profiling reveals extensive heterogeneity in somatic DNA copy number aberrations of canine hemangiosarcoma. *Chromosome Res.* 2014;22:305–319.
24. Poorman K, *et al.* Comparative cytogenetic characterization of primary canine melanocytic lesions using array CGH and fluorescence *in situ* hybridization. *Chromosome Res.* 2015;23:171–186.
25. Mochizuki H, Thomas R, Moroff S, Breen M. Genomic profiling of canine mast cell tumors identifies DNA copy number aberrations associated with KIT mutations and high histological grade. *Chromosome Res.* 2017;25:129–143.
26. Elvers I, *et al.* Exome sequencing of lymphomas from three dog breeds reveals somatic mutation patterns reflecting genetic background. *Genome Res.* 2015;25:1634–1645.
27. Bushell KR, *et al.* Genetic inactivation of TRAF3 in canine and human B-cell lymphoma. *Blood.* 2015;125:999–1005.
28. Sakthikumar S, *et al.* SETD2 is recurrently mutated in whole-exome sequenced canine osteosarcoma. *Cancer Res.* 2018;78:3421–3431.
29. Wang G, *et al.* Actionable mutations in canine hemangiosarcoma. *PloS ONE.* 2017;12.
30. Megquier K, *et al.* Genomic analysis reveals shared genes and pathways in human and canine angiosarcoma. *BioRxiv.* 2019;570879.
31. Amin SB, *et al.* Comparative molecular life history of spontaneous canine and human gliomas. *Cancer Cell.* 2020;37:243–257.
32. Hendricks WP, *et al.* Somatic inactivating PTPRJ mutations and dysregulated pathways identified in canine malignant melanoma by integrated comparative genomic analysis. *PLoS Genet.* 2018;14:e1007589.
33. Wong K, *et al.* Cross-species genomic landscape comparison of human mucosal melanoma with canine oral and equine melanoma. *Nat Commun.* 2019;10:1–14.
34. Lee K-H, Hwang H-J, Noh HJ, Shin T-J, Cho J-Y. Somatic mutation of PIK3CA (H1047R) is a common driver mutation hotspot in canine mammary tumors as well as human breast cancers. *Cancers.* 2019;11:2006.
35. Lorch G, *et al.* Identification of recurrent activating HER2 mutations in primary canine pulmonary adenocarcinoma. *Clin Cancer Res.* 2019.
36. Druker BJ, *et al.* Efficacy and safety of a specific inhibitor of the BCR-ABL tyrosine kinase in chronic myeloid leukemia. *N Engl J Med.* 2001;344:1031–1037.
37. Doherty GJ, *et al.* Cancer treatment in the genomic era. *Annu Rev Biochem.* 2019;88:247–280.
38. Weishaar K, *et al.* c-Kit mutation and localization status as response predictors in mast cell tumors in dogs treated with prednisone and toceranib or vinblastine. *J Vet Intern Med.* 2018;32:394–405.

Adoptive T Cell Therapy

Nicola J. Mason, BVetMed, PhD, DACVIM (Internal Medicine)
Philadelphia, PA, USA

INTRODUCTION

Strategies to manipulate the immune system to treat cancer can be divided into two broad categories: 1) active immunotherapy which describes the use of vaccines, immune modifiers/adjuvants and cytokines; and 2) passive immunotherapy which describes either the use of antibody-based therapies that target malignant cells or modulate the immune responses to them or adoptive transfer of tumor-specific immune cells (most commonly T cells) to eliminate the tumor. Here we will focus on adoptive cell transfer (ACT) and will include the various strategies that are employed to generate large numbers of tumor antigen-specific cells, the remarkable successes that have been achieved with ACT and some of the challenges that remain to improve their feasibility, safety and efficacy. We will also briefly discuss the use of ACT in other disease types including autoimmunity and allergy. Finally, we will discuss future perspectives of ACT and how this approach is set to change the way in which cancer and other disease are managed in the future.

Adoptive cell therapy (ACT) encompasses the intravenous administration of either circulating T cells or tumor-infiltrated lymphocytes (TILs) to a patient to initiate and promote an anti-tumor immune response. While TILs harvested from the patient's own tumor are considered to contain tumor-specific effector cells, circulating peripheral T cells generally require genetic modification to endow them with tumor antigen-specificity. Two approaches are used; genetic modification with a T cell receptor (TCR) such that the TCR-T cell can now recognize tumor antigen presented in the context of MHC or genetic modification with a chimeric antigen receptor (CAR). CARs consist of an antibody-based antigen-binding domain known as a single chain variable fragment (scFv) that is linked to a hinge, (that allows the receptor to move freely on the surface of the cell), a transmembrane domain (to anchor the receptor in the cell membrane) and intracellular signaling domains that initiate T cell activation upon scFv-ligand engagement (**Figure 1**).

Figure 1. Cartoon of a chimeric antigen receptor

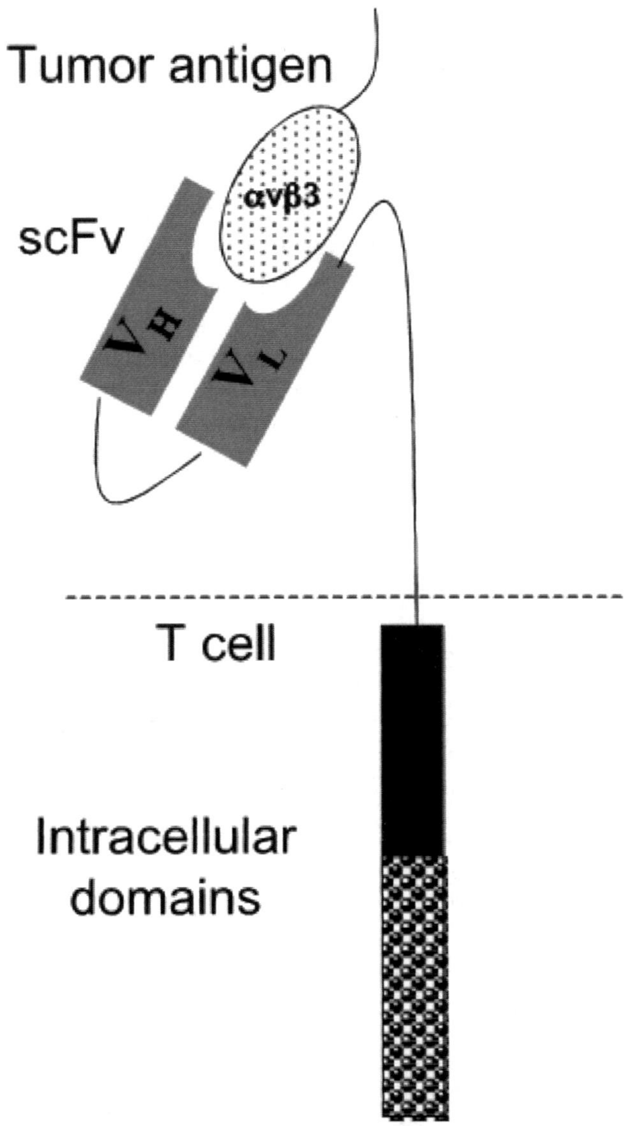

Tumor Infiltrating Lymphocytes

Over 30 years ago, it was discovered that lymphocytes that infiltrate tumors can be isolated from tumor tissue extracts and expanded in the T cell growth cytokine IL-2. Expanded TILs adoptively transferred into mice were found to mediate regression of advanced tumors and micrometastases associated with chemically induced sarcomas, adenocarcinoma, melanoma and bladder carcinoma.[1] These findings led to clinical trials investigating the use of *ex vivo* expanded TILs to mediate tumor regression in patients with metastatic melanoma. In these trials, TILs are isolated from the patient's tumor, T cells are selected, activated and expanded in the presence of IL-2 and then adoptively transferred back into the patient who has undergone pre-conditioning treatment with chemotherapy with or without total body irradiation. Pre-conditioning has been shown to be important for depletion of regulatory T cells and endogenous lymphocytes and enables adoptively transferred cells to compete for the homeostatic cytokines IL-7 and IL-15 and expand to eliminate tumor most effectively. TIL therapy has shown impressive results, for example, in long term follow up in heavily pre-treated human patients with measurable, metastatic melanoma, 22% of patients achieved a complete remission (20/93) and 19 remained in CR beyond 3 years.[2] Interestingly, certain characteristics of the infused T cell product were associated with objective response. These included longer telomeres (younger, less proliferated cells), a greater number of CD8+CD27+ cells (consistent with naïve/central or stem cell memory T cells) and adoptively transferred cell persistence in the periphery at 1 month. Furthermore, the investigators noted that increasing the intensity of lymphodepletion prior to adoptive transfer was associated with increasing objective and complete response, a finding that is echoed in the use of genetically modified T cells (see below). While these responses and other trial responses to TIL in metastatic melanomas, are encouraging, the use of TILs is not without its challenges. For example, TILs cannot be generated from all patients, either because of insufficient starting tumor material or because TILs from certain patients may not expand *ex vivo*; the use of IL-2 in TIL culture can promote the expansion of regulatory T cells which will dampen anti-tumor immune responses and the generation of TIL cultures is labor intensive and expensive.

TCR-Modified T Cells

Genetic modification of autologous circulating T cells represents an alternative ACT strategy that is not dependent on identification and isolation of tumor-specific T cells. In this approach, patients undergo apheresis to procure large numbers of mononuclear cells. T cells are isolated from apheresis products, activated and then transduced with viral vectors or transposons to express either TCR genes or CARs that are specific for the tumor antigen of interest. TCRs are composed of an alpha and beta chain that together recognize their cognate antigen presented in the context of self MHC. Over 400 cancer-related peptides have been identified to date.[3] The alpha and beta genes that encode TCRs against HLA-restricted, tumor associated antigens that include the melanoma antigens MART-1 and gp100,[4] the cancer testes antigen NY-ESO-1,[5] and a p53 epitope expressed on 50% carcinomas have been cloned. Autologous T cells transduced with TCR genes have been employed in the treatment of metastatic melanoma, renal cell carcinoma, colorectal carcinoma and synovial cell sarcoma.[6] In general, only modest responses have been achieved with low affinity TCR that target over-expressed antigens. TCRs that have higher affinity for their peptides hold greater promise but also are more likely to cause on-target, off-tumor side effects[7].

Chimeric Antigen Receptor T Cells

Perhaps the most remarkable ACT clinical results to date have been achieved with autologous CAR-T cells in hematological malignancies such as acute lymphoblastic leukemia (B-ALL), chronic lymphocytic leukemia (B-CLL), B-non Hodgkin lymphoma (B-NHL) and multiple myeloma (MM).[8-11] Indeed, in 2017, the FDA approved Kymriah, a CD19 targeting CAR T cell product for the treatment of children and adolescent patient with refractory or second relapse B-ALL. Approval for adults with relapsed or refractory B-NHL quickly followed and heralded a new era of ACT. As antigen recognition occurs via an antibody fragment within the CAR, there is no requirement for target cells to present the antigen in the context of MHC (as is required for TCR-T cells). Furthermore, as the co-stimulatory domains required to stabilize IL-2 mRNA and amplify the initial activation signal are included within the CAR construct itself, there is no requirement for antigen-presenting cells such as dendritic cells to provide these signals. Following intravenous administration, CAR-T cells come into direct contact with circulating leukemic B cells or neoplastic plasma cells expressing the target antigen and undergo rapid activation and expansion following CAR engagement. They have potent cytotoxic activity and acting as serial killers, they rapidly eliminate tumor cells leading to clinical remission. Importantly, transferred CAR T cells engraft and when remission occurs and the antigenic stimulus is removed, the CAR T cell pool contracts, leaving modified central memory cells to protect against relapse. CAR T cells have been shown to survive in patients for over a decade and their persistence correlates with durable remission.

CHALLENGES OF ACT

Although CAR-T cell therapies have enjoyed unprecedented success in the treatment of hematological malignancies, their benefit in solid tumors to date has been limited. This is in part due to the hostile, immunosuppressive tumor microenvironment frequently associated with solid tumors, that inhibits the effector function of adoptively transferred T cells and their survival within the tumor. Tumor stromal elements including regulatory T cells, myeloid derived suppressor cells, IDO and checkpoint molecules contribute to this hostile environment and must be overcome to enable ACT to achieve the same success seen with hematological malignancies. Another important challenge for CAR T cell therapy is the occurrence of side effects that include cytokine release syndrome (CRS), neurotoxicity and on-target, off tumor cytotoxicity. The severity of CRS is related to tumor burden and can be predicted to occur in patients with advanced disease. The effects of CRS are primarily mediated by IL-6 and the IL-6R antagonist, tocilizumab, has been used successfully in its management. A poorly understood neurotoxicity has also accompanied CART-19 therapy and although the pathophysiology of this toxicity is poorly understood, the symptoms often resolve with symptomatic management and supportive care. Importantly, CAR T cell success is directly related to the fitness of the T cell product.[12] For patients with advanced cancers or those who have been heavily pre-treated with chemotherapy, it may be impossible to generate an autologous CAR T cell product for them due to intrinsic T cell dysfunction and an inability of their T cells to proliferate. Currently, these patients are therefore excluded from CAR T cell trials or treatment.

FUTURE PERSPECTIVES

The advent of CRISPR/Cas9 gene editing technologies has revealed new possibilities for ACT.[13] Perhaps the greatest challenge that now could be overcome with gene editing, is the use of allogeneic T cells for ACT. CRISPR/Cas9 technology acts as a molecular scissor and can precisely remove unwanted genetic elements. As such, CRISPR/Cas9 technology is being employed to eliminate the TCR and MHC of healthy donor, allogeneic cells enabling them to be transferred into an unrelated patient without causing GVHD and without being eliminated by the host's immune response.[14] Furthermore, autologous or allogeneic T cells could also be edited to remove checkpoint molecules that engage with ligands in the solid tumor microenvironment and limit effectiveness of the T cell therapy. Importantly, use of gene-edited, allogeneic T cells to generate large banks of cryopreserved antigen-specific CAR-T cells that could be thawed for use on demand would reduce manufacturing costs and provide an effective product enabling more patients to be treated with this potentially life-saving therapy. While most ACT research is focused on cancer treatments, additional uses of ACT are now being explored. These include the use of *ex vivo* activated and expanded regulatory T cells for the treatment of allergic disease and transplant tolerance and the employment of chimeric auto-antibody receptor T cells (CAAR-T) for specific elimination of auto-reactive B cells in autoimmune disease where the autoantigen is known.[15] CAART cells are engineered to express a chimeric auto-antibody receptor that consists of the known autoantigen itself linked to T cell intracellular signaling domains. When the CAAR engages with B cell receptors expressed on auto-reactive B cells that are specific for the auto-antigen, the T cell is activated and specifically eliminates the auto-reactive B cell. This strategy to precisely remove autoreactive B cell clones rather than enforce global immune suppression has generated considerable excitement and if successful in clinical trials will likely represent a turning point in the treatment of autoimmunity.

Given the remarkable responses and future promise of these ACT approaches in human medicine, attention now turns to their use in canine patients with comparable neoplasms. However, the translation of this technology from human to canine patients is not trivial and requires the development and optimization of canine T cell isolation, activation, expansion, and transduction procedures.[16] Furthermore, single chain variable fragments (scFv) that comprise the target antigen recognition site of the CAR need to be generated from canine specific monoclonal antibodies or scFv phage display libraries and then integrated into canine CAR constructs for *in vivo* use. qRT-PCR primers need to be designed and their use validated for tracking CAR T cell expansion and persistence *in vivo*. With respect to the clinical use of canine CAR T cells, pre-conditioning regimes that aim to lymphodeplete the patient to provide "space" for product engraftment and prevent the development of canine anti-mouse antibodies (CAMA) need to be explored. Pre-conditioning regimes for canine bone marrow transplantation have already been developed and can be employed prior to CAR T cell administration. Once standard operating procedures for the generation and use of canine CAR T cells have been developed and optimized, it is anticipated that comparable results to the human experience may be achieved in canine B cell malignancies using CAR T cell therapy. The promise of a therapy that could induce durable clinical remission in canine B cell lymphomas and leukemias following a single autologous or allogeneic infusion is extremely exciting and might soon be within our reach.

References

1. Rosenberg SA, Spiess P, Lafreniere R. A new approach to the adoptive immunotherapy of cancer with tumor-infiltrating lymphocytes. *Science*. 1986;233:1318–1321.
2. Rosenberg SA, *et al*. Durable complete responses in heavily pretreated patients with metastatic melanoma using T-cell transfer immunotherapy. *Clin Cancer Res*. 2011;17:4550–4557.
3. Vigneron N, Stroobant V, Van den Eynde BJ, van der Bruggen P. Database of T cell-defined human tumor antigens: the 2013 update. *Cancer Immun*. 2013;13:15.
4. Johnson LA, *et al*. Gene therapy with human and mouse T-cell receptors mediates cancer regression and targets normal tissues expressing cognate antigen. *Blood*. 2009;114:535–546.
5. Zhao Y, *et al*. Primary human lymphocytes transduced with NY-ESO-1 antigen-specific TCR genes recognize and kill diverse human tumor cell lines. *J Immunol*. 2005;174:4415–4423.
6. Robbins PF, *et al*. Tumor regression in patients with metastatic synovial cell sarcoma and melanoma using genetically engineered lymphocytes reactive with NY-ESO-1. *J Clin Oncol*. 2011;29:917–924.
7. Sharma P, Kranz DM. T cell receptors for gene transfer in adoptive T cell therapy. *Crit Rev Immunol*. 2019;39:105–122.
8. Maude SL, *et al*. Chimeric antigen receptor T cells for sustained remissions in leukemia. *N Engl J Med*. 2014;371:1507–1517.
9. Porter DL, Levine BL, Kalos M, Bagg A, June CH. Chimeric antigen receptor-modified T cells in chronic lymphoid leukemia. *N Engl J Med*. 2011;365:725–733.
10. Raje N, *et al*. Anti-BCMA CAR T-cell therapy bb2121 in relapsed or refractory multiple myeloma. *N Engl J Med*. 2019;380:1726–1737.
11. Kochenderfer JN, *et al*. Long-duration complete remissions of diffuse large B cell lymphoma after anti-CD19 chimeric antigen receptor T cell therapy. *Mol Ther*. 2017;25:2245–2253.
12. Fraietta JA, *et al*. Determinants of response and resistance to CD19 chimeric antigen receptor (CAR) T cell therapy of chronic lymphocytic leukemia. *Nat Med*. 2018;24:563–571.

13. Zhou X. Empowering chimeric antigen receptor T-cell therapy with CRISPR. *Biotechniques*. 2020.
14. Ren J, *et al*. Multiplex genome editing to generate universal CAR T cells resistant to PD1 inhibition. *Clin Cancer Res*. 2017;23:2255–2266.
15. Ellebrecht CT, *et al*. Reengineering chimeric antigen receptor T cells for targeted therapy of autoimmune disease. *Science*. 2016;353:179–184.
16. Panjwani MK, *et al*. Feasibility and safety of RNA-transfected CD20-specific chimeric antigen receptor T cells in dogs with spontaneous B cell lymphoma. *Mol Ther*. 2016;24:1602–1614.

Listeria-Based Cancer Vaccine Strategies

Nicola J. Mason, BVetMed, PhD, DACVIM (Small Animal Internal Medicine)
Philadelphia, PA, USA

The overarching aim of cancer vaccines is to induce and promote robust tumor-specific, cytotoxic T cell responses that eliminate tumor cells, amplify and broaden the anti-tumor immune response, and promote the formation of memory T cells that ensure durable remissions. However, as our understanding of tumor biology and immunobiology grows and the complex interactions between the tumor and the immune system are recognized, barriers that prevent effective anti-tumor T cell responses become apparent. It is now widely recognized that the tumor microenvironment (TME) plays a pivotal role in suppressing anti-tumor immunity. Tumor stromal elements, suppressor cells, immune-modulatory cytokines and enzymes and checkpoint molecules all serve to inhibit effective T cell responses. As such, vaccines that serve solely to generate tumor specific T cells are unlikely to result in effective therapeutic responses when used as monotherapies. Some biologic vaccine platforms, in addition to their ability to prime and boost tumor-specific T cells, exhibit direct permissive effects on the TME and as such have garnered interest from the medical community. One such platform is the gram-positive, facultative anaerobic, intracellular bacterium *Listeria monocytogenes* (*Lm*).

Lm is a potent stimulator of innate and adaptive immunity. Natural infection occurs following ingestion of contaminated food. The bacteria cross the intestinal epithelium and are rapidly phagocytosed by mononuclear antigen presenting cells (APCs). Pathogen associated molecular patterns (PAMPs) expressed by *Lm* stimulate toll like receptors (TLRs) on APCs and other immune cells leading to NF-κB activation and the release of pro-inflammatory cytokines such as IL-12 which helps orchestrate innate and adaptive immune responses. Most phagocytosed *Lm* are degraded within the phagolysosome, providing antigens that are shuttled into the MHC class II antigen-presenting pathway to prime $CD4^+$ T helper cells. However, a small percentage escape from the phagolysosome into the cytosol by secreting the pore-forming lysin listeriolysin O (LLO) and access the MHC class I processing pathway to prime cytotoxic $CD8^+$ T cells. These unique properties of *Lm* can be exploited to deliver an antigenic payload directly into APCs to prime $CD4^+$ and $CD8^+$ T cells.[1] Tumor-associated antigens (TAA) that are fused directly to a truncated form of LLO (tLLO) can be presented in the context of MHCI and MHCII to prime $CD8^+$ and $CD4^+$ T cells, respectively (**Figure 1**). Finally, the presence of *Lm* in the cytosol activates the STING (stimulator of interferon genes) pathway which leads to the production of type I interferons including IFN-β, which further promotes $CD8^+$ T cell priming.[2] Studies in multiple murine cancer models have demonstrated the safety and effectiveness of a highly attenuated *Lm* strain (*Lm*ddA) to induce antigen-specific cytotoxic T cell responses against tLLO-fused viral antigens such as HPV16-E7 and tLLO-fused self-antigens such as HER2/neu and prostate specific antigen (PSA).[3-5] In both viral or self-antigen targeted settings, these studies showed that induction of antigen-specific immune responses led to the regression of established tumors, prevention of metastatic disease and protection from tumor re-challenge indicating the generation of effective memory T cell responses. Furthermore, the innate immune responses elicited by *Lm*-LLO contribute to anti-tumor activity by supporting the priming, expansion and cytotoxic function of pre-existing, neo-antigen specific T cells effectively broadening tumor-specific T cell responses through epitope spreading.[6]

Figure 1

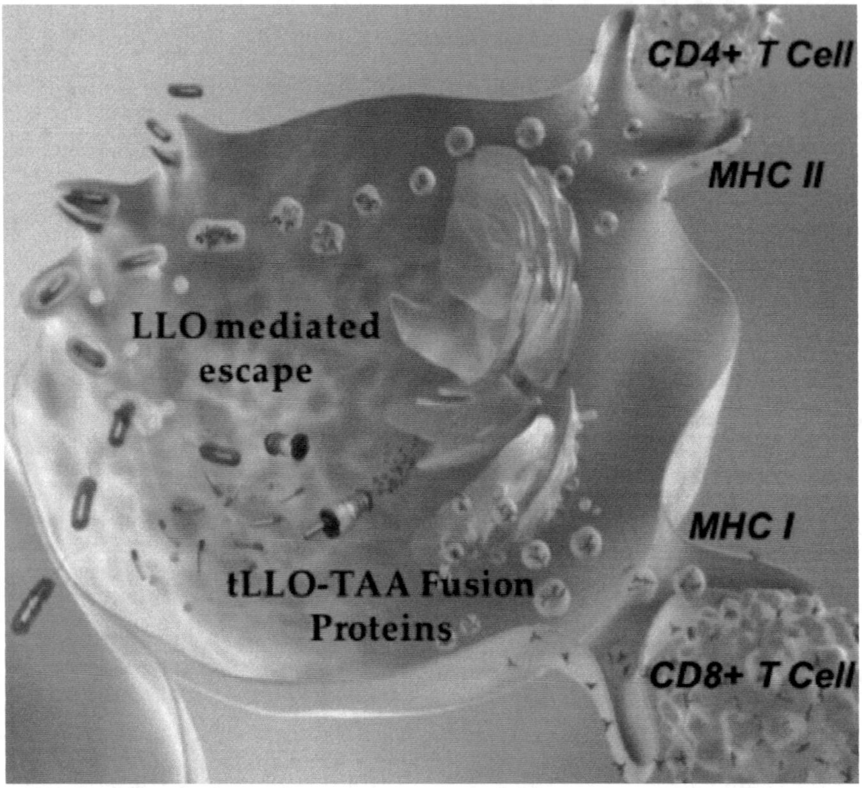

Natural biological properties of *Listeria monocytogenes* enable LLO-fused antigens to be delivered directly into the MHC I pathway. While TAA carried by genetically modified Lm will be processed through the classical MHCII pathway, secretion of LLO by the bacteria within the phagolysosome enables its escape into the cytosol. Here LLO and its fused antigenic payload are effectively processed and presented in the context of MHC I to cytotoxic T cells. Picture courtesy of Advaxis Inc.

THE EFFECTS OF *LISTERIA* ON THE TUMOR MICROENVIRONMENT

In murine models of breast and prostate carcinoma, *Lm*-LLO immune therapies were shown to broaden an effective anti-tumor T cell response in part by reducing the frequency and suppressive capabilities of regulatory T cells (Tregs) and myeloid derived suppressor cells (MDSC) within the TME.[7,8] Interestingly, the reduction in frequency of these immune suppressive cells only occurred within the tumor and not in the periphery (spleen and peripheral blood). Furthermore, this effect seems to be a property of the recombinant *Lm* itself as it was also observed when "control" *Lm*-LLO vaccines carrying irrelevant TAA were administered. Interestingly, while modulation of the TME by *Lm* enabled newly primed and pre-existing tumor-specific T cells to function more effectively, the therapeutic effects of *Lm*-LLO-targeted agents on tumor growth and overall survival in murine models was dependent on the expression of the targeted tumor antigens. This indicates that while the immunomodulatory effects of *Lm* are permissive, specific antigen-targeting and generation of tumor-specific T cells is necessary for the therapeutic effect.[9]

DIRECT TUMOR TARGETING

In addition to its ability to stimulate potent innate and adaptive anti-tumor immunity and modulate the TME, *Lm* has also been shown to directly target tumors leading to clinical responses. *Lm* can directly infect MDSC that traffic to the tumor, bringing the facultative anaerobe into an environment where it can survive.[10] Here, direct tumor cytotoxicity occurs via

NADPH oxidase-mediated production of ROS and intracellular calcium mobilization.[11] Moreover, studies have shown that a subset of MDSC infected with *Lm* can develop an immune stimulating phenotype, producing IL-12 that promotes CD8+ and NK cell responses.[11] These findings confirm the multiple mechanisms by which *Lm* promotes anti-tumor responses and suggests that the accumulation of *Lm* in tumor tissue exerts positive therapeutic effects.

ADXS31-164 in Canine Osteosarcoma

Intravenous administration of a highly attenuated, recombinant, chimeric human HER2/neu (ADXS31-164) targeting *Lm* was evaluated in 18 dogs with histopathologically confirmed osteosarcoma following standard of care amputation and chemotherapy.[12] The pilot study evaluated the safety and ability of ADXS31-164 to induce antigen-specific T cell responses against HER2/neu. Secondary endpoints were disease free interval and overall survival. Dogs recruited to the trial were free of gross metastatic disease 3 weeks after the 4th carboplatin treatment, enabling the evaluation of ADXS31-164 to **prevent** metastases. ADXS31-164 was administered three times at three week intervals. Side effects were transient and low grade, consisting of fever, nausea, and vomiting that occurred on the day of administration. 15/18 dogs generated HER2 specific T cell responses as measured by IFN-γ ELISpot on PBMCs. Disease free interval of vaccinated dogs was 615 days and median overall survival was 956 days. This compared favorably to a historical control group, shown to be free of gross metastatic disease 3 weeks after the 4th carboplatin, where the median overall survival was 423 days. A multi-institutional COTC study enrolling 80 dogs funded by Morris Animal Foundation is currently underway to determine the safety and efficacy of ADXS31-164 in prevention of osteosarcoma metastases. A recent case report described a dog with osteosarcoma who had received 3 doses of a lyophilized version of *Lm*-LLO-HER2/neu and developed an asymptomatic abscess within a metastatic rib lesion three weeks after the last vaccine. Cultures from the abscess revealed the presence of the vaccinal strain of *Listeria*. Removal of the abscess together with the metastatic rib lesion was performed and the dog made an uneventful recovery. The dog remains alive and tumor free greater than 1 year after initial diagnosis.[13] While this report raised concerns about the safety of this approach (see below), particularly for immune compromised patients, it demonstrates the ability of *Lm* to directly target metastatic lesions and in this case provide prolonged tumor free survival of a patient with proven metastatic disease.

The Use of *Listeria* to Target Neo-Antigens

Inducing T cell responses against tumor-associated antigens such as differentiation antigens or self-antigens that are over-expressed by the tumor can be challenging due to mechanisms of immunological tolerance that delete high avidity, self-reactive T cells. Targeting tumor-specific **neo-antigens** that arise from non-synonymous mutations represents an alternative approach that circumvents central tolerance and aims to provide high avidity, functional CD4+ and CD8+ T cells that may effectively eliminate the tumor.[14] Neo-antigen burden is one of the major determinants of tumor immunogenicity and malignancies that exhibit high mutational loads such as melanoma and urothelial carcinoma tend to be most immune responsive. Indeed, tumor neo-epitope burden correlates with tumor T cell infiltration and patient survival[15] and neo-epitope specific, intra-tumoral CD4+ and CD8+ T cells contribute to the clinical responses seen following checkpoint blockade in human patients with melanoma.

With the advent of advanced next generation sequencing (NGS), it is now possible to identify all non-synonymous, expressed mutations within a neoplasm. A subset of these encode immunogenic neo-antigens that are recognized by $CD4^+$ and $CD8^+$ T cells. Immunization against all non-synonymous mutations holds much promise in producing a broad anti-tumor immune response that reduces the chances of antigenic escape, however, it is expensive, requires several months to generate per patient, and has yet to be tested in a clinical setting. As such, efforts to identify non-synonymous somatic mutations that are shared within and across tumor types ("hot-spot" mutations) and that could serve as universal targets in "off-the-shelf" vaccines, are underway.[16,17] The V600E B-Raf mutation is shared across multiple different cancer types in humans and studies in human patients and mice have shown that the mutation is immunogenic and leads to the expansion of functional cytotoxic T cells that can eliminate V600E B-Raf expressing targets. The same mutation is present within up to 87% of canine urothelial carcinoma (UC) cases. Use of the *Lm* platform expressing the V600E B-Raf mutation as a potent neo-antigen targeting immunotherapeutic strategy is hypothesized to enhance V600E B-Raf specific cytotoxic T cell responses and promote epitope spreading leading to tumor regression, prevention of recurrence and durable clinical response.[18,19] The ability of *Lm*-LLO-V600E B-Raf to promote V600E B-Raf specific responses and reduce tumor burden in $V600E^+$ B-Raf UC patients is currently underway in dogs with non-resectable V600E $B\text{-}Raf^+$ UC.

Combination Therapies

Multiple clinical trials using modified *Lm* either alone or in combination with immunomodulatory chemotherapy, radiation, biological response modifiers or checkpoint inhibitors are currently underway in the human oncology clinics.[20] One of the most promising combinations is the use of *Lm* vaccine with checkpoint inhibitors. Checkpoint inhibitors such as anti-PD1 prevent T cell exhaustion and revive tumor specific T cell responses and their combination with a potent vaccine that induces tumor specific T cell responses aims to provide superior anti-tumor effects in cancer patients. Antibodies that inhibit checkpoint molecules are not yet commercially available in veterinary medicine, however it is possible that when available, their impact as a monotherapy might be enhanced by combination with *Lm*-based vaccines.

Safety

Highly attenuated, live recombinant *Lm* vaccines have been shown to be safe and effective in inducing anti-tumor immune responses and therapeutic benefit in human and canine patients. In our experience, an appropriate dose of live, attenuated *Lm*, produces only low grade and transient adverse events.[12] However, there is evidence that *Lm* can accumulate at the tumor site where it can cause abscessation, although to the author's knowledge this has not caused clinical signs of sepsis. Due to concerns over the use of live, attenuated *Listeria* vaccines particularly in immunocompromised patients, researchers have developed a killed but metabolically active (KBMA) *Listeria* platform. *Lm* strains that are deficient in nucleotide repair mechanisms can be inactivated using a DNA crosslinking agent and UV irradiation.[21] The irreversible DNA damage prevents *Lm* from replicating but still capable for a short time of invading cells and generating anti-tumor immunity. This platform has not yet been tested in the human clinic.

In summary, *Listeria* platforms represent a promising strategy to stimulate anti-tumor immunity and enable effective T cell responses that prevent metastatic disease. Furthermore, *Lm* can travel to the tumor within MDSC and exert a direct cytotoxic response that may eliminate the tumor and promote anti-tumor immune memory. Early results with combination approaches in the human clinic are encouraging and it is likely that strategic *Lm* combination therapies will result in further improvements in clinical responses in the near future.

REFERENCES

Additional references are available upon request.

1. Wood LM, Paterson Y. Attenuated *Listeria*: a powerful vector for tumor immunotherapy. *Front Cell Infect Microbiol*. 2014;4:51.
2. Hansen K, *et al*. *Listeria* induces IFNbeta expression through an IFI16-, cGAS- and STING-dependent pathway. *EMBO J*. 2014;33:1654–1666.
3. Hannan R, *et al*. Combined immunotherapy with *Listeria* monocytogenes-based PSA vaccine and radiation therapy leads to a therapeutic response in a murine model of prostate cancer. *Cancer Immunol Immunother*. 2012;61:2227–2238.
4. Lamikanra A, *et al*. Regression of established HPV-16 immortalized tumors *in vivo* by vaccinia viruses expressing different forms of HPV-16 E7 correlates with enhanced CD8(+) T-cell responses that home to the tumor site. *J Virol*. 2001;75:9654–9664.
5. Shahabi V, *et al*. Development of a live and highly attenuated *Listeria monocytogenes*-based vaccine for the treatment of Her2/neu-overexpressing cancers in human. *Cancer Gene Ther*. 2011;18:53–62.
6. Seavey MM, Paterson Y. Antiangiogenesis immunotherapy induces epitope spreading to Her-2/neu resulting in breast tumor immunoediting. *Breast Cancer (Dove Med Press)*. 2009;1:19–30.
7. Wallecha A, *et al*. *Listeria monocytogenes (Lm)*-LLO immunotherapies reduce the immunosuppressive activity of myeloid-derived suppressor cells and regulatory T cells in the tumor microenvironment. *J Immunother*. 2013;36:468–476.
8. Chen Z, *et al*. Episomal expression of truncated listeriolysin O in LmddA-LLO-E7 vaccine enhances antitumor efficacy by preferentially inducing expansions of CD4+FoxP3- and CD8+ T cells. *Cancer Immunol Res*. 2014;2:911–922.
9. Shahabi V, *et al*. Development of a live and highly attenuated *Listeria monocytogenes*-based vaccine for the treatment of Her2/neu-overexpressing cancers in human. *Cancer Gene Ther*. 2011;18:53–62.
10. Quispe-Tintaya W, *et al*. Nontoxic radioactive *Listeria* is a highly effective therapy against metastatic pancreatic cancer. *Proc Natl Acad Sci U S A*. 2013;110:8668–8673.
11. Kim SH, *et al*. High efficacy of a *Listeria*-based vaccine against metastatic breast cancer. *Cancer Res*. 2009;69:5860–5866.
12. Mason NJ, *et al*. Immunotherapy with a HER2-targeting *Listeria* induces HER2-specific immunity and demonstrates potential therapeutic effects in a phase I trial in canine osteosarcoma. *Clin Cancer Res*. 2016;22:4380–4390.
13. Musser ML, *et al*. Vaccine strain *Listeria monocytogenes* abscess in a dog: a case report. *BMC Vet Res*. 2019;15:467.
14. Fritsch EF, *et al*. Personal neoantigen cancer vaccines: the momentum builds. *Oncoimmunology*. 2014;3:e29311.
15. Brown SD, *et al*. Neo-antigens predicted by tumor genome meta-analysis correlate with patient survival. *Genome Res*. 2014;24:743–750.
16. Davies H, *et al*. Mutations of the BRAF gene in human cancer. *Nature*. 2002;417:949–954.
17. Maguire SL, *et al*. SF3B1 mutations constitute a novel therapeutic target in breast cancer. *J Pathol*. 2015;235:571–580.
18. Andersen MH, *et al*. Immunogenicity of constitutively active V599EBRaf. *Cancer Res*. 2004;64:5456–5460.
19. Ilieva KM, *et al*. Effects of BRAF mutations and BRAF inhibition on immune responses to melanoma. *Mol Cancer Ther*. 2014;13:2769–2783.
20. Flickinger JC, *et al*. *Listeria monocytogenes* as a vector for cancer immunotherapy. *Vaccines (Basel)*. 2018;6.
21. Brockstedt DG, *et al*. Killed but metabolically active microbes: a new vaccine paradigm for eliciting effector T-cell responses and protective immunity. *Nat Med*. 2005;11:853–860.

Abstract EN01: Treatment Failure in Hyperthyroid Cats Following Radioactive Iodine (I-131) Injection

Deirdre Mullowney; Yu-Mei Chang; Barbara Glanemann; Harriet Syme
Royal Veterinary College

Radioactive iodine (RAI) is considered the optimal treatment for hyperthyroidism in cats. For convenience, blood sampling is commonly performed immediately before hospital discharge, when total thyroxine (T4) may not have reached its nadir. The purposes of this study were to determine the frequency of, and predictors for, eventual resolution of hyperthyroidism in cats for which T4 remains high at discharge, and to report clinical outcomes for cats that failed initial RAI treatment. Medical records for 959 cats treated with RAI were reviewed. At discharge (15 or 23 days post-injection), 89 cats had T4 >40 nmol/l. Of these, 35 (39%) subsequently had T4 <40 nmol/l (classified as treatment success). Binary logistic regression assessed the utility of T4 (pre-, post- and percentage change), as well as day of sampling and dose of I-131, as predictors of successful treatment. Only the T4 variables were predictive of successful treatment with post-treatment T4 (P<0.001) remaining in the final model. Cats with T4 between 40–100 nmol/l at discharge had a 67.5% chance of treatment success, compared with only 8.2% of cats with higher T4. Of the 54 cats that failed the initial treatment, 16 received a second dose of RAI, 12 underwent thyroidectomy and 21 restarted medical management (some cats received more than one treatment). Re-treatment with RAI was successful in 13 (81%). Four cats had histopathologically confirmed thyroid carcinoma. Cats with T4 >100 nmol/l at discharge have a low probability of becoming euthyroid and may be candidates for immediate re-treatment. Success rates following re-treatment with RAI are good.

Abstract EN29: Day-to-Day Variability of Porcine Lente, Insulin Glargine u300 and Insulin Degludec in Diabetic Dogs

Michelle J. Miller[1]; Jully Pires[1]; Katti Crakes[1]; Rachel Greathouse[1]; Nina Quach[1]; Chen Gilor
[1]Department of Veterinary Medicine and Epidemiology, School of Veterinary Medicine, University of California, Davis; [2]Department of Small Animal Clinical Sciences, College of Veterinary Medicine, University of Florida

The human-recombinant insulin formulations glargine (Toujeo®) and degludec (Tresiba®) are associated with less day-to-day variability and fewer hypoglycemic events compared to Lente-type insulin in diabetic people but are largely unstudied in dogs. A repeated-measured study was used to compare the pharmacodynamics of porcine Lente (Vetsulin®), Toujeo® and Tresiba® in seven diabetic dogs. A continuous glucose monitor (CGM) was placed on dogs throughout the study, recording interstitial glucose concentrations (IG) every 15 minutes. Each dog received the three insulin formulations consecutively. For each insulin, dose escalation was performed using the same algorithm until maximum dose was achieved. Maximum dose was then maintained for 5 days and the CGM data from this period were used for comparison between formulations. To assess day-to-day variability, the mean, SD and coefficient of variation (CV) were calculated at each of the 96 time points for each insulin for each dog and then the mean of these CV's for each dog was compared between insulin formulations. The maximum q12h dose was 0.7 U/kg, 0.5 U/kg and 0.4 U/kg, for Vetsulin®, Toujeo® and Tresiba® respectively. Vetsulin® was associated with increased day-to-day variability compared to Toujeo® and Tresiba® (mean CV 48% vs. 31% and 26% respectively, p=0.002) but with lower mean IG and no difference in IG time spent <70 mg/dL (17%, 11%, 9% respectively, P=0.2). Toujeo® and Tresiba® can maintain lower day-to-day variability in dogs but overall glycemic control might not be better than with Vetsulin®.

Abstract EN03: Iatrogenic Effect of Trilostane (Vetoryl) on Adrenal Steroids Synthesis in Dogs

Luca Giori[1]; Alejandro Esteller-Vico[2]; Jillene Sennon-Greene[2]; Hugo Eiler[2]; Kellie Fecteau[2]
[1]College of Veterinary Medicine, University of Tennessee; [2]BDS Department, College of Veterinary Medicine, University of Tennessee

Trilostane is one of the treatment options for hyperadrenocorticism (HAC). In dogs, clinical signs of HAC often poorly correlate with serum cortisol concentration. We hypothesized that treatment of canine HAC with trilostane results in increased serum concentration of adrenal sex steroids, which can contribute to clinical signs of HAC in dogs. We performed a retrospective analysis of 529 canine blood serum samples submitted to our institution's Endocrinology Laboratory (2015–2019). Serum samples from gonadectomized males and females were evaluated for cortisol, androstenedione, estradiol, progesterone, 17beta-hydroxyprogesterone (17OHP), testosterone and aldosterone at pre- and post-ACTH stimulation. Samples were divided into 3 groups: 0) dogs with HAC, and no reported treatment (n=109); 1) dogs with HAC and receiving trilostane (n=319); and 2) dogs with HAC, with discontinued therapy after poor response (n=101). Group 1 had significantly lower cortisol concentrations pre- and post-ACTH stimulation (mean ± SEM = 4.0±0.2 and 8.3±0.4 µg/dL, respectively) compared to dogs in Group 0 (6.6±0.5 and 21.5±0.9 µg/dL) and Group 2 (5.5±0.3 and 16.3±1.0 µg/dL). Group 1 had significantly higher androstenedione and 17OHP concentrations pre- and post-ACTH stimulation, compared with Groups 0 and 2. No significant differences in steroid concentrations were detected in dogs treated once versus twice a day. Even though cortisol was lower and within the reference interval in Group 1, dogs were reported to have polyuria-polydipsia (39.8%), alopecia (15.9%), and polyphagia (10.3%), among other clinical signs of HAC.

Abstract EN04: Evaluation of a Flash Glucose Monitoring System in Dogs with Rapidly Changing Glucose Concentrations

Carly Patterson; Leigh Howard; Jonathan Lidbury; Shannon Washburn
Texas A&M University

A flash glucose monitoring system (FGMS; FreeStyle Libre 14-day system, Abbott Laboratories) has been evaluated in dogs with uncomplicated diabetes mellitus, as well as dogs with diabetic ketoacidosis. It continuously measures interstitial glucose concentrations for up to 14 days and has proven to be a useful tool for monitoring diabetic dogs that predominantly have hyperglycemia. The utility of this system for detecting hypoglycemia and rapid changes in glucose concentrations has not yet been evaluated in dogs. The objective of this study was to assess the utility of this FGMS in dogs with hypoglycemia and rapid changes in blood glucose concentrations. This IACUC approved study was performed in tandem with a teaching laboratory during which veterinary students administered intravenous regular insulin to 24 fasted dogs and corrected the resulting hypoglycemia by feeding or administering dextrose intravenously; a single subcutaneous dose of dexamethasone was also administered. FGMS sensors were placed 1 hour prior to administration of insulin. A FGMS measurement and a blood sample (collected from a peripheral intravenous catheter) were obtained prior to insulin administration and every 10 minutes over a two-hour period. Blood glucose measurements were promptly made on each sample using a clinical chemistry analyzer (Vitros 4600, Orthos Diagnostics). Glucose concentrations were compared between the two methods for each time point with the clinical chemistry analyzer serving as the reference standard. Agreement between methods was assessed using linear regression and Bland-Altman analysis. The application of the FGMS was uncomplicated and well tolerated by all dogs. However, out of 240 possible readings, 17 sensor errors (7.1%) were reported by the FGMS. When comparing the FGMS and reference standard glucose concentrations, R2 was 0.21, $p>0.001$ and bias (reference standard minus FGMS; 95% CI) was -12.1 mg/dL (39.3 to -63.6 mg/dL). Out of 93 measurements where the reference standard concentration was <60.0 mg/dL, the FGMS measurement was >60 mg/dL for 64 (69%), was >80.0 mg/dL for 18 (19%), and was >100.0 mg/dL for 3 (3%). In this experimental model of dogs with hypoglycemia and rapid changes in blood glucose concentrations, there was limited agreement between FGMS and reference standard glucose measurements and the FGMS failed to reliably detect hypoglycemia. In general, the FGMS measurements were higher than those for the reference standard but this was not consistent, making use of an adjustment factor impossible. These results likely reflect the time it takes for interstitial glucose concentrations to equilibrate with those in the bloodstream.

Abstract EN05: Evaluation for Type 1 Diabetes Associated Autoantibodies in Diabetic and Non-Diabetic Australian Terriers and Samoyeds

Allison L. O'Kell[1]; Mark Atkinson[2]; Paula Henthorn[3]; Rebecka Hess[3]; Clive Wasserfall[2]
[1]College of Veterinary Medicine, University of Florida; [2]University of Florida; [3]University of Pennsylvania

Evidence for an autoimmune etiology in canine diabetes is inconsistent, and could vary based on breed. The objective of this study was to evaluate for the presence of glutamic acid decarboxylase 65 (GAD65), insulinoma-associated protein 2 (IA-2), and zinc transporter 8 (ZnT8) autoantibodies in diabetic and non-diabetic Australian Terriers and Samoyeds, two breeds with comparatively high prevalence of diabetes, in the United States. Banked serum samples from Australian Terriers (n=63 diabetic; n=71 non-diabetic) and Samoyeds (n=32 diabetic; n=42 non-diabetic) previously recruited from United States primary care clinics, breed club events, or academic hospital patient populations were included. Autoantibodies against GAD65, IA-2, and ZnT8 were measured using commercially available kits with specificity for the human proteins. Samples were considered positive if results were above the lower limit of quantification. A Fisher's exact test compared the proportion of dogs in the diabetic and non-diabetic groups within each breed that were positive for each autoantibody. There was no significant difference in the proportion of samples considered positive for GAD65 or ZnT8 autoantibodies in either breed or IA-2 autoantibodies in Australian Terriers ($p>0.05$). The proportion of IA-2 autoantibody positive samples was significantly higher in diabetic versus non-diabetic Samoyeds ($p=0.003$), but substantial overlap was present between diabetic and non-diabetic groups. The present study did not support these autoantibodies as markers of autoimmunity in canine diabetes in the breeds studied here. Future studies investigating alternative markers of autoimmunity in these and other breeds are warranted.

Abstract EN06: The Efficacy, Sensitivity, and Specificity of a Saliva Glucose Monitoring System for Diabetic Canines

Rebecca Silveston-Keith[1]; Delphine Dean[2]; Katherine Hafner[3]; Jeffery Ranta[4]
[1]Accessible Diagnostics; [2]Clemson University; [3]Medical Murray; [4]Coastal Carolina University

Monitoring blood glucose is critical for regulating canine diabetes mellitus (DM) glucose. Insulin regimens prescribed based on in-office glucose curves do not necessarily reflect patients' glucose trends at home as canine blood glucose (BG) varies throughout the day and in response to different environmental stressors including office induced hyperglycemia. The American Animal Hospital Association (AAHA) 2018 Diabetes Management Guidelines for Dogs and Cats recommends incorporating at home BG results and clinical signs with in-practice diagnostic exams for effective management of DM. At home systems such as portable blood glucose meters (PBGMs), continuous glucose monitors (CGMs) and urinary strips are not always used due to owner anxiety in lancing for blood, high cost, or lack of accuracy respectively. Furthermore, owners risk mis dosing insulin, based on this at home data, without veterinarian advice. Decades of research show that partnerships between health professionals and caretakers improve patients' healthcare outcomes. Integration of at-home glucose monitoring into the standard of care provided in the veterinary practice improves the health of the patient if the at home test is effective and a professional can interpret monitoring modalities and blood glucose results in conjunction with clinical signs. The objective of this study was to validate the sensitivity, specificity and effectiveness of a painless, saliva glucose, smart phone enabled at home index test compared to venous blood test for diabetic canines. The side by side testing included 47 DM, and 9 clinically normal, privately owned, canines with informed owner consent under standard fee for service patient clinical care provided by participating veterinary practices. Patient saliva was sampled in duplicate non-invasively and simultaneously with venous blood at two times 60 minutes apart. Automated lab scale blood chemistry analyzers determined the blood glucose. The smart phone proprietary app algorithm, Vet-Tab, determined the corresponding blood glucose from the color change of glucose oxidase reaction with salivary glucose. Pet owners were surveyed about their use behavior for this saliva system, CGMs, PBGM and urine strips. They were also surveyed about their veterinarian relationships in response to the app prompting users to contact their vets for advice when the pet is outside a euglycemic range. The Vet-Tab's salivary system had a positive predictive value for hyperglycemic to hypoglycemic canine venous blood glucose and has a system sensitivity and selectivity is comparable to that of accepted CGMs for diabetic companion animals when the collection and analysis of salivary glucose and BG was done by skilled professionals in a veterinary practice. The Surveillance Error Grid showed that 90% of measurements have only slight risk or no risk for complications due to inaccurate glucose readings in diabetic canines. Owners using Vet-Tab report increased measurement compliance and confidence over other glucose monitoring systems and higher connectivity with their veterinarian. Vet-Tab was shown to be an effective early warning system to supplement in-practice assessments and increase the owner-veterinarians partnership for improved healthcare of the DM patient.

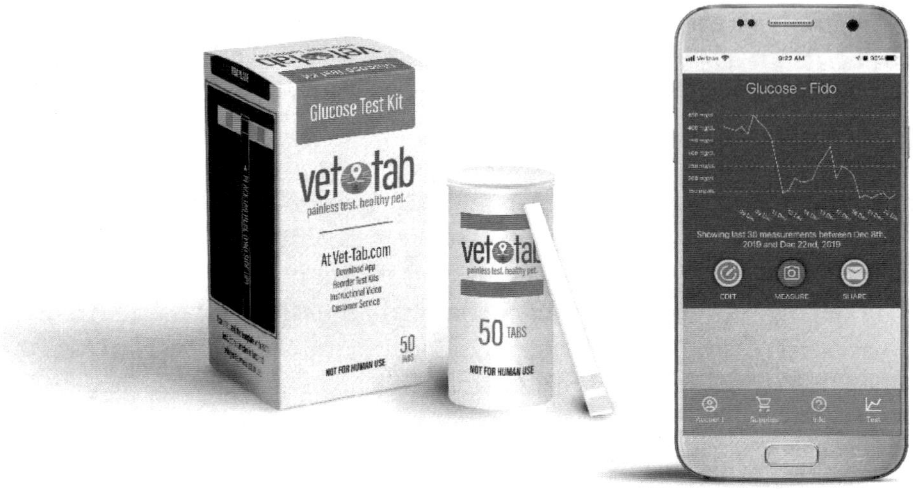
Glucose Test Kit
vettab
painless test. healthy pet.
At Vet-Tab.com
50 TABS
NOT FOR HUMAN USE
Glucose - Fido

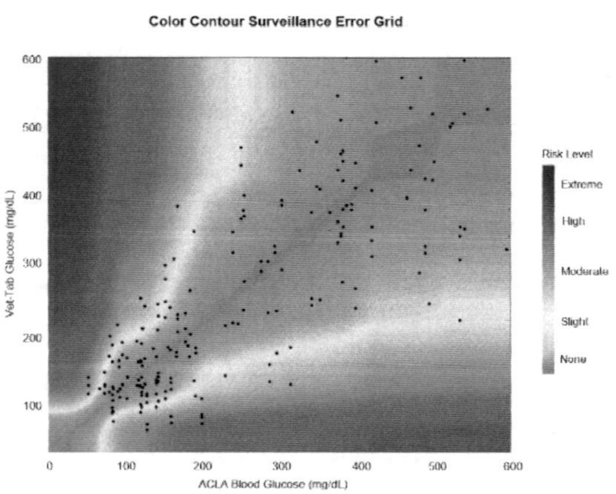
Color Contour Surveillance Error Grid
Vet-Tab Glucose (mg/dL)
ACLA Blood Glucose (mg/dL)
Risk Level
Extreme
High
Moderate
Slight
None
0
100
200
300
400
500
600

Abstract EN07: Organoid Cultures of Follicular-Cell Thyroid Carcinoma: A Novel Canine Model for Translational Thyroid Cancer Research (ESVE Award Winner)

Eve Tièche[1]; Kerstin Hahn[2]; Martina Dettwiler[2]; Federico Massari[3]; Sandra Schallberger[4]; Martin Kessler[5]; Sven Rottenberg[1]; Miguel Campos[1]
[1]Department of Clinical Veterinary Science, Vetsuisse Faculty, University of Bern; [2]Institute of Animal Pathology, Vetsuisse Faculty, University of Bern; [3]DOCVET; [4]Tierklinik Thun; [5]Tierklinik Hofheim

Growing patient-derived tissue in 3-dimensional cell culture systems (organoids) has revolutionized *in vitro* cancer research. In contrast to 2-D cell lines, organoids can be grown more efficiently and conserve important features of the original tumor, such as tissue architecture and cellular heterogeneity. In human medicine, organoid cultures provide a unique platform for personalized cancer therapy. In this study, we aimed to culture and characterize organoids derived from follicular-cell thyroid carcinoma (FTC) in dogs. Tissue samples of follicular type (n=1) and compact type (n=1) FTCs derived from two euthyroid dogs were frozen in DMSO-containing freezing medium within 24 h of thyroidectomy and stored at -150°C until processing. After thawing, the tissue was digested with collagenase IV and dissociated mechanically. Cells were subsequently seeded in Cultrex® Basement Membrane Extract and cultured to organoids in Advanced DMEM supplemented with N-acetylcysteine, B-27 supplement, EGF, Noggin, Rspondin-1 and Rock-inhibitor. Organoids were split and passaged every 9 to 14 days. After 15 to 24 days, organoids were formalin-fixed, pelleted in 2.5% agarose, paraffin-embedded and processed for hematoxylin-eosin staining. Immunohistochemistry (IHC) for thyroid transcription factor-1 (TTF-1), thyroglobulin (Tg), calcitonin, vimentin and Ki-67 was performed on sections of the primary tumors and organoids. Organoids of FTC were cultured efficiently using our protocol. Organoids of both tumors formed follicle-like structures composed of a single epithelial cell layer. These epithelial cells were round to cuboidal, had variably distinct cell borders and abundant eosinophilic to foamy cytoplasm. Anisocytosis and anisokaryosis were not observed in either of the organoid lines but were present in the FTC of compact type. Nuclear expression of TTF-1 in both organoid lines confirmed thyroid origin. The organoids derived from the compact FTC, which had approximately 30% of cells positive to Tg, showed no Tg expression while the organoids derived from the follicular FTC, which had Tg expression in >95% of cells, showed Tg expression in about 50% of the organoids. Vimentin expression was observed in both organoid lines (30–60% of cells) and was higher than in the primary tumors, where only up to 10% of tumor cells were positive. IHC for calcitonin and Ki-67 was negative in both organoid lines. Organoids derived from naturally occurring canine FTC are able to conserve histological and immunohistochemical features of the primary tumors providing an interesting *in vitro* model to better understand the pathogenesis and optimize treatment of thyroid cancer in dogs. The culture protocol likely requires further optimization.

Abstract EN08: Measurement of Pre-trilostane Salivary Cortisol in Dogs with ACTH-Dependent Hyperadrenocorticism

Vanessa Uemura Fonseca[1]; Gabriel Müller[2]; Cíntia Silva[3]; Vinícius Villas Bôas[2]; Paula Papa[4]
[1]School of Veterinary Medicine, University of Santo Amaro and School of Veterinary Medicine and Animal Science, University of Sao Paulo, Sao Paulo, Brazil; [2]Private Practice; [3]School of Veterinary Medicine and Animal Science, University of Sao Paulo; [4]Vetsuisse Faculty, University of Zurich, Zurich, Switzerland and School of Veterinary Medicine and Animal Science, University of Sao Paulo, Sao Paulo, Brazil

Trilostane therapy is used for the management of spontaneous hyperadrenocorticism (HAC) in dogs. The monitoring is often performed by a combination of history, physical examination and assessment of adrenal reserve capacity with an ACTH-stimulation test. Due to the cost and availability of synthetic ACTH for the test, other evaluation methods have been tested, such as urine corticoid:creatinine ratio and baseline cortisol values before and after trilostane administration. The measurement of one or two baseline cortisol pre trilostane has been shown to be promising. However, baseline cortisol can be influenced by stress during blood collection in a veterinary hospital. The cortisol passively diffuses from blood into the saliva, and if the saliva collection takes less than four minutes, there will be a minimal stress effect on the measurement. The goals of this study were: 1- to evaluate the accuracy of pre trilostane salivary cortisol measurements for classifying well and unwell-controlled dogs, and 2- if there is a correlation between serum and salivary cortisol concentrations. Twenty-six client-owned dogs diagnosed with spontaneous ACTH-dependent HAC and on BID trilostane treatment were prospective enrolled. The study was approved by the University of Santo Amaro Animal Care and Use Committee. Three measurements of cortisol were performed, two in the saliva and one in the serum. The first measurement of salivary cortisol was obtained with salivettes at home immediately before the administration of trilostane. And the second was obtained in a veterinary environment immediately after the end of the ACTH-stimulation test, three hours after trilostane. The dogs were classified as well and unwell-controlled, when post-ACTH serum cortisol was between 1.5–5.5 μg/dl and greater or equal than 5.5 μg/dl, respectively. There were 9 well and 17 unwell-controlled dogs. None of the animals had post-ACTH cortisol value below 1.5 μg/dl. The accuracy of salivary cortisol in differentiating well and unwell-controlled dogs determined by the area under the Receiver Operating Characteristic (ROC) curve was 0.79 with p=0.032. The cut-off value less than 0.089 μg/dl for detection of well-controlled showed 71.4% of sensitivity and 86.7% of specificity, whereas a cut-off less than 0.126 μg/dl had 85.7% of sensitivity and 66.7% of specificity. There were differences between salivary cortisol values post-ACTH in well and unwell-controlled group (p=0.026), but no significant difference was observed between salivary cortisol pre-trilostane in well and unwell-controlled group (p=0.087). The Spearman´s rank-order correlation showed that salivary cortisol post-ACTH was moderately correlated with serum cortisol post-ACTH (Rho=0.536, p=0.022). No correlation was observed between salivary cortisol pre-trilostane and serum cortisol post-ACTH (Rho=0.342, p=0.110), and between salivary cortisol pre-trilostane and salivary cortisol post-ACTH (Rho=0.039, p=0.889). These findings suggest that the salivary cortisol measurement may be a potential tool for monitoring HAC. Further research on a larger number of dogs is needed to balance the groups for statistical analyses and confirm the accuracy of this method.

Abstract EN09: Insulin Expression Patterns in Canine Insulinoma

Thomas Schermerhorn; Tera Brandt; Sarah Schneider; Nora Springer
Kansas State University

Canine insulinoma is an insulin-secreting endocrine tumor of pancreatic islet beta cells that causing hyperinsulinemia and hypoglycemia. Little is known about the insulin expression patterns in canine insulinoma or how expression influences hyperinsulinemia. This study used immunohistochemistry (IHC) to compare insulin expression patterns between normal islets and neoplastic tissue in sixteen insulinomas from an archived tissue bank. IHC slides were prepared by the Veterinary Diagnostic Histology Lab and insulin expression quantified using HALO image analysis software. Visual examination showed insulin in insulinoma cells was not concentrated within discreet granules. The mean percentage (%) of insulin staining area (+) within equally-sized regions drawn around normal pancreatic islets (26.8%) and insulinoma tumor tissue (26.0%) was similar. Insulin staining was less intense in insulinomas (23.5% tumor area showed weak staining compared to 14.7% islet area) while 12% of islet area showed moderate or strong staining compared with 2.4% of tumor area. The ratio of insulin expression in tumor and islet (T:I ratio) showed that insulin expression varied between individual tumors (T:I ratio range—0.13 to 3.5); 4 tumors had T:I >1.5 (high insulin content) and 4 had T:I<0.2 (low insulin content). The pattern and intensity of insulin staining is variable in canine insulinoma suggesting insulin content differs between individual tumors. The majority of tumors contained abundant ('high') insulin suggesting over expression could underlie abnormal insulin secretion. Hyperinsulinemia and hypoglycemia associated with tumors with low insulin content suggests another mechanism, such as abnormal glucose sensing, could underlie abnormal insulin secretion.

Abstract EN010: Evaluation of Salivary Vasopressin as an Acute Stress Biomarker in Dogs with Noise Stress

Jeong-Mi Kim[1]; Yi-Kyeong Jeong[2]; Ye-in Oh[2]; Kun-ho Song[2]; Kyoung won Seo[2]
[1]College of Veterinary Medicine, Chungnam National University; [2]Chungnam National University

Stress is associated with various detrimental changes in physiological health, which affect an animal's quality of life. The hypothalamus-pituitary-adrenal (HPA) axis and the sympathetic-adreno-medullar (SAM) axis are two main physiological pathways that contribute to the stress response of an organism. Arginine vasopressin (AVP) is a mediator of the HPA axis and is known to be related to social behaviors and stress. The serum concentration of AVP is higher in more aggressive dogs and humans with post-traumatic stress disorder. Salivary analysis is a non-invasive method to assess stress. The purpose of this study was to evaluate the possibility of using salivary AVP as an acute stress biomarker in dogs. Salivary AVP concentration was measured before and after 30 minutes of vacuum noise exposure. Behavioral assessment, physiologic parameter change assessments, and serum cortisol analysis were conducted in combination. Statistical analysis was conducted twice in the total study population and in more stressed individuals. Based on stress behavior analysis scores, a total of 28 dogs were classified into mildly, moderately, and severely stressed groups. All four physiologic parameters (blood pressure, body temperature, heart rate, and respiratory rate) were significantly increased after noise stimulation in the severely stressed group. Serum cortisol did not show any significant change. Salivary AVP decreased after noise stimulation in the total population. Salivary AVP and blood pressure changes were negatively correlated in the severely stressed group. In conclusion, salivary AVP may be a potential acute stress biomarker in dogs.

Table 1. Characteristics of dog participants

Breed	Number of dogs	Sex of dogs (F/FS/M/MC)
Maltese	8	1/6/0/1
Mongrel	6	0/4/0/2
Shih tzu	2	0/2/0/0
Poodle	2	0/0/0/2
Pomeranian	2	0/1/0/1
Yorkshire terrier	1	0/1/0/0
French bulldog	1	0/1/0/0
Beagle	1	0/1/0/0
Dachshund	1	0/0/0/1
Spitz	1	0/1/0/0
Cocker spaniel	1	0/0/0/1
Border collie	1	0/0/0/1

Abstract EN11: Chronic Low-Dose Rapamycin Does Not Affect Glucose and Insulin Regulation in Middle-Aged, Large Breed Dogs

Jeremy B. Evans[1]; Stephanie Word[1]; Audrey Cook[1]; Matt Kaeberlein[2]; Daniel Promislow[2]; Kate Creevy[1]

[1]Veterinary Teaching Hospital, Texas A&M; [2]University of Washington

Rapamycin is an anti-fungal macrolide that inhibits the mammalian target of rapamycin (mTOR) and is used frequently as an immunosuppressant in human renal transplant recipients. Strikingly, rapamycin administration has been shown to significantly extend the lifespans of numerous laboratory animal models when compared to placebo-treated controls. Due to mTOR's extensive role throughout the body, rapamycin can cause numerous side effects, including derangements in glucose and insulin homeostasis. This study aimed to investigate any potential abnormalities in insulin sensitivity in middle-aged, large breed dogs who received long-term, low-dose oral rapamycin. Seventeen dogs (10 spayed females, 1 intact female, and 6 castrated males) were randomly assigned to receive rapamycin (0.025 mg/kg) or a placebo orally every Monday, Wednesday, and Friday for 6 months. All enrolled dogs underwent complete evaluations (physical exam, chemistry, CBC, urinalysis, and echocardiogram) at baseline, 3 months, 6 months, and 12 months (i.e., 6 months after discontinuation of treatment), and excess blood samples were flash frozen in liquid nitrogen and then stored at -80°C. Samples from baseline and 6 months were analyzed to determine fasting insulin and glucose concentrations and a homeostatic model assessment for insulin resistance (HOMA-IR) was calculated. Mean baseline HOMA-IR scores were not significantly different between control and rapamycin treated groups (2.35±2.1 and 2.14±1.4, respectively; p=0.82), nor at 6 months (2.61±3.1 and 1.99±1.3; p= 0.59). Additionally, the mean change in HOMA-IR over 6 months was not significantly different between rapamycin and control groups (+0.26±2.4 and -0.15±1.5; p=0.67). In this small study of low-dose rapamycin in companion dogs, no overt impact on HOMA-IR scores was detected. Future larger studies are warranted.

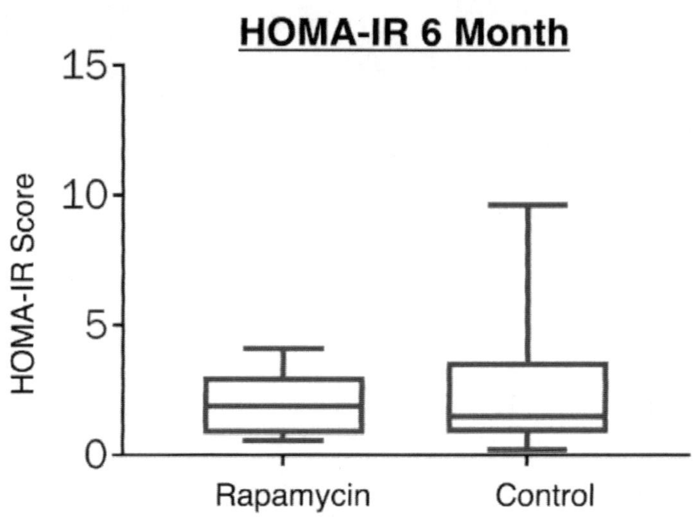

Dog	Treatment	HOMA (BL)	HOMA (6 mo)	Dog	Treatment	HOMA (BL)	HOMA (6mo)
1	Control	2.24	1.58	2	Rapamycin	4.5	1.88
4	Control	1.11	4.17	3	Rapamycin	1.49	0.77
5	Control	1.23	0.67	6	Rapamycin	1.06	0.57
7	Control	5.59	1.87	8	Rapamycin	1.24	3.41
11	Control	1.1	0.22	9	Rapamycin	0.9	0.84
12	Control	5.71	9.62	10	Rapamycin	4.15	2.54
13	Control	0.75	1.34	15	Rapamycin	0.6	1.16
14	Control	1.04	1.4	16	Rapamycin	2.92	2.64
				17	Rapamycin	2.43	4.1

Abstract GI01: Incidence of Bacteremia Secondary to Colonoscopy in Dogs

Martin Granick[1]; Faye Hartmann[1]; Jean-Sébastien Palerme[2]; Jessica Pritchard[1]
[1]University of Wisconsin; [2]Iowa State University

In human medicine bacteremia is a common colonoscopy sequela with a reported incidence of up to 25%. While the majority of these bacteremias are asymptomatic, prophylactic antibiotic guidelines have been created to decrease sepsis risk in certain patient populations. The incidence of transient or clinical bacteremia secondary to colonoscopy has not been characterized in veterinary medicine. We hypothesized that similar to humans, a subset of dogs undergoing colonoscopy would develop bacteremia. Dogs with clinician prescribed colonoscopy were eligible for enrollment. Exclusion criteria included evidence of bowel perforation or concurrent antibiotic administration other than metronidazole. Blood cultures were obtained aseptically from either a cephalic or saphenous vein after anesthetic induction and immediately following colonoscopy. A total of 20 dogs were enrolled between January 2019 and January 2020. Positive blood cultures were detected in 4/20 dogs (20%) pre-endoscopy and 4/20 dogs (20%) post-endoscopy. Blood cultures were positive for one dog both pre- and post-endoscopy although two different microorganisms were cultured. No dogs showed clinical signs consistent with sepsis. There was no difference in the incidence of positive blood cultures pre- and post-colonoscopy. These preliminary results are consistent with other studies documenting bacteremia in 14–52% of clinically well dogs. Additional case recruitment as well as utilizing more sensitive methods to detect bacterial DNA may provide more information on the risk of bacterial translocation secondary to colonoscopy in dogs.

Abstract GI02: Low-Fat Diet Appears to Be Effective Monotherapy in Some Dogs with Protein Losing Enteropathy

Marc Myers; Roger Hostutler; Stephen Martinez; Jonathan Shiroma
MedVet

The efficacy of a low-fat diet as monotherapy or in combined therapy with prednisone for dogs with protein losing enteropathy (PLE) and evidence of lymphangiectasia remains poorly characterized, particularly in non-Yorkshire Terriers. We hypothesized that clinical remission of PLE can be achieved via low-fat diet monotherapy or in conjunction with prednisone in any canine breed. This was a prospective, observational cohort study of 14 dogs with PLE and ultrasonographic evidence of lymphangiectasia. All subjects were placed on dietary therapy and prednisone was added if initial response was deemed inadequate. Dogs were assessed and scored using the Canine Chronic Enteropathy Clinical Activity Index (CCECAI) at study enrollment and at 4 recheck examinations across a 6-month study period, including a final recheck ultrasound at the last examination. Clinical and clinicopathologic variables were collected and dogs were divided into three outcome groups: 1) clinical remission on dietary therapy alone, the "DR" group; 2) clinical remission on dietary therapy plus prednisone, the "DP" group; and 3) treatment failure, the "DF" group. Eleven of 14 dogs (79%) were in clinical remission at the study end date (6 months post-enrollment) based on a CCECAI score of <3, 6 from the DR group, and 5 from the DP group. Dogs from a variety of breeds were represented. Dogs in the DR group had higher median serum albumin (1.6 vs 1.2 g/dL, $p=0.11$) and significantly higher median serum globulins (2.5 vs 1.7 mg/dL, $p=0.03$) on presentation than dogs in the DP group. Four of 11 dogs in clinical remission also had ultrasonographic evidence of resolution of linear striations, 3 from the DR group and 1 from the DP group. Low-fat diet appears to be an effective monotherapy in some dogs with PLE and lymphangiectasia. More severely clinicopathologically affected dogs may be less likely to respond to low-fat diet monotherapy, but results must be interpreted cautiously due to low study enrollment. In dogs that do not respond to low-fat diet monotherapy, the addition of prednisone can also lead to PLE remission although the durability of remission with either an exclusive low-fat diet or low-fat diet in conjunction with prednisone is unknown.

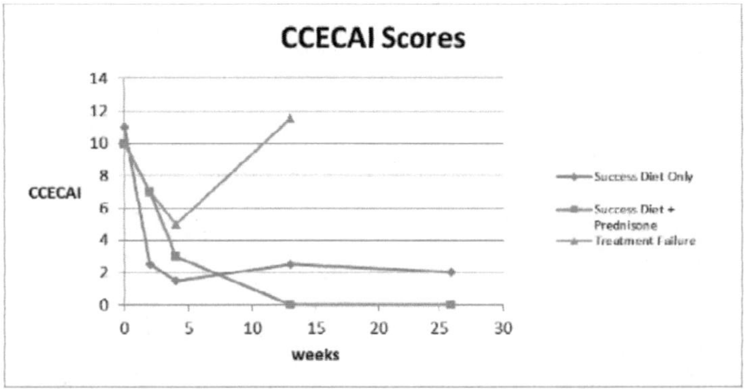

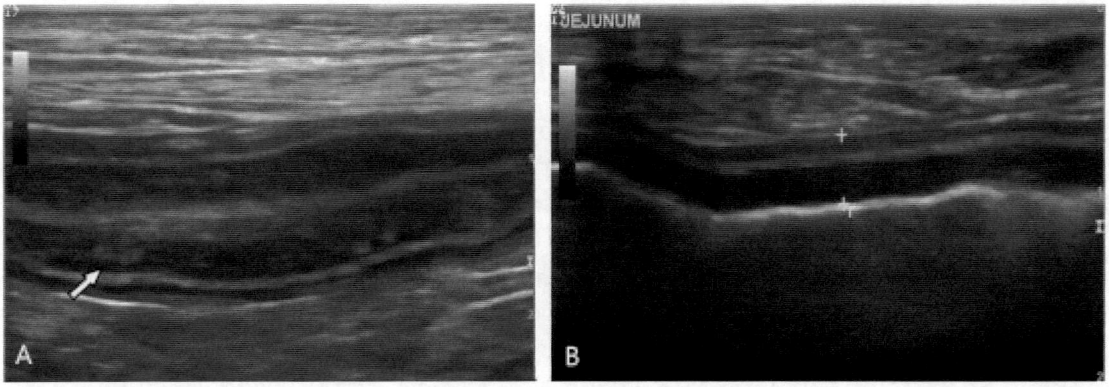
A
JEJUNUM
B

Abstract GI03: Outcome for Cats with Chronic Enteropathy Diagnosed Through Endoscopic Intestinal Biopsies

Kathryn Robb[1]; Betty Chow[2]; Steve Hill[1]; Keith Richter[1]; Joerg Steiner[3]
[1]Veterinary Specialty Hospital by ETHOS Veterinary Health; [2]VCA Animal Specialty & Emergency Center; [3]Gastrointestinal Laboratory, Texas A&M University

Inflammatory bowel disease (IBD) and alimentary small cell lymphoma (LSA) are common forms of feline chronic enteropathy (CE). Immunohistochemistry (IHC) and clonality testing (PCR for antigen receptor rearrangement; PARR) in conjunction with H&E may improve the accuracy of a diagnosis of LSA in patients when compared to H&E alone. There is evidence that using all three modalities (H&E, IHC, and clonality testing) leads to a more frequent diagnosis of alimentary small cell lymphoma than when the diagnosis is based on H&E alone. The goal of this study was to determine if a change in diagnosis, based on the use of IHC and clonality testing, correlates with patient outcome and survival time. A total of 62 cats with clinical signs of chronic enteropathy (>3 weeks duration) who were previously enrolled in a study and had undergone upper and lower gastrointestinal endoscopy with endoscopic biopsies submitted for H&E, IHC, and clonality testing were included in this study. All cases were categorized as "IBD," "possible/probable LSA," or "LSA" based on H&E alone. IHC and clonality testing further classified or reclassified cases as "IBD," "LSA," or "large granular lymphocyte lymphoma" (LGL). All owners were provided with an outcome survey via phone call and/or email. Also, medical records were reviewed to evaluate each patient's health status or cause of death, if known. Necropsy or additional surgical biopsy results were also reviewed if available. Of the 62 cats, 18 (29.0%) were alive at the time of the survey, while 30 (48.4%) were deceased, and 14 (22.6%) were lost to follow up. Estimated median survival times were determined by creating Kaplan-Meier survival curves based on H&E diagnosis alone and for a diagnosis based on the combination of H&E and IHC/clonality testing. The estimated median survival times were 71.9 months for cats diagnosed as "IBD," 15.7 months for cats diagnosed as "possible/probable LSA," and 12.3 months for cats diagnosed as "LSA" when diagnosed by H&E alone. With the combination of H&E, IHC, and clonality testing, the estimated median survival times were 71.9 months for "IBD," 22.8 months for "LSA," and 0.33 months for "LGL." Additional Kaplan-Meier survival curves within each H&E diagnosis by IHC/clonality diagnosis were created. The addition of IHC and clonality testing to patients diagnosed with "IBD" on H&E did not show a significant change in survival time (p=0.593). All patients diagnosed with "LSA" by H&E were confirmed to have "LSA" by IHC and clonality testing. Therefore, the survival time did not change for this group. Patients diagnosed with "possible/probable LSA" were either reclassified as "LSA" (n=12) or "LGL" (n=2). The patients with LGL had a short median survival time. However, the reclassified "LSA" patients had significantly longer survival time (p=0.0001). Our results show that reclassification of feline patients with chronic enteropathy by addition of IHC and clonality testing does not influence the median survival time for patients diagnosed with "IBD" or "LSA" by H&E. IHC and clonality testing are useful for differentiating between "LSA" and "LGL" within the population of cats diagnosed as "possible/probable LSA" by H&E. This differentiation does make a significant difference in survival time.

Abstract GI04: Videofluoroscopic Assessment of Liquid Sildenafil as a Treatment for Canine Generalized Megaesophagus

Susan Mehain; Sarah Guess; Jillian Haines
Veterinary Teaching Hospital, Washington State University

Megaesophagus (ME) is characterized by dysmotility and dilation of the esophagus, causing regurgitation and carrying a poor prognosis. If it can be delivered to the stomach of dogs with ME, sildenafil may cause short-duration relaxation of the gastroesophageal sphincter, thus improving clinical signs in dogs with ME via increased esophageal clearance. The objectives of this study were to determine if liquid sildenafil could be delivered to the stomach of dogs with ME and have significant effects on esophageal clearance, frequency of regurgitation, body weight, and perceived quality of life compared to no treatment or a placebo. In this blinded, randomized, crossover study, 10 healthy, client-owned dogs, previously diagnosed with ME received either liquid sildenafil (1 mg/kg PO q12h) diluted to a total volume of 3–5 mL or a placebo of the same volume for 14 days, followed by a 7-day washout, then the opposite treatment for 14 days. Esophageal clearance time was assessed prior to treatment (baseline), and on day 1 of each treatment period using videofluoroscopy performed over 30 minutes with dogs in an upright position. Dogs were assessed to determine passage of liquid treatment into the stomach, as well as for esophageal clearance of food after receiving liquid (at baseline) or liquid sildenafil/placebo followed by slurry, both containing iohexol contrast. Owners kept logs of regurgitation episodes for 2 weeks before, during the treatment periods, and during the washout periods. Movement of liquid into the stomach was assessed at baseline, and for each treatment, allowing 3 assessments per dog and 30 total assessments. Liquid moved into the stomach successfully 21/30 (70%) times (prior to slurry ingestion = 5, following slurry ingestion = 16) and failed to clear or mixed with the fed slurry 9/30 (30%) times. There were no significant differences found in esophageal clearance between untreated, placebo, or sildenafil treatment; regurgitation episodes between untreated, placebo, or washout periods; quality of life scores between untreated, sildenafil, or placebo; body weight after placebo; or esophageal clearance between untreated, placebo, or sildenafil treatment. Sildenafil did result in significant reductions in regurgitation episodes ($p<0.05$) and increased body weight ($p<0.05$) compared to untreated and placebo. Liquid sildenafil can be successfully delivered to the stomach of dogs with ME during the majority of feedings. Results indicate that there is potential for improved management of dogs with ME treated with long-term liquid sildenafil.

Abstract GI05: Omeprazole Induces Reversible Fecal Dysbiosis in Healthy Dogs

Rachel Pilla[1]; Jan Suchodolski[2]; Jose Garcia-Mazcorro[2]; Joerg Steiner[2]; Jonathan Lidbury[2]; Olivier Dossin[3]
[1]Texas A&M; [2]Gastrointestinal Laboratory, Texas A&M University; [3]Veterinary Clinic Advetia

Omeprazole (OM) is a proton pump inhibitor that is commonly used in small animals. OM treatment has been reported to alter microbial populations in the GI tract in many species, including dogs. However, it is unknown whether these changes are reversible after the cessation of administration. Therefore, the goal of this project was to evaluate the impact of OM on the fecal microbiome in healthy dogs, both during and after omeprazole treatment. Stored samples from a research trial involving 8 healthy dogs receiving OM at 1.1 mg/kg per os q12hrs for 15 days. Fecal samples were collected twice before the beginning of treatment (D-30 and D-15), after the end of treatment (D15), and two weeks after the end of treatment (D29). DNA was extracted, and the qPCR-based dysbiosis index (DI) was used to assess the fecal microbiota. The DI and abundance of select bacterial groups were compared across time points using 1-way RM ANOVA. Significance was set as $p<0.05$. The DI of the two baseline samples (D-30 and D-15) were not significantly different from each other (D-30 mean: -2.9 ± 1.3; D-15 mean: -3.0 ± 1.7; $p=0.994$). After 15 days of OM administration, DI was significantly increased (mean: 1.8 ± 0.7) compared to D-30 ($p<0.001$). When individual genera were considered, the abundance of Turicibacter, a butyrate-producing bacterium, was significantly decreased ($p<0.001$) and that of *Streptococcus* was significantly increased ($p=0.002$). The abundances of *Faecalibacterium*, *E. coli*, *Blautia*, *Fusobacterium*, and *Clostridium hiranonis* were not significantly affected. Two weeks after cessation of treatment, the mean DI (-2.0 ± 1.4) and the mean abundance of *Turicibacter* and *Streptococcus* were no longer significantly different from those of D-30 ($p=0.645$, 0.365, and 0.371, respectively). In summary, the administration of OM to healthy dogs for 15 days induced a significantly increased dysbiosis index. However, the dysbiosis index returned to the reference interval within two weeks after treatment discontinuation, indicating that OM-induced dysbiosis is transient. The effect of OM on bacterial groups not assessed in this index remains to be determined.

Abstract GI06: Evaluation of Feline Gastrointestinal pH and Transit Times

Naila J. Telles[1]; Bradley Simon[2]; Aarti Kathrani[3]; Emily Gould[1]; Elizabeth Scallan[2]; Jonathan Lidbury[1]; Jörg Steiner[1]; Mark Papich[4]; M. Katherine Tolbert[1]
[1]Gastrointestinal Laboratory, Texas A&M University; [2]Texas A&M University; [3]Royal Veterinary College; [4]North Carolina State University

There are knowledge gaps regarding feline gastrointestinal (GI) physiology. As a consequence, veterinarians treat cats as they would a dog, which may be suboptimal. For example, acid suppressants known to be efficacious in dogs are only weakly efficacious in cats. Pharmaceutical companies have encountered difficulties developing oral medications for cats because they have relied on assumptions extrapolated from dogs. However, discrepancies identified between dogs and cats might be explained by differences in physiologic factors, such as intraluminal pH and transit times. Our study objective was to better characterize feline GI transit times as well as GI pH in both the fed and fasted states to determine if critical differences exist between species and contribute to oral drug failure in cats. After withholding food for 20 hours, six healthy, colony-housed, research-bred, spayed female cats had a Bravo pH capsule orally administered followed by 15 mL of water. Five hours after capsule administration, cats were meal fed by offering them their daily allowance of food for 1 hour and maintained in their colony housing. Fasted and fed GI pH, as well as transit times were recorded and compared to those reported for dogs using similar methodology. The median (range) transit times in minutes were: esophageal 10 (1–317), gastric 665 (1–2442), intestinal 1404 (1091–1731), and total transit time 2133 (1421–3850). The median (range) GI pH values were: esophageal 6.8 (6.1–7.3), gastric 2.8 (1.7–4.8), intestinal 8.4 (7.7–8.7), first-hour duodenal 8.3 (7.7–8.7), and last-hour large intestinal 8.5 (7.7–8.8). The median (range) fasted and fed gastric pH values were: 1-hour pre-prandially 3.7 (3.4–7.4), at time of feeding 3.9 (3.6–4.4), and 1-hour post-prandially 3.2 (2.4–4.1). Similar to dogs, GI pH and transit times were highly variable among cats and feeding did not exert a buffering effect on gastric pH. Replicate studies are underway to evaluate intra-individual variability. Colony-housed cats had higher intestinal pH and shorter total GI transit times compared to those previously reported in colony-housed dogs. Based on these results, we believe that differences in intestinal pH and total GI transit times may be contributing reasons for differences in oral drug efficacy between dogs and cats.

Abstract GI07: Bacteria Viability in Stored Canine Feces for Use in Fecal Microbiota Transplantation

Emerald Rodriguez; Mohammad Khattab; Jonathan Lidbury; Joerg Steiner; Jan Suchodolski
Gastrointestinal Laboratory, Texas A&M University, College Station, Texas, USA

Fecal microbiota transplantation (FMT) is an emerging treatment option for dogs and cats with gastrointestinal (GI) disease. The goal of FMT is to re-establish a stable microbial community from a healthy donor in the GI tract of the recipient. Currently, the optimal storage conditions for canine donor feces for use in FMT are unknown. The aim of this study was to determine the effect of different storage conditions on the viability of bacteria in canine fecal samples. Fecal samples from 10 healthy dogs were collected and aliquots of each sample were prepared. One aliquot was immediately treated with propidium monoazide (PMA), followed by DNA extraction to discriminate between DNA from viable versus non-viable bacterial cells. The remaining aliquots were stored for either 1 week at 4°C or 3 months at -20°C or -80°C with or without the addition of 10% glycerol. After storage, these aliquots were also treated with PMA, followed by DNA extraction. The following bacterial groups were evaluated by qPCR in the feces: *Faecalibacterium, Turicibacter, Blautia, E. coli, Streptococcus, Fusobacterium*, and *C. hiranonis*. The abundance of viable bacteria in the stored aliquots was compared to the abundance of viable bacteria in aliquots immediately extracted using repeated measures ANOVA with Dunn's post-test. Significance was set at $p<0.05$. There was a significant decrease in viability of *Fusobacterium* for feces stored at all storage conditions compared to fresh feces ($p<0.002$). For the remaining bacterial groups, storage at 4°C for 1 week did not cause a significant decline in bacterial viability compared to fresh feces ($p>0.108$). After 3 months of storage, freezing with 10% glycerol was superior to freezing without glycerol, as no significant decline in viability was observed with this storage condition at either -20°C or -80°C. In conclusion, storage of fecal samples from healthy dogs resulted in a decline in viability of some bacterial groups. For short-term storage, 4°C without additives appears sufficient, while for longer storage at either -20°C or -80°C, the addition of 10% glycerol is recommended.

Abstract GI08: Clinical Effectiveness of Hydrolyzed Diets for Chronic Gastrointestinal Signs in Cats Under Primary Veterinary Care

Aarti Kathrani; David Church; Dave Brodbelt; Camilla Pegram; Dan O'Neill
Royal Veterinary College

This study included a population of cats with chronic gastrointestinal (GI) signs in primary practice that were prescribed a commercial hydrolyzed diet. The study aimed to identify whether concurrent medical therapy alongside the diet was associated with differing clinical outcomes. Concurrent medical therapies included antibiotics and/or glucocorticoids used to manage their GI signs. Our study included 512,213 cats under veterinary care during 2016 in the UK from the VetCompass database. The medical records of all cats were searched using relevant search terms to identify cats with chronic GI signs receiving hydrolyzed diets. From 5,000 given hydrolyzed diets, 977 met the inclusion criteria. Of 697 cats that received the diet without concurrent antibiotic or glucocorticoid, 66% (457 cats) had a positive outcome, defined as not receiving antibiotic or glucocorticoid for GI signs at subsequent visits nor death from GI signs for a median follow-up time of 818 days (range 184–5,279). In this subpopulation, the following factors increased the odds of a positive outcome in a multivariable model: cats 6 years old or younger when first prescribed the diet compared to cats that were older than 6 years (odd's ratio (OR): 1.55, 95% confidence interval (CI): 1.12–2.17, P=0.009); cats that did not receive antibiotic prior to the hydrolyzed diet being first prescribed compared to those that did (OR: 1.53, 95% CI: 1.09–2.14, P=0.013) and cats that did not receive glucocorticoid prior to the hydrolyzed diet being first prescribed compared to those that did (OR: 2.52, 95% CI: 1.50–4.23, P<0.001). Of the 127 cats that received the diet with concurrent antibiotic and not glucocorticoid, 44% (56 cats) had a positive outcome, defined as not receiving additional antibiotic or glucocorticoid for GI signs at subsequent visits nor death from GI signs for a median follow-up time of 946 days (range 186–3,599). In this subpopulation, cats 6 years old or younger when first prescribed the diet had increased odds of a positive outcome compared to cats that were older than 6 years (OR: 2.27, 95% CI: 1.09–4.75, P=0.029). Of the 153 cats that received the diet with concurrent glucocorticoid with or without antibiotic, 35% (53 cats) had a positive outcome, defined as tapering and discontinuation of the glucocorticoid within 3 months without subsequent need of this medication or antibiotic for GI signs for a median follow-up time of 1,082 days (range 213–3,888). In this subpopulation, the following factors increased the odds of a positive response in a multivariable model: cats 6 years old or younger when first prescribed the diet compared to cats that were older than 6 years (OR: 2.81, 95% CI: 1.37–5.75, P=0.005) and cats that did not receive glucocorticoid prior to the hydrolyzed diet being first prescribed compared to those that did (OR: 2.07, 95% CI: 1.00–4.25, P=0.049). Further studies are needed to determine whether the use of antibiotics or glucocorticoids prior to the use of a hydrolyzed diet with or without concurrent medical therapy affects the response to these diets or whether the results of our study are due to the prescribing habits of primary care veterinarians or the severity of the GI signs.

Abstract GI09: Diagnostic Value of Fecal Bacteriologic Culture and Dysbiosis Index in Dogs with Chronic Diarrhea

Melanie Werner[1]; Jan Suchodolski[2]; Jonathan Lidbury[2]; Joerg Steiner[2]; Katrin Hartmann[1]; Stefan Unterer[1]
[1]Clinic of Small Animal Internal Medicine, Ludwig Maximilians University of Munich, Germany; [2]Gastrointestinal Laboratory, Texas A&M University

Although the clinical usefulness for fecal cultures has been determined to be low in human patients, bacteriologic culture of feces is still frequently performed in the diagnostic work up of dogs with diarrhea. Recently PCR-based methodologies have been established to assess the fecal microbiome. This study aimed to assess the diagnostic yield of fecal cultures in dogs with chronic diarrhea. Fecal culture was performed in 18 dogs with chronic diarrhea and 18 healthy control dogs. Dogs were excluded if they received antibiotics or probiotics in the 4 weeks prior to presentation. Total bacteria and 7 bacterial groups (*Faecalibacterium, Fusobacterium, Turicibacter, E. coli, Streptococcus, Blautia, C. hiranonis*) were analysed by qPCR in all dogs to calculate the fecal dysbiosis index (DI). Results were compared between diseased and healthy dogs. Exact $\chi 2$ test was used to compare categorical variables and Mann-Whitney test to compare the DI. Fecal culture failed to detect any significant difference in the presence of specific enteric pathogens as well as in the assessment of the composition of the gram-negative and gram-positive flora ($p \geq 0.289$). However, the DI was significantly different between dogs with chronic diarrhea (mean [SD]: 0.94 [3.80]) and the healthy control group (mean [SD]: -2.96 [2.77]; $p=0.0002$). In contrast to the PCR-based fecal dysbiosis index, fecal cultures have a low value to differentiate between normo- and dysbiosis. Thus, fecal cultures do not appear to be useful to identify causative agents and to guide treatment decisions in dogs with chronic diarrhea.

Abstract GI10: The Effect of Combined Carprofen and Omeprazole Administration on Gastrointestinal Permeability and Injury in Dogs

Susan M. Jones[1]; Ann Gaier[1]; Hiroko Enomoto[1]; Patricia Ishii[2]; Rachel Pilla[2]; Josh Price[3]; Jan Suchodolski[2]; Joerg Steiner[2]; Mark Papich[1]; Kristen Messenger[1]; M. Katherine Tolbert[2]
[1]College of Veterinary Medicine, North Carolina State University; [2]College of Veterinary Medicine & Biomedical Sciences, Texas A&M University; [3]University of Tennessee

Proton pump inhibitors (e.g., omeprazole) are commonly administered concurrently with nonsteroidal anti-inflammatory drugs (NSAIDs; e.g., carprofen) to reduce the risk of injury to the gastrointestinal (GI) tract. However, evidence to support this practice is weak, and it may exacerbate NSAID-induced enteropathy. Our objective was to evaluate the effect of carprofen alone or with omeprazole in dogs. We hypothesized that the co-administration of omeprazole and carprofen would significantly increase GI permeability and dysbiosis index compared to no treatment or carprofen alone. Serum LPS, plasma iohexol, fecal dysbiosis index, and fecal calprotectin concentration were used to assess GI permeability and injury in six healthy, adult Beagle dogs in a prospective, 3-period design. In the first seven-day sham period, dogs received no intervention (baseline). During the second period, dogs received 4 mg/kg of carprofen q24hrs orally for seven days. In the third period, dogs received 4 mg/kg of carprofen q24hrs and 1 mg/kg of omeprazole q12hr orally for seven days. GI permeability assays were performed at the end of all three periods. Data were analyzed using two-within subjects factor repeated measures mixed-model ANOVA with Tukey-Kramer *post hoc* tests ($p<0.05$). Serum LPS and plasma iohexol concentrations did not differ between treatments. Fecal calprotectin concentrations significantly differed between treatments ($p=0.034$). The fecal dysbiosis index varied over time based on the treatment received ($p=0.031$). Co-administration of omeprazole and carprofen significantly increased fecal dysbiosis index and calprotectin in dogs. Although more studies are warranted, these results suggest that omeprazole can worsen NSAID-induced enteropathy in otherwise healthy dogs.

Abstract GI11: Objective Evaluation of Deglutition in Healthy Cats Using a Free-Feeding Videofluoroscopic Swallow Study Protocol

Megan E. Grobman[1]; Elizabeth Luciani[2]; Carol Reinero[2]
[1]Auburn University; [2]University of Missouri

Background: Videofluoroscopic swallow studies (VFSS) represent the criterion standard for diagnosis of cause of dysphagia in veterinary medicine. Physiologically relevant objective VFSS parameters in healthy cats are lacking. The absence of species-specific swallowing parameters represents a significant limitation in assessing cats with dysphagia. **Objectives/Hypothesis**: Using a standardized unrestrained, free-feeding VFSS protocol, the study aims were to apply objective swallow metrics to healthy cats and to determine species differences by comparing to objective swallow metrics in healthy dogs. We hypothesize that species-specific swallowing parameters can be identified by free-feeding VFSS in healthy cats. Animals: Healthy adult cats (n=13) and dogs (n=24) **Methods**: Free-feeding VFSS utilized fourteen objective metrics of swallowing compared across species and three food consistencies (liquid, puree, and kibble). Objective VFSS metrics included assessment of the oral, pharyngeal, and esophageal stages of swallow. Differences in species and consistency were evaluated by Mann-Whitney Rank Sum Test and One-way ANOVA on Ranks respectively with a $p<0.05$ significance level. **Results**: VFSS features attributed to pathologic states, including reflux (n=2) and macro-aspiration (n=1) were observed in healthy, asymptomatic cats. Statistically significant differences were detected between species (≥5 metrics) and consistencies (≥5 metrics) ($p<0.05$). **Conclusions**: Free-feeding VFSS data was successfully collected in healthy cats with results inclusive of reflux and macro-aspiration. Significant differences with food consistency support the need for standardized recipes for consistent evaluation. Objective swallowing metrics are not directly translatable across species, requiring species specific cut-offs when using objective swallowing metrics.

Abstract GI12: Mycophenolate Mofetil Effect on the Gastrointestinal Microbiome in Dogs

Kenjiro Fukushima[1]; Jason Gagne[2]; Jonathan Lidbury[3]; Lia McCoy[4]; Jorg Steiner[3]; Jan Suchodolski[3]; Michael Lappin[4]
[1]Veterinary Teaching Hospital, Small Animal Medicine, Colorado State University; [2]Purina Nestle; [3]Texas A&M University; [4]Colorado State University

Mycophenolate mofetil (MMF) is an effective immunosuppressant in dogs, but gastrointestinal (GI) toxicity is common. Based on our previous study, 24.4% of 135 client owned dogs experienced GI toxicity (median dose 17.5 mg/kg/day, median time to onset 10 days). Postulated mechanisms for MMF induce GI toxicity include GI dysbiosis, direct cellular toxicity from MMF metabolites, and impaired proliferation of enterocytes. In one study, antibiotics lessened signs of GI toxicity induced by MMF. Based on these data, our hypothesis was that MMF induced GI toxicity in dogs is due to intestinal dysbiosis. The purpose of this study was to determine MMF effects on clinical signs and the fecal microbiome. A total of 15 young adult beagles were assigned into 2 groups (control: n=6, MMF: n=9). Dogs in MMF group 2 were administered MMF (20 mg/kg/day) for 14 days. Appetite, attitude, dehydration, and fecal consistency were scored by blinded personnel until day 28 and fecal microbiome was analyzed using quantitative PCR on samples from days 0, 14, and 28. The MMF group had significantly higher fecal scores than the control group and the MMF group showed significantly higher fecal scores during treatment period compared to baseline. On microbiome analysis, no significant differences were seen between the MMF and control groups. Based on our data, MMF induced diarrhea was common but was not associated with dysbiosis as determined with the here analyzed bacterial taxa. Further study is warranted to determine the cause of MMF induced GI toxicity and whether it can be prevented.

Abstract GI13: Effect of Fasting on the Gastrointestinal Panel in Healthy Dogs

Alexander Saver[1]; Jörg Steiner[2]; Scott Hetzel[3]; Jessica Pritchard[1]
[1]School of Veterinary Medicine, University of Wisconsin-Madison; [2]College of Veterinary Medicine, Texas A&M University; [3]Department of Biostatistics and Medical Informatics, University of Wisconsin-Madison

The gastrointestinal panel, comprising of serum concentrations of cobalamin (vitamin B12), folate (vitamin B9), pancreatic lipase immunoreactivity (cPLI), and trypsin-like immunoreactivity (cTLI), is an essential tool in the diagnosis and management of small intestinal and pancreatic disease in dogs. However fasting requirements for these tests from commercial laboratories range from 6–18 hours and are not evidence-based. Unnecessary fasting in veterinary patients results in reduced patient welfare, logistic challenges for clinicians and, in the acute hospital setting, delayed return to feeding. The objective of this study was to determine evidence-based fasting requirements for analytes part of the gastrointestinal panel in clinically healthy dogs. We hypothesized that there would be no clinically relevant difference in these analytes between a commonly utilized overnight fast (12 hours) and various postprandial timepoints (1, 2, 4, & 8 hours post-feeding). Eleven healthy dogs with no history of gastrointestinal disease completed the study protocol. Dogs were fasted overnight (12 hours) and a baseline serum sample was collected. They were then fed and repeated serum samples were collected at 1, 2, 4, & 8 hours postprandially. Analytes in these samples were compared to baseline using a linear mixed-effects model. Cobalamin was statistically significantly decreased at 4 hours (472 pg/ml; p=0.0080) and 8 hours (476 pg/ml; p=0.0040), compared to 12 hours (526 pg/ml) postprandially, but this difference was not clinically relevant. No difference was seen for serum folate, cPLI, or cTLI concentration. In healthy dogs, fasting didn't result in clinically relevant differences in gastrointestinal panel analytes. This evaluation should be repeated in dogs with gastrointestinal disease to determine if fasting is necessary in this population of dogs.

Abstract GI14: Adverse Effects and Impact on Microbiome in Healthy Dogs Treated with Omeprazole

Kathryn Hogan; Thurid Johnstone; Caroline Mansfield
University of Melbourne

The primary aim of this study is to identify the impact of omeprazole on the fecal microbiome of dogs using the fecal dysbiosis index (FDI). Secondary aims were to ascertain if any adverse clinicals signs occurred because of omeprazole use and if any changes were dose dependent. Twenty healthy, client-owned dogs, over 1 year of age with a normal body condition score, receiving a consistent diet and regular parasite control were recruited to participate in a prospective, randomized trial. Additional inclusion criteria were that dogs had not received any antibiotics, corticosteroids, or non-steroidal anti-inflammatories in the preceding 3 months, and did not have any history of significant or recurrent gastrointestinal disease. Other long-term medications including fluoxetine, zylkene, and oclacitinib maleate were permitted. Dogs were randomly allocated via Excel® spreadsheet random distribution function, to either a 1 mg/kg or 0.5 mg/kg group. Dogs received omeprazole twice daily for 7 days, then once daily for a further 7 days. Fecal collection was performed by the owners or the principal investigator at day 0 and at the end of weeks 1, 2, 4, and 6; samples were immediately refrigerated and transported chilled to the University of Melbourne, mixed with 95% ethanol and stored before being shipped on ice to the University of Texas A&M for FDI testing. Owners also kept a daily log during the first 2 weeks of including fecal scoring, using the Purina fecal index chart. Twenty dogs commenced the trial, however, only 13/20 completed the full 6 weeks of monitoring. Of those that did not complete the study 6/7 were withdrawn in the first week due to adverse effects. A further 1 dog discontinued the trial after finishing the first 2 weeks of omeprazole due to requirement for NSAIDs treatment. Of those dogs withdrawn for adverse effects 3/6 had diarrhea only, 2/6 had both diarrhea and vomiting, and 1/6 had vomiting only. A further 2/13 dogs that did complete the trial had observable changes in fecal consistency and an additional 2/13 had 1 or more episodes of vomiting while receiving omeprazole. In total 10/20 dogs commencing the trial showed evidence of adverse effects; 6/10 of the dogs that had adverse effects received omeprazole at 1 mg/kg. The most substantial change in FDI was noted at the end of the first week of omeprazole dosing with 7/15 dogs in dysbiosis and 12/15 dogs trending towards dysbiosis as defined by a positive increase from individual baseline, in FDI score. Significant inter-individual variation did occur. Due to the large drop-out rate as a result of adverse effects, the high treatment group was not continued and so there are insufficient numbers to determine if statistically the higher dose leads to a greater dysbiosis than the lower dose. While there was significant intra-individual variation, this study demonstrates that the use of twice daily omeprazole in healthy adult dogs causes substantial changes in the fecal dysbiosis index with 12/15 dogs trending towards dysbiosis. Additionally, it highlights the frequency of adverse effects with 50% of dogs in this trial developing vomiting, diarrhea, or both. Omeprazole is frequently used by veterinarians for the management of vomiting and diarrhea, even in the absence of suspected gastric or esophageal ulceration. Given the findings of clinical and microbiome impacts in our study of healthy dogs, treatment with omeprazole in dogs with gastrointestinal signs may inadvertently prolong recovery times and potentially worsen outcome. Further investigation is required to confirm if similar changes are noted in unwell dogs, and if there are longer-term metabolic impacts in healthy dogs.

Abstract GI15: Defining Healthy. The Utility of Building a Companion Animal Fecal Microbiome Reference Dataset

Dawn D. Kingsbury[1]; Jess Jarett[1]; Carlton Osborne[1]; Alex Martin[1]; Jonathan Eisen[2]; Holly Ganz[1]
[1]AnimalBiome; [2]UC Davis

We conducted a participatory research project on the microbiome of cats and dogs where participants submitted fecal samples for bacterial characterization, and built a biobank of >5,800 samples. From these, we curated a subset of 274 fecal samples from apparently healthy cats and dogs. We used this dataset to determine if there is a core fecal microbiome in pet dogs and cats, and whether it might provide a useful frame of reference for companion animal medicine. Fecal samples were submitted and participants answered detailed questionnaires about the pet's health. The microbiome was characterized at the genus level with Illumina sequencing of PCR amplified 16S rRNA genes. From this dataset, we selected 83 cats and 191 dogs as healthy reference animals. Inclusion criteria included a healthy body condition score, the absence of any clinical signs, no diagnosis of disease or behavior problem, and no antibiotic treatment in the preceding twelve months. Ages ranged from 1–12 yr for cats and from 1–11 yr for dogs. Because pets were not examined by a veterinarian for this project, we implemented more conservative screening criteria to eliminate as many occultly unhealthy animals as possible. All genera present in at least 45% of healthy dogs (n=15 genera) or 66% of healthy cats (n=17 genera) were considered part of the normal core microbiome (n=14 shared genera). The distribution of specific genera of veterinary importance was also assessed, including *Escherichia-Shigella*. Low abundances of these taxa are common in healthy animals, and high abundances are often observed in animals with clinical signs. Median values best reflected the most frequently observed relative abundances, and low and high thresholds for each genus were based on the upper and lower 2.5th and 10th percentiles of the normal range for the genus. In dogs, a median of 87.9% (±9.88%) of the total microbiome is represented by the core microbiome of 15 genera, and in cats 73.8% (±9.70%) of the total microbiome is represented by the core microbiome of 1 7 genera. Next, we demonstrated the relevance of the normal taxa ranges as defined by the healthy reference dataset by comparing the clinical signs of animals within the normal range and above the normal range for *Escherichia-Shigella* from the entire database. Both cats and dogs with *Escherichia-Shigella* values in the 90th percentile or above (4.4% for cats, 5.4% for dogs) were significantly more likely to have diarrhea reported than animals below the 90th percentile (n=85 and 64.7% vs. n=565 and 43% for cats, p=0.0003; n=476 and 50.2% vs. n=1356 and 30.5% for dogs, p=2.5 x 10–14 , Fisher's exact test). This research supports the presence of a taxonomically defined core microbiome in healthy pet cats and dogs. Our discovery-based approach determines which bacteria are contributing to imbalances, enables identification of new biomarkers of disease, and can help inform therapeutic recommendations.

Abstract GI16: Relationship Between Anemia, Iron Status, and Cobalamin Status in Cats with Chronic Gastrointestinal Disease

Maria Jugan[1]; Adam Hunt[2]
[1]College of Veterinary Medicine, Kansas State University; [2]College of Veterinary Medicine, University of Tennessee

Anemia is common in humans with chronic gastrointestinal disease, resulting from gastrointestinal bleeding, iron deficiency, and hypocobalaminemia, and increases all-cause mortality in cats. Objectives of this study were to investigate prevalence of anemia in cats with chronic gastrointestinal disease and association with cobalamin and iron status. Twenty client-owned cats with chronic (≥4 weeks) gastrointestinal disease were enrolled prospectively. CBC, serum iron, ferritin, total iron binding capacity (TIBC), serum cobalamin, and serum methylmalonic acid (MMA) tests were performed. The Mann-Whitney U test was used to compare cats with low vs. normal iron and normal vs. elevated MMA. Spearman's test was used to correlate MMA, iron, and CBC parameters. Prevalence of anemia, iron deficiency, and elevated MMA were 15%, 35%, and 40%, respectively. Iron deficiency was characterized as functional based on decreased iron, decreased percent saturation, and normal ferritin. Hematocrit ($p=0.03$), reticulocyte hemoglobin content (CHr; $p=0.02$), and ferritin ($p<0.01$) were lower in cats with decreased iron vs. normal iron. TIBC was lower in cats with increased MMA vs. normal MMA ($p=0.02$). MMA negatively correlated with hematocrit ($r=-0.45$). Iron positively correlated with hematocrit ($r=0.5$), MCV reticulocyte ($r=0.52$), CHr ($r=0.71$), % Sat ($r=0.79$), and ferritin ($r=0.45$). Functional iron deficiency was common in cats with chronic gastrointestinal disease. Cats with lower iron and higher MMA had lower hematocrit. Future studies are needed to determine impact of iron status on survival in cats with chronic gastrointestinal disease.

Abstract GI17: Metronidazole-Induced Dysbiosis Alters Fecal but Not Serum Amino Acid Profiles in Healthy Dogs

Amanda B. Blake[1]; Frédéric Gaschen[2]; Erin Olson[2]; Jonathan Lidbury[1]; Jörg Steiner[1]; Jan Suchodolski[1]
[1]Gastrointestinal Laboratory, Texas A&M University; [2]School of Veterinary Medicine, Louisiana State University

Amino acids are mainly absorbed in the proximal small intestine and can affect a multitude of metabolic processes in the host, such as protein synthesis, formation of neurochemical messengers, and important energy metabolism pathways. Unabsorbed amino acids can be fermented by gut bacteria. Numerous studies have shown that oral antibiotics affect the gut microbiota. With growing interest in the role of amino acids in gastrointestinal disease and the high prevalence of antibiotic usage in these patients, it is important to understand how antibiotics affect amino acid profiles. The objective of this study was to determine if metronidazole-induced dysbiosis alters serum or fecal amino acid profiles in healthy dogs. Baseline fecal and serum samples were collected from 16 healthy pet dogs (D0), that were subsequently treated with metronidazole at 15 mg/kg per os q12h for 14 days. Additional serum samples were collected on day 14 after the last dose of metronidazole (D14) and on day 42 (D42). Naturally passed fecal samples were collected on days 7, 14, 28, and 42 (D7, D14, D28, D42). Fecal and serum samples were analysed for untargeted metabolomics analysis. Kruskal-Wallis tests followed by Dunn's multiple comparison tests were used to assess differences compared to baseline. Spearman's rank adjusted by Benjamini & Hochberg FDR were used to analyse correlations between fecal amino acid abundance and fecal bacterial taxa obtained by 16S rRNA gene sequencing. Statistical significance was set at $q<0.05$. Nine essential and 10 non-essential proteinogenic amino acids were identified by untargeted metabolomics. None of the amino acids were altered in the serum during or after metronidazole administration (D14 or D42) when compared to baseline values. In the feces, lysine was significantly decreased on D14 ($q=0.041$) and tyrosine was significantly decreased on D7 ($q=0.002$) and D14 ($q=0.041$). Additionally, fecal glycine, glutamic acid, and cysteine were significantly increased during metronidazole administration ($q<0.029$). Across all time points, the abundance of *Enterococcus* and *Dorea* correlated with tyrosine ($\rho=-0.702$ and $\rho=0.645$, respectively; $q<0.001$), and *Streptococcus* abundance positively correlated with glutamic acid ($\rho=0.612$; $q<0.001$). Significant alterations of some amino acids were observed in the feces, but not in serum during metronidazole-induced dysbiosis in healthy dogs. This suggests that short-term gastrointestinal microbial dysbiosis caused by metronidazole administration does not affect serum amino acid metabolism in healthy dogs.

Abstract GI18: Chronic Enteropathy Has Bigger Impact Than Diet or Country of Residence on Canine Fecal Dysbiosis

Rachel Pilla[1]; Melissa Guard[2]; Stefan Unterer[3]; Karin Allenspach[4]; Linda Toresson[5]; Aurelien Grellet[6]; Leda MO Barros[7]; Joerg Steiner[8]; Jonathan Lidbury[8]; Jan Suchodolski[8]
[1]Texas A&M; [2]Texas A&M University; [3]Ludwig Maximilian University of Munich; [4]Iowa State University; [5]Evidensia Specialist Animal Hospital; [6]NeoCare, UMR INRA/ENVT, University of Toulouse; [7]FMVZ, University of Sao Paulo; [8]Gastrointestinal Laboratory, Texas A&M University

Alterations of the gut microbiome are often present in dogs with chronic enteropathies (CE). A recently published qPCR-based assay termed the fecal dysbiosis index (DI) can assess fecal dysbiosis in dogs. While previous metagenomic studies have revealed that some changes in gastrointestinal microbiome composition can be induced by diet, it is unclear how those changes impact the keystone bacteria assessed by the DI. The GI microbiota in humans has also been reported to be affected by country of residence, a factor which has not yet been investigated in dogs. Therefore, the goal of this study was to evaluate the impact of CE, dietary macronutrient composition, and country on the fecal DI. A total of 271 fecal samples were selected from previous clinical trials, including 185 from healthy control (HC) dogs from Brazil, Finland, Germany, Italy, Sweden, and the USA, and 86 dogs with CE from Brazil, France, Sweden, UK, and the USA. DNA was extracted and the fecal DI was determined. Data on diet at the time of sampling was available for 76 HC and 45 CE dogs. Mann-Whitney and Kruskal-Wallis tests with multiple comparisons were performed to compare the DI between groups and countries, respectively. The relationship between the proportion of macronutrients (protein, fat, and carbohydrate percentages of metabolizable calories, fiber as percentage of dry matter) and fecal DI were evaluated with Spearman's rank correlation. Statistical significance was set at $p<0.05$. Dogs with CE had a significantly higher DI (median: 3.2) compared to HC (median: -3.7, $p<0.001$). When dietary factors were considered, the fecal DI was not significantly correlated with dietary protein (CE $p=0.948$; HC $p=0.316$), fat (CE $p=0.887$; HC $p=0.461$), or carbohydrate (CE $p=0.448$; HC $p=0.503$) content. Dietary fiber percentages also did not correlate with DI in CE ($p=0.156$) or HC ($p=0.590$). While dietary composition did not affect fecal DI, when individual bacterial taxa were considered, two correlations were identified within the CE groups. Dietary fat showed a weak positive correlation with the abundance of *Turicibacter* ($p=0.019$, $r=0.359$), and carbohydrate content showed a weak positive correlation with *Fusobacterium* abundance ($p=0.016$, $r=0.369$). Country had no influence on DI within the group of dogs with CE ($p=0.065$), and within HC, only dogs from Italy had a significantly higher DI (median: 0.2) compared with other countries (median for Brazil: -2.3; Finland: -4.0; Germany: -4.1; Sweden: -4.5; and USA: -3.9; $p=0.002$). In conclusion, dogs with CE had a significantly higher DI than HC, the median of which was outside of the reference interval. While the abundance of individual bacterial taxa were weakly correlated with dietary factors, a composite index such as the DI is less susceptible to such effects. Indeed, we found no correlation between dietary protein, fat, or carbohydrate percentages of metabolizable calories, or dietary fiber percentages, with fecal DI. Interestingly, fecal samples from healthy dogs living in Italy were different than all other countries, which possibly arose from factors unaccounted for in this study. Overall, these factors are unlikely to affect the use of the DI as a diagnostic and monitoring tool.

This page intentionally left blank.

Abstract GI19: Comparison of the Fecal Microbiota Between Healthy Dogs and Dogs With Diabetes Mellitus

Fabio Alves Teixeira[1]; Bettina Di Donato[2]; Patricia Eri Ishii[3]; Thaila Putarov[4]; Aulus Carciofi[4]; Juliana Jeremias[5]; Mariana Queiroz[2]; Cristiana Pontieri[5]; Joerg Steiner[3]; Jonathan Lidbury[3]; Jan Suchodolski[3]; Marcio Brunetto[2]; Marcia Gomes[2]
[1]School of Veterinary Medicine and Animal Science, University of São Paulo, São Paulo, Brazil; [2]School of Veterinary Medicine and Animal Science, USP, Brazil; [3]College of Veterinary Medicine and Biomedical Sciences, Texas A&M University, College Station, TX, USA; [4]School of Agricultural and Veterinarian Science, Sao Paulo State University; [5]Nutritional Development Center, Grandfood Industry and Commerce LTD (Premier Pet)

There is evidence that both humans and mice with diabetes mellitus (DM) have gastrointestinal dysbiosis. This study aimed to compare the fecal microbiota of diabetic dogs and healthy control dogs. Twenty-one dogs were selected for this study: 8 healthy adult beagles from a research facility and 13 privately owned diabetic dogs of various breeds. The criteria for inclusion of diabetic dogs included: glycemia fluctuation between 90 and 300 mg/dL, body condition score (4–5/9), and same body weight for 60 days. Diabetic dogs received a high-fiber diet to facilitate glycemia control for 60 days and healthy beagles received the same diet for 45 days prior to assessment. Fecal samples were collected and were immediately frozen at -80°C. Microbiota was assessed with 16S rRNA gene sequencing and a qPCR based fecal dysbiosis index (DI). The abundance of bacterial groups and DI between the two groups of dogs were compared using Welch's two sample t-tests with significance set at $p<0.05$. There was no significant difference in the DI between diabetic (mean: -0.69±4.34) and healthy beagles (mean: -1.02±1.76; $p=0.81$). Sequencing data revealed that diabetic dogs had lower abundance of Lactobacillus gender bacteria and a higher abundance of Ruminococcus when compared to healthy dogs. This was the first dog microbiota profile study comparing healthy and diabetic dogs with a standardized diet. 16S rRNA gene sequencing showed similar findings to those from studies of humans and mice in that certain bacterial genera were altered in diabetic dogs. Further research is needed to determine if these alterations in the gastrointestinal microbiota play a role in the pathogenesis of diabetes in dogs.

Abstract GI20: Understanding the Failure of Oral Acid Suppressants in Cats

Katie Tolbert[1]; Desola Odunayo[2]; Phillip Ryan[2]; Shanna Hillsman[2]; Gina Galyon[2]; Silke Hecht[2]; Joerg Steiner[1]; Josh Price[2]
[1]Texas A&M University; [2]University of Tennessee

A sustained increase in gastric pH is central to the successful treatment of upper gastrointestinal (GI) erosion/ulceration and esophagitis in cats. Thus, acid suppressant drugs, such as proton pump inhibitors (PPI; e.g., omeprazole) and histamine-2 receptor antagonists (e.g., famotidine) are commonly prescribed. The requirement for frequent administration and the need to give omeprazole before a meal in cats likely leads to disruption of the human-animal bond, poor owner adherence, and treatment failure. Thus, veterinarians are increasingly prescribing newer PPIs that are longer acting and can be given with food. The efficacy of these novel gastroprotectants has heretofore not been comparatively evaluated and described. Accordingly, our central objective was to evaluate the efficacy of orally administered esomeprazole, lansoprazole, and dexlansoprazole in increasing the intragastric pH of cats. In a randomized, crossover study, 12 cats received 1 mg/kg esomeprazole or lansoprazole or 6 mg/kg dexlansoprazole orally twice-daily for 4 days. Intragastric pH was compared within and between treatments over time using repeated measures mixed model ANOVA. Significant differences in mean percentage time (MPT) intragastric pH ≥3 were found between treatments ($p=.002$) and over time ($p=.039$) but not treatment by time. *Post hoc* tests revealed that cats treated with lansoprazole were found to have lower MPT intragastric pH ≥3 than cats treated with dexlansoprazole or esomeprazole ($p=.027$, for each). Only esomeprazole achieved the pH goal for the treatment of duodenal ulceration as defined for people and this effect was not observed until treatment day 4. The difference between cats and dogs in their response to oral acid suppressants might be explained by differences in feline GI pH and motility and their impact on drug delivery. Preliminary data from a subsequent study on feline gastric pH and gastrointestinal transit times will be included to support this conclusion.

Abstract GI21: Phenotypic and Functional Characterization of Adult Intestinal Organoids From Dogs With Inflammatory Bowel Disease

Sichao Mao[1]; Chelsea Iennarella-Servantez[1]; Todd Atherly[1]; Dana Borcherding[1]; Yoko Ambrosini[2]; Laura Kurr[1]; Agnes Bourgois-Mochel[1]; Yeon-Jung Seo[1]; Jonathan Mochel[1]; Karin Allenspach[1]; Albert Jergens[1]
[1]Iowa State University; [2]University of Texas at Austin

Canine inflammatory bowel disease (IBD) refers to a group of chronic gastrointestinal (GI) disorders of unknown cause and pathogenesis. With the aim of modeling spontaneous GI disease in dogs and humans, intestinal stem cell (ISC)-derived organoids are emerging as a promising *ex vivo* system for studying IBD pathogenesis. Previous studies from our laboratory have validated the cultivation of 3D canine intestinal organoids (enteroids and colonoids) from crypt base columnar (CSC) cells. In collaboration with the US FDA, the aim of this investigation was to compare the similarities and differences in phenotypic and functional characterization between organoids derived from healthy dogs and dogs with IBD. We chose to focus on ileal-derived enteroids as a translational *ex vivo* model due to their unique role in drug absorption and as a primary site of human IBD (i.e., Crohn's ileitis) development. ISCs isolated from endoscopic biopsies of two healthy dogs and two dogs with active IBD were differentiated into intestinal organoids. Ileal organoids and matching primary tissues were probed by RNA *in situ* hybridization and immunohistochemistry for phenotypic changes in IBD. A panel of six phenotypic markers identified different epithelial cell lineages (LGR5: intestinal stem cell, ALP: enterocyte, PAS: goblet cell, NEUROG3: enteroendocrine cell), evidence of epithelial barrier integrity (ZO-1) and cell proliferation (Ki-67). Phenotypic marker expression was then quantitated by light microscopy and ImageJ. In addition, functional features of IBD organoids were investigated by cystic fibrosis transmembrane conductance regulator (CFTR) organoid swelling assay to measure chloride-channel-water conductance across epithelial cells. Forskolin-induced swelling of enteroids was quantitated using an inverted microscope and ImageJ. The increase in surface area after forskolin treatment in organoids from dogs with IBD was compared to results obtained in organoids derived from healthy dogs. Data were analyzed using R version 3.5. Pairwise comparisons were performed to determine group differences. Statistical significance level was set at $P<0.05$. IBD organoids showed different phenotypic features as compared to healthy organoids, including increased expression of ALP, NEUROG3, LGR5, ZO-1, and PAS; however, Ki-67 expression was found to be decreased in IBD enteroids. Among these markers, expression of NEUROG3 and PAS exhibited significant difference ($P<0.05$). Furthermore, similar trends in phenotypic changes were observed between IBD organoids and inflamed primary tissues for LGR5, NEUROG3, and ZO-1 expression. Forskolin significantly ($P<0.05$) increased surface area of IBD organoids after 1 and 4 hr vs. controls indicating functional CFTR in IBD organoids. To conclude, this study provides the first report of phenotypic changes in intestinal organoids from dogs with IBD. Overall, our results show that IBD organoids show different phenotypic features as compared to healthy organoids. In addition, our study shows that ileal organoids derived from dogs with IBD recapitulate both the phenotypic and physiological features of diseased tissue compared to healthy tissue, demonstrating its utility as an *ex vivo* model for investigating pathomechanisms and pharmacotherapy in dogs with IBD.

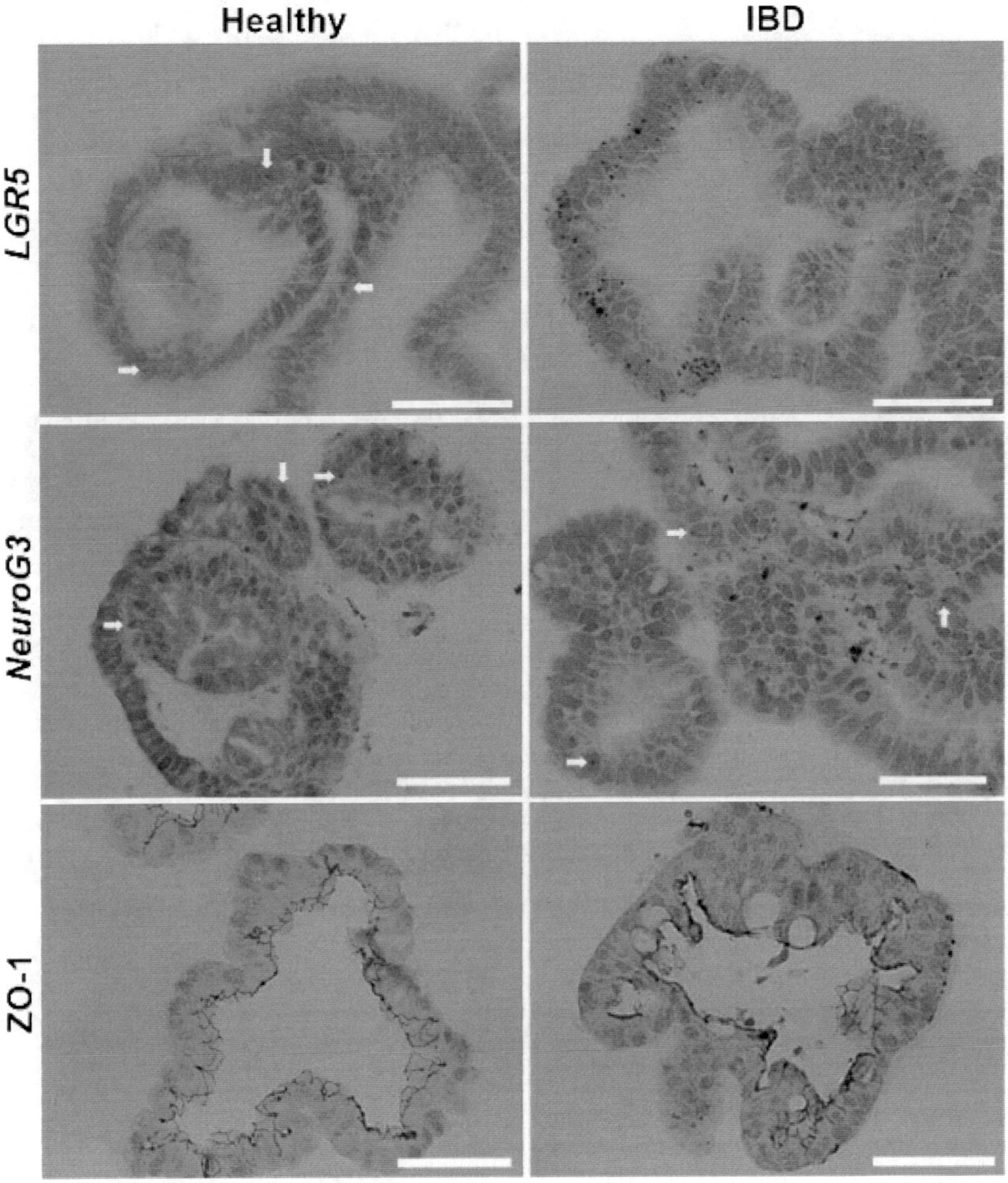

Healthy
IBD
LGR5
NeuroG3
ZO-1

Abstract GI22: Increased Mean Platelet Volume in Dogs with Canine Parvoviral Enteritis

Monique Engelbrecht[1]; Amelia Goddard[2]; Vanessa McClure[2]; Paolo Pazzi[2]
[1]Onderstepoort Veterinary Academic Hospital, University of Pretoria; [2]University of Pretoria

Bacterial translocation from the damaged intestinal tract, reported in canine parvoviral (CPV) enteritis, is thought to be responsible for the systemic inflammatory response resulting from coliform septicaemia, which could ultimately progress to septic shock and death. Alterations in platelet indices, specifically mean platelet volume (MPV), are a consistent finding in critically ill people and dogs with and without sepsis. Increased MPV has been reported to be an indirect indicator of platelet activation and of bone marrow response in people and dogs with sepsis. The study aim was to compare admission MPV and platelet volume distribution width (PVDW) in dogs with CPV enteritis to that of healthy aged matched control dogs. Forty-eight dogs with CPV enteritis and 18 healthy age-matched control dogs were included. CPV infection was confirmed with electron microscopy and concurrent blood-borne infections were excluded using PCR. EDTA whole blood samples were analysed on an automated cell counter, ADVIA 2120, within 30–60 minutes from collection. There was no significant difference for platelet count between the groups. The MPV for CPV infected dogs (median: 14.0; IQR: 12.2–15.1) was significantly higher compared to controls (11.3; IQR: 10.3–13.1, $P=0.002$). The PVDW for CPV infected dogs (66.9; IQR: 64.2–68.8) was significantly higher compared to controls (63.3; IQR: 60.2–65.1, $P<0.001$). These findings suggest that significant platelet activation is present in dogs with CPV enteritis which may play a role in the disease outcome, similar to people with sepsis. Further studies are required to investigate the prognosticating ability of MPV in dogs with CPV enteritis.

Abstract GI23: Proteomic Characterization of Feline Gastric Fluid to Detect Protein Biomarkers of Reflux and Aspiration

Megan E. Grobman[1]; Hansjorg Rindt[2]; Carol Reinero[2]
[1]Auburn University; [2]University of Missouri

Background: Reflux and aspiration of gastric contents are sources of respiratory disease in humans despite 50% of healthy adults aspirating without apparent clinical consequence. In cats, analogous information is lacking. Understanding the frequency of reflux and aspiration in healthy cats is critical to determining their significance in patients with respiratory disease. **Objectives/Hypothesis**: To investigate the prevalence of extra-esophageal reflux (EER) in healthy cats using proteomic characterization of feline gastric fluid and orphopharnygeal (OP) swabs before and after feeding. We hypothesize that healthy cats will have evidence of EER characterized by detectable gastric and intestinal proteins on OP swabs. Animals: 17 healthy companion and research cats. **Methods**: Gastric fluid (GF) (n=4) and OP swabs (n=13), pre- and 30 min post-feeding, were evaluated by liquid chromatography mass spectrometry (LCMS). Differential protein abundance (DA) was evaluated by Student T test ($p<0.01$). Coefficients of variation were calculated for proteins with DA. **Results**: Individual proteins (n=1802) were identified in OP swabs. Significant DA was found ($p<0.01$): pre-feeding vs. post-feeding (n=126), GF vs. pre-feeding (n=267), GF vs. post-feeding (n=217). Marked between-cat variation was observed for proteins with DA. Proteins specific to the stomach or intestines were found in OP swabs from all cats with no differences between pre- and post-feeding samples. **Conclusions**: Gastrointestinal proteins in the OP of healthy cats suggests EER occurs as part of their swallowing physiology. Quantification of GI proteins in the OP may be necessary to determine pathology rather than the presence or absence of such markers.

Abstract GI24: Assessment of Intestinal Permeability in Dogs with Chronic Enteropathy

Patricia Eri Ishii[1]; Ana Rita Carvalho Pereira[2]; Fabio Teixeira[3]; Robert Kyle Phillips[1]; Rafael De Miranda[4]; Franz Yoshitoshi[4]; Paula Giaretta[5]; Ricardo Duarte[6]; Jonathan Lidbury[7]; Joerg Steiner[7]; Jan Suchodolski[7]
[1]Texas A&M University; [2]GastroVet; [3]School of Veterinary Medicine and Animal Science, University of Sao Paulo; [4]Endoscopet Medicina Veterinaria; [5]Universidade Federal de Minas Gerais; [6]Faculdades Metropolitanas de Sao Paulo; [7]College of Veterinary Medicine and Biomedical Science, Texas A&M University

Increased IP has been hypothesized to play a role in the pathogenesis of chronic enteropathy (CE) in dogs. However, due to a lack of an optimal marker, the mechanism by which IP affects or influences CE in dogs remains unclear. Recently, iohexol has been described as an IP marker in healthy dogs, laboratory rats, and humans. The purpose of this study was to assess IP in dogs with CE. Thirty dogs were selected for this study: 15 healthy controls and 15 dogs with CE. Among the 15 dogs with CE, 5 had been classified as having food-responsive enteropathy (FRE), 4 with antibiotic-responsive enteropathy (ARE), and 6 with steroid-responsive enteropathy (SRE). Dogs were fasted for 12 hours before oral administration 2.0 ml/kg of Omnipaque-350 (GE Healthcare). Blood samples were collected from the jugular vein 2 hours after iohexol administration, centrifuged, and frozen. An enzyme-linked immunosorbent assay (FIT-GRF™ Iohexol kit, BioPAL) was used to quantify iohexol in serum. Post-administration serum iohexol concentration were significantly higher in dogs with CE (median: 0.084 µg/mL range: 0.05–0.303 µg/mL) compared to healthy control dogs (median: 0.07 µg/mL range: 0.052–0.118 µg/mL; p=0.042). Dogs with SRE had significantly higher post-administration iohexol concentrations (median: 0.122 µg/mL range: 0.069–0.303 µg/mL) when compared to healthy control dogs (p=0.013). Similarly to what has been shown in humans, this study suggests that a subset of dogs with CE have increased IP when compared to healthy dogs. This finding reinforces the need for further studies to clarify the role of IP on the pathogenesis of gastrointestinal diseases.

Abstract GI25: Effects of Clinical Characteristics and Lifestyle Factors on Fecal Canine S100/Calgranulin Concentrations

Romy M. Heilmann[1]; Melissa Guard[2]; Linda Toresson[3]; Stefan Unterer[4]; Aurelien Grellet[5]; Niels Gruetzner[6]; Jan Suchodolski[7]; Joerg Steiner[7]
[1]College of Veterinary Medicine, University of Leipzig; [2]College of Veterinary Medicine and Biomedical Sciences, Texas A&M University; [3]Evidensia Specialist Animal Hospital; [4]Faculty of Veterinary Medicine, Ludwig-Maximilians-University; [5]Université de Toulouse, ENVT; [6]University of Veterinary Medicine; [7]College of Veterinary Medicine and Biomedical Sciences, Texas A&M University

Fecal S100/calgranulin (S100A12 and calprotectin) concentrations are useful markers of gastrointestinal inflammation in dogs, and the expression of S100A12 relative to that of calprotectin (fCalR) could be useful to further characterize intestinal diseases. In people, fecal S100/calgranulin concentrations are affected by age, obesity, diet, and other lifestyle factors. Knowledge about the effects of such factors on fecal S100/calgranulin concentrations in dogs is currently lacking. However, this knowledge is an important prerequisite for further clinical validation of these two biomarkers. The aim of the study was to evaluate the effect of several patient clinical characteristics and lifestyle factors on fecal S100/calgranulin concentrations and fCalR in a large cohort of healthy dogs. Single spot fecal samples from 181 healthy pet dogs and data derived from a standard questionnaire served to evaluate the effect of age, sex, reproductive status, body weight and body condition, breed type and size, vaccination, endoparasite treatment, diet, environment, and travel history on fecal S100/calgranulin concentrations and the fCalR using a multi-variate mixed effects statistical model and by calculating the effect sizes (ES), with statistical significance set at $P<0.05$. Univariate analysis showed a significant association of intact reproductive status (particularly in female dogs) with higher fecal S100A12 but lower fecal calprotectin and fCalR, and of small breed size with higher fecal S100A12 and calprotectin concentrations but lower fCalR. Pure-bred type was linked to higher fecal S100A12 and lower fCalR, and recent vaccination (particularly with a canine parvoviral vaccine) to higher fecal S100A12 concentrations. Trends for associations were seen: diet with fecal S100A12 and calprotectin, and recent vaccination with fCalR. In multivariate models, small breed size was linked to higher fecal S100A12 and calprotectin (both of crucial clinical relevance), whereas recent vaccination affected only S100A12 concentrations (clinically moderately important increase based on ES); but no effect remained significant for fCalR. Some clinical characteristics (i.e., breed size, recent vaccination, and reproductive status in female dogs) can affect fecal S100/calgranulin concentrations, and these biomarkers should be interpreted in light of those confounding factors. The utility of population-based reference intervals for fecal canine S100/calgranulin concentrations might be improved through stratification by sex/reproductive status and breed size. Fecal canine S100/calgranulin concentrations are not confounded by age, body condition, deworming, diet, environment, or travel history. This study provides an important basis for further evaluating the clinical utility of fecal S100/calgranulin concentrations as gastrointestinal disease markers in dogs.

Abstract GI26: Association of Hypercobalaminemia with Pathological Findings in Dogs and Cats

Romy M. Heilmann[1]; Lea Sielski[2]; Stefanie Kather[2]; Franziska Dengler[2]; Anika Jirasek[3]
[1]College of Veterinary Medicine, University of Leipzig; [2]College of Veterinary Medic, University of Leipzig; [3]Idexx Laboratories Germany

Hypocobalaminemia is common in dogs and cats with exocrine pancreatic insufficiency and chronic enteropathies and also has prognostic value. Hypocobalaminemia has been extensively studied, but increased serum cobalamin concentrations are mostly attributed to oversupply (supplementation) and have thus received little attention in small animal medicine. Current research in human medicine associates hypercobalaminemia with severe systemic illnesses, e.g. neoplastic, hepatic, and renal diseases. Hypercobalaminemia (without prior supplementation) in cats might also be linked to solid neoplasms and hepatopathies. The aim of this retrospective study was to determine the prevalence of hypercobalaminemia and its association with pathological findings in dogs and cats. Medical records were evaluated of all dogs and cats that presented to the UL-VMTH between 2007–2019 and had serum cobalamin concentration measured. Repeated measurements from the same patient, patients that had received supplemental cobalamin, and patients with incomplete medical records were excluded from further analyses. The proportions of patients with hypercobalaminemia (serum cobalamin concentration >908 ng/L in dogs and >1,334 ng/L in cats) was calculated, and the association of hypercobalaminemia with patient characteristics and pathological findings (diagnosis, affected organ system) was evaluated by χ^2 test. Statistical significance was set at $P<0.05$. Of the 654 dogs included in the analysis, 3% (n=21) were hypercobalaminemic (serum cobalamin concentration 914–3,561 ng/L, median: 1,307 ng/L), 29% (n=189) were hypocobalaminemic (<251 ng/L), and 68% (n=441) had a serum cobalamin concentration within the reference interval (251–908 ng/L). Ten (48%) of the hypercobalaminemic dogs had chronic gastrointestinal signs, associated with an inflammatory cause (hepatopathy, enteropathy, and/or pancreatitis) or food intolerance, and 2 hypercobalaminemic dogs (10%) were diagnosed with hypoadrenocorticism. Adrenal pathology was significantly associated with hypercobalaminemia ($P=0.014$). No breed predisposition existed for hypercobalaminemia, but small breeds were overrepresented (57%). Of the 315 cats included in the analysis, 11% (n=34) were hypercobalaminemic (serum cobalamin concentration 1,370–3,107 ng/L, median: 1,713 ng/L), 29% (n=92) were hypocobalaminemic (<270 ng/L), and 189 (60%) were normocobalaminemic (270–1,334 ng/L). Of the hypercobalaminemic cats, 65% (n=22) had a chronic enteropathy, 24% (n=8) acute/chronic pancreatitis, and 18% (n=6) a cholangiohepatopathy. Neoplasia was diagnosed in 2 hypercobalaminemic cats (6%), and 1 cat was hyperthyroid. Hypercobalaminemia was not associated with a specific diagnosis or affected organ system in cats. This study suggests that hypercobalaminemia (without prior supplementation) occurs infrequently in cats, and less often in dogs. Our results confirm the prevalence of hypercobalaminemia in cats reported from a previous study. Hypercobalaminemia in dogs and cats appears to be linked to severe inflammatory, immune-mediated, and neoplastic conditions. Whether hypercobalaminemia in dogs and cats is a benign finding and reflects an increased cobalamin-binding capacity (transcobalamin II) or the presence of autoantibodies against transcobalamin interfering with cobalamin binding in serum warrants further study.

This page intentionally left blank.

Abstract GI27: 1,2-o-dilauryl-rac-glycero-3-glutaric acid-(6′-methylresorufin) Ester (DGGR) Lipase Assay is Not Specific for Feline and Canine Pancreatic Lipase

Sue Yee Lim[1]; Panagiotis Xenoulis[2]; Evangelia Stavroulaki[2]; Jonathan Lidbury[1]; Jan Suchodolski[1]; Frédéric Carrière[3]; Jörg Steiner[1]
[1]Texas A&M University; [2]University of Thessaly; [3]Aix-Marseille University

Measurement of pancreatic lipase is important for the diagnosis of feline and canine pancreatitis. Recent studies have claimed that lipase assays using 1,2-o-dilauryl-rac-glycero-3-glutaric acid-(6′-methylresorufin) ester (DGGR) as substrate are more specific for measuring pancreatic lipase than traditional lipase assays. However, the specificity of this assay for pancreatic lipase has not been demonstrated. Intravenous heparin administration releases hepatic and lipoprotein lipases into the bloodstream. This study aims to evaluate post-heparin plasma lipase activity using a DGGR-based assay and pancreatic lipase immunoreactivity (PLI) concentration in cats and dogs. We hypothesize that hepatic and lipoprotein lipases released after heparin administration can hydrolyze DGGR and contribute to plasma lipase activity measured with a DGGR-based assay. Heparin was administered to six cats and six dogs. Blood was collected at baseline and at 10, 20, 30, 60, and 120 minutes after heparin administration. Lipase activity was measured using a DGGR-based assay, and PLI concentration was measured by Spec fPL (for cats) or cPL (for dogs). Plasma lipase activity as measured by the DGGR-based assay increased significantly 10 minutes after heparin administration in both cats (P=0.003) and dogs (P=0.006) and returned to baseline by 120 minutes. In contrast, PLI concentrations showed no significant changes after heparin administration. In conclusion, DGGR is not only hydrolyzed by pancreatic lipase but also by hepatic lipase and/or lipoprotein lipase in cats and dogs. Because these extrapancreatic lipases are present naturally in cats and dogs, they may contribute to a lack of analytical specificity for the DGGR-based assay.

Abstract GI28: Markers of Endothelial Activation and Inflammation in Canine Parvoviral Enteritis

Brogan K. Atkinson[1]; Sune Pretorius[2]; Zandri Whitehead[2]; Amelia Goddard[2]; Paolo Pazzi[2]
[1]Onderstepoort Veterinary Academic Hospital, University of Pretoria; [2]University of Pretoria

Canine parvovirus (CPV) is a common cause of enteritis, immune suppression, and systemic inflammation in dogs. Markers of endothelial activation, such as intercellular adhesion molecule 1 (ICAM-1), vascular cell adhesions molecule 1 (VCAM-1) and high mobility group box 1 protein (HMGB-1), provide insight into the state of the endothelium during inflammation. Cytokines known to activate the endothelium in response to inflammation include tumour necrosis factor alpha (TNFα), interleukin 1 beta (IL-1β), and IL-4. This study aimed to determine if blood concentrations of endothelial markers and cytokines were increased in CPV enteritis compared to healthy controls, and whether correlation existed between endothelial markers and cytokines. Thirty dogs with naturally occurring CPV enteritis and ten age-matched healthy control dogs were included. Endothelial markers (ICAM-1, VCAM-1, and HMGB-1) were performed on stored EDTA plasma and cytokines (TNFα, IL-1β, and IL-4) were performed on stored serum using commercially available canine-specific ELISA kits. In dogs with CPV enteritis ICAM-1 was significantly lower (5.9 (IQR: 4.3–8.3) compared to controls (8.0 (IQR: 6.9–10.3) p=0.008). No significant difference was found for VCAM-1, HMGB-1, and IL-1β. TNFα and IL-4 were below detection limit in all samples. A strong correlation between HMGB-1 and IL-1β (r: 0.501, p=0.005) was identified. Despite evidence of systemic inflammation in this condition, our results suggest lack of endothelial activation and specific cytokine production based on blood concentrations. Endothelial activation, leukocyte adhesion, and subsequent transmigration through the epithelial is likely to be affected in CPV. The significance of these findings requires further investigation.

Abstract GI29: The Impact of Sampling Method on Gut Microbial Community Profiles in Dogs and Cats

Stacie Summers[1]; Allysa Galloni[2]; Craig Webb[2]
[1]Oregon State University; [2]Colorado State University

Feces, rectal swabs, and colonic mucosal biopsies can be used to characterize the microbial community profiles of the descending colon using 16S ribosomal RNA (16S rRNA) gene sequencing. To date, there are no studies comparing the potential bacterial variation between these three sampling methods in dogs and cats. The aim of this study was to compare the microbial communities between feces, rectal swabs, and colonic biopsies in healthy cats and dogs as well as cats and dogs diagnosed with a chronic enteropathy (CE). Three sample types (feces, rectal swab, and endoscopic distal colonic mucosal biopsies) were collected from clinically healthy dogs (n=9), dogs with CE (n=8), and cats with CE (n=7). A rectal swab and feces were collected from clinically healthy cats (n=8). The 16S rRNA V4 region was amplified and sequenced (Illumina) for all samples using primers 515F-806R. Sequences were processed and analyzed using Quantitative Insights Into Microbial Ecology (QIIME) v1.7.0. A Friedman test with Dunn's *post hoc* analysis or Wilcoxon matched-pairs signed rank test was used to determine differences in alpha diversity indices among feces, rectal swabs, and when available, colonic biopsies for the 4 study groups. Differences in microbial community profiles between the 3 sample types were evaluated within each study group using Analysis of Similarities (ANOSIM) with Bray-Curtis dissimilarity matrix. Chao1 was significantly higher in colonic biopsies when compared to feces in healthy dogs (P=0.029), dogs with CE (P=0.037), and cats with CE (P=0.049), however, no differences were found when colonic biopsies were compared to rectal swabs. Several notable findings were found with ANOSIM. For healthy dogs, the microbial communities were not significantly different among the three sampling methods. For healthy cats, the fecal microbial community was significantly different when compared to rectal swabs (R=0.815; P=0.001). For dogs with CE, the fecal microbial community was significantly different when compared to the colonic biopsy samples (R=0.251; P=0.022). For cats with CE, the fecal microbial community was significantly different when compared to rectal swabs (R=0.272; P=0.001) and colonic biopsy samples (R=0.192; P=0.020). In conclusion, the mucosa of the distal colon exhibits higher bacterial richness compared to feces. Rectal swabs may be an alternate option to collection of fecal samples in dogs and to collection of colonic biopsies in both dogs and cats.

Abstract GI30: Spontaneous Gastroduodenal Perforations in Five Cats

Fernanda Vieira Amorim da Costa[1]; Izadora Zardo[2]; Bárbara Rivas[2]; Gabriela Schaefer[2]; Luciana Sonne[2]; Saulo Pavarini[2]; Marcele Bandinelli[2]
[1]Universidade Federal do Rio Grande do Sul; [2]UFRGS

Gastroduodenal perforations can be classified as spontaneous when there is a perforation with no history of recent gastrotomy and/or anastomosis, external or iatrogenic gastric trauma, gastrointestinal foreign body or volvulo-gastric dilation. Spontaneous gastroduodenal perforations (SGP) are normally associated with pre-existing gastroduodenal ulcers or neoplasia. In cats, SGP has been uncommonly reported in the literature. The aim of this study is to describe cases of PGE in cats seen at UFRGS's Feline Medicine Service between March 2018 and November 2019, and its possible etiologies. Five adult cats were included, four of them were mixed breed and one was Persian, three females and two males, one of them was not neutered. All patients had a history of recent surgical procedure, with an average of eight days post-surgical time until the diagnosis of GP was made. The surgical procedures included: osteosynthesis of the mandible and tibia, hemilaminectomy, enucleation, and ovariosalpingohisterectomy. These patients were submitted to a similar anesthetic protocol, using methadone as a pre-anesthetic medication (mean of 0.3 mg/kg), and for anesthetic induction, propofol to the effect. Surgical maintenance was performed with isoflurane and fentanyl bolus when necessary. Intraoperative hypotension was related in one cat. Four patients received anti-inflammatory drugs, meloxicam was prescribed for three patients, at doses between 0.05 mg/kg to 0.1 mg/kg, for a maximum of 4 days, and one of them received a single dose of dexamethasone (0.25 mg/kg). Clinical presentation for SGP was diverse. Some patients had typical signs of gastrointestinal tract (vomiting, abdominal distension, and diarrhea), but most clinical signs were non-specific, including dehydration, prostration, pale mucous membranes, anorexia or hyporexia, weight loss, hypothermia, tachycardia, hypotension, tachypnea, and hyperthermia. On hematological analysis, 80% of the cats had anemia, leukocytosis, and neutrophilia. Left deviation, lymphopenia, and monocytosis were also observed, but, metamyelocytes, myelocytes, neutropenia, and leukopenia were present in only one patient. The most common finding on blood biochemistry was hypoalbulminemia, presented in four cats. Abdominal ultrasound was performed in one patient, that showed hyperechoic mesenteric fat, gastric wall thickening (0.46 cm), presence of free echogenic peritoneal fluid and reduced motility. Diagnosis was confirmed by exploratory surgery in three cats and necropsy in two. Gastric perforations in the pyloric antrum were more common (4/5), with just one perforation in the proximal duodenum. Three patients underwent necropsy, including one that died after exploratory surgery, and histopathological examination showed transmural necrosis at the perforation site. All cats had diffuse peritonitis and the necropsy findings did not show any predisposing lesion for the injury. Gastric perforation represents a medical emergency and the condition is particularly difficult to diagnose accurately. In our study, the mortality rate for cats was 80%. Although several etiologies for SGP have been suspected, it was not possible to identify the exact origin of the perforation as in the cases mentioned in this study. GP secondary to NSAID or corticosteroid treatment has already been described in cats, and the administration of these drugs may have masked clinical signs associated with acute peritonitis. A history of recent surgery and SGP have only been associated four times in the literature, it is believed that hemodynamic and pharmacological alterations promoted by surgery and anesthesia can lead to mucosal hypoperfusion and consequent predisposing the lesion. Our records showed that SGP arising from surgical procedures are not uncommon as shown in the literature.

Abstract GI31: Fecal Microbiome and Metabolomic Changes in Dogs Receiving Antibiotics Followed by Placebo or Synbiotics

Jacqueline C. Whittemore[1]; Tamberlyn Moyers[2]; Joshua Price[1]; Jan Suchodolski[3]
[1]University of Tennessee, Knoxville; [2]College of Veterinary Medicine, University of Tennessee; [3]Texas A&M

Antibiotic-associated gastrointestinal signs occur in ≤100% of dogs administered enrofloxacin with metronidazole; signs partially are mitigated by synbiotics. The objective of this study was to compare the fecal microbiome and metabolome of 24 healthy dogs administered enrofloxacin (10 mg/kg qd) and metronidazole (12.5 mg/kg BID), followed 1 hour later by placebo or synbiotics (Proviable®-Forte with Proviable®-SB), for 21 days with reevaluation 8 weeks thereafter. Fecal samples were collected on days 5–7 (baseline), 26–28, and 82–84. Sequencing of 16S rRNA genes for operational taxonomic units (OTUs) was performed and mass spectrometry used to determine metabolomic profiles. P<0.05 was considered significant, with Benjamini & Hochberg's False Discovery Rate used to adjust for multiple comparisons. Alpha and beta diversity differed significantly from baseline during treatment and on days 82–84. At the genus level, significant group-by-time interactions were noted for 15 OTUs, including *Adlercreutzia, Bifidobacterium, Slackia, Turicibacter, Clostridium,* [*Ruminococcus*], Erysipelotrichaceae_g_, [*Eubacterium*], and Succinivibrionaceae_g_. Group and time effects were present for an additional 6 OTUs, including *Collinsella*, Ruminococcaceae_g_, and *Prevotella*. Metabolite profiles differed significantly by group-by-time, group, and time for 28, 20, and 192 metabolites, respectively. Short-chain fatty acid, bile acid, indole, sphingolipid, polyamine, and cinnaminic acid metabolites were affected, with some changes persisting through days 82–84 and differing between groups. Antibiotic administration causes sustained dysbiosis in dogs with similar changes in the microbiome and metabolome to those found in cats. Significant group-by-time interactions were noted for a number of OTUs and metabolites, potentially contributing to decreased antibiotic-induced gastrointestinal effects in dogs administered synbiotics.

Abstract GI32: Concurrent Gastrointestinal Signs in Hypothyroid Dogs

Eleonora Gori[2]; Giada Paolinelli[2]; Paola Gianella3; Alessio Pierini[2]; **George Lubas**[1]; Veronica Marchetti[2]
[1]Deptartment of Veterinary Sciences, University of Pisa; [2]University of Pisa; [3]University of Turin

Few observations about prevalence and features of gastrointestinal (GI) signs in hypothyroid dogs (hypoT-dogs) are available. The study aimed (1) to evaluate concurrent GI signs in hypoT-dogs; (2) to analyze clinico-pathological and ultrasound features of hypoT-dogs with and w/out GI signs, and (3) to analyzed GI signs follow-up after thyroid hormone replacement therapy (THRT). Medical records of hypoT-dogs from two Veterinary Teaching Hospitals were retrospectively reviewed. Dogs were classified as hypothyroid if TT4 or fT4 were low/normal with normal/high TSH or inadequate TSH-stimulation test response. Clinical history, GI signs (vomiting, diarrhea, constipation), hematobiochemical parameters and abdominal ultrasound were collected. HypoT-dogs were divided based on the presence of at least one GI signs (GI group and not-GI group). Twenty-seven GI dogs had 3–4 weeks recheck from the beginning of THRT and information on GI signs were recorded. A total of 166 dogs were included (GI group, n=45, 27%; not-GI group, n=121, 73%). GI dogs showed nausea (42%), vomiting (40%), constipation (22%), large bowel diarrhea (40%), small bowel diarrhea (4%), and aspecific diarrhea (40%). No significant difference between GI and not-GI groups on hematobiochemical parameters was found. GI group had significantly higher frequency (20%) of large intestine involvement than not-GI group at the ultrasound (P=0.03; Chi-square test). Twenty-one out of 27 GI dogs had a resolution of GI signs at recheck (P=0.0001; McNemar test). Most of hypoT-dogs had concurrent GI signs mainly due to large bowel involvement. After THRT beginning the concurrent GI signs in hypoT-dogs seem to be reduced.

Abstract GI33: Feasibility and Complications of Video Capsule Endoscopy in 38 Dogs with Suspected Gastrointestinal Bleeding

Jenny Stiller[1]; Alice Defarges[2]; Brigitte Brisson[2]; Alexa Bersenas[2]; David Pearl[2]
[1]Ontario Veterinary College, University of Guelph; [2]University of Guelph

This study aimed to assess feasibility and complications of video capsule endoscopy (VCE) in dogs with occult or overt gastrointestinal bleeding (GIB). From August 2017 to November 2019, a total of 38 dogs (23.1 kg ± 11.2) were examined by VCE because of suspected occult (26) and overt (12) GIB. The ALICAM® was administered orally (28) or by endoscopic deployment (10) after 12 hours of fast. Preparation included enemas (29), administration of polyethylene glycol (19). A standard scoring system was used for GI visibility. All ALICAM® were administered uneventfully and excreted spontaneously. Median transit time of the capsule from administration to excretion was 30.5 hours (range 2.5 hours-8 days). Complications included incomplete studies due to temporary gastric retention (14), lost VCE (1), and unrelated sudden death (1), or euthanasia (1). GI visibility was good to excellent in the stomach and small intestine, limited to poor in the colon. Bleeding lesions were identified in 22 dogs (10 with overt GIB, 12 with occult GIB), and included gastric ulcerations (13 dogs), intestinal ulcerations (6 dogs) and gastric/intestinal/colonic vascular ectasia (2 dogs); intestinal lymphangiectasia was identified on ALICAM® in 2 dogs subsequently confirmed by endoscopic biopsies. In 2 dogs with overt GIB, a bleeding lesion was not identified. VCE is a safe procedure and can be used to diagnose a variety of bleeding lesions in the GI tract of dogs. In ALICAM delivered per os, prolonged gastric retention leading to incomplete studies was frequently noted. Further studies are needed to improve study completion.

Abstract GI34: Use of a Synbiotic for Treating Antibiotic-Induced Diarrhea in Cats

Jeremy A. Kiene[1]; Kelsey Dobesh[1]; Michael Lappin[2]
[1]Center for Companion Animal Studies, Department of Clinical Sciences, College of Veterinary Medicine and Biomedical Sciences, Colorado State University; [2]College of Veterinary Medicine and Biomedical Sciences, Colorado State University

Diarrhea is a common and potentially significant complication in cats undergoing antibiotic therapy. Data supporting the benefit of supplementing probiotics or synbiotics to cats when antibiotic-induced diarrhea has started is lacking. This study examines effects of supplementing a synbiotic (*Enterococcus faecium* SF68 and psyllium) to cats with pre-existing antibiotic associated diarrhea. Sixteen healthy, young-adult research colony cats fed a commercial dry food were administered amoxicillin-clavulanate (Clavamox) at 62.5 mg/cat twice daily. Using a standardized fecal scoring system, trained, masked personnel scored feces every twelve hours, with scores >4 considered diarrhea. Clavamox was discontinued if fecal scores were >4 for 2 days. Cats were then randomized into 2 treatment groups, one supplemented with the synbiotic in 15 g of canned food once daily, the other supplemented with the palatability enhancer from the synbiotic in 15 g of canned food once daily. Diarrhea was less severe overall in the synbiotic group, with 24.5% of fecal samples scored >5, compared with 48.9% in the control group ($p=0.0132$). Time to diarrhea resolution was 4.5 periods from initiation of treatment in the synbiotic group, compared with 6.5 periods in the control group ($p=0.472$). Clinical resolution of diarrhea was achieved in all cats in the synbiotic group. Two cats in the control group had a rescue protocol instituted for diarrhea that failed to resolve during the study period. These results indicate that supplementing this synbiotic to cats with diarrhea provoked by administration of Clavamox reduces severity of diarrhea and may decrease time to resolution.

Abstract GI35: Comparing Adipose-Derived Mesenchymal Stem Cells to Prednisolone for the Treatment of Feline Inflammatory Bowel Disease

Craig B. Webb[1]; Tracy Webb[2]
[1]Veterinary Teaching Hospital, Colorado State University; [2]Colorado State University

A 2015 proof-of-concept study confirmed the safety and potential benefit of treating feline chronic enteropathy with adipose-derived mesenchymal stem cells (fMSC). As a follow-up, this study was designed to compare the efficacy of fMSC to standard prednisolone therapy in confirmed inflammatory bowel disease (IBD). Cats were screened for significant concurrent diseases (history, PE, fecal, CBC, chemistry, urinalysis, TT4, TX A&M GI panel) and food-responsive diarrhea (2-week diet trial). IBD was confirmed by multi-site histopathology, IHC, PARR, and flow cytometry. Enrolled cats were randomly assigned to the fMSC or prednisolone group. All cats received appropriate placebo (IV saline injection or oral liquid vehicle), and owners were blinded to the grouping. fMSC treatment consisted of 2 injections (2 x 106 cells/kg) of allogeneic, fMSC from a single specific pathogen free donor separated by 2-weeks. Prednisolone treatment consisted of 1–2 mg/kg per os daily, tapered according to clinical response. Owners were asked to make no other changes for the first 2-months at which time they either continued with no changes to the 6-month recheck (cat stable/owner satisfied) or "failed," were unblinded, and changes made as necessary (cat not doing well/owner seeking alternate therapy). Six prednisolone and 6 fMSC cats completed the study. All 6 prednisolone cats were female-spayed, mean age 8.3 yrs (range 2–14), mean body weight (BW) 3.6 kg (range 2.5–4.8), and mean pre-treatment FCEAI score 3.4 (range 1–6). The 6 fMSC cats included 3 female-spayed and 3 male-castrated, mean age 8 yrs (range 4.5–13), mean BW 4.9 kg (range 4–5.9), and mean pre-treatment FCEAI score 3.7 (range 2–5). One cat in each group failed at the 2-month recheck. For the remaining cats, at the 6-month recheck, the mean FCEAI score for the prednisolone group was 3.7 (range 0.5–9) and 1.0 (range 0–1.5) for the fMSC group. These results suggest that this fMSC protocol is as effective in the treatment of feline IBD as a standard course of prednisolone.

Abstract GI36: Antibiotic Administration Results in Long-Term Changes to the Immature Feline Fecal Microbiome

Evangelia Stavroulaki[1]; Jonathan Lidbury[2]; Joerg Steiner[2]; Jan Suchodolski[2]; Panagiotis Xenoulis[3]
[1]University of Thessaly; [2]Gastrointestinal Laboratory, Texas A&M University; [3]Clinic of Medicine, Faculty of Veterinary Science, University of Thessaly

Antibiotic treatment (AT) during early life profoundly affects gastrointestinal (GI) microbial composition and function and delays the developmental progression of the microbiome in humans and experimental animals. In humans, it has been shown that antibiotic induced intestinal dysbiosis during infancy and childhood is associated with increased risk for several GI and non-GI related disorders later in life. The aim of this study was to investigate the effect of AT on specific bacterial groups that are potentially associated with GI dysbiosis in kittens. Naturally passed feces were collected from 23 healthy control kittens (Group 1), and 42 kittens that received antibiotics. AT kittens were presented for evaluation and treatment of upper respiratory tract disease and were randomly assigned to receive either amoxicillin/clavulanic acid (Group 2; n=22) for 20 days or doxycycline for 28 days (Group 3; n=20) as part of their standard treatment. Feces were collected on days 0 (baseline), 20 or 28, respectively (last day of AT), 60, 120, and 300. Kittens were approximately 2 months of age at enrollment and received the same diet and antiparasitic treatment during the study period. Kittens were excluded if they received medications known to affect the GI microbiota prior to or during the study period, or if they had or developed significant disease that might have affected the GI microbiota. DNA was extracted from each fecal sample and qPCRs were performed for total bacteria, *Turicibacter* spp., *Faecalibacterium* spp., *Streptococcus* spp., *Escherichia coli* (*E. coli*), *Blautia* spp., *Fusobacterium* spp., *Clostridium hiranonis* (*C. hiranonis*), and *Bifidobacterium* spp. Data were assessed for normality and appropriate statistical analyses were used for independent measurements. Statistical significance was set at $p<0.05$ and correction for multiple comparisons was performed where appropriate. No differences in bacterial abundances were identified among groups on day 0. On day 20, the abundance of total bacteria, *Streptococcus* spp., and *C. hiranonis* were decreased in Group 2 compared to Group 1 ($p=0.001$, 0.045, and 0.005, respectively). *Faecalibacterium* spp., and *Blautia* spp. were decreased in Group 2 compared to Group 1 or Group 3 ($p=0.008$, and 0.001, respectively). On day 20 *E. coli* was increased in Group 2 ($p<0.001$) and on day 60 in Group 3 ($p=0.008$) compared to Group 1. On day 120, *Faecalibacterium* spp. and *Bifidobacterium* spp. were increased in Group 2 compared to Group 1 ($p=0.019$, and 0.049, respectively) On day 300, total bacteria were increased in both AT groups compared to Group 1 (Group 2: $p=0.038$; Group 3: $p=0.033$). In conclusion, similar to humans, both amoxicillin-clavulanic acid and doxycycline appear to influence bacterial taxa commonly associated with dysbiosis in kittens, with changes persisting for at least 10 months after their discontinuation.

Abstract GI37: Coagulation Status, Fibrinolysis, and Platelet Dynamics in Dogs with Chronic Inflammatory Enteropathy

Sara A. Wennogle[1]; Christine Olver[2]; Sarah Shropshire[2]
[1]Colorado State University; [2]Department of Clinical Sciences, Colorado State University

The hypercoagulable state associated with canine chronic enteropathy is poorly understood. The objective of this study was to utilize thromboelastography (TEG) to assess coagulation status and fibrinolysis and platelet aggregometry to assess platelet function in dogs with chronic inflammatory enteropathy (CIE) compared to healthy controls (HC). Twenty-six dogs with CIE were prospectively enrolled. Data from 14 HC was available for comparison. All dogs had tissue factor (TF) + tPA (tissue plasminogen activator) TEG performed. 10/26 CIE dogs and all HC had adenosine diphosphate (ADP) and arachidonic acid (AA) whole blood impedance platelet aggregometry performed. Standard clinical, clinicopathologic, and coagulation data were available for all dogs and correlated to TEG values. Dogs with CIE had higher maximum amplitude (MA)-TF-tPA ($p<.001$), longer clot lysis times (CLT) ($p=.008$), and lower LY30 ($p=.002$) and LY60 ($p=.002$) values when compared to HC, suggesting hypercoagulability and hypofibrinolysis. Although serum albumin concentrations had a weak (rho value, -.46) correlation ($p=.02$) with MA-TF-tPA, 10/16 (63%) normoproteinemic and 9/10 (90%) hypoproteinemic dogs were considered hypercoagulable based on TEG using MA >60 mm to define hypercoagulability. Serum albumin concentrations were also weakly correlated to LY30 (rho value, .44; $p=.03$) and LY60 (rho value, .46; $p=.02$). Serum 25(OH)D concentrations were moderately (rho value, -.6375) correlated ($p<.001$) to MA-TF-tPA. Plasma fibrinogen and antithrombin concentrations were moderately (rho value, .55; $p=.004$) and weakly (rho value, -.42; $p=.03$) correlated to MA-TF-tPA. Platelet aggregometry was not different between groups. Both normoproteinemic and hypoproteinemic dogs with CIE were considered hypercoagulable and hypofibrinolytic based on TEG when compared to HC. Humans with inflammatory bowel disease have been demonstrated to have anti-tPA antibodies that are suspected to contribute to a hypofibrinolytic state and recently, anti-thrombotic actions of vitamin D have been evaluated. More work is needed to determine if these factors could play a role in the coagulation status of dogs with CIE, and whether they could become targets for therapy in the future.

Abstract GI38: Enhanced Gut Microbial Fermentation and Metabolism by Different Starch-Rich Products in a Canine Gastrointestinal Model

Sergi Segarra[1]; Jonas Ghyselinck[2]; Pieter Van den Abbeele[2]; Massimo Marzorati[3]
[1]Bioiberica SAU; [2]ProDigest BVBA; [3]CMET, Faculty of Bioscience Engineering, Ghent University

The purpose of the study was to evaluate the prebiotic effects of three different starch-rich products on gut microbial activity using an *in vitro* canine gastrointestinal model. The simulator of the canine intestinal microbial ecosystem (SCIME) platform was used to simulate the colonic environment in a single stage reactor setup and assess the effects of three products (2201, 2202, and 2203) using a freshly prepared fecal inoculum from a healthy adult Beagle dog as a single donor. All tested products contained the same source of chondroitin sulfate, β-glucans and mannanoligosaccharides; but a different source of starch: high amylose maize starch (2201), tapioca dextrin (2202), and tapioca maltodextrin (2203). A sugar-depleted nutritional medium was used as negative control (blank). Products were incubated for 48 hours at 39°C, under shaking (90 rpm) and anaerobic conditions at a concentration of 5 g/L. Each condition was performed in triplicate. Changes in microbial fermentation (pH and gas production), and metabolic activity [levels of lactate, ammonium, and short-chain fatty acids (SCFA)] were analyzed after 0, 6, 24, and 48 hours. Comparisons between groups overtime for the different parameters were performed using unpaired two-tailed Student's T-test. Differences were considered statistically significant if P value was <0.05. All three tested products were well fermented by the canine microbiota and led to significantly lower pH and ammonium levels, and to significantly higher production of gas and total SCFA (approximately, 2-fold higher vs. blank). Product 2201 yielded highest lactate, acetate and butyrate concentrations and a more marked pH decrease than the other two products, while products 2202 and 2203 led to highest propionate concentrations. Fermentation of product 2201 slightly lowered branched SCFA concentrations compared to control, while products 2202 and 2203 showed slightly higher branched SCFA concentrations than the control. Product 2201 also appeared to exert a faster effect, especially during the first 6–24 hours. In conclusion, the three tested combinations of chondroitin sulfate, β-glucans, mannanoligosaccharides and starch were well fermented by the gut microbiota of the Beagle dog and enhanced the *in vitro* production of health-related metabolites, hence confirming a remarkable prebiotic effect. Moreover, the combination including maize as source of starch led to superior results. Although further investigations are warranted, these results point towards the potential clinical usefulness of such combinations in dogs with chronic enteropathies.

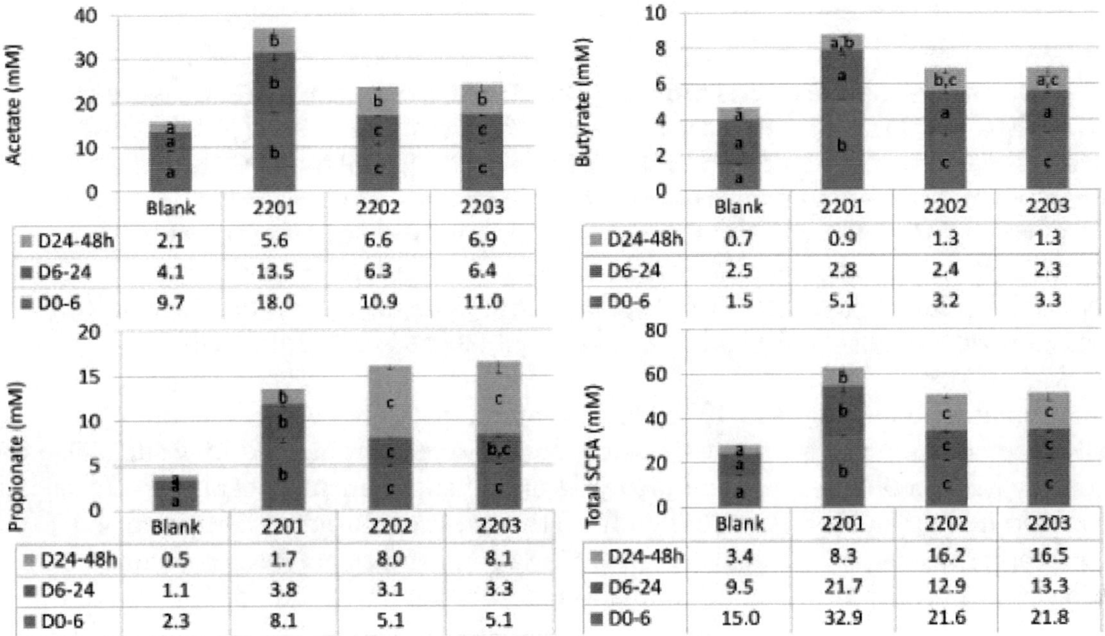

Acetate (mM)	Blank	2201	2202	2203
D24-48h	2.1	5.6	6.6	6.9
D6-24	4.1	13.5	6.3	6.4
D0-6	9.7	18.0	10.9	11.0

Butyrate (mM)	Blank	2201	2202	2203
D24-48h	0.7	0.9	1.3	1.3
D6-24	2.5	2.8	2.4	2.3
D0-6	1.5	5.1	3.2	3.3

Propionate (mM)	Blank	2201	2202	2203
D24-48h	0.5	1.7	8.0	8.1
D6-24	1.1	3.8	3.1	3.3
D0-6	2.3	8.1	5.1	5.1

Total SCFA (mM)	Blank	2201	2202	2203
D24-48h	3.4	8.3	16.2	16.5
D6-24	9.5	21.7	12.9	13.3
D0-6	15.0	32.9	21.6	21.8

Abstract GI39: Increased Expression of the Ileal Cobalamin Receptor in Hypocobalaminemic Dogs with Idiopathic Inflammatory Bowel Disease

Stefanie Kather[1]; Johannes Kacza[2]; Helga Pfannkuche[3]; Gotthold Gäbel[3]; Jörg Steiner[4]; Franziska Dengler[3]; Romy Heilmann[1]
[1]Small Animal Clinic, Veterinary Teaching Hospital, Institute of Veterinary Physiology, University of Leipzig; [2]Saxon Incubator for Clinical Translation, Bioimaging Core Facility, University of Leipzig; [3]Institute of Veterinary Physiology, University of Leipzig; [4]Gastrointestinal Laboratory, Texas A&M University

Idiopathic inflammatory bowel disease (IBD) in dogs can be associated with cobalamin deficiency. Hypocobalaminemia is also a risk factor for negative outcome in dogs with IBD, and affected dogs may not respond to treatment unless receiving supplemental cobalamin. Thus, it has been proposed that the uptake of cobalamin from the intestinal lumen is compromised as a result of ileal receptor deficiency in dogs with IBD. However, expression of the cobalamin receptor in dogs with chronic intestinal inflammation has not yet been studied. The aim of this study was to quantify the expression of the cobalamin receptor subunits amnionless (AMN) and cubilin (CUBN) in ileal biopsies from dogs with IBD in comparison to healthy dogs. Endoscopic biopsies from the canine ileum were evaluated from dogs that were assigned to one of the following study groups: (1) dogs with IBD and severe hypocobalaminemia (n=6), (2) dogs with IBD and suboptimal serum cobalamin status (n=7), (3) dogs with IBD and normocobalaminemia (n=7), and (4) healthy control dogs (n=9). Formalin-fixed and paraffin-embedded sections of tissue sections were stained for AMN and CUBN using polyclonal primary and fluorescence-labelled secondary antibodies. AMN and CUBN expression was quantified using confocal laser scanning microscopy. Receptor expression was compared among the groups of dogs and was also correlated with clinical patient data (i.e., age, sex, canine chronic enteropathy clinical activity index [CCECAI] score, serum cobalamin and folate concentrations, and histopathologic criteria). Statistical significance was set at $P<0.05$. Ileal mucosal expression of AMN and CUBN was significantly higher in the group of severely hypocobalaminemic dogs with IBD (median: 5.0 and 4.4, respectively) compared to healthy control dogs (median: 3.4; $P=0.007$ and median: 3.4; $P=0.034$, respectively). There was no significant difference among any of the other groups of dogs. AMN expression in ileal biopsies was significantly correlated with age ($P=0.001$), sex ($P=0.030$), CCECAI score ($P=0.036$), the severity of ileal lacteal dilatation ($P=0.039$), duodenal macrophage infiltration ($P=0.027$), and with serum folate ($P=0.004$), but not serum cobalamin concentrations ($P=0.420$). CUBN expression was correlated with AMN expression ($P<0.0001$), age ($P=0.020$), ileal macrophage infiltration ($P=0.020$), and serum folate concentration ($P=0.008$), but also not serum cobalamin concentration ($P=0.399$). Expression of the cobalamin receptor subunits AMN and CUBN appears to be altered in severely hypocobalaminemic dogs with IBD. Contrary to the previously proposed pathogenetic mechanism, cobalamin receptor downregulation does not appear to be the primary cause of severe hypocobalaminemia in canine IBD. Additionally, there was a relation between cobalamin receptor expression and several patient characteristics: more severe clinical signs (CCECAI score >9), histopathologic presence of lacteal dilatation and macrophage infiltration, older age (>7 years), and male sex (regardless of neuter status) were associated with a higher AMN expression. Higher CUBN expression was associated with older age (>7 years) and presence of macrophage infiltration.

The underlying mechanisms and clinical implications of these findings, and also the effect of selected patient characteristics on AMN and CUBN expression, warrant further investigation.

Abstract HM01: Immune Profiles of Cocker Spaniels and Old English Sheepdogs, Breeds Predisposed to Autoimmune Blood Disorders

Michael Barchilon[1]; Austin Viall[1]; Jordan Gagne[1]; Emily Phalen[1]; Paula Boggiatto[2]; Robert Schaut[3]; Unity Jeffery[4]; Marjory Brooks[5]; Dana LeVine[1]
[1]Iowa State University; [2]National Animal Disease Center; [3]Elanco Animal Health; [4]Texas A&M University; [5]Cornell University

Cocker Spaniels (CS) and Old English Sheepdogs (OES) are predisposed to immune-mediated hemolytic anemia (IMHA) and immune thrombocytopenia (ITP). Decreased regulatory T-cells (Treg) in ITP and elevated macrophage/monocyte recruiting cytokines and keratinocyte chemoattractant (KC)) in canine IMHA (monocyte chemoattractant protein-1 (MCP-1) are described. We aimed to determine if healthy CS and OES have altered immune cell and cytokine profiles versus other breeds, hypothesizing decreased Tregs, increased MCP-1 and increased KC. We prospectively enrolled healthy CS (21), OES (23) and age/sex-matched dogs of other breeds (42). Circulating Tregs, CD4+ and CD8+ T-cell percentages were determined flow cytometrically, serum cytokines assayed via multiplex, and leukograms performed. CS, OES, and matched controls were compared with paired parametric/non-parametric tests. We found no differences in Tregs between CS (mean 5.6%, 95% CI 4.4–6.8%) and CS-controls (5.7%, 4.6–6.8%) or between OES (median 6.0%, 4.8–7.7%) and OES-controls (median 5.5%, 4.8–6.6%). CD4+ and 8+ percentages were similar between groups. We identified several cytokines elevated in CS versus controls: Interleukin-7 (CS median 142.7 pg/ml, 52.8–256.6; control 72.0 pg/ml, 50.0–110.2; p=0.035), MCP-1 (CS median 399.3 pg/ml, 343.7–481.0; control 348.2, 318.6–379.6; p = 0.022), and KC (CS mean 683.5 pg/ml, 522.4–844.5; control 445.8, 319.6–572.0; p=0.0005). CS had increased percent (CS mean 5.2%, 4.1-6.3; control 3.6%, 2.9-4.2; p=0.016) and absolute (CS mean 613/µl, 448–778; control 334/µl 265–403; p=0.0029) monocytes. Our data combined with previous canine IMHA cytokine studies suggest that CS may have a breed-related state of monocyte/macrophage activation enhancing susceptibility to autoimmune blood disorders.

Abstract HM02: Comparison of Direct Venipuncture Versus Peripheral Catheter Samples for Serum Biochemistry Testing in Dogs

Aria L. Guarino; Allison O'Kell; Andrew Specht; Sarah Beatty
University of Florida Veterinary Hospitals

Drawing blood from dogs for serum biochemistry (SB) analysis from a freshly placed peripheral intravenous catheter (PIVC) rather than direct venipuncture (DV) has the potential to reduce patient morbidity and improve efficiency. However, whether PIVC blood collection provides reliable SB results has not been assessed in dogs. The purpose of this study was to determine whether there is clinically acceptable agreement between SB results from samples collected from a PIVC at the time of placement and samples collected contemporaneously by DV in dogs. This was a prospective study involving 61 dogs presenting for clinical illness or elective procedures that required PIVC placement. Samples were collected by DV and through a freshly placed PIVC in randomized order and processed immediately. Bland-Altman analysis was used to determine whether there was clinically acceptable agreement between the two sampling methods based on application of a priori limits of agreement (LoA) from previously published recommendations. Most analytes had results that were within the LoA 100% of the time. Analytes with results within the LoA 95–99% of the time (between 1-3 cases outside LoA) included alkaline phosphatase, phosphorus, glucose, and bicarbonate. The remaining analytes (aspartate aminotransferase, total bilirubin, and potassium) had results within the LoA 92%, 90%, and 83% of the time respectively. Based on this data, it appears that blood samples collected from PIVC, while not entirely equivalent to those collected by DV, generally provide results within previously published guidelines for clinically acceptable agreement. However, some values such as potassium require more cautious interpretation.

Abstract HM03: Evaluation of Hemostasis in Hyperthyroid Cats

Audrey Keebaugh[1]; Stefanie DeMonaco[1]; David Panciera[1]; Jonathan Abbott[2]; Katie Boes[1]; Giulio Menciotti[1]
[1]Virginia-Maryland College of Veterinary Medicine; [2]College of Veterinary Medicine, University of Tennessee

Hyperthyroid cats are predisposed to thrombus formation. The mechanism for thrombogenesis is currently unknown, but could be associated with altered hemostasis as seen in hyperthyroid humans. The purpose of this study was to evaluate markers of hemostasis in hyperthyroid cats compared to healthy cats, and in hyperthyroid cats before and after treatments with radioactive iodine (RIT). Twenty-four cats with hyperthyroidism (elevated serum T4 concentration) and 12 healthy euthyroid cats >8 years of age were studied. Blood was collected by jugular venipuncture for measurement of prothrombin time (PT), activated partial thromboplastin time (aPTT), fibrinogen, antithrombin (AT), and D-dimers. An echocardiogram was performed in all cats and healthy cats with abnormal echocardiograms were excluded. A normal echocardiogram was defined as two-dimensional (2D) and M-mode left ventricular free wall and interventricular septal wall thickness at end-diastole of <6 mm, 2D left atrium to aortic ratio of <1.5, no or only trivial insufficiencies of the pulmonic and tricuspid valves, and no insufficiency of the aortic and mitral valves. Measurements of hemostasis were evaluated again >6 months after RIT in 8 hyperthyroid cats. The Wilcoxon signed-rank test was used to compare measurements of hemostasis between hyperthyroid and euthyroid controls, pre and post RIT values, and in hyperthyroid cats with normal versus abnormal echocardiograms. Linear regression analysis was used to test for correlation of serum T4 and fibrinogen and serum T4 and antithrombin. The median serum T4 was 134 nmol/L (range 49–509 nmol/L [reference interval 16–37.7 nmol/L]) in hyperthyroid cats. Median follow up serum T4 was 23.3 nmol/L (range 16.3–48.9 nmol/L). Fibrinogen and AT were significantly greater (P<0.001) in hyperthyroid cats compared to healthy controls. After RIT, there was a significant decrease in fibrinogen and antithrombin (P<0.01 and P=0.02, respectively) and a significant prolongation of PT (P=0.03). Fibrinogen and antithrombin had a strong positive correlation with serum T4 value (r=0.79; 95% CI 0.62–0.89 and r=0.71; 95% CI 0.49–0.84, respectively). Significant differences were not detected when hemostatic markers from hyperthyroid cats with normal or abnormal echocardiograms were compared. These results provide evidence of altered hemostasis in hyperthyroid cats that is independent of cardiac abnormalities. These differences resolved after radioiodine therapy.

Abstract HM04: Effect of Duration of Canine Blood Storage on Red Blood Cell Alloimmunization and Compatibility Testing

Alison Thomas-Hollands; Rebecka Hess; Nicole Weinstein; Kimberly Marryott; Samantha Fromm; Nicole Chappini; Mary Beth Callan
University of Pennsylvania

Longer duration of RBC storage has been associated with increased RBC alloimmunization in a mouse transfusion model, as well as development of spurious major crossmatch incompatibilities with equine blood. The objectives of this prospective study were: 1) to determine if longer duration of storage of canine pRBCs is associated with increased RBC alloantibody formation in transfused dogs, and 2) and if storage duration has an effect on compatibility testing. Transfusion-naïve dogs in need of a RBC transfusion were included. Pre- and post-transfusion plasma/serum samples were stored at -80°C until fresh RBCs from the original blood donor were available. Major crossmatches were performed using standard gel columns, with and without addition of canine antiglobulin reagent. Whole blood segments from pRBC units were used for major and minor crossmatches (gel column) after storage for 0, 7, 14, 21, 28, and 35 days. Pre- and post-transfusion samples were available for 27 dogs that received a median of 1 pRBC transfusion (range, 1–4). Post-samples were obtained a median of 24 days (range, 6–388) after the first pRBC transfusion. Eleven of 27 (41%) dogs developed major crossmatch incompatibilities, but 5 of 27 (19%) dogs had detectable RBC alloantibodies only with the addition of antiglobulin. Transfused pRBC units had been stored a median of 24 days (range, 1–32). There was no association between duration of pRBC storage and development of major crossmatch incompatibilities. Serial major (n=66) and minor (n=69) crossmatches were performed weekly with pRBC units (n=24) stored for 35 days. There were no significant changes in compatibility over time. In conclusion, RBC alloantibody formation is common in dogs post-transfusion but not associated with longer duration of pRBC storage. Unlike equine blood, storage of canine pRBC units for 35 days does not result in spurious changes in crossmatch compatibility over time.

Abstracts HM05: Cryopreserved Platelets Versus Lyophilized Platelets for Management of Thrombocytopenia Associated Bleeding in Dog

Robert Goggs[1]; Benjamin Brainard[2]; Janine Calabro[3]; Karyn Harrell[4]; Philip Bergman[5]; Tracy Mills[5]; Dana LeVine[6]; Richard Stone[7]; Benjamin Davidson[7]; Christine Iacovetta[8]; Lauren Harris[8]; John Gicking[8]; Brenda Fulcher[9]; Teresa Lightfoot[9]; Meredith Miller[10]; John Loftus[10]; Jennifer Kishbaugh[11]; Anne Hale[11]
[1]Cornell University; [2]University of Georgia; [3]Friendship Veterinary Hospital; [4]North Carolina State University; [5]VCA Clinical Research, VCA Animal Hospitals; [6]Iowa State University; [7]Blue Pearl Pet Hospital; [8]Blue Pearl Pet Hospital; [9]Blue Pearl Science; [10]Cornell University College of Veterinary Medicine; [11]BodeVet Inc.

Lyophilized platelet products have long shelf-lives and can be easily transported, stored, and administered in various settings. Thrombocytopenia is common in canine critical care, but limited availability of fresh platelet concentrates in veterinary medicine necessitates more frequent use of DMSO-stabilized cryopreserved canine platelets. A novel trehalose-stabilized lyophilized canine platelet product was recently developed, but its efficacy in dogs with clinical bleeding is unknown. This multicenter, randomized, non-inferiority clinical trial compared DMSO-stabilized cryopreserved canine platelets (CP) with a trehalose-stabilized lyophilized canine platelet product (LP) for the control of life-threatening thrombocytopenic bleeding in dogs. Dogs with platelet counts $<50\times10^9$/L and evidence of active bleeding using a standardized bleeding assessment tool (DOGiBAT) were randomized to receive 3×10^9 platelets/kg of LP or CP. Primary outcome measures were DOGiBAT score, platelet count, need for additional red cell transfusion and all-cause mortality. In total, 91 dogs were enrolled; 53 received LP and 38 received CP. Baseline demographics and clinical characteristics of both groups were comparable. At 1-hour post-transfusion the LP were superior for the change in DOGiBAT score, and non-inferior at 24-hours post-transfusion. The LP were non-inferior to the CP for change in platelet count, need for additional red blood cell units and for survival to discharge. LP were superior for change in hematocrit at 1-hour post-transfusion, and non-inferior at 24-hours. No adverse effects were noted in either group. In conclusion, a novel lyophilized canine platelet product is non-inferior to DMSO-stabilized cryopreserved platelets for management of thrombocytopenic bleeding in dogs.

Abstract HM06: Inhibition of Myristoylated Alanine-Rich C Kinase Substrate (MARCKS) Decreases Canine Clot Retraction and Platelet Aggregation

Allison J. Rowland; Samuel Jones
North Carolina State University

Thrombosis in canines is a recognized consequence of many disease processes and some treatments and is associated with increased morbidity and mortality. Use of antithrombotic therapy is increasing though treatment options are limited, require frequent monitoring, and may cause unwanted bleeding. A better understanding of canine platelet function may elucidate new methods of treatment. Myristoylated alanine-rich C kinase substrate (MARCKS) protein helps to regulate cellular events utilizing dynamic actin reorganization and is essential for platelet function. This study was performed to determine the role of MARCKS in canine platelet function. Blood was collected from healthy dogs for whole blood clot retraction (n=4), optical platelet aggregometry using platelet rich plasma (n=6), or isolated platelet adhesion assay using a luminescence assay (n=7). A MARCKS-specific inhibitor peptide called myristoylated n-terminal sequence (MANS) was used to block MARCKS function and the negative control peptide was random n-terminal sequence (RNS) which has a rearranged form of the same 24 amino acids as MANS. Each assay was performed on samples exposed to phosphate-buffered saline (unexposed control), RNS (negative control), and MANS (treatment). Gross visual change in clot size was recorded for the whole blood clot retraction assay. Results of aggregometry and adhesion assays were analyzed using a one-way ANOVA with Tukey's multiple comparisons test. All whole blood clot retraction samples had decreased clot retraction in the MANS treated group compared with the unexposed and negative controls. For the optical platelet aggregometry there was a significant decrease in the percent aggregation in the MANS treated group as compared with the unexposed and negative controls (p=0.023 and p=0.0073, respectively). Results of the platelet adhesion assay showed a significantly decreased adhesion in the MANS treated group as compared with the unexposed control group (p=0.036) though no significant difference between the MANS and RNS groups (p=0.566) was found. These results support that MARCKS protein is essential in canine clot retraction and platelet aggregation. Inhibiting MARCKS is a potential therapeutic target for the treatment of thrombosis in canines.

Abstract HM07: Neutropenia in Dogs Receiving Vincristine for Treatment of Immune-Mediated Thrombocytopenia

Kathryn LaQuaglia; James Robertson; Katharine Lunn
College of Veterinary Medicine, North Carolina State University

Myelosuppression, specifically neutropenia, is a recognized adverse effect of vincristine when used in multidrug chemotherapy protocols. The incidence of neutropenia, or other indicators of myelosuppression, in dogs receiving vincristine for treatment of immune-mediated thrombocytopenia (ITP) has not been reported. A retrospective cohort study was performed to determine the incidence of neutropenia, and evaluate risk factors, in dogs receiving vincristine for treatment of ITP. Client-owned dogs with severe thrombocytopenia (platelet count ≤15,000/μL), presumptively diagnosed with ITP, over a 15-year period were included. Cases were excluded if they were diagnosed with neoplasia, presented with neutropenia, or were treated with vincristine prior to admission. Administration of immunomodulatory agent(s) or human intravenous immunoglobulin, vincristine dose, presence of hyperbilirubinemia, and vector-borne disease status were evaluated as potential risk factors, using logistic regression models. 143 dogs were included in the study, of which 126 received vincristine; 19 of them became neutropenic. Neutropenia was identified between 2 and 14 days (median 5 days) after vincristine administration, and resolved between 1 and 8 days (median 3 days) following nadir. Furthermore, of 36 dogs with ITP with an initial regenerative anemia, 28 received vincristine, and the anemia became non-regenerative in 23 of these dogs. Of the risk factors evaluated, cyclosporine administration was significantly associated with development of neutropenia in dogs receiving vincristine ($p=0.00001$). These results suggest that alternative immunomodulatory agents, delay in the initiation of cyclosporine treatment, or vincristine dose reductions for dogs chronically receiving cyclosporine should be considered when using vincristine for treatment of ITP.

Abstract HM08: Evaluation of Hematocrit in Juvenile Shelter Dogs Presenting for Routine Ovariohysterectomy or Neuter

Kate KuKanich; Joshuah Klutzke; Nora Springer; Butch KuKanich
Kansas State University

Hematological parameters are established for healthy juvenile dogs. Applicability of these reference values to a shelter population of dogs with potential comorbidities is unknown. Shelter dogs (77 female, 61 male) of varied breeds presented to the Kansas State University Junior Surgery Laboratory. All dogs received physical examinations including dental aging, CBC, 4Dx, and flea comb, and dogs with diarrhea had fecal float +/- parvovirus test. Dogs were grouped into 3 categories: Puppies (P) with no adult incisors (<3 months old); Mideruption (M) with some adult incisors (3-6 months old); or Adult (A) with all adult canines (>6 months old). A one-way ANOVA was used to compare hematocrit (HCT) between groups. There were 34 P, 22 M, and 82 A dogs, with mean calculated HCT 35.6% (range 27-46), 37% (31-48%), 45.8% (34-59%), respectively. Nine dogs had positive 4Dx results [3 *Anaplasma* (HCT 36-56%), 5 *Ehrlichia* (HCT 34-47%), 1 heartworm (HCT 50%)]. Twelve dogs had fleas or flea dirt [10 A (HCT 37-47%), 1 M (HCT 37%), and 1 P (HCT 30%)]. Two dogs had hookworms (HCT 29-35%), 3 had roundworms (HCT 27-34%), 1 had tapeworms (HCT 35%), 1 had coccidia (HCT 37%), and 1 was parvovirus positive (HCT 40%). Hematocrit differed significantly between adults and puppies ($p<0.01$) and between adults and mideruption dogs ($p<0.01$), whether infectious cases were included or excluded. Juvenile shelter dogs have lower HCT than adults, similar to prior reports. Further investigation into underlying cause and influence of infectious disease is warranted.

Abstract HM09: Point of Care Assessment of Fibrinolysis Using the Viscoelastic Coagulation Monitor

Armelle deLaforcade; Elizabeth Rozanski
Cummings School of Veterinary Medicine at Tufts University

Hyperfibrinolysis (HF) has been increasing identified as pathologic process in a variety of critical illnesses in dogs. However, HF is challenging to detect using conventional assays. Thromboelastography (TEG), modified with the addition of Tissue Factor (TF) and tissue plasminogen activator (TPA) has been described as a sensitive marker of fibrinolysis. However, TEG is not widely available in practice, and when available, requires specialized training and equipment to accurately perform the assay. The Viscoelastic Coagulation Monitor (VCM) has recently been introduced as a cartridge based viscoelastic test which removes the requirement for specialized equipment and training. The VCM is marketed as a point of care test with fresh whole blood. The goal of this study was to evaluate the effect of the addition of TPA to fresh whole blood in VCM cartridges. VCM produces results similar to conventional TEG, but are named Clot Time (CT), defined as when clot initially forms; Clot formation time (CFT) defined as when there is 10% amplitude of the clot from baseline; Angle defined as angle of the clot curve, maximum clot formation (MCF) defined as the firmness of the clot; A10 and A20 are the amplitude of the clot at 10 and 20 minutes after CT; while LI (lysis index) 30 and LI 45 represent the amplitude of the clot at 30 and 45 minutes after clot time as a percentage of the MCF. Healthy dogs were enrolled. A 2 ml sample of blood was collected. 500 ul was immediately placed in a VCM cartridge for point-of-care assessment of coagulation per the manufacturer's instruction. 400 ul of FWB was additionally placed into a vial containing 25 IU TPA, inverted 5 times, and then placed into a VCM cartridge. Results for VCM parameters were compared between treatment using a paired T-test, with a p value of <0.05 considered significant. Eleven dogs were enrolled. There was no difference between CT, CFT, Alpha, A10, and A20 following the addition of TPA. However, the LI 30 and LI 45 were significantly affected by the addition of the TPA. The addition of TPA to FWB results in accelerated fibrinolysis as assessed by the VCM analyzer. Further investigation of the modified assay in dogs with evidence of hemorrhage is warranted.

Abstract HM10: Reactive Oxygen Species Production and Biomarkers of Oxidative Stress in Anemic Dogs

Andrew Woolcock; Andrea Santos; Priscila Serpa; George Moore; John Christian
Purdue University

Oxidative stress has a role in the pathophysiology of multiple disease processes including anemia, and may contribute to ongoing red blood cell injury in hemolytic states. ROS are increased and antioxidant systems are depleted in people with autoimmune-hemolytic anemia (AIHA), as well as anemia of chronic inflammatory disease. The role of oxidative stress in canine anemia is minimally understood. Gluthatione peroxidase is reduced in anemic people, and was identified to be reduced in anemic dogs, but no association with hemolysis was noted. A novel assay for direct measurement of intraerythrocytic ROS by flow cytometry was recently validated, and shows promise for identification of ROS in canine disease states, including anemia. The objective of this study was to evaluate intraerythrocytic ROS production, as well as glutathione and vitamin E concentrations in anemic dogs. We hypothesized that anemic dogs would have increased ROS detected by flow cytometry, and decreased concentrations of glutathione and vitamin E when compared to controls. We further hypothesized that markers of oxidative stress would be different in dogs with hemolytic anemia when compared to dogs with non-hemolytic anemia. Dogs were recruited from the hospitalized population of the veterinary teaching hospital. Dogs were eligible for inclusion to the study if their hematocrit or packed cell volume were <30%. Once included, medical records were reviewed to determine the most likely cause for the anemia. Causes were classified as either non-hemolytic or hemolytic. A second population of healthy control dogs was recruited as well. Forty-eight anemic dogs were included in the study, with 11 diagnosed with immune-mediated hemolytic anemia, and 37 diagnosed with non-hemolytic causes for their anemia. 20 healthy dogs were included in the study. Data were non-parametrically distributed with summary descriptive statistics presented as median [range] and assessed with the Wilcoxon rank sum test. Median GSH:GSSG values from plasma and hemolysate samples of anemic and non-anemic, healthy dogs, further divided into those with hemolytic and non-hemolytic anemias. Healthy dogs were found to have significantly higher RBC GSH:GSSG when compared to anemic dogs of both causes (Table 1, $p<0.0001$). Dogs with hemolytic anemia had lower GSH:GSSH compared to dogs with non-hemolytic anemia, but this difference did not reach significance ($p=0.081$). Dogs with hemolytic anemia had significantly higher plasma GSH:GSSG when compared to dogs with non-hemolytic anemia and healthy dogs (both $p<0.0001$). There was no difference in intracellular ROS or vitamin E between groups. This study demonstrated a significant depletion of glutathione in anemic states, especially in dogs with hemolytic anemia. While the use of flow cytometry to label intracellular ROS did not find differences between groups, this method still shows promise. Prospective investigation of oxidative stress in immune-mediated hemolytic anemia is warranted, with an evaluation of the effect of antioxidant supplementation on these parameters.

Table 1. Median GSH:GSSG values from plasma and RBC hemolysate samples of anemic and non-anemic, healthy dogs, further divided into those with hemolytic and non-hemolytic anemias. Values with matching superscript letters represent significant differences.

	Plasma GSH:GSSH		RBC GSH:GSSG	
	Median	**Range**	**Median**	**Range**
Healthy	2.80[a]	0.09-29.91	7.22[c,d]	0.51-29.7
Non-Hemolytic Anemia	1.29[b]	0.01-7.09	4.11[c]	2.23-26.37
Hemolytic Anemia	7.96[a,b]	0.42-15.68	2.39[d]	0.4-30.83

Abstract HM11: Identification of Five New Feline Erythrocyte Antigens Based on the Presence of Naturally Occurring Alloantibodies

Marie Binvel[1]; Julie Arsenault[1]; Boris Depré[2]; Marie-Claude Blais[1]
[1]Faculté de médecine vétérinaire, Université de Montréal; [2]Adomvet Urgences vétérinaires

Since the discovery of the feline Mik antigen, several studies have described blood incompatibilities unrelated to the AB system. Based on the presence of naturally occurring alloantibodies (NOAb), the purpose of this study was to begin mapping the corresponding feline erythrocyte antigens (FEA) behind these incompatibilities. By groups of six, 274 transfusion-naïve cats were prospectively evaluated for the presence of alloantibodies using a gel column crossmatch test. When non-AB-related alloantibodies were detected in a cat, its plasma was used to assess the presence or absence of the corresponding antigen in cats included thereafter in the study (blood typing). Plasma from 18 of 263 type-A cats (6.8%) caused 49 incompatible results out of 1257 crossmatches performed (3.9%). Presence of NOAb was statistically associated with an age of <2 years ($P=0.04$). Using 7 of the 18 alloantibodies, systematic blood typing was performed, and results compared. Based on agreement analysis, 3 NOAb appeared to correspond to the same antigen (k ^{3} 0.64); hence the identification of 5 different putative FEA (numbered in order of identification). FEA 1, 4, and 5 were most frequent with a prevalence of 84%, 65%, and 96%, respectively. Only FEA 1 was significantly associated with NOAb ($P=0.005$), which were observed in 8 of 43 FEA 1-negative cats (19%). This study represents a first step of FEA's identification outside the AB system. Because of its prevalence and association with NOAb, FEA 1 may correspond to the lost Mik antigen. Banked alloantibodies will facilitate future studies, notably regarding their clinical relevance.

Abstract HM12: Comparison of Dogs Treated for Primary Immune-Mediated Hemolytic Anemia in Tuscany, Italy and Texas, USA

George Lubas[1]; Unity Jeffery[2]; Chiara Alaimo[3]; Giulia De Feo[3]; Alessandra Gavazza[4]
[1]Department of Veterinary Sciences, University of Pisa; [2]College of Veterinary Medicine and Biomedical Sciences; [3]University of Pisa; [4]University of Camerino

Large-scale studies are needed to determine optimal treatment for canine immune-mediated hemolytic anemia (IMHA). Multicenter studies reduce the time for case enrollment, but differences in disease severity and outcome could complicate their interpretation. This retrospective study compared clinical characteristics between dogs treated for IMHA by veterinary teaching hospitals in Tuscany, Italy and Texas, USA between 2010 and 2018. As the two institutions used different diagnostic criteria, Tuscany cases (n=48) were included if flow cytometry was positive for anti-erythrocyte antibodies and/or marked spherocytosis was reported. Texas cases (n=43) were included if there was clinical evidence of hemolysis and marked spherocytosis and/or a positive saline agglutination test. Cases were excluded if a secondary IMHA was evidenced. Categorical data was compared by Chi-squared test and continuous data by Mann-Whitney test. The two populations did not differ significantly in age, but Texas dogs were more commonly neutered and had lower bodyweight. Chihuahuas and mixed breed dogs were most common in Texas and mixed breed dogs and Cocker Spaniels in Tuscany. Median hematocrit, erythrocyte regeneration response, occurrence of spherocytes and hyperbilirubinemia were not significantly different. Hemolyzed plasma and bilirubinuria were more common in Texan cases. CHAOS scores were not significantly different between groups. Texan cases were more likely to be hospitalized at initial presentation, but there was no difference in survival to discharge between the two locations. Dogs treated for primary IMHA in a European or North American location were broadly similar, suggesting cases from both locations could be combined in future clinical trials.

Abstract HM13: Chronic Low-Dose Rapamycin Does Not Cause Red Blood Cell Microcytosis in Middle-Aged, Large Breed Dogs

Jeremy B. Evans[1]; Ashley Morrison[1]; Unity Jeffery[1]; Daniel Promislow[2]; Matt Kaeberlein[2]; Kate Creevy[1]
[1]Veterinary Teaching Hospital, Texas A&M; [2]University of Washington

Rapamycin is a macrolide antibiotic that inhibits the mammalian target of rapamycin (mTOR), a protein complex involved in multiple metabolic pathways. Due to its ubiquitous nature, disruption of the mTOR complex via rapamycin has numerous systemic effects, including immunosuppression, anti-proliferative effects, and extension of lifespan in laboratory models. However, side effects, such as dyslipidemias and red blood cell microcytosis have been reported in humans and mice. This study aimed to determine the prevalence of progressive red blood cell microcytosis in middle-aged, large breed dogs who received long-term, low-dose oral rapamycin. Seventeen dogs (10 spayed females, 1 intact female, and 6 castrated males) were randomly assigned to receive rapamycin (0.025 mg/kg) or a placebo orally every Monday, Wednesday, and Friday for 6 months. All enrolled dogs underwent complete evaluations (physical exam, chemistry, CBC, urinalysis, and echocardiogram) at baseline, 3 months, 6 months, and 12 months (i.e., 6 months after discontinuation of treatment), and erythrocyte mean corpuscular volume (MCV) was recorded for each visit. There was no significant difference in MCV between the control and rapamycin treated groups at baseline (71.4 fL vs. 72.0 fL, respectively; p=0.56), 3 (70.9 fL vs. 70.9 fL; p=0.99), 6 (70.1 fL vs. 69.8 fL; p=0.74), or 12 months (69.3 fL vs. 70.4 fL, p=0.14). Additionally, there were no significant differences in the change in MCV between groups over the 6 month period (-1.3 fL [control] vs. -2.2 fL [rapamycin]; p=0.18). In the cohort receiving rapamycin, there was a significant reduction in mean MCV from baseline to 6 months (72.0 fL vs. 69.8 fL; p=0.03), but there was no significant difference in MCV between 6 and 12 months (69.8 fL vs. 70.4 fL; p=0.54). In the control group, a significant decrease in MCV was observed from baseline to 12 months (71.4 fL vs. 69.3 fL; p=0.04), a finding that reflects the need for more study subjects in order to better assess the potential of rapamycin-induced changes in MCV in healthy dogs.

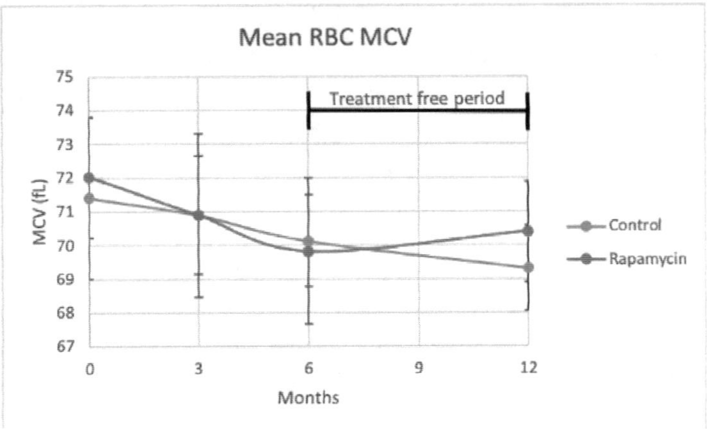

Treatment	0mo MCV (fL)	3mo MCV	6mo MCV	12mo MCV	Treatment	0mo MCV	3mo MCV	6mo MCV	12mo MCV
Control	72.1	72	71	67.9	Rapamycin	72.9	73	72.4	72
Control	73.5	70.7	70.8	70.2	Rapamycin	75.7	72.9	73.1	72.7
Control	74.8	74.1	72.8	71.5	Rapamycin	69.3	68	67.9	68.5
Control	70.8	69.5	69.3	68.8	Rapamycin	71.7	67.6	69.5	69.1
Control	67	68.6	68.5	67.8	Rapamycin	73.3	72	71.6	70.4
Control	69.5	70.3	69.4	70.1	Rapamycin	71.5	68.4	66.5	69
Control	71.8	72.1	69.3	69.5	Rapamycin	71.7	72.9	68.6	70.5
Control	71.8	69.9	69.9	68.7	Rapamycin	70.8	73.4	69.5	71
Mean	71.4	70.9	70.1	69.3	Rapamycin	71.3	69.8	69.3	N/A
					Mean	72	70.9	69.8	70.4

Abstract HP01: Evaluation of Coagulation in Dogs with Gallbladder Mucoceles

Michelle Pavlick[1]; Cynthia Webster[2]; Dominique Penninck[2]; Armelle DeLaforcade[2]
[1]Newtown Veterinary Specialists; [2]Tufts University

Gallbladder mucoceles (GBM) are a common biliary disorder in dogs and are associated with inflammatory and thrombotic complications. Limited information is available on their coagulation status. The aim of this study was to assess coagulation in dogs with GBM. Twenty-three dogs with GBM identified on ultrasound were prospectively enrolled. Blood was collected at the time of GBM identification for determination of complete blood count, biochemical panel, packed cell volume, prothrombin time (PT), activated partial thromboplastin time (aPTT), factor VIII activity, fibrinogen, D-dimers, thromboelastrography (TEG), protein C activity (PC), antithrombin activity (AT), and von Willebrand's factor activity (vWF). Overall, when compared to the hospital generated reference range population, dogs with GBM had significantly decreased K values and increased angle, MA and G ($p<0.001$). Based on G value, 19/23 dogs (82.6%) were classified as hypercoagulable and 4/23 (17.4%) were classified as normocoagulable. Four of 23 (17.3%) of dogs were classified as hyperfibrinolytic based on increased Ly 60 values. Plasma based coagulation tests showed coagulation changes consistent with hypercoagulablitity such as hyperfibrinogenemia (8/23), thrombocytosis (9/23) and increased D-dimers (4/13). Increased PC activity (20/24) and AT activity (9/23) were seen. The PT was normal 21/22 dog but the aPTT was prolonged in 8/22. vWF was not increased in any dog while Factor VIII was increased in 3/23 dogs. In conclusion, dogs with ultrasonographically identified GBM have changes in whole blood kaolin-activated TEG supporting a hypercoagulable state although traditional plasma-based coagulation testing might suggest that a more complex state of hemostasis exists.

Abstract HP02: Dogs with Chronic Hepatitis Have Altered Amino Acid Profiles Compared to Healthy Dogs

Robert Kyle Phillips[1]; Amanda Blake[2]; Yuri Lawrence[3]; Jan Suchodolski[2]; Jörg Steiner[2]; Jonathan Lidbury[2]
[1]GI Lab, Texas A&M University; [2]Texas A&M University; [3]Austin Veterinary Emergency and Specialty Center

The liver plays a central role in protein metabolism, and abnormalities of serum amino acid (AA) concentrations contribute to the development of sequelae to chronic hepatitis (CH), such as hepatic encephalopathy. Consequently, measurement of serum AA concentrations may better our understanding of the pathogenesis of CH in dogs, as well as serving as potential diagnostic and/or prognostic biomarkers. The purpose of this study was to compare serum AA profiles between dogs with CH and healthy control dogs. Serum samples were collected from 10 dogs with histologically confirmed CH and from 38 healthy control dogs. Serum AA were measured with a Biochrom 30+ (Biochrom Ltd., Cambridge, UK) amino acid analyzer. The concentration of each AA between groups was compared using a Mann-Whitney test and corrected for multiple comparisons using the Benjamini-Hochberg procedure. Significance was set at $q<0.05$. Twenty-eight AA were measurable in the serum samples. Serum concentrations of 13 AA were significantly increased in dogs with CH: glutamic acid, phenylalanine, asparagine, ornithine, serine, lysine, glutamine, histidine, citrulline, methionine, α-aminobutyric acid, tyrosine, and aspartic acid. The serum concentration of hydroxyproline was significantly decreased in dogs with CH. Serum concentrations of the following AA were not significantly different between groups: alanine, valine, 1-methylhistidine, proline, isoleucine, phosphoserine, leucine, threonine, glycine, arginine, tryptophan, α-aminoadipic acid, 3-methylhistidine, and hydroxylysine. Dogs with CH had altered serum AA concentrations compared to healthy control dogs. Serum concentrations of 13 of 28 measured AA were increased in dogs with CH, which may be attributable to increased protein turnover and/or diminished AA catabolism. Further investigation to determine if these alterations constitute a unique serum AA profile for patients with CH that is distinguishable from dogs with other forms of hepatic disease is warranted.

Abstract HP03: Gastroduodenal Ulceration in Canine Liver Disease

Kirsten Cooke; Allison O'Kell; Alexander Gallagher
College of Veterinary Medicine, University of Florida

Although liver disease is frequently cited as a cause of gastroduodenal ulceration (GDU) in dogs, studies regarding GDU and esophageal varices (EV), another possible sequelae of liver disease, are limited. The objective of this study was to document the presence of GDU and EV in dogs with liver disease. Dogs that underwent liver biopsy and/or CT angiography to diagnose congenital or acquired liver disease were enrolled. All dogs had gastroduodenoscopy performed with photographic and video documentation in a standardized fashion. Lesions (hemorrhage, erosions, ulcers) in the esophagus, stomach, and duodenum were scored based on a grading scale and presence of EV was recorded. Dogs were categorized into 4 groups according to the cause of liver disease (inflammatory disease, cirrhosis, congenital, other) and the total endoscopic score was compared among the groups using a Kruskal-Wallis test. The presence or absence of ulcers and/or erosions was compared among the groups using a Fisher's exact test. Forty dogs were enrolled in the study with the following distribution: 12 congenital, 13 inflammatory, 3 cirrhosis, and 9 other. Four dogs had GDU (11%), nine dogs had erosions and/or GDU (22.5%), and 1 dog had EV (2.5%). There was no difference in total endoscopic score between the groups (p=0.21) or in the proportion of dogs with ulcers and/or erosions versus those without (p=0.25). GDU and EV were not common in this population of dogs with liver disease. Further studies are warranted to confirm these findings in larger numbers of dogs with specific disease etiologies. GDU and EV were not common in this population of dogs with liver disease. Further studies are warranted to confirm these findings in larger numbers of dogs with specific disease etiologies.

Abstract HP04: Intracellular Distribution of Copper in Liver Specimens from Cats

Punyamanee Yamkate[1]; David Twedt[2]; Jan Suchodolski[1]; Joerg Steiner[1]; Jonathan Lidbury[1]
[1]Gastrointestinal Laboratory, College of Veterinary Medicine, Texas A&M University; [2]College of Veterinary Medicine and Biomedical Sciences, Colorado State University

The intracellular distribution of copper in the liver has been investigated in many species, including rats, sheep, humans, and dogs. Intracellular copper distribution has been reported to differ in patients with copper hepatopathy compared to those of healthy subjects, providing important insight into the pathogenesis of this condition. However, hepatic copper distribution has not previously been reported in cats. Therefore, the aim of this study was to assess the intracellular copper distribution in liver specimens from cats with copper concentrations within or above the reference interval. Thirteen frozen liver specimens collected from cats were available for this study, 6 with copper concentrations within and 7 with copper concentrations above the reference interval (<180 µg/g dry weight). Each specimen of frozen liver tissue was divided into two equal portions. The first portion was kept for copper quantification. The other portion was homogenized and was separated into fractions (nuclear, large granule, microsomal, and cytosolic) using differential centrifugation. Copper concentrations in liver specimens and liver fractions were measured using flame atomic absorption spectroscopy. The distribution of copper among fractions was calculated and compared between liver specimens with copper concentrations within or above the reference interval using Mann Whitney test. Statistical significance was set as $p<0.05$. For liver specimens with copper concentration within the reference interval, median (range) percentage of copper in each of the 4 fractions was as follows: nuclear, 23.8% (11.6–41.6); large granule, 11.6% (9.4–27.8); microsomal, 4.2% (2.2–8.7); and cytosolic, 58.6% (44.5–62.8). For liver specimens with copper concentration above the reference interval, median (range) percentage of copper in each of the 4 fractions was as follows: nuclear, 20.2% (11.2–35.8); large granule, 19.1% (2.6–26.2); microsomal, 8.2% (1.3–9.7); and cytosolic, 56.1% (41.8–79.2). There was no significant difference in copper distribution for each fraction between cat liver specimens with hepatic copper concentrations within or above the reference interval ($p>0.36$). Our findings indicate that, similar to other species, intracellular copper is predominantly found in the cytosolic and nuclear fractions in liver specimens from cats. The distribution of copper in liver specimens from cats with copper concentrations above the reference interval was not significantly different to those with copper concentrations within the reference interval. Consequently, further investigation of hepatic intracellular copper distribution in a greater number of liver specimens from cats is warranted.

Abstract HP05: Approach to the Diagnosis of Hepatocutaneous Syndrome in Dogs: A Retrospective Study and Systematic Review

Karah C. Burns DeMarle[1]; Lluis Ferrer[2]; Dominique Penninck[1]; Cynthia Leveille-Webster[1]
[1]Cummings School of Veterinary Medicine, Tufts University; [2]Autonomous University of Barcelona

Hepatocutaneous syndrome (HCS) is a rare and often fatal disease seen in dogs characterized by skin lesions (superficial necrolytic dermatitis [SND]) and associated hepatopathy. Although various combinations of clinical signs, biochemistry tests, imaging findings and skin and hepatic histopathology have been used to diagnose this syndrome, there is currently no consensus on which combination would enable the most timely and non-invasive method for diagnosis. In order to accomplish this, medical records were reviewed retrospectively for dogs with skin biopsy proven HCS (n=24) and data were compared to cases found by systematic review of the literature (n=105). The most consistent findings were ulcerative-crusted lesions affecting paw pads or mucocutaneous junctions (103/103, 100%), marked plasma hypoaminoacidemia (50/50, 100%) and the presence of a "honeycomb-like" appearance to the liver on ultrasound (62/63, 98%). Six out of 23 (26%) of the dogs in the retrospective study had marked keratinocyte apoptosis, a finding associated with the development of diabetes mellitus. A literature search confirmed that the marked plasma hypoaminoacidemia seen in dogs with HCS is unique and has not been reported in other hepatic disease (n=6 manuscripts) or in non-hepatic disorders (n=5 manuscripts). An algorithm for diagnosing HCS was developed which involves sequential identification of characteristic skin lesions, plasma hypoaminoacidemia and the presence of a honeycomb-like pattern of the liver on ultrasound. This diagnostic approach simplifies the diagnostic pathways used in the literature and permits earlier identification of HCS cases before the painful lesions prompt the compassionate decision for euthanasia.

Abstract HP06: Validation of an In-Clinic Assay for the Measurement of Canine and Feline Bile Acids

Elizabeth Schooley; Eric Steva
IDEXX Laboratories Inc.

This study evaluates IDEXX Catalyst® Bile Acids (CatalystBA) for canine and feline bile acids measurement. Serum samples, originally submitted for clinical purposes, from 73 dogs and 45 cats were obtained from IDEXX Reference Laboratories (IRL). Samples were provided in accordance with IRL Terms and Conditions. Samples were analyzed using CatalystBA, once on a Catalyst One® analyzer, once on a Catalyst Dx® analyzer and twice using the Diazyme Total Bile Acids Assay (Enzyme Cycling Method) run on the Beckman Coulter AU5800 analyzer (RefMethod). Each CatalystBA result was paired with the corresponding mean RefMethod result. Testing was completed at IDEXX. Precision was evaluated with a 10-day protocol run across 4 Catalyst instruments. RefMethod median bile acids was 22.3 µmol/L; range: 1.0 to 155.8 µmol/L. CatalystBA median bile acids was 21.0 µmol/L; range: 1.1 to 177.8 µmol/L. No statistical difference existed between concentrations obtained by the two methods (p=0.86; Wilcoxon signed ranks). Method comparison results are reported with 95% confidence limits in parentheses. Passing-Bablok regression analysis: intercept -1.04 µmol/L (-1.49 to -0.74); slope 1.05 (1.02 to 1.09); Tau 0.88. Pearson's correlation coefficient (r): 0.99. Results were assigned to one of three categories (table 1). Calculation of percent concordance between the assays showed excellent agreement (92.6%) on result classification. Results of precision testing are outlined in table 2. This study confirms minimal bias, good precision and concordance, as well as excellent correlation with the RefMethod and provides confidence that CatalystBA shows strong performance for in-clinic measurement of canine and feline bile acids.

Table 1–Canine and Feline Categories for Concordance Calculation

Canine – 90.6% concordance					Feline – 96.2% concordance				
		RefMethod (µmol/L)					RefMethod (µmol/L)		
CatalystBA (µmol/L)		<30	30-40	>40	CatalystBA (µmol/L)		<15	15-30	>30
	<30	63	6	0		<15	46	2	0
	30-40	1	12	2		15-30	1	10	0
	>40	0	4	50		>30	0	0	20

Table 2–Precision Results

Mean Concentration (µmol/L)	Standard Deviation (µmol/L)	Replicates
14.9	1.5	318
23.8	2.1	318
79.1	3.4	320

Abstract ID01: Canine Leishmaniasis in North America: Imported and Autochthonous Cases, 2006–2019

Taylor Gin[1]; Edward Breitschwerdt[2]; Barbara Qurollo[3]
[1]College of Veterinary Medicine, Texas A&M University; [2]NCSU; [3]Department of Clinical Sciences, NCSU

Sand fly transmission of *Leishmania infantum*, a protozoa that causes a potentially life-threatening infectious disease in dogs and humans in many parts of the world, has not been definitively documented in the United States or Canada. Infection with *L. infantum* has been reported commonly in foxhounds throughout North America, but only rarely in other dog breeds. A convenience sample of dogs tested by *L. infantum* immunofluorescent antibody (IFAT), *Leishmania* PCR, or both diagnostic assays was established for this study by review of a database containing canine vector-borne diseases (CVBD) diagnostic testing results performed between 01/04/2006 and 05/22/2019 at the North Carolina State University (NCSU), College of Veterinary Medicine (CVM), Vector-borne Disease Diagnostic Laboratory (VBDDL). We report demographics and travel history obtained through email or phone contact with the primary veterinarian for dogs positive for *L. infantum* based on positive serology and/or PCR. Over this time period, 126/1,961 (6.4%) of dogs tested by the NCSU-CVM-VBDDL were serologically (n=100) or PCR (n=60) positive for *L. infantum*. Information related to travel history was available for 70/126 (55.6%) dogs. Sixty (85.7%) dogs had history of travel outside of the US or Canada, whereas 10 (14.3%) dogs had no travel history outside of North America, including 3 PCR positive dogs. Importantly, 116/126 (92%) of *L. infantum* positive dogs were breeds other than foxhounds. *L. infantum* was detected in many non-foxhound breeds, some of which had no reported travel history outside of the US or Canada. Increased *Leishmania* diagnostic testing and disease surveillance in dogs are essential for monitoring potential emergence of this zoonotic infectious disease in North America.

Abstract ID02: *Heterobilharzia Americana* Infection in Dogs: Clinical Features and Outcome in 60 Cases (2010–2019)

Amber M. Graham[1]; Valentina Moshnikova[2]; Amy Davenport[3]; Lindsey Gilmour[1]; Michelle Fabiani[3]; Audrey Cook[1]
[1]Texas A&M University; [2]University of Minnesota; [3]Gulf Coast Veterinary Specialists

Fecal PCR testing has simplified the diagnosis of *Heterobilharzia americana* in dogs. The purpose of this retrospective study was to provide updated information regarding the clinical features of this infection. Medical records at two hospitals were searched for patients with schistosomiasis, and data were collected regarding signalment, history, diagnostic results, treatment, and outcome. Sixty dogs were diagnosed using fecal PCR (n=49), PCR and histology (n=3), histology (n=5), or fecal floatation (n=3). Mean age was 7.5 years (range: 0.6–17.2) and median body weight was 23.2 kg (range: 3.4–49). Forty-one dogs presented for vomiting +/- diarrhea; eight for weight loss +/- anorexia. Laboratory data were available for 54 dogs: hypercalcemia was noted in 4/54; azotemia in 10/52; hypoalbuminemia in 14/54; hyperglobulinemia in 18/54. Eosinophil count was >500/µL in 22/52. Increased ALP (n=13; median: 306 IU/L; range: 161–1795) and ALT (n=7; median: 335 IU/L; range: 131–1447) activities were noted in 14/47. Transabdominal ultrasonography revealed pin-point hyperechoic foci in the small intestine +/- liver +/- mesenteric lymph nodes in 38/59 dogs; effusion was noted in 14/59. Fifty-six dogs received oral praziquantel for 2–3 days (median dose: 27 mg/kg q 8 hr). Concurrent medications included fenbendazole (40/56) and anti-inflammatory doses of prednisone (12/56). Follow-up PCR was negative in 11 dogs; 5 were PCR positive and retreated. Six-month survival data was available for 34 treated dogs: 25 were alive, 2 died acutely during treatment, 7 died or were euthanized due to their infection (n=3) or of unrelated causes (n=4). Clinical features of schistosomiasis are variable and non-specific. It is noteworthy that <10% of dogs were hypercalcemic and two died acutely during treatment. Schistosomiasis should be considered in dogs in endemic areas with gastrointestinal signs, weight loss, elevated liver enzymes or consistent ultrasound changes.

Abstract ID03: Evaluation of Canine Parvovirus Neutralizing Antibody (KIND-030) as a Prophylactic and Therapeutic Treatment in Puppies

Ellen R. Ratcliff[1]; Laurie Larson[2]; Allyson Avenatti[1]; Stephanie Pierce[1]; Tianhua Hu[1]; Melinda Poole[1]
[1]Kindred Biosciences, Inc.; [2]School of Veterinary Medicine, University of Wisconsin-Madison

The primary purpose of this proof-of-concept study was to evaluate the effectiveness of canine parvovirus (CPV) neutralizing antibody (KIND-030) as a prophylactic and therapeutic treatment in puppies. KIND-030 is a monoclonal antibody. It is under clinical investigation and has not received approval by the USDA. Eight purpose-bred, CPV-2 seronegative, healthy beagle puppies aged 11 weeks were randomly allocated to four open-label groups. No treatment beyond study drug was permitted. Groups 1 (n=2, IV 5 mg/kg) and 2 (n=2, SC 5 mg/kg) received treatment with KIND-030 on Day 1. All four groups were inoculated with a challenge dose of 106 TCID50 dose of virulent CPV-2 on Day 4. Group 3 (n=2) received KIND-030 IV 5 mg/kg upon detection of CPV-2 in feces as confirmed via cage-side ELISA. Group 4 (n=2) did not receive KIND-030 treatment. **Survival**: Groups 1 and 2 puppies remained healthy throughout the study. Group 3 puppies survived and recovered during the study period. Group 4 puppies were severely affected by CPV infection and euthanized for humane reasons before study completion. **SNAP Test**: In Groups 1 and 2, none of the puppies had a positive SNAP test during the study. In Group 3, both puppies had positive SNAP tests on Day 7. Group 4 puppies had a positive SNAP test on Day 7 (n=1) and Day 8 (n=1). **Adverse Events**: There were no adverse events and no injection site reactions in Groups 1 and 2. Following KIND-030 treatment, Group 3 did not exhibit any remarkable adverse events. Group 4 puppies demonstrated symptoms consistent with CPV infection. In this proof-of-concept study, prophylactic administration of KIND-030 prior to challenge protected 100% of puppies from CPV-2 infection. In addition, 100% of puppies given therapeutic administration of KIND-030 after confirmed CPV infection survived and recovered.

Abstract ID04: A Novel *Rickettsia* Species Infecting Febrile Dogs in the United States

Barbara A. Qurollo[1]; James Wilson[1]; Edward Breitschwerdt[1]; Nicholas Juhasz[1]; Henry Marr[1]; Joao Filpe de Brito Galvao[2]; Carmela Pratt[3]
[1]College of Veterinary Medicine, North Carolina State University; [2]VCA Arboretum View Animal Hospital; [3]Oklahoma Veterinary Specialists

In the United States, tick-borne *Rickettsia rickettsii*, causative agent of Rocky Mountain Spotted Fever (RMSF), is the only known cause of spotted fever group (SFG) rickettsioses in dogs. SFG *Rickettsia*, including *R. parkeri*, *Rickettsia 364D* and *R. rickettsii* are well-documented causes of human SFG rickettsioses. If, or to what extent, other SFG *Rickettsia* are pathogenic in dogs is unclear, but serosurveys report high SFG *Rickettsia* seroprevalence in dogs in North America. In this study, we describe a novel *Rickettsia* sp. infecting three dogs with febrile illness and hematological abnormalities. All dogs were *R. rickettsii* IFA seroreactive and identical *Rickettsia* DNA sequences were amplified from their diagnostic blood specimens. Case 1 had acute onset fever with a rapid response to doxycycline. Case 2 had acute onset fever and neutrophilic polyarthritis. Case 3 had acute onset fever and a concurrent disease process that contributed to protein losing nephropathy. Case 3 was also febrile and seroreactive to *Rickettsia* by IFA one year prior to documentation of the novel *Rickettsia* sp. infection. Geographically, these cases were distributed among four states, including Tennessee and Arkansas (travel history case 1), Illinois (case 2) and Oklahoma (case 3). Tick exposure was documented in all three dogs and illness occurred during summer months. The novel *Rickettsia* sp. was confirmed by PCR amplification and sequencing of several *Rickettsia* genus targets. A multi-locus *Rickettsia* spp. phylogenetic tree using 2,576 nucleotides concatenated from 5 regions within 3 genes (ompA, gltA, and 17 kDa) and 2 intergenic spacer regions (23S-5S and mmpA-purC) was assembled using the maximum-likelihood method and Tamura-Nei model. The novel *Rickettsia* sp. was most similar to *R. heilongjiangensis* and *R. massiliae*, tick-borne human rickettsial pathogens. These three canine rickettsioses cases underscore the potential for canine infection with a previously undescribed vector-borne pathogen in the United States and the utility of dogs as sentinels for the identification of emerging and potentially zoonotic tick-borne pathogens.

Abstract ID05: Flea-Borne Bacterial Pathogens from Free-Roaming Cats and their Fleas

Erin W. Lashnits[1]; Charlotte Manvell[1]; Hanna Berman[1]; William Swain[2]; Kelli Ferris[1]; Edward Breitschwerdt[1]; Benjamin Callahan[1]
[1]North Carolina State University; [2]University of Wyoming

Free-roaming cats live at the interface between humans, their companion animals, local wildlife, and the parasites that infest these animals, creating novel opportunities for cross-species pathogen transmission. The cat flea (*Ctenocephalides felis*) is the most common ectoparasite of free-roaming cats and is the vector of transmission for multiple zoonotic bacterial pathogens, including *Bartonella*, hemotropic *Mycoplasma*, and *Rickettsia* species. The goal of this study was to determine the prevalence of flea-borne bacterial pathogens in free-roaming cats and their fleas across different geographic locations using both targeted real-time PCR (qPCR) and 16S rRNA next generation sequencing (NGS). Fleas were collected from free-roaming cats at spay and neuter clinics in northern California, Louisiana, and North Carolina. Tissue samples (ear tips and reproductive tissues including ovary, testicle, uterus, placenta, and fetus) were concurrently collected in North Carolina. Real-time PCR was performed on DNA extracted from whole washed fleas and tissue samples, with primers for *Bartonella* ssrA and hemotropic *Mycoplasma* spp. 16S. Positive results were confirmed with Sanger sequencing and species identity determined based on alignment with the NCBI Local Alignment Search Tool (BLAST). A subset of 45 fleas from North Carolina (n=33) and California (n=12) also had whole-flea eubacterial community composition determined using NGS targeting the V3-V4 region of the 16S rRNA gene. By qPCR, 28% of *C. felis* (33/142) contained *Bartonella* spp. DNA. *Bartonella* was most common in fleas from North Carolina (20/28 positive, 71%), moderately common in fleas from Louisiana (13/44 positive, 30%), and relatively rare in fleas from northern California (2/67 positive, 3%; chi-squared test of independence for each proportion $p<0.01$). Based on 16S-NGS on whole washed fleas, the most common pathogenic bacteria found were *Bartonella* and *Rickettsia* genera; a large proportion of fleas also carried *Wolbachia* spp. Microbiota community composition in this pilot set of fleas was associated with geographic region, confirming results from targeted qPCR. The sensitivity and specificity of 16S-NGS (with targeted qPCR as the reference standard) for *Bartonella* spp. was 76% and 92% respectively. Based on targeted PCR and 16S-NGS, no fleas contained hemotropic *Mycoplasma* spp. DNA. For 68 cats with ear tip tissue collected (all from North Carolina or Virginia), 21% were *Bartonella* PCR positive (most commonly *B. clarridgeiae*) and 16% were hemotropic *Mycoplasma* spp. PCR positive (most commonly *M. haemominutum*). Of the 68 cats sampled, only 21 had fleas collected: 14% (3/21) of these cats with fleas were *Bartonella* PCR positive, compared to 20% (5/25) of cats with no fleas collected ($p=0.71$). In reproductive samples (173 tissue samples from 49 cats), *Bartonella* spp. DNA was qPCR amplified from 18% and hemotropic *Mycoplasma* from 8%. Our findings document geographic differences among flea-borne bacterial pathogens, with *Bartonella* spp. most common in fleas from North Carolina cats. Hemotropic *Mycoplasma* spp. were not documented in fleas, despite infection being found in cats. Further investigations are needed to elucidate factors associated with the incidence of these important and potentially fatal flea-borne diseases that affect both cats and humans worldwide.

Abstract ID06: The "Backyard" Identified as an Environmental Reservoir of blaNDM-5 *E. coli* During a Veterinary Outbreak

Stephen Cole; Jaclyn Dietrich; Shelley Rankin
University of Pennsylvania

Healthcare associated outbreaks of carbapenem-resistant Enterobacteriaceae (CRE) have been associated with environmental reservoirs that include handwashing sinks, shower drains and medical equipment. There are limited reports of outbreaks of CRE in veterinary settings and little is known about the veterinary hospital epidemiology of these organisms. Results of point prevalence surveys (PPS) during the response to an outbreak of blaNDM-5 positive *Escherichia coli* at a referral veterinary hospital in 2018–2019 suggested possible ongoing in-facility transmission to animals. Since CRE is spread via the Oral-Fecal route, we hypothesized that the outdoor elimination area or "backyard" that was shared by many patients served as a reservoir for infection. Samples were collected from an outdoor elimination area using sterile electrostatic cling cloths (Swiffer). The shoes of the sampler were tested before and after collection. Cloths were incubated at 37°C in 100–200 mL of buffered peptone water (BPW) in a WhirlPak for 16–24 hours and 10 uL subcultured to selective, chromogenic agar plates (CARBA, BioMerieux). 0.5 mL of the BPW enrichment b was added to 4.5 mL of tryptic soy broth supplemented with a 10 ug meropenem disc and incubated and subcultured as described above. Presumptive isolates were confirmed as *E. coli* with the Vitek2 GN Card and shown to produce a carbapenemase with the modified Carbapenem Inhibition Method (mCIM). Bacterial DNA extraction was performed and the isolates were tested with a loop mediated isothermal PCR assay that detects the NDM gene. Positive sites identified in the elimination area included multiple sections of the synthetic turf, metal and composite wood siding, a hose and "pooper scooper" that were stored on the turf. The metal door jamb and a rug leading from the area back into the clinical areas of the hospital were also positive. In addition, the shoes of the collector were negative prior to collection but positive following collection. All 10 environmental isolates were identified as *E. coli* and were positive for the NDM gene. This study suggests that the elimination area may have served as an environmental reservoir for CR *E. coli* during an outbreak at a veterinary hospital. To what degree patient-to-patient transmission played a role cannot be determined within the 16 patients with NDM-5 *E. coli* identified by point prevalence testing. This study also demonstrated that the shoes of animal handlers can be contaminated. This could reintroduce CRE from outdoor elimination areas into the hospital environment and potentially expose more animals and people. Shared elimination areas should be considered high risk areas for nosocomial transmission of antibiotic resistant bacteria and that risk should also be considered for the people that accompany animals into this area.

Abstract ID07: Knowledge, Attitudes and Influencers of Dog-Owners Surrounding Antimicrobials and Antimicrobial Stewardship in North America

Madeleine R. Stein[1]; Michelle Evason[2]; Jason Stull[2]; J McClure[2]; J. Scott Weese[3]
[1]University of Prince Edward Island; [2]Atlantic Veterinary College; [3]Ontario Veterinary College

Antimicrobial resistance (AMR) in companion animals is a growing concern. Research on pet-owner knowledge, attitudes, and influencers (KAIs) surrounding antimicrobials will elucidate education needs. Study objectives were to: 1) Quantify the individual influences of antimicrobial cost, method of administration, and drug importance in human medicine on dog-owner preferences, and 2) Determine KAIs of dog-owners surrounding antimicrobials and antimicrobial stewardship. Data were collected through an online survey targeting three dog-owner participant groups. These consisted of individuals residing in 1) Canada, 2) United States, and 3) any country recruited through social media. Conjoint analyses were used to quantify the influence of antimicrobial cost, method of administration and drug importance in human medicine. Descriptive statistics were used for data evaluation. A total of 809 surveys were completed (US participants, n=315; Canadian participants, n=298; social media group, n=196). Conjoint analysis results quantification revealed antimicrobial cost as accounting for 47% of dog-owner preferences, followed by method of administration (31%) and drug importance (22%). All groups preferred drugs that cost $25 and were administered once by injection. US and Canadian participants were more likely to prefer drugs considered "very important" in human medicine, whereas social media recruited participants were more likely to prefer drugs that were "not at all important." Most respondents (86%) reported AMR as important in human medicine and 29% believed pet antimicrobial use posed a risk for humans. Our research reveals cost as most important in dog-owner antimicrobial preferences and provides practical information for antimicrobial stewardship initiatives.

Abstract ID08: Serum 25-Hydroxyvitamin D and Canine Infectious Respiratory Disease Complex in Shelter Dogs

Andrew Sun[1]; Cody Blakeman[1]; Jared Jaffey[1]; Nancy Bradley-Siemens[1]; Michael Lappin[2]; Jennifer Hawley[2]; Randy Ringold[3]; Rachael Kreisler[1]
[1]College of Veterinary Medicine, Midwestern University; [2]College of Veterinary Medicine, Colorado State University; [3]VDI Laboratory

The background is as follows. Hypovitaminosis D is associated with increased susceptibility to acute respiratory infections and repletion decreases this risk as well as severity of illness in humans. Canine infectious respiratory disease complex (CIRDC) is common in animal shelters. These respiratory infections negatively impact dogs and consumes a substantial amount of shelter resources. The role of vitamin D in shelter dogs with CIRDC is unknown. Objectives are as follows. (1) Compare serum 25-hydroxy (OH)-vitamin D concentrations between shelter dogs with respiratory signs associated with CIRDC and shelter dogs without respiratory signs. (2) compare serum 25(OH)D concentrations between shelter dogs polymerase chain reaction (PCR) positive for *Bordetella bronchiseptica* (BB), *Mycoplasma cynos* (MC), canine herpesvirus-1 (CHV-1), and canine distemper virus (CDV). The methods are as follows. Serum 25(OH)D concentrations were measured with a commercially available chemiluminescent immunoassay kit and nucleic acids of common CIRDC agents were amplified by use of a commercially available service (Antech Diagnostics). Samples were obtained from left-over blood samples taken by a shelter during a CIRDC outbreak. Unpaired two-tailed t-tests were used to compare square root transformed 25(OH)D concentrations with values reported back transformed and $p<0.05$ considered significant. The results are as follows. One hundred and forty-six shelter dogs were included. The median length of stay prior to sampling was 46 days (IQR 28–73). Thirty-six (25%) shelter dogs exhibited respiratory signs at the time of inclusion, while 51 (35%) were sampled after resolution of respiratory signs and 4 (3%) before signs. Five dogs did not have available PCR results. Forty-seven (32%) of dogs were PCR positive for at least one pathogen with 32 (23%), 11 (8%), 6 (4%), and 1 (1%) positive for MC, BB, CHV-1 and CDV, respectively. Only two dogs were positive for more than one pathogen. The mean 25(OH)D concentration for dogs with respiratory signs at the time of sampling was 59.3 ng/ml (SD 1.5), while it was 66.6 ng/ml (SD 1.4) for dogs without, which was significantly different, $t(142)=2.0317$, $p=0.0440$. Dogs that were CHV-1 positive had significantly lower serum 25(OH)D concentrations with a mean of 50.6 ng/ml (SD 1.8) than dogs that were CHV-1 negative which had a mean of 65.5 ng/ml (SD 1.4), $t(137)=1.9814$, $p=0.0495$). There was no difference for dogs positive for BB or MC and dogs negative for those pathogens, $p=0.82$ and $p=0.47$, respectively. The conclusions/clinical importance are as follows. These results suggest that vitamin D has a potential clinical role in shelter dogs, particularly those with viral infections, and provides a rationale for future studies to investigate if repletion of vitamin D can decrease the prevalence, morbidity, or both of CIRDC in shelter dogs.

Abstract ID09: Markers of Inflammation and Cytokine Concentrations During Experimental *Babesia rossi* Infection of Beagle Dogs

Brogan K. Atkinson[1]; Andrew Leisewitz[2]; Varaidzo Mukorera[2]; Peter Thompson[2]
[1]Onderstepoort Veterinary Academic Hospital, University of Pretoria; [2]University of Pretoria

Babesia rossi (*B. rossi*), the most virulent of canine *Babesia* parasites, causes severe clinical disease and death in dogs in sub-Saharan Africa. Complications and multiple organ dysfunction seen with babesiosis are likely caused by the effects of an unfocused and excessive inflammatory response. The aim of this study was to investigate markers of inflammation and cytokine kinetics from the point of inoculation with *B. rossi* throughout the course of clinical disease and determine if infectious dose influenced rate and severity of disease progression. This experimental study was performed on 5 healthy sterilised male beagle dogs. Three dogs were given a high infectious dose and 2 dogs a low infectious dose of parasite. CBC, C-reactive protein, albumin and lactate were determined daily. Cytokines were quantified on stored plasma using a canine specific cytokine magnetic bead panel (Milliplex©). Dogs were monitored daily until predetermined end points for treatment were reached. Post inoculation, initial parasitaemia occurred on day 1 and day 3 in the high and low dose groups respectively. The rate of increase in parasitaemia in the high dose group was significantly higher than seen in the low dose group. Significant differences between the high and low dose groups were found in body temperature, interferon gamma (INFγ), keratinocyte chemoattractant (KC), INFγ-induced protein 10 (IP10), interleukin 8 (IL8), IL15, IL18, CRP, albumin, segmented and band neutrophils at certain time points. Significant differences were noted for pulse, glucose, lactate, granulocyte-macrophage colony-stimulating factor (GM-CSF), IFNγ, KC, IP10, IL2, IL6, IL7, IL8, IL10, monocyte chemoattractant protein 1 (MCP1), tumour necrosis factor alpha (TNFα), segmented neutrophils, band neutrophils, monocytes, CRP and albumin at certain time points within the high dose group and baseline. Significant difference was also noted for pulse, glucose, IFNγ, KC, IP10, IL8, segmented neutrophils, CRP and albumin between low dose group and baseline at certain time points. The majority of the variables evaluated were higher when comparing the high dose to low dose group and both groups to the baseline with the exception of albumin throughout the study and segmented neutrophils during infection, which were consistently lower. These findings suggest that in *Babesia rossi* infection, initiation of inflammation occurs before the onset of clinical disease and infectious dose affects the evolution of inflammation and the course and severity of disease.

Abstract ID10: Effect of Probiotic Containing *Aspergillus*-derived Ingredients on Serum and Urine Galactomannan Antigen Assay in Dogs

Krystle L. Reagan[1]; L. Joseph Wheat[2]; Jane Sykes[1]
[1]University of California, Davis; [2]MiraVista Diagnostics

The Platelia *Aspergillus* galactomannan antigen (GMA) enzyme-linked sandwich immunoassay (ELISA) assay is used to support a diagnosis of systemic aspergillosis in dogs. In humans, false-positive results are associated with administration of medications derived from molds. Several probiotic supplements contain fungal organisms or enzymes derived from molds that may cross-react with the GMA ELISA. We sought to determine the effect of administration of an oral probiotic containing *Aspergillus*-derived ingredients on serum and urine GMA levels in dogs by conducting a prospective, cross-over study. Serum and urine galactomannan indices (GMI) were measured in 10 healthy dogs before (day 0), after 1 week (day 7) of probiotic administration, and 2 weeks after the probiotic was discontinued (day 21). Median (range) serum GMI were 0.19 (0.08–0.62), 0.22 (0.07–1.15) and 0.17 (0.14–0.63) at day 0, 7 and 21 respectively. Two of 10 dogs developed positive GMI (≥0.5) results after probiotic administration, and one dog initially had a positive GMI at day 0 that then returned to normal at days 7 and 21. No significant treatment effect was noted in the study period. Median (range) urine GMI results were 0.06 (0.04–0.22), 0.07 (0.05–0.41) and 0.06 (0.03–0.16) at day 0, 7 and 21 respectively. No positive urine GMIs were noted. Administration of a probiotic containing *Aspergillus*-derived ingredients did not reliably result in elevated serum or urine GMA levels. Three dogs had low-grade positive GMI, one at each study time point, that did not correlate to probiotic administration.

Abstract ID11: Feline Parvovirus Seroprevalence is High in Cats from Disease Outbreak and Non-Outbreak Regions in Australia

Elizabeth L. Jenkins[1]; Conor Davis[2]; Maura Carrai[2]; Michael Ward[2]; Susan O'Keeffe[3]; Martine van Boeijen[4]; Louise Beveridge[5]; Costantina Desario[6]; Canio Buonavoglia[6]; Julia Beatty[2]; Nicola Decaro[6]; Vanessa Barrs[2]
[1]University Veterinary Teaching Hospital Sydney; [2]Sydney School of Veterinary Science, University of Sydney; [3]School of Veterinary and Lifesciences, Murdoch University; [4]Perth Cat Hospital; [5]Bedford-Dianella Vet Centre; [6]Department of Veterinary Medicine, University of Bari

Multiple, epizootic outbreaks of feline panleukopenia (FPL) caused by feline parvovirus (FPV) were recorded in shelter-housed and owned cats in eastern Australia between 2014 and 2018. Most affected cats were unvaccinated. We hypothesised that low population immunity was a major driver of re-emergent FPL. The aim of this study was to (i) determine the prevalence and predictors of seroprotective titres to FPV among shelter-housed and owned cats, and (ii) compare the prevalence of seroprotection between a region affected by FPL outbreaks (Sydney, eastern Australia), and a region with no recent history of FPL outbreaks (Perth, western Australia). FPV antibodies were detected by haemagglutination inhibition assay on sera from 523 cats and titres ≥1:40 were considered protective. Socioeconomic indices based on postcode and Australian Bureau of Statistics census data were included in the risk factor analysis. The overall prevalence of protective FPV antibody titres was 94.3% and was similar between cats from outbreak (94.3%) and non-outbreak regions (94.2%). On multivariable logistic regression analysis vaccinated cats were 29.94 times more likely to have protective FPV titres than cats not known to be vaccinated. Cats from postcodes of relatively less socioeconomic disadvantage were 5.93 times more likely to have protective FPV titres. The predictors identified for FPV seroprotective titres indicate that support targeted vaccination strategies in regions of socioeconomic disadvantage would be beneficial to increase population immunity. The critical level of vaccine coverage required to prevent FPL outbreaks should be determined to support initiatives to prevent the reemergence of this frequently fatal disease.

Abstract ID12: Point Prevalence Surveys to Investigate a Veterinary Hospital Outbreak of Carbapenem Resistant *Escherichia coli*

Stephen Cole; Jaclyn Dietrich; Donna Oakley; Shelley Rankin
University of Pennsylvania

Reports of Carbapenem-resistant Enterobacteriaceae (CRE) are limited to sporadic cases in animals and there is therefore, a lack of outbreak investigation guidelines for veterinary facilities. As these organisms continue to emerge in veterinary medicine it is critical to evaluate the utility of outbreak investigation strategies. The CDC Facility Guidance for Control of Carbapenem-resistant Organisms suggests that serial point prevalence surveys (PPS) are useful to document ongoing transmission in healthcare facilities. In May 2019, the clinical microbiology lab at Matthew J. Ryan Veterinary Hospital at the University of Pennsylvania (MJR-VHUP) reported a cluster of NDM-5 producing *Escherichia coli* isolated from 15 animals to the Philadelphia Department of Public Health (PDPH). This represented the first report of a CRE outbreak in a US veterinary facility. The PDPH recommended a weekly PPS and due to lack of IACUC approval for the collection of rectal swabs this was performed using free-catch fecal samples for two weeks. Following review by a clinical ethicist, rectal swabs were collected from all hospitalized animals on the day of testing for the next 9 weeks. Specimens were screened using chromogenic agar. The production of carbapenemase was confirmed by mCIM. Epidemiologic data collected included; number of visits to MJR-VHUP and days hospitalized prior to sampling. PPS's were performed on fecal specimens from 16 of 21 animals and 14 of 22 animals in weeks one and two. All tested negative. Review of the protocol by a clinical ethicist stated that "if [PDPH] determine[d] that testing [was] necessary for all hospitalized patients…rectal swabs should be obtained…because it is imperative that sampling be thorough and expedient." Rectal swabs or fecal samples were obtained from 268 patients over 11 weeks. Fourteen animals (13 dogs and 1 cat) tested positive (5.2%). The average number of visits for positive patients was 10 (vs. 6 for negative) and the average number of days hospitalized prior to sampling was 5 (vs. 2 for negative). Two patients tested negative and then positive in a second PPS. Half (n=7) of the colonized patients were on their first visit to MJR-VHUP and 4/7 were hospitalized less than 48 hours prior to sampling. An important "lesson learned" is that all veterinary facilities should have policies in place to sample animals as part of a public health response. The PPS was a useful tool to document in-facility transmission for two cases, but positive animals detected within 48 hours of admission suggested a potential community source and dictated the need, in this outbreak, for admission and discharge testing.

Abstract ID13: Pilot Pharmacodynamic and Safety Study of Canine-Parvovirus-Neutralizing-Antibody (KIND-030a and KIND-030b) in Purpose Bred Puppies

Ellen R. Ratcliff[1]; Laurie Larson[2]; Allyson Avenatti[1]; Stephanie Pierce[1]; Tianhua Hu[1]; Melinda Poole[1]
[1]Kindred Biosciences, Inc.; [2]School of Veterinary Medicine, University of Wisconsin-Madison

The purpose of this proof-of-concept study was to determine the onset of detection, half-life, safety, and bioavailability of canine parvovirus neutralizing antibody (KIND-030a and KIND-030b). Both are chimeric, high-affinity monoclonal antibodies targeting canine parvovirus (CPV). Six unvaccinated, purpose-bred, female beagle puppies, 10 to 12 weeks of age, with negative CPV titers on Day 0 were randomized 1:1 to a single dose of 5mg/kg of either KIND-030a (n=3) or KIND-030b (n=3). A monovalent CPV vaccine was initially administered when the CPV hemagglutination inhibition (HI) titer dropped to 80 or lower as *in vivo* proof of viral neutralization and administered at least 2 more times. CPV HI titers on Day 0 were <20 prior to dosing in all 6 puppies and increased to greater than or equal to 2560 on Day 1 after KIND-030 administration. The half-lives of KIND-030a and KIND-030b were 10.1 days and 8.6 days, respectively. This half-life roughly matches maternally derived antibody. The duration of immunity, defined as the duration in which the HI titer was maintained at 80 or higher, ranged from 30 to 42 days after one dose of either KIND-030a or KIND-030b. The virus neutralization (VN) results mirrored the HI results. The initial CPV vaccination did not result in seroconversion in 5 out of 6 puppies, potentially due to residual KIND-030 antibodies neutralizing the vaccine. Blocking of vaccinal virus replication by both KIND-030a or KIND-030b was the first *in vivo* correlation to the *in vitro* neutralization data. Subsequently, three puppies seroconverted after the 2nd and one after the 3rd vaccination. The study was terminated on Day 100. In this proof-of-concept study, both KIND-030a and KIND-030b resulted in high titers, were well tolerated in puppies and no adverse events or injection site reactions were reported. This product is under clinical investigation and has not received approval by the USDA.

Abstract ID14: Using Big Data to Investigate Feline Intestinal Parasitism by Geographic Region and Age

Sarah L. Sweet[1]; Evan Hegarty[2]; Donald McCrann[2]
[1]IDEXX Inc.; [2]IDEXX Reference Laboratories

Intestinal parasite risk to pet cats is underrepresented in the literature and is often overlooked by pet owners and veterinarians. The purpose of this study is to examine the risk of intestinal parasitism in cats by presenting the percentage of positive results for fecal samples tested using zinc sulfate flotation by centrifugation and Fecal Dx® Antigen Panel, as well as combined positive results from both methods. A retrospective sample of 89,379 test results conducted using both testing techniques was gathered from IDEXX Reference Laboratory data from the continental United States (U.S.) in 2017 and 2018. Percentages of samples that tested positive were calculated according to testing-method, age, and region for roundworm, hookworm, and whipworm.* Percentages of positive test results are presented in Table 1. Roundworm was the most common intestinal parasite of cats in this study, while hookworm and whipworm were not commonly seen in these cats. This study reinforces the need for regular testing in younger cats, as cats less than 1 year of age were nearly five times more likely to be parasitized than cats 1 year of age or older when using flotation and antigen techniques combined. All U.S. geographic regions besides the West had notable percentages of positive test results, with approximately 1 in 20 cats testing positive for roundworm, hookworm, or whipworm. The percentage of samples that tested positive by antigen or by antigen in combination with flotation was appreciably higher than the percentage of samples that tested positive by flotation alone. This suggests that fecal antigen testing may be a more sensitive method for detection of these parasites, and the Companion Animal Parasite Council (CAPC) recommended combination of flotation with antigen testing may offer the greatest sensitivity. This study demonstrates that pet cats in the U.S. are at risk of intestinal parasitism and should receive regular fecal testing. *Hookworm: *Ancylostoma tubaeforme, Ancylostoma braziliense, Uncinaria stenocephala*; Roundworm (Ascarid): *Toxocara cati, Toxascaris leonine*; Whipworm: *Trichuris serrata* (formerly *Trichuris felis*).

Table 1. Percent of positive test results for intestinal parasites in U.S. feline fecal samples from 2017 and 2018

	Flotation	Antigen	Float+Antigen
Intestinal Parasite			
Roundworm	3.6%	4.9%	5.3%
Hookworm	0.4%	0.8%	0.9%
Whipworm	<0.1%	0.1%	0.1%
Age of Cat			
< 1 year old	8.4%	11.3%	12.2%
≥ 1 year old	1.3%	2.3%	2.6%
Geographic Region			
Midwest	4.9%	6.7%	7.4%
Northeast	4.6%	6.7%	7.1%
South	4.0%	5.6%	6.3%
West	1.2%	2.0%	2.2%

Abstract ID15: Pilot Study to Determine whether *Cytauxzoon felis* has Expanded into Colorado

Pierce Chan[1]; Michael Lappin[1]; Jennifer Hawley[2]
[1]Veterinary Teaching Hospital, Colorado State University; [2]Center for Companion Animal Studies

Cytauxzoon felis is a piroplasm that infects both domestic and wild felids, with bobcats being the definitive host. *Amblyomma americanum* is considered the most important vector of *C. felis* and confirmed cases of cytauxzoonosis in cats overlaps with the distribution of *A. americanum*. The historical range of *Amblyomma* ticks includes the south central and southeastern US, but recent data has shown that the range has been expanding. Therefore, cases could be missed due to historical low prevalence. The purpose of this study was to use a PCR assay to amplify *C. felis* DNA in the blood of cats with suspected infectious hemolytic anemia in Colorado. Remnant whole blood DNA samples previously tested through the CSU Diagnostic Laboratory were stored at -80°C until assayed in this study. Submissions were searched from June 2004 to December 2018 for cats in which an infectious cause of anemia was suspected. Only samples from cats living in Colorado with adequate DNA were assayed. A previously published quantitative PCR assay that co-amplifies the DNA of *C. felis* and *Babesia* spp. mitochondrial DNA was used for this study. Other PCR assay results that were previously performed for *Bartonella* spp. (29/98 samples) and *Ehrlichia* spp. (23/98 samples) were included in the analysis for this study. Samples were also tested for haemoplasmas (*Mycoplasma hemofelis*/*M. turicensis* and *M. haemominutum*) if not previously performed. Overall, DNA of an infectious agent was amplified from 13/98 (13.2%) samples. *Bartonella* spp. DNA was amplified from 1/29 (3.4%) samples and haemoplasma spp. DNA was amplified from 12/97 (12.3%) cats. *Babesia* spp., *Ehrlichia* spp., and *C. felis* DNA were not amplified in any sample. *Cytauxzoon felis* DNA was not amplified in any sample, although 13% of samples tested positive for an infectious agent. The results of this pilot study show no evidence that the range of *C. felis* has expanded to include Colorado.

Abstract ID16: Prevalence of *Ehrlichia canis* Antibodies in Dogs in Baja Sur, Mexico

Michael R. Lappin[1]; Cody Minor[2]; Arianne Morris[1]; Melissa Brewer[1]; Jennifer Hawley[1]; Amy Rex[3]
[1]Colorado State University; [2]Unknown; [3]Vet Clinic

Rhiphicephalus sanguineous is the most common tick in Baja Sur, Mexico and is known to be a vector for multiple pathogens including *Ehrlichia canis*, *Babesia vogeli*, *Anaplasma platys*, and *Rickettsia rickettsii*. The objectives of this study were to determine the prevalence of *E. canis* antibodies in dogs not currently treated with an acaracide and those provided a collar with 10% imidacloprid + 4.5% flumethrin (Seresto®; Bayer Animal Health) and tested again 180 days later. The study was performed in the Todos Santos area over two, 180-day periods. Owners were offered free testing for tick borne agents as well as flea and tick control products. A subset of owners with dogs that were negative for *E. canis* antibodies (IFA <1:80 or SNAP4DX Plus negative) and negative for *E. canis* DNA by PCR assay were given a collar and offered a financial incentive to return for recheck examinations and testing. Dogs with lost collars replaced within 7 days remained in the study. *Ehrlichia canis* antibodies were detected in 64.8% (79 of 122) and 43.2% (121 of 280) of the dogs on Day 0 of Period 1 and Period 2, respectively. Nineteen negative dogs in Period 1 and 60 negative dogs in Period 2 returned for all rechecks. Of these 79 dogs, three had *E. canis* antibodies (2 concurrently PCR positive) in the 180-day samples. For 2 dogs, the collar was missing for greater than 7 days and for one dog, the owner stated that the collar had been lost but was unsure whether it was replaced within 7 days. Antibodies against *E. canis* are common in dogs that are not treated with acaracides in this region of Mexico.

Abstract ID19: Etiology and Seasonality of Fecal Enteropathogens from Diarrheic Cats: A Retrospective Study of 1620 Cases

Ye-In Oh[1]; Kyoung-Won Seo[2]; Sung-Soo Kim[1]; Sanghyuk Seo[1]; Jaegon Ah[1]; Do-Hyung Kim[3]; Doo-Sung Cheon[4]
[1]VIP Animal Medical Center, Chungnam National University; [2]Chungnam National University; [3]Knotus Inc.; [4]Postbio Inc.

Diarrhea is one of the most common clinical symptoms encountered in cats and can be caused by infectious pathogens. In this retrospective study, we aimed to identify the etiology, clinical features, and seasonality of enteropathogens in diarrheic cats using quantitative polymerase chain reaction (qPCR). Comprehensive diarrheal qPCR panels were performed on fecal samples from 1620 cats with diarrhea in Republic of Korea from April 2016 to April 2019. Among 1620 diarrheic cats, the positive infection rate was 82.59%. Bacterial, viral, and protozoal infections accounted for 49.3%, 37.57%, and 13.13% of cases, respectively. Feline enteric coronavirus (FECV) was the most common pathogen (29.37%), followed by *Clostridium* (C.) *perfringens*, *Campylobacter* (C.) *coli*, feline parvovirus, and *Tritrichomonas foetus*. The seasonality of enteropathogens was observed with peaks as follows: bacterial infections peaked in October, viral infections peaked in November, and protozoal infections peaked in August. Viral and protozoal infections showed differences in prevalence according to patient age. Investigation of infection patterns revealed that the ratios of single infections, mixed infections, and co-infections were 35.72%, 9.87%, and 54.41%, respectively. FECV was predominant in single infections. The most common infection patterns were a combination of *C. perfringens* and *C. coli* in mixed infections and a combination of *C. perfringens* and FECV in co-infections. Infection patterns differed according to the enteropathogen species, seasonality, and age distribution in cats. The results of this study could be helpful in the diagnosis, treatment, and prevention of spreading infection in feline infectious diarrhea. In addition, continued monitoring of feline enteropathogens is required.

Abstract ID20: Vaccination History in Dogs with Acute Respiratory Disease Suspected to be Associated with Parainfluenza Virus

Cindy Sotelo[1]; Michael Lappin[2]; Ashley Gange[3]; Tracey Jensen[3]
[1]Colorado State University; [2]Internal Medicine Service, VTH, Colorado State University; [3]Wellington Veterinary Hospital

Parainfluenza virus is a cause of the canine infectious respiratory disease complex (CIRDC) and vaccines containing the virus are available for parenteral or intranasal administration. Parenteral parainfluenza containing vaccines are frequently recommended every 3 years after the initial vaccination series since it is usually combined with canine distemper virus and canine parvovirus which induce long term sterilizing immunity. A group of dogs were evaluated in a clinic in North central Colorado for acute clinical signs of disease consistent with CIRDC. Each dog had samples evaluated at commercial laboratories for nucleic acids of the common bacterial and viral agents associated with CIRDC. The 19 dogs were all sampled between December 20, 2018 and January 8, 2019 and were all suspected to be involved in the same CIRDC outbreak. Nucleic acid amplification rates for *Mycoplasma* spp., parainfluenza, *Bordetella bronchiseptica,* and adenovirus were 84%, 52%, 21%, and 5%, respectively. Of the 10 dogs with parainfluenza virus RNA, 90% were also positive for nucleic acids of *Mycoplasma* (8 dogs), *B. bronchiseptica* (2 dogs), or adenovirus (1 dog). All of the dogs positive for parainfluenza RNA had been administered an oral *B. bronchiseptica* vaccine with the time since the last dose ranging from 1.5 weeks to 23 months. Of the 10 dogs positive for parainfluenza virus RNA, the time since administration of the last parenteral vaccine containing this virus ranged from 16 months to 7 years. The patient that had parainfluenza RNA isolated on swab with no other pathogen isolated received parainfluenza vaccination (along with distemper and parvo) 1 year and 4 months prior to the respiratory panel. Parainfluenza virus was suspected as the cause of this CIRDC outbreak since none of the dogs had been recently vaccinated with products containing this virus to explain the positive molecular assay results and there was one patient positive for parainfluenza virus RNA alone. If this hypothesis is correct, it suggests that use of parenteral parainfluenza vaccines at an interval at greater than 16 months may not provide complete protection.

Abstract ID21: The Role of *Streptococcus equi* Subspecies *Zooepidemicus* and Influenza Viruses in URIs in Shelter Cats

Anna Winner; Hayley Clark; Jennifer Hawley; Michael Lappin
College of Veterinary Medicine and Biomedical Sciences, Colorado State University

Feline respiratory disease complex (FRDC) is characterized by clinical signs of respiratory and/or ocular disease in susceptible cats and is a prevalent problem in overcrowded settings like animal shelters. Included in the complex are a variety of pathogens, particularly feline herpesvirus-1 (FHV-1), feline caliciviruses (FCV), *Bordetella brochiseptica*, *Mycoplasma* spp., and *Chlamydia felis*. Recently, a number of cats have been shown to be infected by influenza A viruses or *Streptococcus equi* subspecies *zooepidemicus*, but little is known about the incidence of these potential pathogens in cats in shelters in Colorado. This study aims to determine the estimated prevalence rates for influenza A viruses and *S. zooepidemicus* to determine whether these are additional major contributing pathogens to FRDC. All cats in the study were housed in shelters in north-central Colorado and had acute clinical signs of FRDC. Archived DNA and RNA extracts (stored at -80°C) from 2008 (52 cats) and prospective samples collected from 24 cats in 2019 were included. All samples were assayed for nucleic acids of influenza A viruses and for *S. zooepidemicus*. In the experiment completed in 2019, while none of the samples from the 21 cats were positive for RNA of influenza viruses or DNA of *S. zooepidemicus*, nucleic acids of *C. felis* (4.2%), FHV-1 (8.3%), *Mycoplasma* spp. (20.8%), or FCV (37.5%) were amplified from many cats. In the experiment completed in 2008, while none of the samples from the cats tested were positive for RNA of influenza viruses, DNA of *S. zooepidemicus*, or *C. felis*, nucleic acids of FCV (2.1%), *Mycoplasma* spp. (58.1%), or FHV-1 (74%) were amplified from many cats. Both single and co-infections were detected in both sample sets. While influenza virus and *S. zooepidemicus* infections have been detected in dogs in the shelters used in this study, the infections did not appear to be associated with clinically ill cats during the study periods.

Abstract ID22: Evaluation of the Clinical Performance of a Point-of-Care *Coccidioides* Antibody Tests in Dogs

Krystle L. Reagan; Ian McHardy; George Thompson; Jane Sykes
University of California, Davis

Point-of-care (POC) *Coccidioides* antibody assays may provide veterinarians with a more rapid, patient-side diagnosis when compared with the serologic reference standard, immunodiffusion. The objective of this study was to determine the sensitivity and specificity of a POC lateral flow assay (LFA), sona *Coccidioides* (IMMY, Norman, OK), using the immunodiffusion assay (Coccidioidomycosis Serology Laboratory, University of California, Davis, CA) as the reference standard. Sera specimens were collected from client-owned dogs both prospectively and from stored specimens. Animals were classified as *Coccidioides* antibody positive (CP, n=36) based on a positive quantitative IgG immunodiffusion. Animals were *Coccidioides* antibody negative (CN, n=12) based on a negative immunodiffusion test. The sensitivity and specificity of the POC assay were calculated by comparing LFA results to the reference standard results. The sona *Coccidioides* antibody LFA assay correctly classified 32/36 CP specimens and 12/12 CN specimens resulting in a sensitivity of 88.9% (CI; 74.7–95.6%) and specificity of 100% (CI; 75.8–100%) as compared to immunodiffusion. A receiver operator characteristic curve was constructed, and the area under the curve was calculated to be 0.944 (CI; 0.880–1.000). Four false negatives were observed, all of which had low quantitative IgG immunodiffusion titers of 1:4. This POC assay is an alternative and rapid diagnostic tool to the traditional immunodiffusion assay. False negatives were observed in dogs with low quantitative immunodiffusion titers, therefore confirmatory testing is recommended.

Abstract ID24: *In Vitro* Inhibitory Effects of Polyammonium Bisulfate and Nanosulfur Against Canine Isolates of *Pythium insidiosum*

Cara E. Martin; Stuart Walton; Domenico Santoro; Justin Shmalberg
University of Florida College of Veterinary Medicine

Pythium insidiosum, an aquatic oomycete, causes life-threatening dermatologic and gastrointestinal disease in dogs. Infection is characterized by rapid clinical progression with a high mortality rate due to extensive, but typically localized, spread of disease. There is no consensus on standard treatment recommendations but complete surgical resection of affected tissues, where possible, and long-term adjunctive antifungal therapy are commonly advised. Variable response to treatment has prompted the use of other compounds and combination protocols as alternative options for the management of *Pythium insidiosum* infections in animals. Polyammonium bisulfate molecular clusters (PBMS) and nanosulfur are new compounds successfully used in agriculture to treat algae, fungi, and bacterial infections in plants and water reservoirs which are theorized to exert effects on oomycetes. The primary aim of this study was to investigate the *in vitro* activity of PBMS and nanosulfur against clinical isolates of *Pythium insidiosum* acquired from canine patients. Four isolates of *Pythium insidiosum* were obtained from cultures of infected canine tissue. Each isolate was incubated for 24 hours at 37°C in triplicate on 20% clarified v8 agar with and without antimicrobials (nanosulfur and PBMS). Three two-fold dilution of antimicrobials (nanosulfur and PBMS) were added to the agar immediately before plating. Nanosulfur was used at concentrations of 500, 1,000, and 2,000 μg/mL. Three solutions of PBMS containing 5% copper, 9.8% zinc, and 5% zinc with 2.5% copper were used at concentrations between 0.25 to 5 μg/mL of Cu and 0.5 to 9.8 μg/mL of Zn. Isolates not exposed to antimicrobials served as positive controls. After 24 hours, each isolate was re-implanted onto new plates to simulate repeat dosing. Radial growth of the colony was recorded in mm at 24, 48, and 72 hours. Colony diameter was calculated as the mean of two perpendicular measurements with the diameter of the mycelial plug subtracted. The median of the triplicates was used for statistical analysis. Kruskal-Wallis followed by the Dunn's multiple comparison test were performed. A p value of <0.05 was considered statistically significant. Colony growth was uniform among all the isolates. Compared to the positive control, PBMS (9.8% Zn) at 9.8 μg/mL significantly inhibited the growth after 48 hours (40.5% diameter reduction) and 72 hours (29.2% diameter reduction) ($p=0.005$ and $p=0.02$, respectively). Compared to the positive control, nanosulfur significantly inhibited the growth at 2,000 μg/mL after 72 hours (38.9% diameter reduction), at 1,000 μg/mL after 24 (43.8% diameter reduction) and 72 hours (38.9% diameter reduction), and at 500 μg/mL after 24 (46.9% diameter reduction), 48 (40.5% diameter reduction), and 72 hours (36.1% diameter reduction) ($p<0.05$). Radial growth of *Pythium* isolates was significantly inhibited by all concentrations of nanosulfur tested and by 9.8 μg/mL of PBMS 9.8% Zn. Re-implantation of isolates was designed to mimic more realistic circumstances in which antimicrobials would be administered repeatedly, assuming adequate absorption into infected tissue. Inhibition of growth appeared to be most pronounced at earlier time points. Results suggest that nanosulfur and PBMS with Zn may be promising adjunctive therapies for the treatment of *Pythium insidiosum*. Additional investigation is necessary to determine whether nanosulfur or PBMS 9.8% Zn improve growth inhibition when used in combination protocols with other previously-studied agents such as mefenoxam, caspofungin, itraconazole, and terbinafine.

This page intentionally left blank.

Abstract IM01: RNA Sequencing of Dogs with Primary Immune-Mediated Hemolytic Anemia

Corie Borchert[1]; Aimee Brooks[2]; Adam Herman[3]; Steven Friedenberg[1]
[1]University of Minnesota, Twin Cities; [2]Purdue University; [3]Minnesota Supercomputing Institution

The immunologic and cellular mechanisms that trigger the onset of immune-mediated hemolytic anemia (IMHA) in dogs are largely unknown. Given the significant morbidity and mortality caused by IMHA in dogs, further investigation into its underlying etiology is warranted. For many diseases, transcriptome-wide gene expression studies have proven to be a powerful tool to inform our understanding of disease pathophysiology and identify novel therapeutic targets. In this study, we hypothesized that RNA sequencing of dogs with IMHA would reveal marked dysregulation of genes and gene pathways related to erythropoiesis and immune system function. We collected whole blood samples in an RNA stabilizing medium from 18 dogs with a new diagnosis of primary IMHA and an equal number of breed- and age-matched controls from two tertiary care referral institutions. Total RNA was extracted from each sample, depleted of hemoglobin and ribosomal RNA using sequence-specific capture probes, barcoded, pooled, and sequenced in a 50 base-pair paired-end read configuration to a target depth of 60 million reads per sample. Analysis of differentially expressed genes between cases and controls revealed greater than 8-fold (B-H $p<0.05$) upregulation of many genes related to erythrocyte production, the complement system, and neutrophil function in affected dogs. In particular, several genes encoding for erythrocyte membrane proteins which have been associated with hemolytic anemia in humans were significantly upregulated, as well as scramblases which translocate phospholipids across erythrocyte cell membranes. Unexpectedly, many genes and pathways related to lymphocyte function were markedly downregulated, which could point to differences in the activation states of circulating vs. organ-resident lymphocytes in IMHA patients. Gene co-expression network analysis identified groups of genes that are strongly associated with changes in reticulocyte ($p=5\times10\text{-}6$), neutrophil ($p=6\times10\text{-}5$), and eosinophil ($p=2\times10\text{-}6$) count in affected dogs. Further studies are warranted to validate these findings in a larger population of affected dogs, and to determine which dysregulated genes or pathways might be potential targets for therapeutic intervention.

Abstract IM02: Proteomic Analysis of Canine Vaccines

George E. Moore[1]; Jackeline Franco[2]; Uma Aryal[2]; Harm HogenEsch[2]
[1]College of Veterinary Medicine, Purdue University; [2]Purdue University

Canine vaccines provide protection against important zoonotic and non-zoonotic infectious disease by stimulating immune responses to vaccine antigens. Unintended or adverse immune responses may develop, however, against protein antigens that are present in the vaccine but do not originate from the pathogen(s) for which the vaccine was designed. Dogs with adverse reactions to vaccines have demonstrated immunoreactivity to proteins of bovine origin, e.g., from fetal calf serum, but specific protein components of canine vaccines have not been elucidated. The objective of this study was to use proteomic analysis to qualitatively and quantitatively identify protein components of commonly used canine vaccines. The vaccines analyzed were purchased commercially and produced by 4 major manufacturers. The vaccines included 8 rabies vaccines (four with 1-year duration of immunity and four with 3-year duration of immunity), 9 multivalent distemper-adenovirus 2, parvovirus, parainfluenza, and leptospirosis vaccines, and 7 Lyme (*Borrelia burgdorferi*) vaccines. For vaccines designed to be reconstituted, the lyophilized and liquid components were analyzed separately. This resulted in 29 products for proteomic analysis; all vaccines, lyophilized, and liquid components were processed in duplicates. Proteins were acetone precipitated, reduced with 10 mM dithiothreitol, alkylated with 20 mM iodoacetamide and then digested with a trypsin and Lys-C mix before liquid chromatography-tandem mass spectrometry (LC-MS/MS) analysis using the UltiMate 3000 RSLC nano System coupled to the Orbitrap Fusion Lumos Mass Spectrometer. LC-MS/MS data were searched against *Bos taurus* protein database using MaxQuant (version 1.6.3.3) for protein identification and label-free relative quantitation (LFQ intensities). At MS/MS spectral counts >5, there were 357 different proteins identified (molecular weight range 5–587 kDa). All rabies vaccines had significantly (>2–8 fold) more proteins identified than were present in the other types of vaccines. Bovine serum albumin (BSA-69 kDa) was the most common protein identified but varied in quantity by vaccine and manufacturer. Other commonly identified proteins, also varying by vaccine and manufacturer, included α-2-HS-glycoprotein (38 kDa), α-1-antiproteinase (46 kDa), and α-fetoprotein (68 kDa). Proteomic analysis of vaccines identified specific, yet diverse, proteins that may initiate potentially beneficial or adverse immune responses following vaccination.

Abstract IM03: Gene Expression of Immunoinflammatory and Immunological Status of Obese Dogs Before and After Weight Loss

Vivian Pedrinelli[1]; Thiago Vendramini[1]; Henrique Macedo[1]; Andressa Amaral[1]; Mariana Rentas[1]; Matheus Macegoza[1]; Rafael Zafalon[1]; Lígia Mesquita[1]; Julio Balieiro[1]; Karina Pfrimer[1]; Raquel Pedreira[2]; Cristiana Pontieri[2]; Cristina Gomes[1]; Marcio Brunetto[1]
[1]University of Sao Paulo; [2]Grandfood Industry and Commerce Ltd

Adipose tissue actively participates in inflammation and immunity, and several defense cells of the organism may, therefore, be involved in the diversity found between obese and ideal weight individuals. Thus, the present study aimed to evaluate the gene expression profile of immunoinflammatory response and the lymphoproliferation of obese dogs before and after weight loss. Eight female dogs, neutered, aged between 1 and 8 years, obese, with body composition determined by the deuterium isotope dilution method were included. The obese dogs were enrolled in a weight loss program and after losing 20% of their initial weight became a second experimental group. A third experimental group consisted of eight female dogs, neutered, aged between 1 and 8 years and with ideal composition and body condition score ideal. Procedures were previously approved by the Animal Use Ethics Committee (AUEC) of the School of Veterinary Medicine and Animal Science of the University of São Paulo (protocol number 4668091214). Gene expression of immunoinflammatory cytokines (resistin, leptin, adiponectin, TNF-α, IL-6, IL-8, and IL-10) was assessed by qRT-PCR and immunity was assessed by lymphoproliferative response using the flow cytometry technique. The obese group presented increased gene expression of resistin, adiponectin, and IL-8 in relation to the weight loss group. Weight loss resulted in an increase in the lymphoproliferation rate (18.48%) compared to obese dogs at the beginning of the study (10.71%) (p=0.004). These results indicate that weight loss modulates the immunoinflammatory response of obese dogs and may present important benefits to health and longevity of dogs.

Abstract IM04: Effect of Distemper-Adenovirus-2-Parainfluenza-Parvovirus Vaccination on Platelet Numbers and Development of Anti-Platelet Antibodies in Healthy Dogs

Maggie Williams; Michael Lappin; Nida Chornarm; Matt Khorsand; Melissa Brewer; Jennifer Hawley; Sarah Shropshire
Colorado State University

Immune thrombocytopenia (ITP) can occur when platelets are destroyed as a result of antibody-mediated mechanisms. In humans, ITP has been associated with the measles, mumps, and rubella (MMR) vaccine. Canine distemper, adenovirus-2, parainfluenza, and parvovirus (DA2PP) vaccines contain a similar paramyxovirus virus to those in MMR. A link between vaccination with DA2PP and thrombocytopenia has been proposed although a recent retrospective study failed to find a correlation with clinical ITP and recent vaccination. There is increasing concern for a link between vaccination and immune-mediated conditions in dogs but a causal relationship between vaccination and ITP, determined by anti-platelet antibodies (APA), has not been investigated. The objective of this study was to determine if administration of modified live DA2PP vaccines induces a thrombocytopenia and development of APA in a population of healthy, client-owned dogs. Thirty client-owned healthy adult dogs were randomly divided into 1 of 3 DA2PP vaccine groups (n=10/group). On day 0, all dogs had a physical examination, CBC, chemistry, and APA flow cytometry followed by DA2PP vaccination. On day 10, repeat platelet count, mean platelet volume (MPV), and APA flow cytometry were performed. On day 0, one dog was positive for APA (26.1% IgG) with normal platelet count and was excluded from the study; all other dogs were negative for APA (<10% IgG) and had normal platelet counts. On day 10, no dogs were thrombocytopenic or had clinical signs of bleeding. From day 0 to 10, platelet counts increased by a mean of 43 k/uL (range -36 to 214; SD 48.7, p<0.001) and percent IgG increased by mean of 1.72% (range -4.13 to 6.72, SD 2.9, p=0.0037). One dog became weakly positive for APA (10.7% IgG; weak positive range 10–14%) but had an increase in platelet count and the remaining dogs were negative for APA. There was no significant change in MPV. In this population of healthy, client-owned dogs, vaccination with commercial DA2PP vaccines was not associated with development of thrombocytopenia or clinically relevant APA formation. Additional longitudinal studies evaluating platelet count and APA after vaccination are warranted.

Abstract NM01: Comparison of Essential Nutrients in Plant-Based Pet Foods with Nutrient Requirements of Dogs and Cats

Sarah A.S. Dodd[1]; Anna-Kate Shoveller[1]; David Ma[1]; Andrea Fascetti[2]; Zhenshou Yu[2]; Adronie Verbrugghe[1]
[1]University of Guelph; [2]University of California–Davis;

Some pet owners choose to feed plant-based diets to their facultative carnivorous dog or obligate carnivorous cat, either as part of (7.9% of pet owners) or the complete (1.3% of pet owners) diet. Previous evaluations of essential nutrient content of these diets have been limited, thus the objective of this study was to investigate the dietary content of multiple essential nutrients in plant-based diets. Proximate analyses, amino acids, fatty acids, minerals (not including selenium and iodine), and vitamins A, B_{12} and D_3 were measured in accordance with AOAC methods in single samples of all plant-based pet foods commercially available in Canada at the time the study was conducted. Depending on analytical variance, analyses were performed in singlet (proximates, minerals, vitamins), duplicate (fatty acids), or triplicate (amino acids). Twenty-six (18 dog, 5 cat, 3 both; 18 dry, 8 wet) products were included. All foods produced in the USA (11) and 2 of 4 foods produced in Canada had an AAFCO adequacy statement on their label. No foods produced in Europe (11) had an AAFCO statement on their label, though indications of completeness and/or balance in accordance with FEDIAF were present. Essential nutrient analyses of all diets were compared to NRC nutrient requirements for maintenance or growth, for their intended species and life-stage as labelled. Twenty percent (4 of 20) of canine and none (0 of 8) of feline adult or all life stage diets met all requirements for adult maintenance. No canine (0 of 7) or feline (0 of 5) growth or all life stage diets met all requirements for growth. One canine growth diet met requirements for adult maintenance. Estimates of nutrient density were corrected to account for inter-laboratory variation and those failing to comply with recommended allowance are presented in Table 1. Most (22/26, 85%) diets failed to meet the nutrient profile for which they were intended. Of greatest concern were failures to meet the calcium, phosphorus and vitamin D3 requirements of growing dogs and cats, and sulfur amino acid requirements of both growing and adult dogs and cats. Formulation of plant-based diets for pets must focus on inclusion of these essential nutrients to meet the requirements of the animals for which they are intended to be fed, and animals being fed plant-based diets should be closely monitored for signs of malnutrition, particularly associated with sulfur amino acid and bone metabolism, such as skeletal deformities, abnormal growth, and failure to thrive.

Table 1. Range of percent (%) of recommended allowance met for species and lifestage intended. Bolded values represent percentages below or above recommended allowances. Zero values indicate nutrients below detection levels, which were below the minimum requirement.

Nutrient	Canine maintenance	Canine growth <14 weeks old	Canine growth >14 weeks old	Feline maintenance	Feline growth
Methionine	**65**–398	129–375	173–505	260–**865**[*]	106–334
Methionine + cystine	**39**–276	**83** - 256	110–338	177–566	**67**–219
Taurine	ND	ND	ND	**77**–189	**84**–211
Total fat	155–457	100–288	100–288	**95**–162	**95**–162
Linoleic acid	131–**1360**[**]	131–**894**[+]	131–**894**[+]	255–**1936**[++]	603–**1936**[++]

Arachidonic acid	ND	ND	ND	108–276	**32**–83
EPA+DHA	**0–83**	**0–33**	**0–33**	**0–55**	**0–55**
Calcium	**96**–473	**39**–182	**39**–182	160–490	**58**–178
Phosphorus	177–496	**63** - 112	**63** - 112	242–412	**81**–149
Sodium	**54**–1603	123–315	123–315	251–1062	225–516
Chloride	**73**–2942	**78**–485	**78**–485	387–1524	**87** -163
Zinc	**74**–399	**93**–359	**93**–359	120–485	119–479
Vitamin A	**29**–459	**84**–459	**84**–459	**75**–696	**75**–696
Cholecalciferol	**0**–450	**0**–564	**0**–564	**0**–446	**0**–693
Vitamin B_{12}	**0**–19735	**0**–19735	**0**–19735	**0**–31862	**0**–31862

ND=recommended allowance not determined
*Exceeds safe upper limit (SUL) by 13%; **Exceeds SUL by 130%; +Exceeds SUL by 51%; ++Exceeds SUL by 94%

Abstract NM02: Evaluation of Nutrient Content and Caloric Density in Commercially Available Foods Formulated for Senior Cats

Stacie Summers[1]; Jonathan Stockman[2]; Anais Sanchez Rodriguez[2]; Lei Zhang[2]
[1]Oregon State University; [2]Colorado State University

Despite the lack of determined nutritional guidelines for senior cats, multiple commercial cat food products that are marketed for senior cats are available to owners. Although these products adhere to the Association of American Feed Control Officials (AAFCO) adult guidelines, the variability in the protein, fat, caloric density, and mineral content of these diets are unknown. The aim of the study was to measure the caloric density and caloric distribution of protein, fat, and select minerals in commercially available foods labeled for senior cats (≥7 years) and compare results to adult cat foods. Thirty-one senior cat foods and 59 adult cat foods (both canned and dry) were purchased from pet stores. Two lots were available and purchased for 26/31 senior cat foods. Food analysis was performed using official methods of the Association of Official Agricultural Chemists International. All food samples were analyzed for complete proximate (moisture, crude protein, crude fat, crude fiber, ash) and select minerals (phosphorus, calcium, potassium, sodium). Measured crude protein, fat, mineral percentages were converted to g/1,000 kcal metabolizable energy (ME) using modified Atwater factors for protein, fat, and carbohydrate to allow for comparisons among study foods. Caloric density was calculated based on dry weight. Descriptive statistics are presented as median and range. An unpaired t test or Mann-Whitney test and one-way ANOVA or Kruskal-Wallis test was used to compare caloric density and mineral contents between groups. A p-value<0.05 was considered significant. The average of the 2 lots were used for statistical analysis. All senior cat foods were formulated to meet the AAFCO nutrient profile for adult maintenance. For the adult cat foods, 31 diets had adequacy statements for AAFCO nutrient profile for adult maintenance, and 28 diets had statements for all life stages. No significant difference in the analyzed caloric density or mineral content of senior cat foods was found when compared to adult cat foods and cat foods formulated for all life stages. In conclusion, senior cat foods had highly variable caloric densities and mineral content. Caution should be taken when making broad recommendations for the use of commercially available senior foods in cats.

	Adult foods n=59	Senior foods n=31	P-value
Caloric density kcal/100 grams	413 (359–635)	406 (337–505)	0.204
Protein g/1,000 kcal ME	100 (50–173)	94 (77–139)	0.147
Fat g/1,000 kcal ME	48 (26–90)	46 (33–79)	0.452
Phosphorus g/1,000 kcal ME	3.1 (0.9–5.8)	3.2 (1.5–4.4)	0.636
Calcium g/1,000 kcal ME	3.9 (0.8–8.7)	3.7 (1.9–5.7)	0.768
Calcium:phosphorus ratio	1.3 (0.8–1.7)	1.3 (0.9–1.6)	0.605
Magnesium g/1,000 kcal ME	0.27 (0.06–0.54)	0.28 (0.12–0.56)	0.591
Potassium g/1,000 kcal ME	2.2 (1.2–3.9)	2.3 (1.7–3.6)	0.131
Sodium g/1,000 kcal ME	1.1 (0.5–3.9)	1.1 (0.7–2.7)	0.913

Canine Obesity Knowledge Among Brazilian Owner's Is Associated to Overweight Risk in Dogs

Mariana Y. Porsani; Fabio Teixeira; Vinicius Oliveira; Gabriel dos Santos; Tatiane Pooli; Pamela Vasconcerva; Marcio Brunetto
School of Veterinary Medicine and Animal Science, USP, Brazil

This study aimed to assess the canine obesity knowledge and body condition score (BCS) perception among dog owners in Brazil.
A face-to-face questionnaire was applied to owners (n=926). The dog's BCS (1 to 9) were evaluated by veterinary nutricionists and owners. The information was associated with the chi-square test, and the odds ratio test (OR).
Dogs (n=926) were divided into two groups: ideal weight (IW) (BCS 4 and 5/9)=51.8% (480/926); and overweight (OvW) (BCS ≥6/9) 48.2% (446/926), equally, owners were classified according to their dog's group: 62.5% of owners of IW dogs (300/480) and 74.4% of OvW dogs (332/446) believed that snacks are related to their dog's weight gain, this answer was associated with BCS group ($p<0.001$) and resulted in OR 1.747 to dogs be overweight. Acknowledge that obesity is a health problem was reported by 97.1% of IW owners and 80.9% OvW, it was associated with BCS ($p <0.001$) and resulted in OR 2.06 of being overweight. The majority (97.1%) of owners would be willing to enroll their dog's into a weight loss program if recommended by a veterinarian, however, 50.5% of OvW and 63.6% of IW underestimate their dog´s BCS ($p<0.001$). Belief that supplements are necessary for weight loss resulted in OR 1.537 and it was associated with overweight dogs.
The owner's knowledge regarding obesity was associated with a higher risk of dogs being overweight.

Abstract NU01: Urine-Specific Gravity is Less Predictive than Nutritional Factors for Management of Urolithiasis in Cats

Jennifer M. MacLeay[1]; Mikaleigh Woodward[2]; Stephen Davidson[1]; Kathy Gross[1]
[1]Hill's Pet Nutrition, Inc.; [2]KSU

Urolithiasis due to calcium oxalate (CaOx) or struvite (MAP) are common causes of lower urinary tract disease in cats. A common recommendation is to monitor urine-specific gravity (USG) as a surrogate for relative supersaturation calculation (RSS) and calcium oxalate titration testing (COTT) as these tests are not commercially available. However, the refractometer is limited in sensitivity to small molecules that may have a profound impact on urine performance. In this study a meta-analysis of 41, two or four week long monadic feeding tests was undertaken to determine the correlation between RSS MAP, RSS CaOx or COTT and USG. All methods were approved by an internal animal care and use committee. Tests encompassed 32 dry, 7 wet and 2 combined wet and dry foods. Each test was comprised of 10 or 12 healthy adult cats 1–11 years old. At the end of each test freely voided urine was collected over 24 hours using a closed system with inert litter box beads. Urine pH was measured using a calibrated meter and USG by refractometer. The COTT was performed by titrating whole urine with sodium oxalate until precipitation occurred according to the method described by Davidson *et al.* (*JVIM* 2015). RSS CaOx and MAP were calculated using EQUIL software from urinary concentrations of NH_4^+, Ca+2, Cl^-, citrate, creatinine, Mg^{+2}, oxalate, PO_4^{-3}, K^+, Na^+, SO_4^{-2} and urine pH. Correlation tests were performed using Proc Corr in SAS. Significant differences were set at $p<0.05$. Across these 41 studies, age and USG were not correlated. RSS for MAP and CaOx were significantly correlated with USG but the r values were low (MAP $r=0.19$, $p<0.0001$ and CaOx $r=0.11$, $p=0.02$). COTT was not significantly correlated with USG ($r=-0.04$, $p=0.44$). Across the entire data set, USG ranged from 1.016–1.084 and fold differences in urine analyte concentrations were broad; Ca 86x, Cl 31x, Mg 80x, P 35x, K 15x, Na 132x, oxalate 8x, Sulfate 15x, citrate 373x, pH 0.68x, and creatinine 6x. It was observed that minimum values for dry and wet foods were similar for USG, RSS MAP, RSS CaOx and COTT. Further, USG for the 7 wet formulas were not uniformly low (1.020, 1.027, 1.028, 1.035, 1.037, 1.044, 1.046) and lastly, 2 foods with different nutritional profiles were fed as both dry and wet forms and the wet forms had reduced USG (wet, dry food A 1.044,1.061 and food B 1.020,1.054), RSS CaOx and COTT but RSS MAP was similar. In conclusion, a meta-analysis of 41 cat urinary studies showed that nutrition has a profound impact on the concentration of urine constituents and use of USG as a surrogate for RSS or COTT is not supported as dry foods can induce urine with low RSS and COTT. Practitioners must continue to rely upon information from manufacturers concerning how any food impacts urine mineral concentration, pH and, in turn, stability. However, a subset of this data suggests monitoring USG is most useful after instituting nutritional therapy with a food shown to reduce oxalate risk.

Abstract NU02: Voided Urine Collection Methods for Urine Culture in Female Dogs with Lower Urinary Tract Signs

Lynn Little[1]; Michael Childress[1]; Ben Olson[2]; Audrey Ruple[1]; Sarah Steinbach[1]; Lynn Guptill[1]
[1]Purdue University; [2]Pieper Memorial

Cystocentesis is currently the only recommended collection method for urine culture in female dogs, presenting a diagnostic challenge in patients for which cystocentesis is contraindicated, such as transitional cell carcinoma or thrombocytopenia. We hypothesized that culture results of clean, midstream voided urine samples would predict culture results of samples obtained via cystocentesis or cystoscopy if quantitative interpretive cut-offs were applied. Fifty-three female dogs experiencing lower urinary tract signs were prospectively enrolled with informed client consent at Purdue University's Veterinary Teaching Hospital. All dogs had 3 urine samples collected and submitted for urine culture within 24 hours: standard mid-stream voided (SMV), clean mid-stream voided (CMV), and a sterile sample obtained by cystocentesis or cystoscopy (CC). One or more cultures were positive for bacterial growth in 37/53 (70%) patients, however, only 40% (21/53) of CC cultures were positive. Comparing CMV and CC samples, growth of the same organism(s) or no growth on either culture occurred in 64% of samples (34/53). Comparing SMV and CC samples, the same result occurred in only 27/53 (51%). When SMV and CMV samples with quantitative cultures (35/53) were assessed for significant bacteriuria (≥105 CFU/ml based on current guidelines), the CC sample result could be predicted from the SMV sample in 80% (28/35) of patients and the CMV sample in 83% (29/35). While sterile sampling methods remain the gold standard method for urine culture in female dogs, when interpretive cut-offs are applied, voided samples may be an acceptable alternative in patients for which cystocentesis is contraindicated.

Abstract NU03: Reticulocyte Indices for the Assessment of Iron Status in Cats with Chronic Kidney Disease

Adeline Betting; Thierry Francey; Ariane Schweighauser
Vetsuisse Faculty, University of Bern

Gastrointestinal bleeding, erythropoietin therapy, and chronic hematuria can lead to iron deficiency (ID) in cats with chronic kidney disease (CKD), especially in those with subcutaneous ureteral bypasses (SUB). However, the assessment of the iron status is poorly defined in this species. The reticulocyte indices, reticulocyte hemoglobin content (CHr) and reticulocyte mean cell volume (MCVr), have been proposed to reflect the iron status of humans and dogs, but they have not been evaluated extensively in cats. The objective of this study was therefore to assess their diagnostic value for ID in cats with CKD with or without SUBs. Sixty-two cats were prospectively enrolled (15 CKD, 31 SUB, 16 healthy controls). All cats were characterized with a physical examination, abdominal radiographs, blood and urinalysis, and a complete iron panel (total serum iron, ferritin, total iron binding capacity). The transferrin saturation (TSAT) was calculated and used to define ID as TSAT<20%. ID was diagnosed in 9/62 (15%) cats, including 3/15 (20%) cats with CKD, 5/31 (16%) cats with a SUB and 1/16 (6%) healthy cats; 11/62 (18%) cats were anemic. Median TSAT was 26.4% (range, 16.1–46.8%) in cats with CKD, 26.1% (10.0–41.3%) in cats with SUBs, and 28.1% (18.1–52.2%) in healthy cats. Cats with ID had a lower CHr (15.6 pg; interquartile range, 14.7–17.0) than cats without ID (17.7 pg; 16.6-18.7; p=0.016), with a large overlap between the groups. The correlation between CHr and TSAT was moderate (r=0.37). A ROC curve analysis yielded a moderate discriminatory value for CHr in the diagnosis of ID (AUC=0.75; p=0.006) and compared favorably to the currently available variables MCV (AUC=0.58; p=0.458) and MCHC (AUC=0.71; p=0.023). It was however a weaker predictor of ID than the total iron concentration (AUC=0.97, p<0.001). MCVr was not different between the groups (p=0.226). Based on these results, CHr and MCVr do not seem to add diagnostic value to the currently available hematology variables in cats with CKD. Despite its limitations, total serum iron concentration was the best alternative to a complete iron panel in our study population.

Abstract NU04: Association Between Hyperlipidemia and Lower Urinary Tract Calcium Oxalate Stones in Dogs: Initial Results

Mathieu V. Paulin[1]; Marilyn Dunn[2]; Catherine Vachon[2]; Bérénice Conversy[2]
[1]Faculty of Veterinary Medicine, University of Montreal, Canada; [2]Montréal's Faculty of Veterinary Medicine in Saint-Hyacinthe

In people, obesity and metabolic syndrome have both been associated with formation of calcium oxalate stones (CaOxS). The goal of our study is to evaluate the association between obesity and hyperlipidemia (total cholesterol-TC and triglycerides-TG) and CaOxS formation in dogs. Dogs presenting for lower urinary tract stone removal with confirmed CaOxS (S-dogs) were prospectively enrolled. Control dogs were deemed healthy based on blood work, urinalysis and imaging ruling out CaOxS (SF-dogs). Body condition scores (BCS, 1–9 scoring scale), serum TC and TG levels were recorded. Statistical analyses included-[A] or not-[B] Schnauzers and comorbidities (Mann-Whitney-Wilcoxon and Fisher's exact tests). Seventy-one S-dogs and 33 SF-dogs (76/36% of males, median ages 9/6.5 years old, median weights 10.5/7 kg, respectively) were included in the study. The S-group was composed of Miniature Schnauzers (20%), Shih Tzus (17%), Yorkshire Terriers (14%) and 12 other breeds, and the SF-group of 15 breeds without any Schnauzers. In the S-group, comorbidities included hyperparathyroidism (3 dogs), hypothyroidism (2), hyperadrenocorticism (2), and diabetes mellitus (1). Calcium oxalate stones were primarily removed by minimally invasive procedures. Body condition score was significantly higher in S-dogs than SF-dogs ($p=0.02$–[A]). Thirty-eight percent [8/21] of hypertriglyceridemic S-dogs were Schnauzers. Hypertriglyceridemia (>2.38 mmol/L) was significantly more frequent in S-dogs ($p=0.0001$–[A]; 20 versus 0%, $p=0.005$–[B]). Median TC ($p=0.01$–[A]; 6.5 versus 5.6 mmol/L, $p=0.03$–[B]; Figure 1) and TG levels ($p<0.0001$–[A]; 0.91 versus 0.49 mmol/L, $p=0.001$–[B]; Figure 2) were significantly higher in S-dogs. Our study suggests that BCS and hyperlipidemia are significantly associated with lower urinary tract CaOxS in dogs.

Figure 1: Total cholesterolemia in dogs with calcium oxalate lower urinary tract stones (S-dogs) and healthy dogs (SF-dogs) - medians are reported

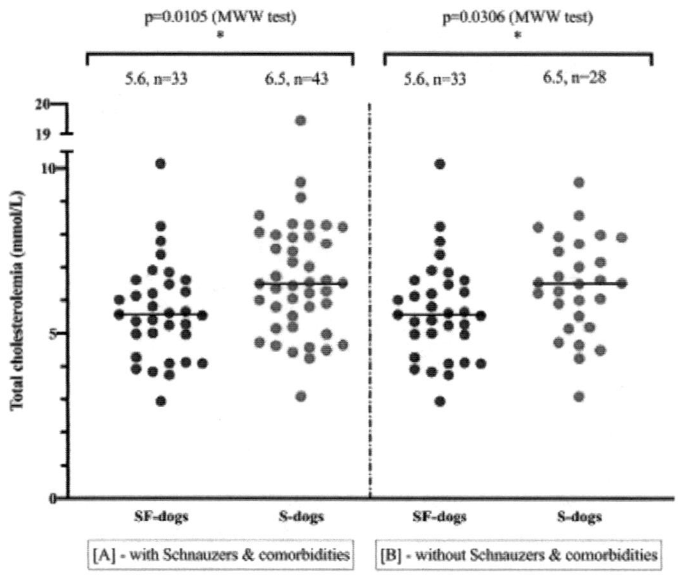

Figure 2: Triglyceridemia in dogs with calcium oxalate lower urinary tract stones (S-dogs) and healthy dogs (SF-dogs) - medians are reported

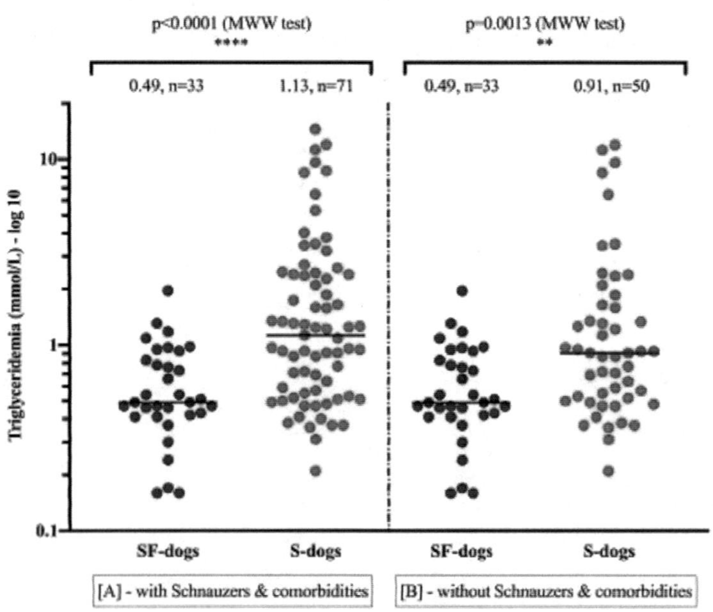

Abstract NU05: Predicting Early Risk of Chronic Kidney Disease in Dogs

JoAnn Morrison[1]; David Aucoin[2]; Richard Bradley[3]; Geert De Meyer[3]; James Kennedy[4]; Minseung Kim[5]; Yiannis Kokkinos[5]; Theodoros Panagiotakos[5]; Ilias Tagkopoulos[5]; Phil Watson[3]
[1]Banfield Pet Hospital; [2]Antech Diagnostics; [3]Waltham Petcare Science Institute; [4]Mars Advanced Research Institute; [5]PIPA LLC

Study purpose: Advanced machine learning methods combined with large sets of health screening data provide opportunities for diagnostic value in human and veterinary medicine. The purpose of this study was to derive a model to predict the risk of dogs developing chronic kidney disease (CKD) using data from electronic health records (EHRs) collected during routine veterinary practices at primary care hospitals in the United States. Methods: A total of 55885 dogs that visited Banfield Pet Hospitals between February 1, 1996 and March 31, 2018 were included in the study. Longitudinal EHRs from Banfield Pet Hospitals were extracted and randomly split into 2 parts. The first 66% of the data were used to build a prediction model, which included feature selection and identification of the optimal neural network type and architecture. The remaining unseen EHRs were used to evaluate the model performance. Results: The final model was a recurrent neural network (RNN) with 6 features (creatinine, blood urea nitrogen, urine-specific gravity, urine protein, weight and age). When predicting CKD near the point of diagnosis, the model displayed a sensitivity of 91.4% and a specificity of 97.2%. Model sensitivity decreased when predicting the risk of CKD with a longer horizon, having 69.2% sensitivity 1 year before diagnosis and 44.9% 2 years before diagnosis, but with specificity remaining around 97%. Conclusions reached: The use of models based on machine learning can support veterinary decision making by improving early identification of canine CKD. Considering the prognosis of CKD in canines, earlier identification could allow for greater opportunities for interventions and positive impacts on prognosis and patient outcomes.

Abstract NU06: Alterations of Serum Amino Acid Profiles in Cats with Chronic Kidney Disease

Stacie Summers[1]; Amanda Blake[2]; Jessica Quimby[3]; Jonathan Lidbury[2]; Joerg Steiner[2]; Jan Suchodolski[2]
[1]Oregon State University; [2]Texas A&M Gastroenterology Lab; [3]The Ohio State University

Chronic kidney disease (CKD) is common in senior cats and is associated with dysrexia, weight loss, cachexia, and accumulation of harmful metabolites of protein fermentation by colonic bacteria. Amino acid derangements may be a therapeutic target in CKD cats to reduce loss of muscle mass. The objective of our study was to measure serum amino acid concentrations in cats with CKD and compare results to healthy senior cats. Serum was collected from healthy senior cats (≥8 years; n=16) and CKD cats (International Renal Interest Society (IRIS) stage 1, n=2; stage 2, n=9; stage 3, n=11; stage 4, n=4). After deproteinization and filtration, the serum was analyzed on a Biochrom 30+ Amino Acid Analyzer to obtain concentrations of amino acids and other nitrogenous compounds. Unpaired student t test or Mann-Whitney test was used to compare amino acid concentrations between two study groups. A Spearman rank was used to assess correlation between serum amino acid and creatinine concentrations. Cats with CKD had significantly decreased concentrations of the essential amino acids leucine ($p=0.010$), methionine ($p=0.014$), phenylalanine ($p=0.004$), threonine ($p=0.003$), tryptophan ($p=0.002$), and valine ($p=0.040$), and the nonessential amino acids serine ($p=0.031$), tyrosine ($p=0.001$), proline ($p=0.005$), and asparagine ($p=0.046$) when compared to healthy senior cats. The essential amino acid taurine ($p=0.002$) and the nonessential amino acids citrulline ($p=0.002$) and aspartic acid ($p=0.037$) were increased in CKD cats when compared to senior cats. 3-methylhistidine, an indicator of skeletal muscle protein breakdown, was significantly increased in CKD cats ($p<0.0001$) when compared to senior cats and was positively correlated with serum creatinine ($p<0.0001$; $\rho=0.858$). In conclusion, the serum amino acid profile is altered in cats with CKD. 3-methylhistidine may be a biomarker of skeletal muscle protein breakdown in CKD cats but since the amino acid is excreted by the kidneys further evaluation is needed.

Abstract NU07: Feline Ureteral Obstruction: A Case-control Study of Risk Factors (2016–2019)

Alex Kennedy; Joanna White
Small Animal Specialist Hospital, Sydney

Ureteral obstruction (UO) in cats causes acute kidney injury and typically requires surgical intervention. Further information is required about potentially modifiable risk factors to inform prevention strategies. A case-control study was performed to assess risk factors associated with feline UO. Cases were defined as cats with i) ureteral obstruction confirmed with pyelography or ii) a creatinine concentration>140 mcg/ml with both ureteral obstruction and pyelectasia ≥5 mm on abdominal ultrasound. Controls were defined as cats without evidence of ureteral obstruction on history, physical examination and abdominal ultrasound. Age, sex, breed (domestic or pedigree), diet (predominantly dry, mixed or predominantly wet food), housing (indoors or mixed) and total calcium were evaluated for their association with for UO using multivariable logistic regression. A receiver operator characteristics (ROC) curve was created to evaluate the predictive ability of the final model. One hundred and sixty-eight cats (28 cases, 140 controls) were included. Neither age, sex, total calcium, breed nor housing were significantly associated with UO (Table 1). Diet was significantly associated with UO. Compared to cats eating a predominantly wet food diet, cats fed a predominantly dry food diet were 15.9 times more likely to develop a UO (confidence interval 2.9–295, p=0.009). There was no difference in the association between diet and UO in cats fed a mixed diet compared to cats fed a predominantly wet food diet (p=0.25). The area under the curve of the ROC curve was 72%. While the study is limited by the accuracy of owner recollection of diet, changes in dietary formulation could provide a simple and economical method to reduce the risk of UO.

Table 1. 168 cats (28 ureteral obstruction and 140 controls): demographic data and results from bivariate logistic regression

	Ureteral obstruction	No ureteral obstruction	Coefficient (SE)	P value	Odds ratio (CI)
Age (years) median (Q1–Q3)	5.75–10.875	5.000–13.000	-0.733	0.46375	-
Sex (number of cats) male:female	13:15	61:79	-0.278	0.781	-
Breed (number of cats) pedigree:domestic	13:15	63:77	0.139	0.89	-
Housing (number of cats) indoor only: indoor/outdoor	16:10	86:49	0.21	0.834	-
Diet (number of cats) mainly wet: mixed: mainly dry	1:10:17	28:82:30	2.603	0.00925	15.87
Total calcium (mmol/L) median (Q1–Q3)	2.212–2.284	2.09–2.4	0.812	0.417	-

CI= confidence interval
OR=odds ratio

Abstract NU08: The Probability of Persistence of an Increased SDMA in Cats and Dogs

Rebekah Mack-Gertig; Evan Hegarty; Donald McCrann
IDEXX

Symmetric dimethylarginine (SDMA) has been shown to be a sensitive and early marker of decreased glomerular filtration rate (GFR); however, some practitioners are uncertain how to interpret mild increases in SDMA concentrations above the reference interval (RI) as these may be perceived as inconsequential. The goal of this retrospective study was to assess the probability that an increased IDEXX SDMA® (>14 μg/dL) will be persistent, i.e., above the RI on follow-up testing. Chemistry panel results from 1,578,808 cats and 3,643,452 dogs were collected from a 2-year period beginning in July 2015. To meet inclusion criteria, animals were required to have a minimum of 3 chemistry results: the first with an SDMA concentration within the RI (0–14 μg/dL), followed by a second result with increased SDMA concentration, followed by a third SDMA result within 14 days to 18 months. In the 16,670 cats and 16,712 dogs that met these criteria, 53% of cats and 42% of dogs with mildly increased SDMA concentrations (15–19 μg/dL) on their second result were found to have increased SDMA concentrations on the third result. For comparison, the results of 45,169 cats and 117,225 dogs in which the first two results had SDMA concentrations within the RI were examined. The probabilities of a third result above the RI were found to be 14% in cats and 7% in dogs, markedly lower than those with increased SDMA results. This suggests that even a mildly increased SDMA infers 4 to 6 times the probability of an increased test on follow-up compared to test results within the reference interval. With higher SDMA concentrations on the second result, higher persistence probabilities were noted: SDMA was increased on the third result in 76% of cats and 75% of dogs when SDMA concentrations were 20–24 μg/dL and 85% of cats and 82% of dogs when SDMA concentrations were 25+ μg/dL. This study demonstrates that patients with a mild increase in SDMA had a much higher probability of being above the RI on follow-up testing than those with SDMA concentrations within the RI on second result, and that with higher SDMA concentrations there is a greater probability of persistence. These data suggest that even a mildly increased SDMA result is a sign of ongoing GFR impairment, indicating the need for appropriate follow-up testing of SDMA and other renal biomarkers.

Abstract NU09: Association Between Ultrasound and Renal Biopsy Features in Dogs with Protein-losing Nephropathy

Lucy Kopecny; Carrie Palm; Kelsey Brust; Michelle Giuffrida; Larry Cowgill; Eric Johnson
University of California–Davis

Advances in categorization of canine glomerular disease have occurred with comprehensive pathologic evaluation, however, corresponding ultrasonographic features are not known. This study described ultrasonographic features in dogs with protein-losing nephropathy (PLN) and evaluated associations between ultrasonographic and renal biopsy findings. Dogs with PLN undergoing renal biopsy with evaluation by light, electron and immunofluorescence microscopy from 2008–2018 were included. Medical records and renal biopsy reports were reviewed. Archived ultrasound images obtained within 30 days of biopsy were reviewed by two radiologists. Descriptive statistics were calculated. Continuous variables were assessed for normality. Ultrasound variables were compared using Fisher's exact or Wilcoxon rank sum tests. P<0.05 was statistically significant. Seventy dogs were included. At biopsy, median creatinine, BUN, albumin and urine-protein-to-creatinine ratio were 3.4 mg/dL (range, 0.5–15.2; reference range [RR] 0.8–1.5), 75 mg/dL (range, 7–245; RR 11–33), 2.4 g/dL (range, 1.0–3.8; RR 3.4–4.3) and 11.2 (range, 0.4–79.3; RR<0.2), respectively. Morphologic diagnoses included membranoproliferative glomerulonephritis (n=19), focal segmental glomerulosclerosis (n=19), amyloidosis (n=13), membranous glomerulopathy (n=12), mesangioproliferative glomerulonephritis (n=4), mixed glomerular disease (n=3) and end-stage renal disease (n=2). Histopathologic abnormalities within the tubulointerstitial space were associated with increased cortical echogenicity (p=0.008). Gastric wall thickness >5 mm was associated with histopathologic diagnosis of acute, versus chronic disease (p=0.004). No significant associations were detected between: a) immune and non-immune-mediated glomerulopathy, b) acute and chronic disease or c) presence and absence of tubulointerstitial disease regarding corticomedullary ratio, medullary echogenicity or abdominal fluid. These findings suggest ultrasonographic features are poorly associated with specific pathologic patterns in canine PLN.

Abstract NU10: Comparison of Dipstick Salivary and Blood Urea Nitrogen Measurement to Diagnose Azotemia in Dogs

J.D. Foster
Friendship Hospital for Animals

Rapid detection of azotemia may be very helpful in emergency situations. Dipstick measurement of blood urea nitrogen (BUN) has been previously shown to have 85–95% sensitivity in identifying azotemia in dogs. Salivary urea nitrogen concentration correlates well with serum urea nitrogen, suggesting dipstick salivary urea nitrogen measurement may be an acceptable method of identifying azotemia. Dogs were prospectively enrolled to receive concurrent saliva and whole blood urea nitrogen dipstick measurement, stratified into four categories based upon the reagent area color change. Measurements were compared to serum BUN measurement via a dry chemistry machine. Azotemia was defined as BUN>30 mg/dL measured by the reference method. The rate of correct classification and sensitivity, specificity, and predictive values for predicting azotemia were calculated for dipstick methods. 126 dogs were enrolled in this study, 38 (30%) dogs were azotemic. Saliva and blood dipstick testing correctly classified 44% and 56% of dogs, respectively, into the appropriate category. Saliva dipstick results within categories 3 or 4 (BUN>30 mg/dL) was 81.6% sensitive, 69.3% specific, and had a 53.5% positive predictive value and 89.7% negative predictive value for identifying azotemia. Blood dipstick results within categories 3 or 4 was 94.7% sensitive, 78.4% specific, and had a 65.6% positive predictive value and 97.2% negative predictive value for identifying azotemia. Both blood and saliva dipstick measurements had poor accuracy in determining serum BUN. Blood dipstick measurement had better performance than saliva. Saliva dipstick testing may be best used to exclude azotemia if results are within category 1 or 2.

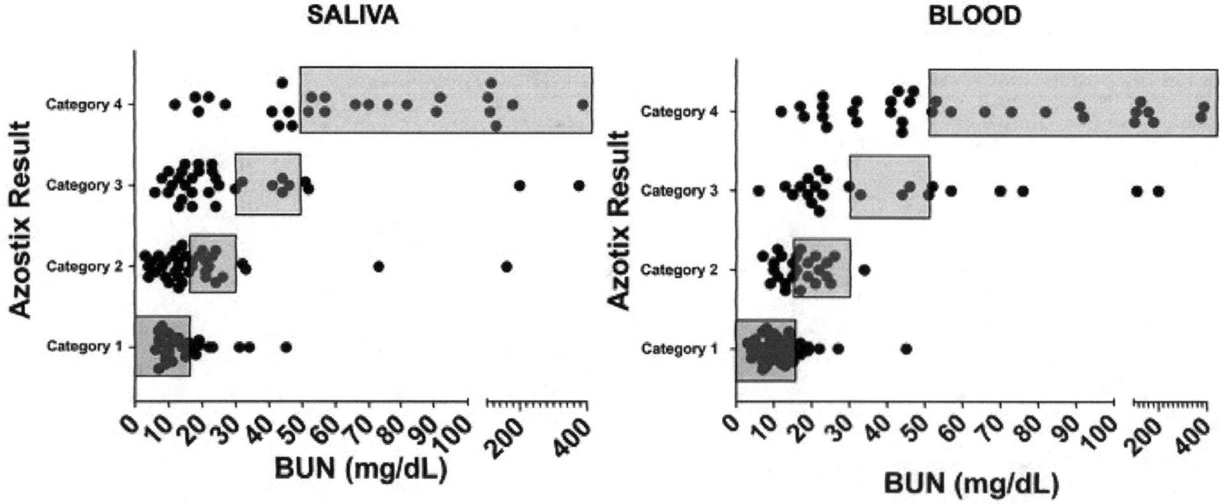

This page intentionally left blank.

Abstract NU11: Agreement of Renal Biomarkers: Longitudinal Evaluations of Increased SDMA and Creatinine in Cats and Dogs

Rebekah Mack-Gertig; Evan Hegarty; Donald McCrann
IDEXX

Symmetric dimethylarginine (SDMA) and creatinine (Cr) are commonly used renal biomarkers. Previous studies have supported SDMA as an earlier and more sensitive indicator of decreased glomerular filtration rate (GFR) than Cr. However, in practice it may be difficult to see how often an increased SDMA is the first sign of impaired renal function. This retrospective study aimed to assess this by addressing the question: when SDMA is increased, how often is Cr increased on the same chemistry or within 1 year of the increased SDMA? And how do these compare to the same statistics for increased Cr? Chemistry panel results from 2,208,408 cats and 5,085,278 dogs were collected from a 2-year enrollment period beginning in July 2015. For inclusion, animals were required to have a chemistry with IDEXX SDMA® and Cr results both below their upper reference limits (SDMA: ≤14 µg/dL, Cr: cats ≤2.3 mg/dL, dogs ≤1.5 mg/dL). Following this chemistry, animals were included in either the SDMA cohort and/ or the Cr cohort if they had two sequential increased SDMA result s (>14 µg/dL) or two sequential increased Cr result s (cats>2.3 mg/dL, dogs>1.5 mg/dL), respectively, between 14 days and 18 months apart. These and all subsequent chemistry results from up to 3.5 years of available data were used to determine expected percentage of agreement between markers using the Aalen Johansen estimator. For the SDMA cohort, agreement was also calculated by initial increased SDMA concentration grouped by severity. Of the initial population, 9,707 cats and 8,040 dogs were enrolled for the SDMA cohort, and 2,603 cats and 2,220 dogs were enrolled for the Cr cohort. Results are shown in Table 1. When SDMA was increased, 30% of cats and 28% of dogs had concurrently increased Cr, with Cr agreement increasing to 57% of cats and 54% of dogs by 1 year. For Cr, 75% of cats and 69% of dogs had concurrently increased SDMA, with SDMA agreement increasing to 93% of cats and 87% of dogs by 1 year. The greatest diagnostic utility was seen in patients with initial increased SDMA results in the 15–19 µg/dL range, which comprised a majority of cases in this study and were likely animals whose declining renal function was detected early. SDMA was the only indicator of renal decline in 80% of these cats and 82% of these dogs at presentation. This study supports that an increased SDMA is an early indicator of renal impairment and often precedes an increase in Cr.

Table 1. Percentage of Cats and Dogs with agreement of Renal Markers at First Increase and by 1 Year.

Study Cohort	Cats		Dogs	
	At Presentation	1 Year	At Presentation	1 Year
	Increased Creatinine		Increased Creatinine	
SDMA	30%	57%	28%	54%
Initial Increased SDMA Result (µg/dL)				
15-19	20%	50%	18%	45%
20-24	39%	66%	39%	71%
25+	76%	89%	77%	91%
	Increased SDMA		Increased SDMA	
Creatinine	75%	93%	69%	87%

Abstract NU12: Changes in Symmetric Dimethylarginine and Glomerular Filtration Rates in Dogs Receiving Oral Prednisone

Jean-Sebastien Palerme; Jonathan Mochel; Austin Viall; Shane Murphy; Jessica Ward
Iowa State University

Symmetric dimethylarginine (SDMA) is a renal biomarker that correlates well with glomerular filtration rates (GFR) in dogs with renal disease and is thought to be superior to other markers such as creatinine since it is less influenced by non-renal factors such as muscle mass. However, since SDMA is a product of proteolysis, it is possible that in catabolic conditions, such as hypercortisolism, SDMA levels might be increased independently of GFR. This study tested the hypothesis that SDMA would correlate poorly with GFR in healthy dogs receiving various doses of exogenous glucocorticoids. Eight healthy purpose-bred research beagles were studied. Dogs underwent 5 sequential 5-day treatment periods with intervening 9-day washout periods: control (no prednisone) and 4 prednisone treatments (0.5, 1, 2, 4 mg/kg/day q24h). Samples from days 1 and 5 of each treatment were submitted for minimum database, urine protein creatinine ratio, GFR by iohexol clearance, free and bound cortisol levels, and SDMA. While GFR increased significantly in dogs receiving 4mg/kg/day prednisone (13.86 vs. 8.66 ml/kg/day; p=0.039), creatinine, BUN, and SDMA did not differ from controls at any prednisone dose. Contrary to BUN and creatinine, SDMA did have a weak correlation with prednisone doses (r=-0.28, p=0.040) and total cortisol concentrations (r=0.31, p=0.022). No significant correlation existed between GFR and any of the renal biomarkers. These results suggest that SDMA levels might be affected by exogenous glucocorticoids.

Abstract NU13: Evaluation of Feline Urine Concentrations of Amoxicillin and Clavulanate

Kate KuKanich[1]; Butch KuKanich[1]; Kallie Woodruff[1]; Mark Papich[2]; Zackery Bieberly[1]; Joshuah Klutzke[1]; Ron Orchard[1]; Lucy Schermerhorn[1]
[1]Kansas State University; [2]North Carolina State University

The aim was to measure feline urine concentrations of amoxicillin and clavulanate after oral administration of amoxicillin-clavulanate, and to suggest updated feline urine-specific susceptibility testing breakpoints for oral amoxicillin and amoxicillin-clavulanate. The Clinical and Laboratory Standards Institute (CLSI) uses a feline breakpoint based on plasma drug concentrations (≤0.25 μg/mL), but a canine urine-specific breakpoint exists (≤8 μg/mL). Eleven healthy non-azotemic research cats were administered three 62.5 mg doses of amoxicillin-clavulanate orally every twelve hours. Following the third dose, urine was collected from litter pans or via cystocentesis (the last urine collection) over the next 28 hours, recording each urination time and volume. A minimum of 3 urine samples were collected per cat. Liquid chromatography with mass spectrometry determined the urine concentrations of amoxicillin and clavulanate. The mean half-life of amoxicillin in urine was 1.8 hours, and the mean recovery was 29%. The mean urine amoxicillin concentrations were 929 μg/mL (0–6 h) and 491 μg/mL (6–12 h). Amoxicillin concentrations were >8 μg/mL in urine collected prior to 12 hours from 10/11 cats, and >2 μg/mL in the remaining cat. Clavulanate was detected in all urine samples. This study documented that oral amoxicillin-clavulanate produced urine amoxicillin concentrations above the wild-type cutoff (8 μg/mL) for *Escherichia coli* in 10/11 non-azotemic cats. Because urine-specific susceptibility testing breakpoints can be determined using urine concentrations, this information should allow new CLSI feline uropathogen susceptibility breakpoints for amoxicillin and amoxicillin-clavulanate in non-azotemic cats, specifically increasing the urine breakpoint from ≤0.25 μg/mL to ≤8 μg/mL.

Abstract NU14: The Clinical Presentation and Outcome Using Various Treatment Modalities in 11 Dogs with Proliferative Urethritis

Max Emanuel[1]; Taryn Donovan[1]; Ken Lamb[2]; Allyson Berent[1]; Chick Weisse[1]
[1]Animal Medical Center; [2]Lamb Consulting

The background is as follows. Proliferative urethritis (PU) is an uncommon inflammatory and infiltrative disease of the urethra in female dogs, often associated with a urinary tract infection (UTI). PU typically presents with evidence of a urethral obstruction (UO). The objective is as follows. To identify clinical features in dogs with PU and document outcomes after different treatment modalities. The animals are 11 client-owned dogs. The methods are as follows. Medical records of dogs with a histopathologic diagnosis of PU from 2011–2020 were retrospectively evaluated, including information on clinical pathology, imaging, and histopathology. The outcomes of various treatment modalities were recorded and compared. Long-term urethral patency (>6 months) was considered a treatment success. The results are as follows. All dogs were female and 8/11 (73%) had a history of UTI(s). Ten of 11 survived to discharge and were used for long-term data collection. Seven of 10 dogs (70%) were treated with an effacement procedure (balloon dilation [BD] and/or stent) and 6/7 (86%) had urethral patency long term. Seven of 10 had UO recurrence after their first procedure, including 3/3 (100%) that did not have effacement and 4/7 that did (57%), at a median time of 101 and days, respectively. After effacement, the duration of patency was longer for those treated with a stent (retrievable/permanent) than BD alone (median 843 and 452, respectively). The conclusions and clinical importance are as follows. PU is a recurrent disease often associated with a UTI. The best chance long term urethral patency was after lesion effacement, either with BD or stenting (retrievable/permanent). Future prospective studies are needed to determine the impact of various immunosuppressive agents.

Abstract NU15: Symmetric Dimethylarginine and Creatinine Concentrations in Cats with Ureteral Obstruction Before and After Decompression

Jessica Himelman[1]; Allyson Berent[2]; Kendall Wilson[2]; Chick Weisse[2]; Donald Szlosek[3]; Michael Coyne[3]; Donald McCrann[3]; Rachel Murphy[3]
[1]Veterinary Emergency and Referral Group; [2]The Animal Medical Center; [3]IDEXX Laboratories Inc.

The purpose of this study is to evaluate symmetric dimethylarginine (SDMA) and creatinine concentrations in cats presenting with ureteral obstruction(s) (UO), and to determine if pre-decompression values are predictive of underlying renal function after decompression is accomplished with a subcutaneous ureteral bypass (SUB) device. Twenty-five cats admitted for treatment of UO(s) were prospectively evaluated. Serum SDMA and creatinine concentrations were assessed at presentation, 24 hours, 1 week, 1 month, and 3 months post-decompression. Urinalysis and urine culture were assessed at presentation, 1 month, and at 3 months post-decompression. Cats were excluded if: they had bacterial growth on urine culture, evidence of re-obstruction during the study period, additional surgical procedures during the study period, or if all samples were not able to be obtained at designated time points. A significant correlation was observed between the initial SDMA value and the initial creatinine value pre-decompression ($\rho=0.871$, $p<0.001$). Both the initial mean creatinine, and the initial mean SDMA values pre-decompression were shown to significantly decrease on all time points when controlling for individual patient effects ($p<0.001$, $p<0.001$, respectively). No significant correlation was observed between the initial SDMA value pre-decompression and the 3-month post-decompression SDMA value (Figures 1 and 2, $\rho=0.270$, $p=0.760$). SDMA appears to be a useful marker of post-renal azotemia in cats with UO. Pre-decompression SDMA should not be used as a means to predict underlying renal disease and long-term renal recovery in patients undergoing SUB device placement for decompression of their UO(s).

Figure 1. Box and whisker plot of SDMA values pre-decompression (initial), and post-decompression

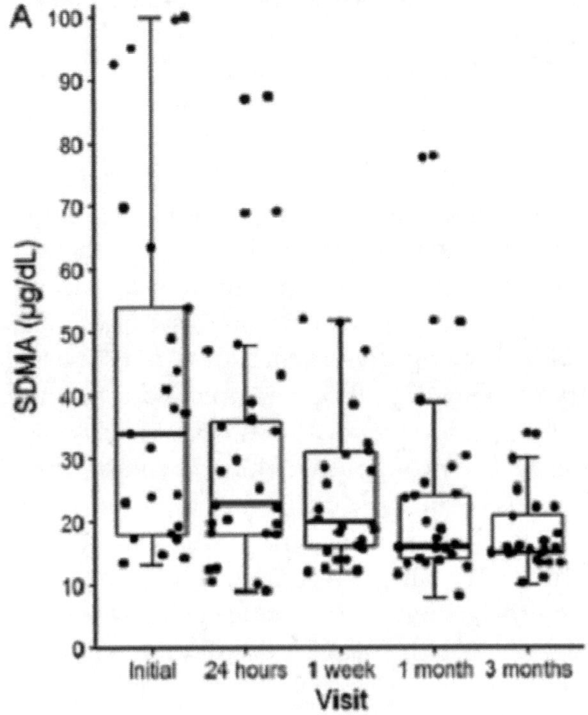

Figure 2. Scatter plot between initial SDMA value pre-decompression and 3 months post-decompression

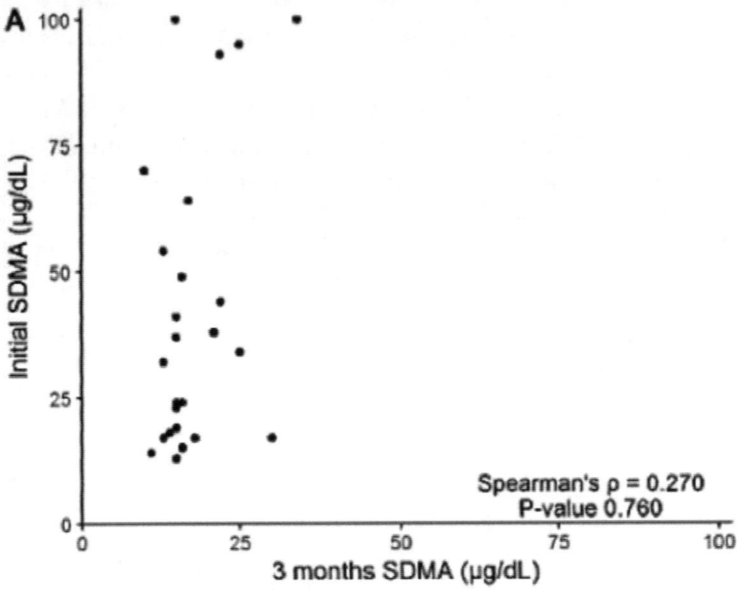

Abstract NU16: Identification of a Potential Marker of Immune Complex Mediated Glomerulonephritis in Canine Urine

Jessica A. Hokamp; Rachel Cianciolo; Jesssica Quimby; Andrew Reed
The Ohio State University

The most common causes of glomerular disease in dogs are immune complex mediated glomerulonephritis (ICGN), amyloidosis, and focal segmental glomerulosclerosis. A diagnosis of active ICGN prompts immunosuppressive therapy which is not recommended in other glomerular diseases. Renal biopsy is required for accurate diagnosis of glomerular disease in dogs, but there is significant interest in discovery of specific disease markers in samples collected by minimally invasive means (urine, blood). The objective of this study was to use liquid chromatography tandem mass spectrometry (LC-MS/MS) for proteomic discovery of high molecular weight protein(s) that will accurately distinguish dogs with ICGN from dogs with non-ICGN diseases. Urine from dogs with membranoproliferative glomerulonephritis (MPGN, a type of ICGN) (n=9) and non-ICGN diseases (n=6, specifically, amyloidosis and focal segmental glomerulosclerosis) was analyzed with Tris-Acetate gel electrophoresis to separate high molecular weight (MW) proteins. A distinct high MW band, present in MPGN cases but absent to faint in non-ICGN cases, was excised from the gel lanes. Identical regions were excised from the lanes of non-ICGN cases. These bands were trypsin digested followed by peptide extraction. Liquid chromatography tandem mass spectrometry (LC-MS/MS) was performed on the extracted peptides to acquire MS/MS data. Sequence data was searched against the Canis lupus NCBI protein database. Total normalized spectral counts were used to quantify relative protein abundances. The excised band of interest contained a large protein, de-identified as "Protein A", in 6 out of 9 (67%) of the MPGN cases and 1 out of 6 (17%) of the non-ICGN cases. The mean number of peptides matching Protein A was significantly higher in dogs with MPGN (10.7 peptides) vs. non-ICGN (0.24 peptides) (two-sample Wilcoxon rank-sum test, $p=0.03$). Preliminary data suggests that urinary "Protein A" might be a marker of canine MPGN but additional work is needed to confirm these results in a greater number of cases. Future studies will also be expanded to include other forms of ICGN (membranous and mesangioproliferative glomerulonephritis).

Abstract NU17: Influence of a Meal on Serum Concentrations of Gut-Derived Uremic Toxins in Healthy Adult Cats

Stacie Summers[1]; Jessica Quimby[2]; Linxing Yao[3]; Corey Broeckling[3]; Michael Lappin[3]
[1]Oregon State University; [2]The Ohio State University; [3]Colorado State University

The main gut-derived uremic toxins indoxyl sulfate (IS), p-cresol sulfate (pCS), and trimethylamine-n-oxide (TMAO) are metabolites of dietary protein. This study assessed the short-term effect of a meal on serum IS, pCS, and TMAO in healthy adult cats. Nine healthy research-bred cats were enrolled, and jugular catheters were placed for repeat blood sampling. Cats were fed the same commercial dry cat food (8.74 g protein per 100 kcal) for at least 8 weeks prior to enrollment and during the study. After a 12-hour fast, a blood sample was collected at hours 0800, 1000, 1200, 1400, 1600, 1800. The next day after a 12-hour fast the cats were offered food at hour 0600 and a non-fasted blood sample was collected at hours 0800, 1000, 1200, 1400, 1600, and 1800. Total IS, pCS, and TMAO serum concentrations were determined by liquid chromatography-tandem mass spectrometry. A paired t test or Wilcoxon matched-pairs signed rank test was used to compare serum concentrations between the fasted and non-fasted state. A one-way ANOVA with Geisser-Greenhouse correction or Friedman test was used to compare serum concentrations between the collection time points throughout the day. A p value ≤ 0.05 was considered significant. Serum pCS concentrations were significantly lower in the non-fasted state when compared to the fasted state at hours 1000 ($p=0.004$) and 1200 ($p=0.008$). Serum IS concentrations were significantly lower in the non-fasted state when compared to the fasted state at hours 1000 ($p=0.006$), 1200 ($p=0.005$), 1400 ($p=0.050$), and 1800 ($p=0.026$). Serum TMAO concentrations were significantly lower in the non-fasted state when compared to the fasted state at hours 0800 ($p=0.005$), 1000 ($p<0.001$), 1200 ($p<0.001$), 1400 ($p<0.001$), 1600 ($p<0.001$), and 1800 ($p<0.001$). Serum concentrations differed between multiple collection time points for IS in the non-fasted ($p=0.001$) and pCS in the fasted state ($p<0.001$). In conclusion, feeding may reduce serum concentrations of pCS, IS, and TMAO over a 12-hour period in cats. To compare serial measurements, it would be prudent to collect samples at the same time of day and consistently in either a fasted or non-fasted state.

Abstract NU18: Dietary Potassium Chloride Promotes Urine Dilution and Lowers Relative Supersaturation in Dogs and Cats

Esther Bijsmans; Vincent Biourge; **Yann Quéau**
Royal Canin

Prevention of calcium oxalate (CaOx) urolithiasis is challenging, and crystallization risk can be decreased by balanced dietary minerals in combination with urine dilution. Sodium chloride is often used as a strategy to promote urine dilution. The effect of potassium chloride (KCl) salt on urine parameters is less known. This study aimed to investigate the effects of increasing levels of KCl on urine dilution and struvite and CaOx relative supersaturation (RSS) as a measure of crystallization risk. Three dry extruded maintenance diets were manufactured using the same base formula, but supplemented with KCl to reach K^+ of 1.56 g/Mcal (Diet A), 3.35 g/Mcal (Diet B) and 4.62 g/Mcal (Diet C). Macronutrients and minerals did not differ between diets, including sodium, which was 1.0 g/Mcal in all. Diets were fed in a cross-over fashion to 9 healthy adult cats and 5 healthy adult Miniature Schnauzers for 7 days, followed by 3 days of urine collection. Water intake and urine volume were recorded. Urinary ions were measured on the pooled urine sample using ionic chromatography. Struvite and CaOx RSS were calculated using SUPERSAT software. Data was log-transformed if needed depending on the residuals to meet statistical assumptions of the model. Mixed effect models were used to investigate the effect of diet on water intake, urine volume, urine-specific gravity, struvite, and CaOx RSS, using animal as a random term. Data is presented as mean±SD except for struvite which is presented as median [25th, 75th percentile]. Post-hoc comparisons were performed using Tukey's HSD. Results are presented in the table. In conclusion, increasing dietary KCl leads to an increased water intake, increased urine volume and decreased USG in healthy dogs and cats, in a linear fashion. Struvite RSS significantly decreases in cats, but no significant difference is found in dogs. Median struvite RSS is undersaturated on all diets. CaOx RSS is significantly lower on the diets with the highest potassium content compared to the diets with the lowest potassium content in both dogs and cats. These results suggest that KCl can be used effectively to dilute urine and decrease crystallization risk in both dogs and cats.

	Cat (n=9)				Dog (n=5)			
	Diet A K+ 1.56 g/Mcal	**Diet B** K+ 3.35 g/Mcal	**Diet C** K+ 4.62 g/Mcal		**Diet A** K+ 1.56 g/Mcal	**Diet B** K+ 3.35 g/Mcal	**Diet C** K+ 4.62 g/Mcal	
Water intake (ml/kg BW/day)	22.4±5.3	25.9±6.0	28.5±7.2	A < C *P* < 0.01	33.7 ±3.0	38.7±1.9	46.1±1.9	A < B < *P* < 0.00
Urine volume (ml/kg BW/day)	8.5±2.5	13.1±1.7	15.4±2.1	A < B < C *P* < 0.0001	20.2±2.84	24.0±2.77	33.8±2.5	A < B < *P* < 0.00
USG	1.070±0.008	1.065±0.007	1.060±0.005	A < B < C *P* < 0.0001	1.055±0.008	1.048±0.006	1.041±0.004	A < B < P < 0.0
Struvite RSS	0.34[0.19,0.74]	0.09[0.05,0.19]	0.07[0.02,0.20]	A > B & C *P* < 0.001	0.28[0.18,0.36]	0.50[0.19,1.62]	0.21[0.11,0.43]	*P* = 0.0
CaOx RSS	2.55±0.76	1.95±0.60	1.65±0.52	A > B & C *P* < 0.01	8.76±0.1.40	7.05±1.89	5.66±0.95	A > B > *P* < 0.0

Abstract NU19: The Effect of Attenuating Dietary Phosphate Restriction in Cats with Azotemic CKD and Ionised Hypercalcemia.

Rebecca F. Geddes; Henk Van den Broek; Yu-Mei Chang; Rosanne Jepson; Jonathan Elliott
Royal Veterinary College

Dietary phosphate restriction may contribute to hypercalcemia development in some cats with CKD. Optimal dietary management strategy for these individuals is unclear. A database of cats diagnosed with azotemic CKD from 1/1/14–7/1/18 was searched for cats with ionised hypercalcemia at time of CKD diagnosis (ionised [calcium] on iSTAT (iCa) >1.40 mmol/l; "baseline hypercalcemics"), or after feeding a clinical renal diet ([iCa] >1.50 mmol/l once or persistently >1.40 mmol/l; "RD hypercalcemics"). Of 48 hypercalcemic cats identified, 28 were transitioned onto Royal Canin Senior Consult Stage 2 (RCS2; 1.5 g/Mcal phosphorus; Ca:P ratio 1.3), a moderately phosphate-restricted diet. Twenty cats were excluded as they lacked baseline [iCa] data (n=3), received prednisolone (n=3) or were not offered RCS2 (n=14). Variables over time were assessed using linear mixed models within groups, with visit 1 (v1) taken as the date RCS2 was initiated. Data are presented as median [25th,75th percentile]. Eighteen of 28 cats transitioned onto RSC2 had follow-up data. Nine were baseline hypercalcemics with v1 iCa 1.47[1.42,1.55] mmol/l, and 9 were RD hypercalcemics (hypercalcemia identified 148 [110,236] days after CKD diagnosis) with v1 [iCa] 1.52[1.51,1.62] mmol/l). Blood iCa (coefficient (β), -0.030; 95% confidence interval; -0.048 to -0.012; p=0.003) significantly decreased when RD hypercalcemics were fed RCS2, with [iCa] decreasing to <1.4 mmol/l in 8/9 cats after 67 [56,112] days. Plasma phosphate did not change. These variables did not change in baseline hypercalcemics (follow-up 273[147,325] days on RCS2 diet). Attenuation of dietary phosphate restriction could normalise [iCa] in cats that develop hypercalcemia whilst eating a clinical renal diet.

Abstract NU20: Urinary Fibrinogen and the Development of *Enterococcus* spp. Bacteriuria in Dogs

Michael Wood; Adam Lepold; Dahlia Tesfamichael
University of Wisconsin-Madison

In humans, increased urinary fibrinogen concentration induces *Enterococcus* spp. bacteriuria. Commonly this fibrinogenuria occurs secondary to urinary catheter induced urothelial trauma. Routine catheterization of canine patients is relatively uncommon; however, the prevalence of *Enterococcus* spp. in dogs with recurrent bacteriuria (17–25%) is increased when compared to dogs with all forms of bacteriuria (8.8–11.3%). This study hypothesizes that previously identified *Enterococcus* spp. bacteriuria risk factors in dogs (uroliths, lower urinary tract (LUT) neoplasia, and anatomic abnormalities of the LUT) will similarly be associated with increased urinary fibrinogen concentrations. To test this hypothesis urine culture negative samples were collected from 46 risk factor dogs including 8 dogs with uroliths, 27 with LUT neoplasia, and 11 with LUT anatomic abnormalities. Healthy control dogs numbered 21. Urine fibrinogen concentrations were measured with an optimized canine urine-specific fibrinogen ELISA and normalized by calculating a urine fibrinogen:creatinine ratio. In total 46 risk factor dogs had a median urinary fibrinogen:creatinine ratio of 2.87 compared to 0.17 in the healthy group ($p<0.05$). The LUT neoplasia group had the highest median fibrinogen:creatinine ratio (5.52; $p<0.05$) followed by uroliths (0.72; $p<0.05$) and LUT anatomic abnormalities (0.48; $p>0.05$). This study concludes that urinary fibrinogen concentrations are elevated in dogs with LUT neoplasia and uroliths as compared to healthy dogs. This hyperfibrinogenuria may be predisposing these dogs to *Enterococcus* spp. Bacteriuria; however, other factors are also likely important given that increased urinary fibrinogen was not identified in dogs with LUT anatomic abnormalities.

Abstract NU21: The Effects of Calcifediol Supplementation on the Renin-Angiotensin-Aldosterone System in Dogs with Chronic Kidney Disease

Matthew S. Miller[1]; Jessica Quimby[1]; Catherine Langston[1]; Marisa Ames[2]; Valerie Parker[1]
[1]The Ohio State University; [2]Colorado State University

Increased renin-angiotensin-aldosterone system (RAAS) metabolite concentrations instigate pro-fibrotic and pro-inflammatory changes within the kidney, potentially leading to worsening chronic kidney disease (CKD). While vitamin D metabolites have been shown to ameliorate activation of RAAS, CKD leads to decreased serum vitamin D metabolite concentrations. We hypothesize that supplementation with calcifediol (25-hydroxyvitamin D) will decrease circulating RAAS mediators in dogs with CKD. In a pilot study, 10 dogs with International Renal Interest Society (IRIS) stage 2 and 3 CKD received an oral extended release calcifediol supplement for 3 months. Dogs were evaluated at baseline, then after 3 months of calcifediol supplementation, and again 2 months later after discontinuation of calcifediol. Serum concentrations of 25-hydroxyvitamin D [25(OH)D] and RAAS mediators (angiotensin I/II/III/IV/1-5/1-7, and aldosterone) were evaluated at all time-points. RAAS mediators were evaluated using a novel liquid-chromatography-tandem mass spectroscopy assay (RAS-Fingerprint™). The ratio of angiotensin II to angiotensin I served as an indicator of angiotensin-converting enzyme (ACE) activity. Six dogs were included in analysis after exclusion of four dogs for use of RAAS-inhibitors (enalapril n=2, telmisartan n=1) or lack of study completion (n=1). Friedman's repeated-measures test with Dunn's post-test was utilized for analysis. With calcifediol supplementation, 25(OH)D concentration significantly increased (*p=0.03), and ACE activity significantly decreased (**p=0.01) (Figure 1). Serum concentrations of variables (median, range) in 6 dogs at baseline, after 3 months of calcifediol supplementation, and 2 months after calcifediol discontinuation are listed in Table 1. There was no significant difference in angiotensin II, angiotensin 1-5, and angiotensin 1-7 at any time-point. In conclusion, short-term calcifediol supplementation in this small subset of CKD dogs affected ACE activity.

Variable	25(OH)D (ng/ml)	ACE activity	Angiotensin I (pmol/L)	Angiotensin II (pmol/L)	Aldosterone (pmol/L)
Baseline	43.2 (33.8–58.3)	0.67 (0.61–1.28)	31.3 (13.2–75.9)	22.4 (17.0–46.3)	29.2 (7.5–85.8)
Month 3	249.9 (204.2–310.3)*	0.55 (0.37–1.00)**	55.5 (29.2–111.4)	31.1 (12.1–60.7)	126.9 (7.5–301.2)
Month 5	68.6 (49.4–107.6)	0.63 (0.58–1.32)	31.2 (20.2–77.3)	25.7 (14.7–46.6)	62.0 (25.0–122.6)

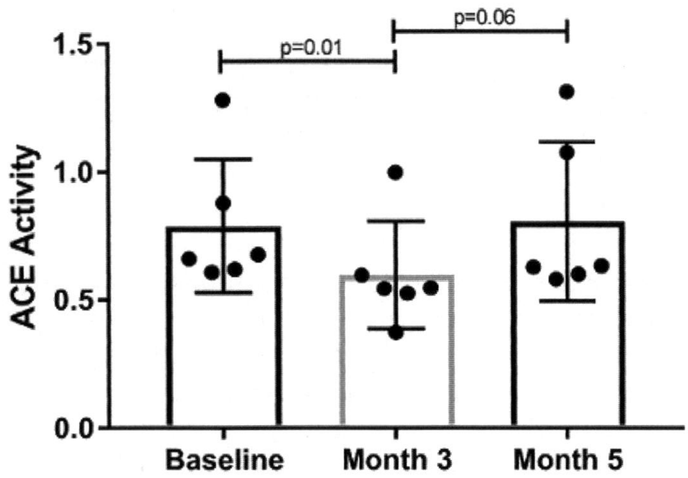
p=0.01
p=0.06
ACE Activity
1.5
1.0
0.5
0.0
Baseline
Month 3
Month 5

Abstract NU22: The Use of a 3D-Ultrasound Device to Investigate Urinary Retention in Hospitalized Dogs

Edward J. Vasquez[1]; Allison Kendall[1]; Sarah Musulin[2]; Shelly Vaden[2]
[1]North Carolina State University; [2]College of Veterinary Medicine, North Carolina State University

Urine residual volume (URV) is the volume of urine remaining in the bladder immediately after the completion of voiding and is a clinically important measurement for assessing bladder function. URV>1 ml/kg can lead to serious clinical consequences such as detrusor atony and urinary tract infection. Multiple factors, such as anesthesia and surgery have been implicated to cause postoperative urinary retention in people, which has led to routine bladder monitoring. Monitoring for urinary retention in hospitalized canine patients is not routinely performed yet urine retention may result in clinical repercussions. URV can be measured directly via urethral catheterization or indirectly via 2D ultrasound formulas. However, these techniques may impose risks such as sedation-related adverse events or catheter-associated urinary tract infection, and require appropriate operator skill and equipment. Clinical application of a 3D ultrasound device has been used for volumetric assessments of the canine urinary bladder and is routinely used for fast, 'bedside' estimation of urinary bladder volumes in people. This study aims to investigate the degree of urinary retention in hospitalized dogs using a 3D ultrasound device, and to describe various factors that could be contributing to urinary retention. We hypothesized that the majority of hospitalized dogs would develop a degree of urinary retention, and that various factors such as anesthetic events or use of opioids would contribute. A total of 25 hospitalized dogs were enrolled in this prospective, observational study. All dogs were hospitalized for more than 24 hours, and had no apparent concurrent urinary or neurologic disease that would impact their ability to ambulate or voluntarily urinate. Pre- and immediately post-void bladder volumes were measured with a 3D ultrasound device within 12 hours of presentation; subsequent measurements were obtained throughout the duration of hospitalization at approximately the same time each day. Urine retention was observed in all hospitalized dogs as manifested by increased URV. This was evident regardless of the length of hospitalization; however, the majority of dogs experienced the greatest degree of urinary retention on the second day. Of the 25 dogs enrolled, 18 dogs had an anesthetic event during their hospitalization. Of these dogs, 15/18 (83%) had URV above the reported normal reference range (0.2–0.4 ml/kg) with an average of 4.34 ml/kg (0.5–13.4 ml/kg). URV is an important clinical measurement that directly assesses bladder function and urine retention in dogs. We have determined that a 'cage-side' 3D ultrasound device can be used to safely and efficiently measure daily urinary bladder volume in hospitalized dogs. Daily monitoring of hospitalized dogs, particularly those who have undergone an anesthetic event, could help in early identification and intervention of patients that are retaining urine, thereby preventing the adverse effects of urinary retention. Further studies are warranted to determine if urinary retention is persistent after discharge in those dogs that had increased URV during hospitalization.

Abstract NU23: Urethral Tissue Engineering for Canine Urinary Incontinence

Hilla Chen[1]; Anna Shipov[1]; Shai Sherbo[2]; Keren Sinik[2]; Ishay Attar[2]; Gilad Segev[1]
[1]Koret School of Veterinary Medicine, The Hebrew University of Jerusalem;
[2]BioChange Ltd.

Urinary incontinence is a common medical concern in female dogs, most commonly caused by urethral sphincter mechanism incompetence (USMI). Disadvantages of medical therapy include the need for lifelong treatment, potential side effects and incomplete response in some dogs. VetFoam™ is an injectable scaffold made of natural materials, biocompatible and biodegradable. In this study, scaffold and saline (control) were locally injected into the urethra and sub-cutaneous tissue of a pig. Injection sites were sampled 15, 30, 60, 90 or 140 days (skin and urethra, respectively) post-injection. Masked histopathological evaluation using semi-quantitative scoring method (grade 0–4) was used to describe collagenesis along with scaffold resorption. VetFoam™ was than examined clinically by endoscopically guided injections to the proximal urethra of 15 client-owned, incontinent females dogs diagnosed with USMI. Continence score (CS) was assigned (1–5 scale) before and after the procedure and monthly thereafter. Histopathology of porcine tissue showed new tissue formation within ~90 days, while the implanted scaffold was undetectable by that time. Collagenesis grade was 3–4 in VetFoam™ injected tissue compared to 0 in control sites. No inflammation, necrosis, cavity formation or edema were documented. Median post-procedural CS of dogs increased from baseline (1.5 vs. 4, respectively, $p<0.001$), mean continence duration was 11.3±11.0 months. VetFoam™ promote collagenesis and thus has potential to maintain continence longer compared with other bulking agents. The clinical effect demonstrated in dogs, exceeded the time to full scaffold degradation, hence VetFoam™'s efficacy is likely the outcome of tissue remodeling, rather than of a bulking effect.

Abstract NU24: Influence of Laboratory Parameters on Survival in Dogs with Chronic Kidney Disease

Vivian Pedrinelli; Fabio Teixeira; Mariana Porsani; Andressa Amaral; Marcia Kogika; Marcio Brunetto
University of Sao Paulo

Chronic kidney disease (CKD) is one of the most common diseases in dogs. Factors such as creatinine, phosphorus, and albumin serum concentrations may influence survival of dogs with this disease. However, information on the influence of other laboratory parameters on survival of these animals is scarce. The aim of this study was to evaluate retrospectively the relationship between survival in dogs with CKD and common laboratory parameters evaluated in this disease. A total of 116 dogs with CKD stages 2 to 4 were included. Kaplan-Meier curves and log-rank test were used to compare survival of dogs according to stage of disease, serum phosphorus, albumin, and hematocrit. P values <0.05 were considered significant. Stage of disease ($p<0.0001$), serum phosphorus ($p=0.0005$), and hematocrit ($p=0.0001$) influenced on survival. Dogs with serum phosphorus below 5,5 mg/dL had an increase of 4.21 times in the survival, and dogs with hematocrit higher than 37% presented an increase of 4.19 times in the survival. Albumin concentration, however, did not influence survival in dogs with CKD ($p=0.2260$). According to the results observed in this study, early diagnosis, along with proper laboratory evaluation, are determinant strategies to extend survival in dogs with CKD.

1. IRIS (2019) available at www.iris-kidney.com/guidelines/staging.html.

Abstract NU25: Reduced Dietary Sodium and Calcium Improves Urine Stability in Dogs

Jennifer M. MacLeay; Stephen Davidson; Kathy Gross
Hill's Pet Nutrition, Inc.

Calcium oxalate urolithiasis is common in dogs and humans and recurrence rates are 40% within 2 and 10 years, respectively.[1,2] Nutritional recommendations in people to reduce risk commonly includes increased water intake and reduced dietary protein and oxalate.[3,4] In veterinary medicine salt may be recommended to increase thirst and water intake, whereas it is not recommended in human medicine[5,6] as dietary sodium increases urinary calcium.[7] Increased dietary sodium associated with increased urine calcium has also been shown in cats.[8] Increased urinary calcium can lead to urine instability as measured by a urine calcium oxalate titration test (COTT), which has also been used to evaluate risk for calcium oxalate recurrence in humans[9] and the test has been adapted for use in cats.[10] Foods rich in antioxidants and fatty acids and reduced in sodium have shown significant improvements in COTT and urine calcium concentration in cats.[11,12] In this study, a dry dog food with added fatty acids and antioxidants and relatively lower in sodium and calcium (Food A; Na 0.3%, Ca 0.7% DMB) was compared with a higher sodium and calcium food (Food B; Na 0.5%, Ca 1.6% DMB), and urine parameters related to risk for calcium oxalate were compared. All procedures were approved by an internal animal care and use committee and dogs had opportunities to exercise indoors and outdoors throughout the study. In succession, ten healthy adult dogs consumed a test (A) and a control (B) food for 14 days and on the last day of each trial all urine voluntarily voided over a 24-hour period was collected into a closed system where urine was maintained at body temperature until laboratory testing. Urine pH was measured using a calibrated meter. The COTT was performed by titrating whole urine with a sodium oxalate solution and monitored at 585 nm until precipitation occurred according to the method described by Davidson.[10] RSS for calcium oxalate and struvite were calculated using EQUIL software using urinary concentrations of NH_4^+, Ca^{+2}, Cl^-, citrate, creatinine, Mg^{+2}, oxalate, PO_4^{-3}, K^+, Na^+, SO_4^{-2} and urine pH. A one tailed T-Test was performed and significance was set at $p<0.05$. Dogs consuming Food A had significantly reduced urine sodium and calcium, urine-specific gravity, and RSS Struvite compared to Food B. The urine was more stable as evidenced by a significantly reduced calcium oxalate titration test result. In this study, differences in urine composition that impacted urine stability were better reflected by the COTT test than RSS calculation for calcium oxalate. Similar to people and cats, feeding foods with higher calcium and sodium levels may result in greater risks for calcium oxalate stones as measured by a urine stability test.

1. Lulich. *JVIM* 1992; 2. Singh. *Mayo Clin Proc* 2015; 3. Worcester. *NEJM* 2010; 4. Queau. *VCNA* 2019; 5. Delaney and Fascetti. Nutritional Management of Uroliths. 2016. www.vetmed.ucdavis.edu/hospital/animal-health-topics/uroliths v4-04. Accessed Jan 16, 2020; 6. Ticinesi *et al. NDT*; 7.Nouvenne *et al. AJCN* 2009; 8. Gluhek. *JVIM* 2012; 9. Laube. *Urol Res* 2000; 10. Davidson. *JVIM* 2015; 11. MacLeay. *JVIM* 2015; 12. MacLeay. *JVIM* 2018

Abstract NU26: Fecal Primary and Secondary Bile Acids in Cats with Chronic Kidney Disease

Stacie Summers[1]; Jessica Quimby[2]; Jenessa Winston[2]
[1]Oregon State University; [2]The Ohio State University;

Cats with chronic kidney disease (CKD) have an intestinal dysbiosis. Primary bile acids are converted to secondary bile acids by intestinal microbiota and fecal concentrations may be altered with dysbiosis. To date, fecal bile acids have not been quantified in CKD cats, therefore the objective of this study was to compare fecal bile acid concentrations in cats with CKD and compare to a group of healthy senior cats. Voided feces was collected from healthy senior cats (≥8 years; n=10) and cats with CKD (IRIS stage 2, n=17; IRIS stages 3 and 4, n=12). Major primary and secondary bile acids in feces were measured by liquid chromatography with tandem mass spectrometry. Fecal bile acids were compared between groups with Mann-Whitney test and Kruskal-Wallis test with Dunn's multiple comparisons test. A Spearman rank was used to assess correlation between serum creatinine and fecal bile acid concentrations. A p value ≤0.05 was considered significant. The primary bile acids cholic acid and chenodeoxycholic acid and the secondary bile acids lithocholic acid and deoxycholic acid were not significantly different between healthy senior cats and CKD cats. The secondary bile acid ursodeoxycholic acid (UDCA) was significantly lower in CKD cats (median, 12.9 ng/mg; range, 0–166.0 ng/mg) when compared to healthy senior cats (median, 32.2 ng/mg; range, 9.4–56.2 ng/mg; p=0.051). Stages 3 and 4 CKD cats (median, 6.5; range, 0–166 ng/mg) had significantly lower fecal UDCA concentrations when compared to healthy senior cats (p=0.043). No significant difference in fecal UDCA concentrations was found between healthy senior cats and stage 2 CKD cats (p=0.696) and between stage 2 CKD cats and stages 3 and 4 CKD cats (p=0.386). Fecal UDCA concentrations weakly correlated with serum creatinine concentrations (p=0.041; ρ=-0.328). In conclusion, fecal UDCA concentrations are significantly decreased in cats with CKD, especially cats with late-stage disease.

Abstract NU27: Urinary 15-F2-Isoprostanes in Dogs with Transitional Cell Carcinoma and Other Lower Urinary Tract Disease

Andrew Woolcock; Adrienne Cheney
Purdue University

Transitional cell carcinoma (TCC) is the most common type of urinary bladder tumor in dogs. Signs of lower urinary tract disease like dysuria from tumor size/obstruction or secondary urinary tract infections may prompt evaluation in these cases, but often this leads to a late diagnosis of the tumor. In people, oxidative stress is implicated in the carcinogenesis of tumors of the stomach, colon, liver, and urogenital tract. By-products of oxidative damage, such as the F2-isoprostanes, can be easily measured in urine. In people, urinary F2-isoprostanes are increased in men with lower urinary tract diseases and prostatic carcinoma, with carcinoma patients demonstrating the highest magnitude increase in urinary F2-isoprostanes. Urinary F2-isoprostanes have not been evaluated in canine urogenital tumors. The objective of this study was to measure urinary F2-isoprostanes in dogs with TCC as well as in dogs demonstrating signs of lower urinary tract disease (LUTD). We hypothesized that urinary F2-isoprostanes would be increased in dogs with TCC and LUTD when compared to healthy controls. We further hypothesized that urinary F2-isoprostanes would be increased in dogs with TCC when compared to dogs with LUTD. Dogs with newly diagnosed, histologically confirmed, TCC of the urinary tract were eligible for inclusion in the study, and were excluded if they had received any non-steroidal anti-inflammatory drugs (NSAID), corticosteroids, or cytotoxic chemotherapeutic agents within 30 days of presentation. Dogs with other, non-neoplastic, lower urinary tract disease (LUTD) were eligible for inclusion as well, with LUTD defined by the presence of dysuric clinical signs including stranguria, pollakiuria, or urinary incontinence. Urine was collected by mid-stream voided collection, sterile urethral catheterization, cystocentesis, or collection through cystoscopy. Gas chromatograpy-negative ion chemical ionization-mass spectrometry (GC-NICI-MS) was performed for quantification of F2-isoprostanes in all samples, with values normalized to urine creatinine. All dogs also had a concurrent urinalysis performed. Isoprostane measurements were parametrically distributed with summary descriptive statistics presented as mean+SD and assessed by one-way ANOVA. All other data were non-parametrically distributed with summary descriptive statistics presented as median [range]. Nonparametric numerical data for groups was assessed with the Wilcoxon rank sum test. Seventy-six dogs were included in the study, with 46 dogs diagnosed with TCC and 30 dogs diagnosed with LUTD. The most common LUTD diagnosed were cystitis (n=16; n=8 bacterial cystitis, n=8 lymphocytic and other sterile cystitis). The urinary F2-isoprostanes did not differ between the dogs with TCC, LUTD, or control dogs (mean+SD, TCC 6.58+6.37 vs. LUTD 5.52+3.13 vs. control 4.23+1.60, p=0.07). Dogs with TCC had higher qualitative measures of proteinuria (p=0.004), microscopic hematuria (p=0.013), and epithelial cells noted on sediment examination (p=0.002) compared to the other two groups. Urinary F2-isoprostanes do not appear to be a useful biomarker in the diagnosis of TCC in dogs, but further research is required to determine if urinary F2-isoprostanes decrease during treatment of lower urinary tract diseases and neoplasia in dogs.

Abstract NU28: Transient Increase in Serum Symmetric Dimethylarginine Level During Anesthesia in Cats Undergoing Routine Dental Procedures

Ira Roth; Sara Gonzalez; Alison Meindl
College of Veterinary Medicine, University of Georgia

Renal function is traditionally monitored by serum levels of creatinine and BUN even though marked renal dysfunction must occur prior to elevation in these markers. Symmetric dimethylarginine (SDMA) is a functional renal biomarker that correlates with glomerular filtration rate (GFR) and has been shown to be more sensitive than creatinine or BUN. The purpose of this study was to evaluate the immediate and residual effect of anesthesia on common renal function biomarkers in 12, client-owned cats that were anesthetized for routine dental procedures. Samples were collected at 3 timepoints: post 8 hour fast, but prior to anesthesia (baseline), post-surgery (extubation), and at a 14-day follow-up, unfasted (day 14). SDMA, creatinine, BUN, and urine-specific gravity (USG) were determined at each time point. Two of the study cats were previously diagnosed with IRIS Stage 2 CKD. Differences in renal biomarkers and USG at each timepoint were tested with paired Wilcoxon Rank Sum tests, with p-values adjusted for multiple comparisons using the Holm-Bonferroni method. All tests in this study used a significance threshold of $p<0.05$. Median SDMA concentrations increased by 4 µg/dl between baseline and extubation (10 µg/dl to 14 µg/dl). This represents a significant increase ($p=0.002$) in SDMA from baseline to extubation. (Figure 1) At the 14 day recheck, SDMA concentrations, though lower than baseline, were not significantly different from baseline concentrations ($p=0.393$). No significant increases were observed in creatinine concentrations from baseline to extubation ($p=0.324$), and while BUN concentrations increased in some cats from baseline to extubation, the increase was not significant ($p=0.152$). USG values did not significantly change from baseline to extubation ($p=0.221$). At the 14 day recheck, creatinine concentrations, BUN concentrations, and USG values were not significantly different from baseline ($p>0.3$). Length of anesthesia, body temperature, and volume of fluids administered during anesthesia were not significantly correlated with changes in SDMA levels from baseline to extubation tested with Spearman's Rank Correlation. In this study, SDMA was the only renal biomarker that increased in cats following routine dental procedures. These data support that GFR decreases in cats undergoing routine dental procedures, likely because of anesthesia, and that SDMA is the only renal biomarker sensitive enough to detect this change. The return of SDMA levels to within the reference interval two weeks post anesthesia suggests that GFR changes associated with anesthesia are transient and returns to pre-anesthesia levels in cats.

Figure 1. Trends in renal markers in felines undergoing routine dental procedures

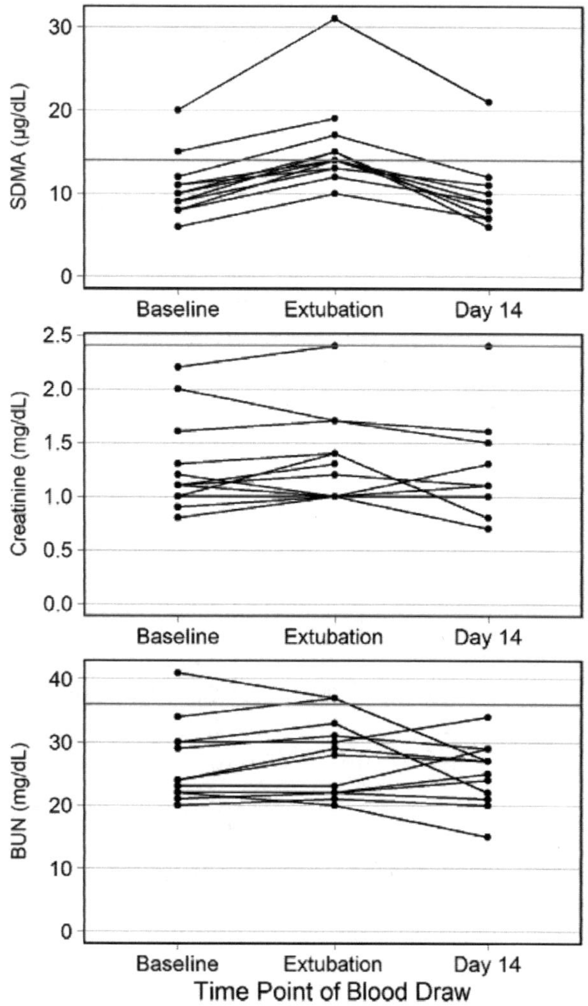

Abstract NU29: Survey of Litter Box Defecation Habits in Apparently Healthy and Chronic Kidney Disease Cats

Sarah Jones[1]; Jessica Quimby[1]; Sarah Caney[2]; Sierra Adams[1]; Stacie Summers[1]; Adam Rudinsky[1]
[1]The Ohio State University; [2]Vet Professionals Ltd

Changes in bowel movements (BM) are an important clinical sign in many diseases, including chronic kidney disease (CKD), and the purpose of this study was to collect information on BM and fecal score (FS) in both apparently healthy and CKD cats. A secondary aim was to assess owner awareness of BM frequency. Owners were asked to complete an initial online questionnaire about their cat's health (to corroborate health status) and litter box habits (including predicted BM frequency and FS). Owners were then asked to clean the box daily for 7 days and report results (observed frequency of BM and FS) daily using Purina Fecal Score (1–7) to assess feces. Differences in BM frequency and FS between apparently normal and CKD cats were compared using Mann-Whitney test and predicted versus observed data were compared using Wilcoxon sign rank test. Difference in percentage of cats defecating more or less than once daily were assessed with Fishers exact test. Survey data from 124 owners of apparently healthy cats and 43 owners of CKD cats who submitted two or more days of daily observations (in addition to the initial questionnaire) were analyzed. Age (years) of healthy cats was reported as 15+ (6.4%), 11–14 (13.2%), 7–10 (20.1%), 1–6 (55.2%), or <1 (4.3%). Age of CKD cats was reported as 15+ (62.8%), 11–14 (23.3%), 7–10 (9.3%), 1–6 (4.7%). Median serum creatinine (reported in n=26 CKD cats) was 2.5 mg/dL (range 1.6 mg/dL to 3.3 mg/dL). Eighty-five percent of apparently healthy cats were observed to defecate one or more times per day and 15% defecated less than once per day. Fifty-eight percent of CKD cats defecated one or more times per day and 42% defecated less than once per day. A significantly higher percentage of CKD cats defecated less than once per day in comparison to apparently healthy cats ($p<0.0001$). Data for predicted and observed BM frequency and FS are presented in Table 1. Observed BM frequency was significantly less in CKD cats compared to healthy cats ($p=0.02$). Owner predicted BM frequency was significantly different than observed BM frequency in healthy cats ($p=0.04$), but not in CKD cats. Owner predicted FS was significantly different than observed FS in healthy cats ($p=0.01$), but not in CKD cats. Observed FS was not significantly different between healthy and CKD cats ($p=0.16$). In conclusion, the observed BM frequency of cats with CKD was less than apparently healthy cats and represents a clinically important variation from normal.

		Predicted Avg BM/day*	Observed Avg BM/day*	Predicted Avg FS	Observed Avg FS
Healthy Cats	Mean±SD	1.16±0.5	1.29±0.6	2.14±0.5	2.08±0.5
	Median (range)	1.0 (0.14 to 3.0)	1.0 (0.33 to 3.3)	2.0 (1.0 to 4.0)	2.0 (1.0 to 4.0)
CKD Cats	Mean±SD	0.88±0.3	1.12±0.6	2.34±1.1	2.19±0.9
	Median (range)	1.0 (0.33 to 2.0)	1.0 (0.25 to 3.5)	2.0 (1.0 to 5.0)	2.0 (1.0 to 5.5)

*Predicted Avg BM/day: 0.33=every third day, 0.25=every fourth day, 0.14=once per week

Abstract NU30: Survey of the Practice of Canine and Feline Urinalyses in the United States and Canada

Nicole H. Gibbs[1]; Johanna Heseltine[2]; Mark Rishniw[3]; Mary Nabity[2]
[1]Texas A&M University, College of Veterinary Medicine & Biomedical Sciences;
[2]Texas A&M University; [3]Veterinary Information Network; Cornell University

Although urinalysis is a common diagnostic tool, no known studies describe factors influencing the performance of urinalyses by small animal practitioners. A survey was developed and posted on the Veterinary Information Network (VIN), enlisting veterinarians who perform urinalyses for dogs and cats. After answering general questions, participants were directed to question banks based on whether urinalyses are performed in-house, by an outside diagnostic laboratory, or using an in-house automated instrument. Participants using multiple methods were directed to questions from each appropriate question bank. A total of 1,059 participants from the United States and Canada completed the survey (86.6% from the United States). Of the practices surveyed, 90.9% were first opinion practices with 70% seeing exclusively dogs and/or cats, and 73.6% of the participants worked in practices with >1 full-time veterinarian. Participants performed a median of eight urinalyses per week. Practitioners performed urinalyses much less frequently than blood work during routine examination. For example, in patients over 7 years of age, the number of veterinarians who always performed urinalyses (n=142) was considerably fewer than the number who always performed a CBC (n=544) or chemistry panel (n=614). The most common factors preventing veterinarians from performing a urinalysis with blood work included clients' financial constraints (n=447), difficulty obtaining urine (n=365), and the veterinarian considering it unwarranted (n=315). Approximately 29% of participants performed urinalyses in-house by manual examinations and by submission to outside laboratories, whereas 11% exclusively sent urinalyses to outside laboratories, 23% exclusively performed in-house urinalyses with manual examinations, and 11% used automated dipstick readers and/or sediment analyzers. For participants who sent urinalyses to diagnostic laboratories and performed in-house manual examinations, the median percentage of samples sent to diagnostic laboratories was 50%. The most common reasons for submission to a diagnostic laboratory included efficiency, more trusted results, and convenience. Speed of obtaining results was the most common reason for performing urinalyses in-house. Of the participants that performed in-house urinalyses, only 57% always performed a sediment examination. Although urinalyses are commonly utilized in veterinary practice, major factors including finances, urine collection, efficiency, and lack of perceived expertise may contribute to urinalyses being an underutilized diagnostic tool.

Abstract NU31: Plasma Copeptin Concentrations in Dogs with Chronic Kidney Disease

Ya li Chang; Ya-Jane Lee; Pei-Shiue Tsai
Institute of Veterinary Clinical Science, National Taiwan University

Arginine vasopressin (AVP or antidiuretic hormone) plays a dual role in physiology. First, AVP is responsible for osmoregulation by facilitating water transport in the collecting duct; second, AVP can constrict arterioles to increase arterial blood pressure. Thus, the abnormal AVP secretion can be the cause and consequence of a declined renal function. However, due to the unstable characteristics of the circulated AVP, the measurement of AVP concentration is unlikely reliable. However, the activation of AVP system stimulates the secretion of copeptin, a C-terminal part of the pre-provasopressin (pre-pro AVP) in equimolarity with AVP. Therefore, copeptin directly reflects AVP concentration and can be used as a surrogate marker of AVP secretion. Recent studies suggested that increased copeptin level was associated with worse outcomes in chronic kidney disease (CKD) including microalbuminuria, hypertension, renal function decline, renal cyst formation and decreased kidney length in humans. Furthermore, copeptin was considered as an early predictor of kidney disease and cardiorenal syndrome in both human and laboratory animals. However, the information of copeptin in kidney diseases remains largely unknown in small animals. The aim of current study was to characterize plasma copeptin in dogs with CKD. Canine plasma samples were collected in the National Taiwan University Veterinary Hospital. Signalment, clinical information including history, physical examination and results of the hematology, plasma biochemistry and urinalysis of each case were recorded. Control groups (11 dogs) was defined as the dogs without any clinical signs associated with any disease, and no remarkable physical examination findings, as well as with the normal values of hematology, plasma biochemistry (BUN<30 mg/dL, creatinine $\leq$1.4 mg/dL) and urinalysis. The dogs with azotemia (creatinine >1.4 mg/dL) and with the history and clinical signs related to renal disease and/or abnormal kidney image (small irregular kidneys and decreased corticomedullary distinction) for at least 3 months were enrolled into CKD group. Excluding criteria included post-renal cases (obstruction in the lower urinary tract, lower urinary tract infection or cystitis), and with concurrent diseases other than kidney diseases. The CKD group was further classified into 4 subgroups based on the IRIS staging system; in total, 5 dogs in stage 1, 15 dogs in stage 2, 10 dogs in stages 3 and 5 dogs in stage 4. In addition, stages 1 and 2 of CKD stage were grouped as early stage, while stages 3 and 4 were late stage. Progression was defined as a 0.5 mg/dL increment in creatinine level or death within the 6 months. Progression day was also recorded. Blood osmolarity was derived from the formula of 2 (Na mmol/L + K mmol/L). Plasma copeptin concentration was determined with the average of duplicated tests by a commercial ELISA kit. Results showed that plasma copeptin concentrations were significantly decreased in CKD group (median [IQR]2.717 [2.08–5.29] pmol/L), as compared with those in the control group (3.994 [3.107–6.278] pmol/L) ($p=0.049$). Linear regression analysis indicated plasma copeptin levels were significantly decreased with the increase of creatinine (B=0.092, $p=0.004$), urine-specific gravity (B=0.033, $p=0.002$), potassium (B=0.136, $p=0.007$), and urine protein-creatinine ratio (B=0.153, $p=0.032$), while it increased with the increments of BUN (B=1.211, $p=0.014$), PCV (B=1.419, $p<0.001$), sodium (B=4.452, $p=0.005$), systolic blood pressure (B=10.912, $p=0.005$), and blood osmolarity (B=9.59, $p=0.006$). However, plasma copeptin levels had no significant difference between progression and non-progression cases in CKD group (p=0.222) and between the cases with and without antihypertensive medicine which included angiotensin converting enzyme inhibitors (ACEi) (p=0.538, n=35), and calcium channel blocker (p=0.281, n=33). In conclusion, similar to results of the studies of human and laboratory animals, plasma copeptin level in canine CKD was related with hypertension and several parameters responsible for osmoregulation in our study; however, plasma copeptin decreased in the CKD cases. Furthermore, an inverse correlation between plasma copeptin and creatinine, proteinuria and urine-specific gravity individually were noticed, which was different from those in human and laboratory animal studies.

Figure 1. Plasma copeptin concentrations in early and late stages of CKD dogs

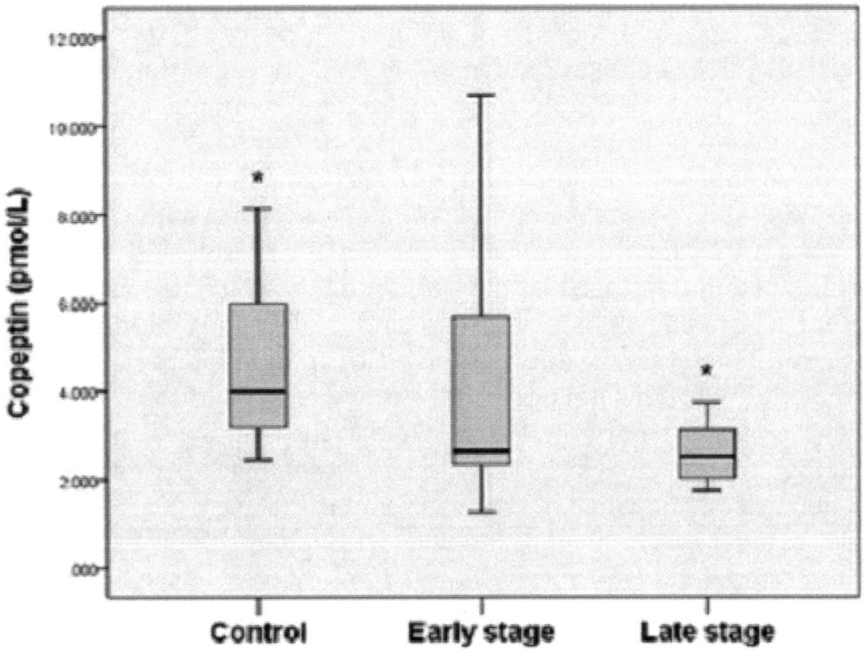

* indicated significant difference (p<0.05) between the groups

Table 1. Simple linear regression analysis of different parameters and plasma copeptin in all cases

Parameters	B value	Adjustment β	95% CI for B value	P value
Age (year-old)	0.022	0.063	-0.085–0.12	0.680
Hematocrit (%)	1.419	0.490	0.662–2.17	<0.001 **
Albumin (g/dL)	0.604	0.083	-0.009–0.015	0.604
Glucose (mg/dL)	0.162	-0.179	-0.513–0.18	0.352
BUN (mg/dL)	1.211	0.358	0.252–2.17	0.014*
Creatinine (mg/dL)	0.092	0.410	0.31–0.15	0.004*
Phosphate (mg/dL)	0.14	0.344	-0.008–0.28	0.063
Sodium (mmol/L)	4.452	0.442	1.402–7.501	0.005*
Potassium (mmol/L)	0.136	0.428	0.039–0.23	0.007*
Chloride (mmol/L)	0.019	0.080	-0.062–0.1	0.640
ALT (U/L)	2.128	0.206	-1.633–5.88	0.257
ALP (U/L)	-3.022	-0.097	-14.429–8.38	0.593

Systolic blood pressure (mmHg)	10.912	0.575	3.666–18.15	0.005*
USG	0.033	0.448	0.013–0.05	0.002*
Blood osmolarity (mOsm/kg)	9.59	0.441	2.992–16.18	0.006*
UPC	0.153	0.363	0.014–0.29	0.032*

ALT, Alanine aminotransferase; ALP, Alkaline phosphatase; BUN, Blood urine nitrogen; USG, Urine-specific gravity; UPC, Urine protein-urine creatinine ratio.
*p<0.05; **p<0.001

Abstract NU32: Plasma Copeptin Concentrations in Cats with Naturally Occurring Chronic Kidney Diseases

Syu-Yin Lin[1]; Wei-Li Hsu[2]; Ya-Jane Lee[3]; Ji-Ming Teh[4]
[1]Institute of Veterinary Clinical Science, National Taiwan University; [2]National Chung Hsing University; [3]Institute of Veterinary Clinical Science, School of Veterinary Medicine, National Taiwan University Veterinary Hospital; [4]National Taiwan University

Arginine vasopressin (AVP) has been reported to exert proproteinuric effects and cause renal function decline in human medicine. And, thus, the elevated levels of vasopressin could potentially serve as a prognostic marker for chronic kidney disease (CKD). It has been demonstrated that vasopressin is negatively associated with estimated glomerular filtration rate (eGFR), and with the development of CKD in both animal and human models via activation of V2 receptors in collecting ducts. However, the small molecular size and unstable characteristic of AVP made the measurement unreliable and that limits its application in clinical practice. The C-terminal glycopeptide, also called copeptin, is produced in an equimolar ratio (1:1) along with AVP as the result of the cleavage of the precursor of AVP prohormone in hypothalamus. Compared to vasopressin, quantification of copeptin is relatively easier due to the high stability when measured by sandwich immunoassay. Thus, copeptin acts as a surrogate marker that reflects the level of AVP release in many human clinical studies. Positive associations between plasma copeptin and renal function decline as well as proteinuria have been shown in human with CKD. However, the research of plasma copeptin and renal diseases in small animal practice is lacking. Thus, the aims of this study were several folds to investigate: (1) whether the increment of plasma copeptin concentration is proportional to renal function in cases with chronic kidney disease (CKD), (2) whether the concentration of plasma copeptin can be used to predict the risk of renal function decline, and (3) whether plasma copeptin concentration is relative to proteinuria. All cases were collected from National Taiwan University Veterinary Hospital. Healthy control group were the cases with no clinical signs related to CKD and with normal reference values of BUN, creatinine and hematological examination as well as urinalysis. The cats with plasma creatinine level ≥1.6 mg/dL and with one of the following conditions including a history of clinical signs (such as polyuria and polydipsia), an abnormal urinalysis (urine-specific gravity <1.030 or proteinuria [UPC >0.5]), abnormal radiology examination (such as small, irregular kidneys or decreased corticomedullary distinction) for at least 3 months were enrolled as CKD cats. However, the cases with one of the following problems were excluded, including abnormal serum total thyroxine concentration, any lower urinary tract disease and any symptoms other than renal diseases such as liver, heart, infectious and/or neoplastic disease. The CKD cats were further classified into different groups based on International Renal Interest Society (IRIS) staging criteria. And CKD cats with an acute increment of plasma creatinine concentration above 0.5 mg/dL within 14 days would be separated and defined as acute-on-chronic injury group (ACKI). A total of seventy cats were enrolled into: 1. healthy control group (n=7); 2. IRIS Stage 2 group (n=21); 3. IRIS Stage 3 group (n=25); 4. IRIS Stage 4 group (n=10); and 5. ACKI group (n=6). The definition of CKD progression was an increment of plasma creatine concentration above 0.5 mg/dL within 90 days. Plasma copeptin concentrations were determined from the averaged results of duplicate plates by a commercial quantitative sandwich ELISA kit. Statistically significant difference of plasma copeptin concentration was neither among the control group, CKD and ACKI groups (medians and interquartile range [IQR]: 4.028 [3.966–5.063], 4.039 [3.522–5.615], 4.575 [3.053–5.758], 4.914 [3.472–6.913] and 5.12 [3.958–7.668] pmol/L, respectively), nor among the different stages of CKD. While age ($p=0.002$), RBC ($p<0.001$), plasma hemoglobin ($p<0.001$), BUN ($p<0.001$), creatinine ($p<0.001$), phosphate ($p<0.001$) concentration, value of hematocrit ($p<0.001$), urine-specific gravity (USG) ($p<0.001$) and urine protein to creatinine ratio (UPC) ($p=0.001$) had significant differences among groups (Table 1). However, in linear regression analysis, plasma copeptin concentrations were significantly decreased with the increment of creatinine($p=0.014$), phosphate ($p=0.024$) concentration and urine protein-to-creatinine ratio (UPC) ($p=0.007$) (Table 2). Furthermore, logistic regression analysis showed that plasma copeptin concentration was not significantly correlated with the progression in feline CKD. In conclusion, plasma copeptin concentrations can be correlated with plasma creatinine, phosphorus and proteinuria in feline CKD, despite without significant increase/decrease in cats with CKD or ACKI. Moreover, our findings implied that plasma copeptin might not be associated with the progression of feline CKD.

Table 1. Clinical characteristics and biochemical parameters in control group, IRIS CKD stages 2~4 and acute-on-chronic kidney injury groups

Parameters	Control		IRIS CKD Stage 2		IRIS CKD Stage 3		IRIS CKD Stage 4		Acute on chronic kidney injury		P value
	Median (IQR)	n	Median (IQR)	n	Median (IQR)	n	Median (IQR)	n	Median (IQR)	n	
Copeptin (pmol/L)	4.03 (3.97, 5.06)	7	4.04 (3.52, 5.61)	21	4.56 (3.05, 5.76)	25	4.91 (3.47, 6.91)	10	5.12 (3.96, 7.67)	6	0.656
Age	2.00^{a,b} (1.00,11.00)	7	9.50^{a} (7.75,13.00)	21	13.00^{a,b} (12.05, 12.65)	25	12.85^{a,b} (5.53,16.85)	10	14.75^{b} (13.85, 17.13)	6	0.002 *
RBC (10^6/uL)	9.87^{c} (7.69, 10.18)	5	8.47^{b,c} (7.07, 9.03)	21	7.75^{a,b,c} (6.51, 8.66)	20	5.57^{a} (4.44, 6.62)	10	5.60^{a,b} (4.46, 6.15)	5	<0.001 **
Hemoglobin (g/dL)	14.90^{a,b} (12.90, 15.40)	5	14.00^{a,c} (12.20, 14.85)	21	12.25^{a,b,c} (10.23, 13.10)	20	9.70^{b,d} (6.65, 11.25)	10	8.80^{c,d} (8.00, 9.50)	5	<0.001 **
Hematocrit (%)	43.10^{c} (37.00–44.60)	5	39.60^{a,c} (34.00, 41.80)	21	35.10^{a,b,c} (26.83, 37.30)	20	27.90^{b} (19.35, 30.83)	10	24.70^{a,b} (21.25, 27.60)	5	<0.001 **
WBC (/uL)	8100 (5300–12050)	5	7000 (5750–9100)	21	8800 (7025, 11325)	20	13100 (10050, 17750)	10	12500 (8450, 17500)	5	0.009
BUN (mg/dL)	20.00^{a,b} (19.00, 24.25)	7	28.00^{a} (21.50, 34.50)	20	56.00^{b,d} (39.00, 64.00)	25	87.50^{c} (74.50, 120.75)	10	120.50^{c,d} (62.00, 233.25)	6	<0.001 **
Creatinine (mg/dL)	1.70^{a} (1.60, 1.90)	7	2.10^{a} (1.80,2.45)	21	3.70^{b} (3.20,4.40)	25	6.30^{b} (5.48, 7.73)	10	7.60^{b} (5.00, 18.18)	6	<0.001 **

Glucose (mg/dL)			123.50 (105.00, 150.25)	14	116.00 (100.00, 127.50)	9	120.50 (107.75, 163.00)	6	148.00 (134.00, -)	3	0.109
Sodium (mmoL/L)			153.60 (152.30, 154.70)	19	155.00 (153.90, 156.15)	22	153.20 (152.20, 155.85)	10	154.30 (151.40, 161.53)	6	0.077
Potassium (mmol/L)			3.86 (3.56, 4.27)	19	3.95 (3.72, 4.20)	22	3.83 (3.52, 4.14)	10	4.24 (3.55, 4.84)	6	0.420
Chloride (mmol/L)			116.80 (113.10, 118.10)	19	117.55 (116.45, 119.90)	22	115.30 (114.10, 118.95)	10	119.00 (117.43, 124.00)	6	0.051
Phosphate (mg/dL)	4.75[a,b] (4.63, 4.80)	4	3.90[a] (3.60, 4.20)	19	4.50[a] (3.85, 5.40)	21	9.35[b] (7.10, 10.53)	10	14.25[b] (8.80, 19.70)	6	<0.001 **
Plasma osmolality (mOsm/L H_2O)			311.60 (308.05, 313.41)	19	314.08 (311.68, 316.33)	22	310.67 (308.33, 315.47)	10	312.4 (306.08, 327.99)	6	0.081
Systolic arterial pressure (mm Hg)			138.00 (128.00, 142.00)	12	151.00 (137.00, 170.50)	17	139.00 (130.00, 150.00)	7	133.00 (118.50, 144.50)	4	0.043 *
USG	1.056[a] (1.045, 1.067)	7	1.014[b] (1.010, 1.030)	13	1.011[b] (1.009, 1.013)	18	1.010[b] (1.007, 1.012)	7	1.009[a,b] (1.008, -)	3	<0.001 **
UPC	0.01[a] (0.01, 1.00)	7	0.14[a] (0.09, 0.20)	8	0.23[a,b] (0.12, 0.51)	11	0.69[b] (0.52, 1.08)	5	1.05[a,b] (0.37, -)	3	0.001 *

1. The parameters were analyzed by Kruskal-Wallis test and Dunn test. The values were shown with median and IQR in brackets.
2. The different superscripts ([a, b, c, d]) indicated various significant differences.
3. ALKP, Alkaline phosphatase; BUN, Blood urine nitrogen; UPC, Urine protein-urine creatinine ratio; USG, Urine specific gravity
4. * p<0.05; ** p<0.001

Table 2. Simple linear regression analysis of different parameters and plasma copeptin in all cases

Parameters	B value	Adjustment β	95% CI for B value	P value
Age (year-old)	0.083	0.053	-0.085–0.12	0.664
Hematocrit (%)	0.020	0.008	-0.656–0.695	0.008
RBC (10^6/μL)	0.014	0.025–0.132	- 0.160	0.846
Hemoglobin (g/dL)	0.014	0.015	-0.224–0.252	0.909
WBC (/μL)	-230.733	-0.114	-755.164–293.699	0.382
Albumin (g/dL)	-0.002	-0.019	-0.036–0.031	0.899
Glucose (mg/dL)	-0.393	-0.058	2.895–2.108	0.751
BUN (mg/dL)	2.289	0.164	-1.123–5.702	0.185
Creatinine (mg/dL)	0.292	0.296	0.062–0.522	0.014*
Ionized calcium (mmol/L)	-0.038	-0.468	-0.086–0.010	0.107
Total calcium (mg/dL)	0.083	0.936	-0.012–0.178	0.064

Phosphate (mg/dL)	0.349	0.291	0.048–0.650	0.024*
Total calcium x phosphate	1.566	0.686	-3.491–6.624	0.314
Sodium (mmol/L)	0.021	0.023	-0.230–0.273	0.866
Potassium (mmol/L)	0.032	0.192	-0.012–0.076	0.152
Chloride (mmol/L)	0.148	0.127	-0.163–0.460	0.345
AST (U/L)	1.063	0.101	-3.587–5.714	0.640
ALT (U/L)	-0.668	-0.008	-32.975–31.638	0.966
ALP (U/L)	-1.310	-0.216	-4.070–1.451	0.334
Systolic arterial pressure (mm Hg)	3.023	0.268	-0.549–6.595	0.095
USG	-0.001	-0.120	-0.003–0.001	0.416
Plasma osmolarity (mOsm/kg)	0.074	0.039	-0.436–0.585	0.771
UPC	0.089	0.031	0.026–0.152	0.007*

1. AST, aspartate aminotransferase; ALT, Alanine aminotransferase; ALP, Alkaline phosphatase; BUN, Blood urine nitrogen; USG, Urine specific gravity; UPC, Urine protein-urine creatinine ratio
2. * p<0.05; ** p< 0.001

Abstract NU33: Evaluation of Ultrasonographic and Biochemical Parameters to Predict Outcome After Treatment of Feline Ureteral Obstructions

Elisa P. McEntee[1]; Allyson Berent[1]; Kenneth Lamb[2]; Alex Le Roux[1]; Chick Weisse[1]
[1]The Animal Medical Center; [2]Lamb Consulting

The background is as follows. To date, no pre-operative prognostic factors have been clearly identified for cats undergoing renal decompression after a diagnosis of a ureteral obstruction (UO). Frequently, judgment of long-term outcome is made based on the ultrasonographic appearance of the kidney, but this has not been well substantiated by any clinical data. The objectives are as follows. To determine whether pre-operative imaging characteristic(s) in cats suffering from a unilateral UO was predictive of renal recovery and long-term outcome after renal decompression. The animals are 34 cats with unilateral, benign UO. The methods are as follows. Pre-operative imaging characteristics (including renal pelvis diameter, parenchymal thickness [transverse and sagittal planes], renal length, and pelvic size:overall renal size) and biochemical data were retrospectively evaluated for all cats diagnosed with a unilateral UO treated with a SUB device. All patients were followed between 3 and 6 months post-operatively to obtain post-operative baseline biochemical data and had confirmed renal decompression based on SUB device flushing and patency confirmation. The results are as follows. No pre-operative imaging characteristics or biochemical findings were found to be significantly correlated with long-term creatinine. The length of the kidney and parenchymal thickness in the sagittal plane were found to be correlated to change in BUN and creatinine with decompression, but not with long-term renal values or IRIS Stage. The conclusions and clinical importance is as follows. Prediction of long-term renal recovery based on pre-operative biochemical data or ultrasound imaging findings should not be made in cats with ureteral obstruction(s) regardless of the severity of renal pelvic dilation (large or small pelvis), kidney size, or thickness of renal parenchyma assessed.

Abstract NU34: High-Protein Diet Does Not Affect Markers of Kidney Function in Cats with Asymptomatic Hypertrophic Cardiomyopathy

Yann Quéau[1]; Esther Bijsmans[2]; Elizabeth Bode[3]; Geoff Culshaw[4]; Jo Dukes-McEwan[3]; Jonathan Elliott[5]; Julie Hamilton-Elliot[3]; Hannah Hodgkiss-Geere[3]; Jeremy Laxalde[2]; Yolanda Martinez Pereira[4]; Paul Motskula[6]; Valentina Palermo[7]; Ingrid van Hoek[2]
[1]Royal Canin; [2]Royal Canin SAS; [3]University of Liverpool; [4]University of Edinburgh; [5]Royal Veterinary College; University of London; [6]Estonian University of Life Sciences; [7]Anderson Moores Veterinary Specialists

Prevalence of asymptomatic hypertrophic cardiomyopathy (aHCM) and chronic kidney disease (CKD) in cats increases with age. Cardiac and renal function are tightly integrated across species. High protein (HP) and phosphorus diets are associated with worse outcome of CKD, however a HP diet restricted in starch and with omega-3 fatty acids supplementation showed reverse remodeling effects of primary echocardiographic variables in cats with aHCM (van Hoek et al, ECVIM 2019). The aim of this study was to determine whether feeding a HP diet has deleterious effects on renal health in cats with aHCM. Forty-four cats with aHCM (ISACHC stage 1b) without signs of kidney disease (aged 5 [1.4–17] years) were included after informed owner-consent and randomly allocated to diet group HP or low protein (LP) (Table 1), in a prospective, double-blind, multicenter study. Both dry and moist forms of the two diets were available (Table 1). Owners could opt for feeding only dry, only moist or mixed feeding according to the cat's preference. Serum creatinine, BUN, phosphate, and fibroblast growth factor-23 (FGF-23) concentrations were measured in non-sedated, fasted cats before and after 6 and 12 months of feeding the allocated diet. Two-way repeated measures ANOVA compared the measured variables with both time (baseline, 6 and 12 months) and group (HP and LP) with a significance level of 5%. Results are expressed as median and range (minimum–maximum). Table 2 shows a significant diet-time interaction for BUN ($P=0.0013$) with significant increase after 6 and 12 months of feeding the HP diet, but no significant change over time for creatinine, phosphate or FGF-23 in either group, and the small number of cats with serum concentration ≥ upper reference range at each timepoint. There was no evidence that feeding a high protein diet over 12 months has detrimental effects on renal excretory function in cats with aHCM in this study. This paves the way to maximising the therapeutic potential of dietary manipulation in aHCM, in cats without pre-existing CKD.

Table 1. Diet characteristics of dry and moist forms of Diet LP and Diet HP

	Diet group LP		Diet group HP	
	dry	moist	dry	moist
Moisture (%)	5.5	79	5.5	80
Protein (g/MCal)	63.6	113.4	125	133.6
Fat (g/MCal)	36.8	38.5	34.8	41.3
EPA+DHA (g/MCal)	0.019	0.05	0.6	1.8
Starch (g/MCal)	113.4	46.5	38.8	19.4
NFE (g/MCal)	117.9	59.5	57.3	32.2
Sodium (g/MCal)	0.64	0.6	0.68	0.6
ME (kcal/kg)	3955	392	3765	383

Table 2. Serum creatinine, BUN, phosphate, and FGF-23 concentrations. Different superscript letters identify significant within- and between-group comparison differences for different time points with square brackets indicating number of cats with serum concentrations ≥ upper reference range

	Group LP			Group HP		
Time (months)	0	6	12	0	6	12
Creatinine (µmol/L)	127 (95-180)^{b} [3 of 21]	127 (82-167)^{ab} [0 of 21]	119 (80-221)^{b} [3 of 21]	120 (96-164)^{ab} [0 of 23]	113 (72-204)^{ab} [1 of 23]	111 (65-167)^{a} [0 of 23]
BUN (mmol/L)	9.5 (7.3-22.5)^{ab} [1 of 21]	8.9 (7.1-14.1)^{ab} [1 of 21]	9.3 (7.7-19.7)^{ab} [5 of 21]	8.9 (6.6-16.1)^{a} [1 of 23]	10.45 (7.3-13.4)^{b} [0 of 23]	10.3 (7.0-13.1)^{b} [0 of 23]
Phosphate (mmol/L)	1.3 (0.7-2)^{ab} [0 of 21]	1.3 (0.9-1.6)^{ab} [0 of 21]	1.3 (0.8-1.8)^{ab} [0 of 21]	1.3 (0.7-2.2)^{a} [1 of 23]	1.3 (1.1-1.7)^{b} [0 of 23]	1.3 (0.9-1.5)^{b} [0 of 23]
FGF-23 (pg/mL)	119 (12.7-741.6)^{ab} [1 of 21]	158.9 (40.1-656.4)^{b} [0 of 21]	113.8 (49.4-1520.8)^{ab} [2 of 21]	89.3 (21.0-3043.1)^{ab} [1 of 23]	73.7 (4.5-2358.1)^{ab} [1 of 23]	76.11 (43.8-4263.6)^{a} [1 of 23]

Abstract NU35: 16S Ribosomal RNA Gene Amplification of Urine from Cats with Suspected Renal Dysfunction

Kenny Siu[1]; Cindy Sotelo[1]; Stacie Summers[2]; Jennifer Hawley[1]; Jesse Buch[3]; Michael Lappin[1]
[1]Colorado State University; [2]Oregon State University; [3]IDEXX

All causes of renal dysfunction in cats have not been ascertained. While some cats with renal diseases are culture positive, the majority are negative. However, it is possible that difficult to culture (*Bartonella* spp.; *Leptospira* spp.) or unculturable bacteria could be playing a causative role. Use of 16S rRNA primers to amplify the microbial DNA in biological samples with sequencing used to determine the genus and species is being used more frequently in veterinary medicine. The objectives of this study were to determine the concentrations of DNA in the urine of client owned cats after extraction using a standardized protocol and to determine amplification rates for 16S rRNA genes. Blood and urine from cats with suspected renal dysfunction (urine specific gravity was <1.035 and symmetric dimethylarginine (SDMA) ≥15 µg/dL) from 2 commercial laboratories (California and Massachusetts) were available for study. Urine total DNA had been stored at 80°C until used in this study. The urine samples had been centrifuged at 16,000 g for 30 minutes at 4°C. The supernatant was removed while preserving the urine pellet undisturbed. Five hundred microliters of sterile phosphate buffered solution (PBS) was added to reconstitute and wash the urine pellet by brief vortexing. The PBS-urine pellet mix was centrifuged for 15 minutes at 21,000 g at room temperature. The supernatant was carefully pipetted off the pellet and the pellet was suspended in 200 µL PBS for DNA extraction using a commercially available kit (QIAamp DNA Mini Kit; Qiagen). The concentrations of double stranded DNA were determined by Qubit (Invitrogen; sensitivity =10 pg/µL to 100 ng/µL) and total nucleic acid concentrations were determined by Nanodrop (ThermoFisher; sensitivity =2 ng/µL). Bacterial DNA was amplified using a previously published real-time 16S rRNA V4 PCR assay. A convenience sampling of 50 cats with suspected renal dysfunction and white blood cells on urine sediment examination (2–5/HPF; n=25), hematuria (≥6/HPF) without pyuria (0–1/HPF; n=18) or proteinuria alone (urine dipstick ≥1+; n=7) was chosen. The median starting volume of urine for pelleting and DNA extraction was 0.95 ml (range, 0.1–1.0 ml). Nucleic acids were detected by the Nanodrop protocol in 48/50 samples (median, 1.26; range 0.04–11.5 ng/µL). Double-stranded DNA was detected by the Qubit protocol in 44/50 samples (median, 0.01; range, 0.005–0.034 ng/µL). None of the samples were positive for 16S rRNAV4 gene sequences. Microbial DNA was not amplified from urine samples suggesting a sterile urine environment in this cohort of cats. However, based on the results from the 2 non-specific nucleic acid quantification techniques, the concentrations of DNA may have been too low for detection of the 16S rRNA gene. False-negative results could have occurred with the PCR assay used. Future studies should use larger urine volumes, use samples that have not been frozen and thawed which may degrade DNA, and evaluate urine specific extraction techniques with more sensitive PCR assays.

Abstract NU36: Expression Patterns of Krüppel-Like Factors (KLFfs) in Renal Tissue in Felines with Kidney Dysfunction

Selena Tavener[1]; Dennis Jewell[2]; Kiran Panickar[3]
[1]Hill's Pet Nutrition, Inc.; [2]Kansas State University; [3]Hills PNC

Krüppel-like factors (KLFs) are a family of zinc-finger transcription factors that play an important role in cellular differentiation and proliferation during embryonic development. In addition, KLFs are also involved in key renal physiological processes including podocyte function, mesangial function, renal fibrosis, and tubulointerstitial inflammation. The aim of the present study was to investigate changes in gene expression associated with fibrosis and inflammation in renal tissue of cats with kidney disease or calcium oxalate stone formers (CaOx) from samples obtained at necropsy (end-of-life samples). At the time of death the circulating levels of creatinine, symmetric dimethyl arginine (SDMA), as well as blood urea nitrogen, all markers of kidney decline in cats, were significantly higher in cats with renal disease (n=11) or stone-forming cats (CaOx, n=12) when compared to controls (n=19). Using RNAseq in renal cortex tissue, we found a significant decrease in KLF15 in cats with kidney disease (-1.99 fold, $p<0.001$) and CaOx (-1.89, $p<0.001$) when compared to controls. Given that KLF15 can attenuate fibrosis by inhibiting the actions of TGF-β, a key regulator of fibrosis, our results indicate that a lower level of KLF15 in kidney disease and CaOx may be contributing to renal fibrosis in cats. Also, as KLF15 is a regulator of podocyte differentiation and is protective against podocyte injury, its decline in cats with kidney dysfunction indicates a potential damage to podocytes and thus the glomerular filtration barrier. Consistent with this there was a decline in podocin and nephrin, markers of podocytes, compared to controls. KLF9 was also decreased in kidney disease (-1.35 fold; ns) and it was significantly decreased in CaOx (-1.59; $p<0.05$). Whether KLF9 plays an important role in maintaining glomerular barrier function is not known, but its role in maintaining intestinal barrier has been reported. A decline in KLF5 was also found in both groups compared to control (both ns). In contrast, KLF6 and KLF10 were modestly increased in both groups compared to controls (both >1.50; ns). KLF6 has been reported to be increased in renal fibrosis. There was little to no change in other KLFs detected including KLF3, KLF4, KLF7, KLF11, and KLF12, in either group compared to controls. Our results indicate the different expression levels of KLFs in renal tissue in cats with kidney disease and CaOx. Taken together, a decline in KLF15, KLF9, and an increase in KLF6 may be important indicators of renal fibrosis and glomerular filtration dysfunction.

Abstract NU37: Intra-Individual Variability of Urinary Calcium to Creatinine and Oxalate to Creatinine Ratios in Cats

Esther Bijsmans; **Gwendoline Chaix**; Laurence le Verger; Vincent Biourge; Yann Quéau
Royal Canin

Spot urinary calcium-to-creatinine ratio (UCa:Crea) and oxalate-to-creatinine ratio (UOx:Crea) have been proposed as measures of calcium oxalate crystallization risk, though few studies have researched normal values in cats. Dijcker (2013) reported (median[range]) UCa:Crea: 0.08[0.04–0.31] and UOx:Crea: 0.05[0.03–0.06] mmol/mmol in 3 stone-formers. This study aimed to determine the intra-individual variability of UCa:Crea and UOx:Crea in healthy cats, with comparison to those stone-forming cats. A base dry extruded diet (Diet A, 0.7% K^+) was supplemented with potassium chloride (Diet B, 1.7% K^+), to achieve greater urine dilution. Eighteen healthy adult cats were fed each diet for 7 days, followed by 3 days of urine collection. Each micturition was collected. Urinary ions were measured using ionic chromatography. The coefficient of variation (CV) for urinary calcium and creatinine was <5%, and <10% for oxalate. Data was log-transformed as needed. Mixed effect models assessed the effect of diet and micturition on UCa:Crea and UOx:Crea, using animal as random term. Intra-individual coefficient of variations (ICV) are reported in the table below. There was no significant difference between diets or micturitions for both UCa:Crea and UOx:Crea ($p>0.10$ for all), but variability was seen. In conclusion, UCa:Crea showed important intra-individual variability between micturitions, but UOx:Crea varied less. Both showed overlap with values reported in stone-formers. Assessment of clinical applicability requires greater patient numbers and determination of discriminating power.

	Diet A			Diet B		
	Median[range] (mmol/mmol)	ICV	Cats with ICV>10%	Median[range] (mmol/mmol)	ICV	Cats with ICV>10%
UCa:Crea	0.04 [0.01-0.13]	19.9%	71%	0.03 [0.02-0.09]	16%	78%
UOx:Crea	0.07 [0.04-0.11]	8.4%	35%	0.08 [0.04-0.11]	10.7%	39%

Abstract NU38: Intra-Individual Variability of Urinary Calcium to Creatinine and Oxalate to Creatinine Ratios in Dogs

Esther Bijsmans; **Gwendoline Chaix**; Laurence le Verger; Yann Quéau; Vincent Biourge
Royal Canin

Spot urinary calcium-to-creatinine ratio (UCa:Crea) and oxalate-to-creatinine ratio (UOx:Crea) have been proposed as monitoring indexes for crystallization risk in dogs with calcium oxalate urolithiasis, although only UCa:Crea is reported to be increased in stone-forming dogs. This study aimed to determine the intra-individual variability of UCa:Crea and UOx:Crea, and compare values to those described in stone-formers. Two dry extruded maintenance diets (A & B) with different macronutrient profiles were fed to 6 healthy adult Miniature Schnauzers (MS) for 7 days, followed by 3 days of urine collection. Each micturition was collected. Urinary ions were measured using ionic chromatography. Coefficient of variation (CV) for urinary calcium and creatinine was <5%, and <10% for oxalate. Data was log-transformed as needed. Mixed effect models investigated the effect of diet and micturition on UCa:Crea and UOx:Crea, using animal as random term. Intra-individual coefficient of variations (ICV) are reported below. No significant difference was found between diets or micturitions for UCa:Crea and UOx:Crea ($p>0.05$). Seventeen (24%) urine samples had higher UCa:Crea than stone-forming MS (0.19 mmol/mmol) (Furrow, 2017). In conclusion, UCa:Crea showed important intra-individual variability in spot urine samples, with values above those reported in stone-forming MS. Its use as a risk index or for monitoring should be applied with caution. Urinary Ox:Crea had lower ICV but its relevance remains to be determined.

Abstract NU39: Tetrasodium EDTA: An Efficient Chemolytic Agent for Dissolution of Feline and Canine Uroliths *In Vitro*

Alice Defarges[1]; Allyson Berent[2]; Gabrielle Monteith[3]; Scott Weese[4]; Hayden Marshall[5]
[1]University of Guelph; [2]Head of Interventional Endoscopy, Animal Medical Center; [3]Statistician Technician, University of Guelph; [4]Full Professor, University of Guelph; [5]DVM student, University of Guelph

The objective was to determine if tetrasodium EDTA (tEDTA) dissolves feline and canine calcium oxalate (FeOx, K9Ox), and struvite (FeSt, K9St) uroliths *in vitro*. 120 (3–5 mm) analyzed cystic FeOx, K9COx, FeSt, K9St uroliths were obtained. Sixty uroliths were soaked in artificial urine, weighed daily and lavaged twice daily with 2% tEDTA (10 treated per type), or artificial urine (5 controls per type) for 5 consecutive days per week for 2 weeks. The same experiments were performed using 1% tEDTA. A general linear model for repeated measures controlling for each initial urolith weight was used to test for significant differences. $P<0.05$ was considered significant. All 2% tEDTA treated uroliths weight decreased significantly compared to controls by day 1. K9Ox, FeSt weight decreased significantly by day 5, K9St and FeOx weight by day 12 when lavaged with 1% tEDTA ($p<0.001$). 1 and 2% tEDTA dissolve 3–5 mm K9St, FeSt, K9Ox and FeOx uroliths *in vitro*, which supports the potential for tEDTA to be evaluated as a topical chemolytic *in vivo*.

Percentage of treated and control uroliths completely dissolved at different time points (days 5, 7 and 10) during the experiments involving 1 and 2 % tEDTA and artificial urine

	Group	Percentage of uroliths dissolved at					
		Day 5		Day 7		Day 10	
tEDTA %		1	2	1	2	1	2
K9Ox	Treated	0	40	0	40	0	100
	Control	0	0	0	0	0	0
FeOx	Treated	0	30	0	30	0	100
	Control	0	0	0	0	0	0
K9St	Treated	0	20	0	20	10	100
	Control	0	0	0	0	0	0
FeSt	Treated	0	100	0	100	100	100
	Control	0	0	0	0	0	0

Abstract NU40: Investigation on Urinary and Serum Alpha Klotho in Dogs with Chronic Kidney Disease

Dasol Park[1]; Hongjae Yi[1]; Jong-Bok Lee[2]; Kyu-Pil Lee[1]; Kunho Song[1]; Kyoung won Seo[1]
[1]College of Veterinary Medicine, Chungnam National University; [2] Yonam College

As a co-receptor for FGF-23, klotho plays a pivotal role in phosphate metabolism. The kidney is not only known to be a main source of soluble alpha klotho, but also a principal regulator for maintaining its concentration. The objective of this study is to measure serum and urinary alpha klotho in CKD dogs and to identify their association with IRIS CKD stages and other parameters known to be associated with CKD. Dogs were diagnosed with CKD and classified into different stages using IRIS guidelines. Control group was composed of healthy dogs without any clinical signs. Serum and urinary alpha klotho were measured by a commercially available canine specific ELISA kit. Calculated urine klotho to creatinine ratio (UrKl/Cr) was used to minimize the effect of USG. UrKl/Cr was significantly lower in stage 3 dogs compared to that of control group ($p = 0.001$; Figure 1). UrKl/Cr of stages 3 and 4 was significantly decreased compared to that of stages 1 and 2 ($p < 0.001$). UrKl/Cr was negatively correlated with sSDMA, BUN, creatinine and phosphorus concentration (Figure 2). UrKl/Cr was decreased in dogs with advanced CKD. In addtion, it was negatively correlated to parameters related to kidney function or damage. Therefore, it supports the possibility that klotho has a potential role in the pathogenesis of CKD and its clinical consequences in dogs. Serum klotho was not significantly related to IRIS CKD stages or other parameters tested in this study.

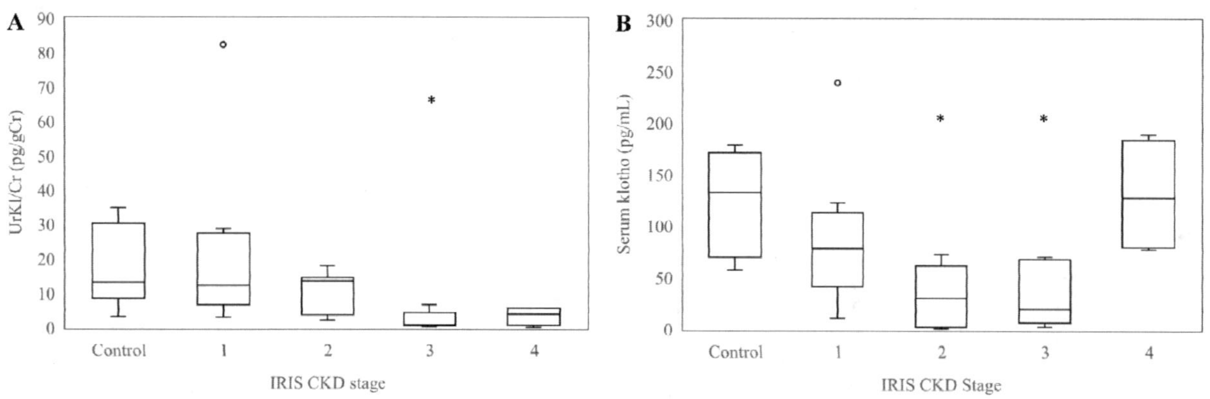

Figure 1. Box and Whisker plot illustrating urinary (A) and serum (B) klotho concentrations based on IRIS CKD stage and healthy control dogs. The boxes represent the 25th and 75th percentiles, and the central lines in the boxes represent the median values. The whiskers represent the range of concentrations. Dots represent outliers. (A) Asterisk represent statistically decreased serum klotho concentration in dogs with IRIS stages 3 compared to control dogs. (B) Asterisks represent statistically decreased serum klotho concentration in dogs with IRIS stages 2 ($p = 0.001$) and 3 ($p = 0.001$) compared to control dogs. UrKl/Cr and serum alpha klotho concentrations were compared between groups by a nonparametric Kruskal-Wallis test. Mann Whitney U-test for nonparametric data was used for confirmation of the difference between groups.

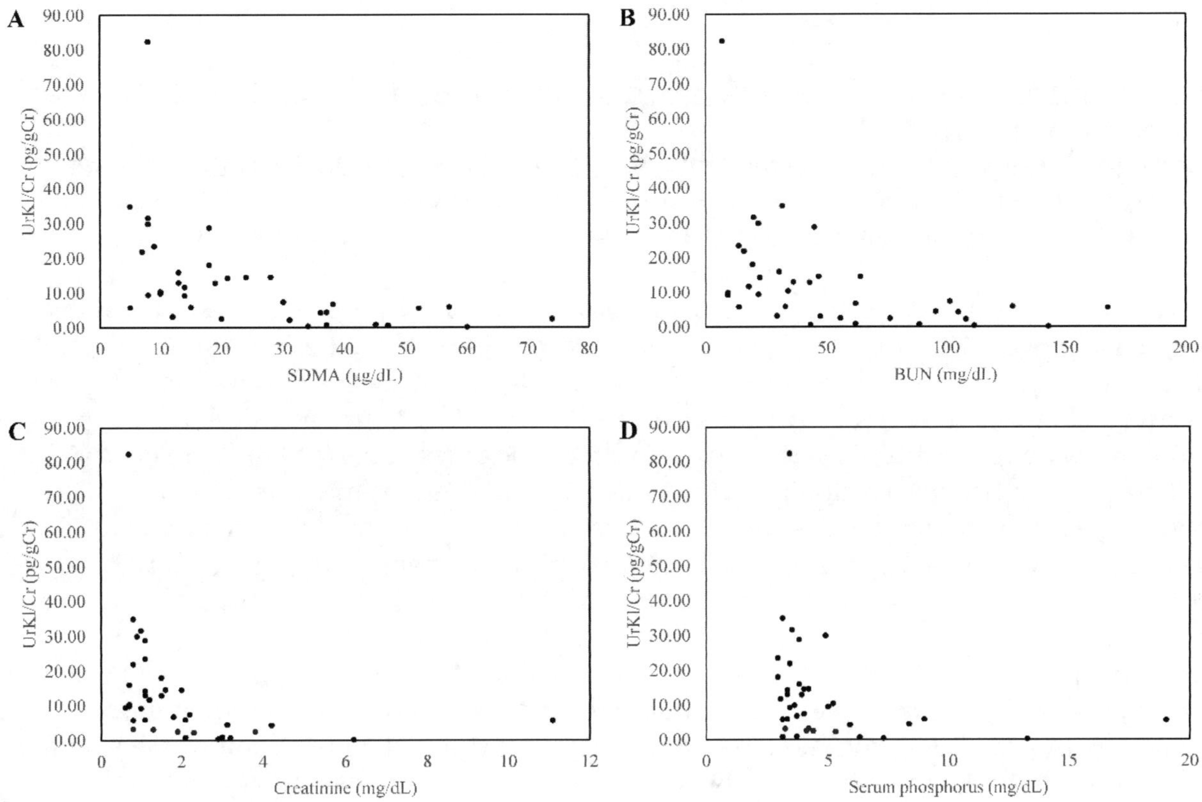

Figure 2. Scatterplot illustrating urine klotho to creatinine ratio (UrKl/Cr) by (A) serum SDMA, (B) BUN, (C) creatinine and (D) phosphorus concentrations. UrKl/Cr was negatively correlated with sSDMA ($r = -0.661$, $p < 0.001$), BUN ($r = -0.631$, $p < 0.001$), creatinine ($r = -0.651$, $p < 0.001$) and phosphorus ($r = -0.449$, $p = 0.005$) concentrations. Spearman's correlation coefficient was used to evaluate the relationships between variables.

Abstract OT01: The Dog Aging Project Study Population–An Initial Look at the Members of the Pack

Lucy Chou[1]; Audrey Ruple[2]; Katie Tolbert[1]; Katee Creevy[1]; Matt Dunbar[3]; Matt Kaeberlein[3]; Daniel Promislow[3]; Dog Aging Project Team[1]
[1]Texas A&M University; [2]University of Purdue; [3]University of Washington

Age is the single greatest risk factor for many diseases in humans, yet most of what we know about the biology of aging we have learned from laboratory-based studies of short-lived model organisms like yeast, worms, mice, and flies. This work has advanced our knowledge of aging in meaningful ways, but we still have gaps in our understanding of the genetic and environmental determinants of healthy aging. Aging in dogs resembles that of humans and they share age as a risk for developing many of the same diseases as humans. Dogs also share a tremendous amount of our genetics and they share our daily environment—including our climate, infectious disease exposure, and chemical exposures. Thus, studying aging in dogs may help us to understand aging in both dogs and humans. We have launched a large-scale longitudinal study of aging in dogs that will utilize a team of citizen scientists to help us learn more about aging in companion dogs. All dog owners in the United States have the opportunity to nominate their dog to this project. The initial call for nominations was overwhelmingly successful, with over 78,000 dogs nominated to the project in less than two months. Most of the dogs included in this initial nomination lived in single dog households (45,976/78,489), but 12.8% of the dogs live in households with three or more dogs. The vast majority of the dogs were indoor dogs, with 99.0% of nominated dogs sleeping in the house at night (77,703/78,489) and many of these dogs were reported to sleep in the owner's room (77.3%). The majority of the dogs nominated were seen by their veterinarian at least one time per year (75,245/78,489), but a small number of dogs were reported to have never been seen by a veterinarian (279/78489). The population of nominated dogs was split evenly between mixed (50.4%) and pure breed dogs (49.6%). The most common pure breeds of dogs reported were, in order of descending frequency, Labrador retriever (4,836), Golden retriever (4,023), German shepherd dog (1,887), Dachshund (1,186), and Australian shepherd (1,132). There were slightly more male dogs (51.7%) nominated than female dogs, but the majority of dogs, regardless of sex, were neutered. The average age of nominated dogs was 6.6 years, with the oldest dogs nominated being 24 years of age. Dogs were nominated from across the United States in similar proportion to the population of people in each state (Figure 1). This initial group of dogs nominated to the Dog Aging Project represent a diversity of breeds and ages and includes both sterilized and intact dogs of both sexes. Also, the geographic diversity of this population is representative of that which we see in our human population. Thus, the Dog Aging Project population will be useful in helping us to better understand the genetic and environmental determinants of healthy aging in dogs and we will be able to make inferences about human populations as well.

Nominations by State
Number Nominations
6000
4000
2000

Abstract OT02: Scoring of Erythema, Excoriation, and Lichenification Severity in Canine Atopic Dermatitis Using Deep Learning

Jung-Hyun Kim[1]; Suk-Jun Lee[2]; Sung-Sik Yoon[2]; Yoon-Mi Kim[1]; Young-Rok Kim[1]; Sang-Won Kim[1]
[1]Konkuk University; [2]Kwangwoon University

The background is as follows. Canine Atopic Dermatitis Extent and Severity Index (CADESI) is the most commonly used for evaluation of the severity of atopic dermatitis in dogs. It requires training and validating of evaluators, and it is still challenging to score the severity of canine atopic dermatitis objectively with CADESI score. Therefore, a new approach to assess the severity of canine atopic dermatitis is needed to solve the problems of unsupervised diagnostic methods and improve the quality of veterinary medical services. Automated standardization of erythema, excoriation, and lichenification severity with clinical images has not been investigated yet. It is expected to contribute to the practical use in veterinary medicine. The objectives are as follows. The aim of this study was to determine whether the deep learning using convolutional neural networks (CNNs) could assess erythema, excoriation, and lichenification severity at the level of competence comparable to veterinarians' scoring. The methods are as follows. We made a standard dataset of 2,000 clinical images of dogs showing atopic dermatitis. These images were scored 0 to 3 for each sign by three veterinarians. We trained four CNNs (ResNet V1, ResNet V2, GoogLnet, and VGG-Net) with the image dataset, and then examined which CNNs was most suitable for erythema, excoriation, and lichenification scoring. In this process, we use 200 cropped images for erythema, and 84 cropped images for excoriation scoring, and 156 cropped images for lichenification scoring. The results are as follows. For erythema, ResNet V1 152 showed the highest accuracy (74.38%), and ResNet V1 50 was the best (80.88%) for excoriation. In case of lichenification, ResNet V1 50, 101, and 152 showed the best performance with accuracy (90.32%). The conclusion and clinical relevance are as follows. These results suggest some CNNs have a performance capacity for erythema, excoriation and lichenification scoring at the level of competence comparable to veterinarians. Finally, deep learning for scoring severity of erythema, excoriation, and lichenification could be a useful tool for the evaluation of canine atopic dermatitis. Further research is needed to improve the diagnostic accuracy, and if scoring system using CNNs is applied to veterinary medical devices, it may be useful for routine veterinary clinical practice.

Abstract P01: An *In Vitro* Evaluation of Intravenous Lipid Emulsion on Three Common Canine Toxicants

Emery A. Jones[1]; Stuart Walton[1]; Jennifer Davis[2]; James Colee[3]
[1]College of Veterinary Medicine, University of Florida; [2]Virginia-Maryland College of Veterinary Medicine; [3]University of Florida

Intravenous lipid emulsion (ILE) therapy is a novel therapy recommended for lipophilic and local anesthetic drug toxicities. There are several proposed mechanisms of action for ILE therapy including the lipid sink theory, an increased mitochondrial recovery, and a direct ionotropic activity. Previous reports of therapeutic success of ILE therapy relied on resolution of clinical signs as opposed to documented decreases in drug concentration levels. The primary aim of this study was to determine whether ILE therapy significantly reduces the concentration of free drug (baclofen, ibuprofen, and bromethalin) in canine whole blood over time. The secondary aim was to determine the percent reduction in drug (baclofen, ibuprofen, and bromethalin) concentration over time with the use of ILE therapy to simulate the direct lipid effects in the lipid sink theory. Seven units (500 ml) of 1.1 negative canine blood were divided into 125 ml bags and randomly assigned to one of 3 (ibuprofen, baclofen or bromethalin) treatment groups or 4 control groups (a positive control for each study group and a negative control group). A measured amount of injectable ibuprofen (200 mg/kg), baclofen (8 mg/kg), or bromethalin (3 mg/kg) was apportioned into each smaller unit of canine whole blood and incubated for 30 minutes at 38.5°C. A known amount of ILE (12.4 ml, Intralipid®) was added to each sample and the solution was vortexed at 215 rpm for 15 minutes at 37°C (98.6°F). Whole blood was sampled at designated time points (0, 15, 30, 60, 180, 360 min), centrifuged, and separated into plasma and RBC fractions. Plasma samples were ultracentrifugation at 22,000 g at 37°C for 10 minutes to separate lipid rich and lipid poor fractions. All samples were stored at -80°C until prior to analysis and allowed to thaw at room temperature. Samples were subjected to protein precipitation followed by mixing, vortexing for 30 seconds and centrifugation at 16,100 x g for 10 minutes. Ultra-high-performance liquid chromatography with tandem mass spectrometry was used to determine the concentrations of ibuprofen, baclofen, bromethalin, desmethylbromethalin (bromethalin metabolite) in the lipid poor sample of canine plasma. Ultra-high-performance liquid chromatography was also used to determine ibuprofen concentration in red blood cells. Change in concentration of drug with the addition of ILE over time was measured and compared to the positive control to assess the binding ability of ILE. Collected data was analyzed using the Tukey-Kramer analysis, least squares mean estimates, and the GLIMMIX procedure. The total change in concentration of each drug was compared to the change in concentration of the control. A significant (p value 0.0006) change in concentration of total drug was only established for bromethalin with a percent reduction of 86%. ILE significantly reduced both the overall bromethalin concentration and bromethalian concentration at all time points when compared to the positive controls. Neither baclofen, ibuprofen, nor desmethylbromethalin had significant changes in concentration. In this single compartment *in vitro* study, ILE therapy may be efficacious for the treatment of bromethalin toxicosis. Furthermore, measurement of desmethylbromethalin, a bromethalin metabolite, did not demonstrate a significant change in concentration, suggesting that the decrease in bromethalin was not due to the conversion of bromethalin to its metabolites. However, it is possible that bromethalin is being converted to other unmeasurable metabolites as opposed to a true response to ILE therapy. These initial results are suggestive that the ILE lipid sink may be beneficial in reducing free bromethalin in canine serum. In the case of a simulated ibuprofen or baclofen overdose, treatment with ILE did not demonstrate a significant decrease in drug concentration. As a single compartment study, this study does not evaluate other proposed mechanisms of action of ILE therapy, and therefore ILE therapy may have other means of significantly decreasing the concentration of drugs such as bromethalin, baclofen, and ibuprofen in cases of toxicosis. Based on the results from this study, the lipid sink mechanism is only responsible for decreases in drug concentration in bromethalin toxicosis. Further *in vivo* study in naturally occurring toxicosis is required.

Abstract P02: Pharmacokinetic Interactions of Carprofen and Omeprazole in Dogs

Ann Gaier[1]; Susan Jones[1]; Hiroko Enomoto[1]; Mark Papich[1]; M. Katherine Tolbert[2]; Kristen Messenger[1]
[1]North Carolina State University; [2]Texas A & M University

Because omeprazole and nonsteroidal anti-inflammatory drugs (NSAIDs) are often administered concurrently, the aim of this study was to determine the effects of omeprazole on the steady-state pharmacokinetics of carprofen in dogs. We hypothesized that omeprazole would change the pharmacokinetics of carprofen in dogs affecting safety or efficacy. Carprofen (50 mg; approximately 4.4 mg/kg) was administered orally to 6 healthy adult Beagle dogs every 24 hours for 7 days alone, then with 10 mg (1 mg/kg) omeprazole administered orally twice daily in a two-period sequential study with a 4-week washout between treatments. Blood samples were obtained on day 7 of each treatment at pre-determined time points for plasma (R-) and (S+) carprofen enantiomer analysis by high performance liquid chromatography (HPLC). Pharmacokinetic parameters were calculated using noncompartmental methods. Paired t-tests or Wilcoxon Rank Sum were used to compare parameters between groups ($p<0.05$ significant). Coadministration of carprofen and omeprazole significantly altered the pharmacokinetics of carprofen for both the R(-) and S(+) enantiomers. Area-under-the-curve (AUC) was lower, and terminal half-life (T½) was shorter following concurrent carprofen and omeprazole administration than carprofen alone, while clearance (Cl/F) was significantly increased. Median S(+) AUC were 184.03 and 132.25 µg*hr/mL ($p=0.0118$), Cl/F were 12.72 and 18.37 mL/hr/kg ($p=0.0246$), and T½ were 8.8 and 6.7 hr ($p=0.0016$) for carprofen only and carprofen plus omeprazole treatments respectively. R(-) AUC was 133.37 and 98.17 µg*hr/mL, Cl/F was 18.41 and 24.82 mL/hr/kg, and T1/2 was 7.13 and 6.0 hr, respectively. A pharmacokinetic interaction does occur between omeprazole and carprofen. The clinical significance of this interaction could involve decreased analgesia or anti-inflammatory activity.

Abstract P03: Clinical Efficacy of a Long-acting Subcutaneous Methadone Solution for Postoperative Analgesia in Dogs

Butch KuKanich; Kate KuKanich; Kara Berke; Emily Klocke; David Upchurch; Brad Crauer; Alyssa Comroe; Maria Jugan; Diane Mason; Gina Jensen; Zack Bieberly; Joshua Klutzke; Kallie Woodruff; Ally Fitzgerald; Ron Orchard; Kelsey Decker
Kansas State University College of Veterinary Medicine

Current injectable opioids for postoperative dogs require frequent injections to maintain analgesia. Fluconazole significantly increases the half-life and duration of methadone. The purpose of this study was to compare the perioperative analgesia of subcutaneous methadone/fluconazole to subcutaneous standard methadone in dogs. We hypothesized subcutaneous methadone/fluconazole would provide 24 hours of analgesia with 2 doses. Forty-one healthy female dogs were enrolled, blocked by body weight, and randomly allocated to one of two treatments. The positive control group (methadone STD) received subcutaneous 0.5 mg/kg methadone q4h. The experimental group (methadone/fluconazole) received subcutaneous 0.5 mg/kg methadone and 2.5 mg/kg fluconazole, repeated once at 6 hours. All dogs received perianesthetic acepromazine, propofol, and isoflurane. Experienced surgeons performed standardized ovariohysterectomies between 8 AM and 12 PM. Perioperative monitoring included the Glasgow Composite Pain Scale (GCPS) and sedation assessed as none, slight, moderate, profound or unresponsive. All dogs were administered oral carprofen starting 24 hours postoperatively per standard practice. There were no significant differences ($p>0.05$) in postoperative GCPS scores between groups. One dog total (1/41) received rescue analgesia (methadone/fluconazole). Two methadone/fluconazole dogs had previous ovariohysterectomies (confirmed by abdominal exploratory). A single episode of vomiting occurred in 3 dogs (methadone STD) and 1 dog (methadone/fluconazole). All dogs were able to walk (moderate sedation or less) by 1 PM (methadone/fluconazole) and 4 PM (methadone STD) the day of surgery. Two doses of subcutaneous methadone/fluconazole provided 24 hours of perioperative analgesia. This should increase compliance and analgesic efficacy in veterinary clinics.

Abstract P04: Pharmacokinetics and Safety of Orally Administered Cannabidiol in Domestic Cats

Aaron Rozental; Stephanie McGrath; Lisa Bartner; Breonna Thomas; Daniel Gustafson
Colorado State University

As cannabidiol (CBD) continues to gain popularity, understanding the science behind its use in domestic animals is essential. The objective of this study was to determine the single-dose pharmacokinetic properties of orally administered highly purified CBD in healthy research cats. Eight cats were included in the study. For each dosing round, the cats were randomly divided into groups of four and assigned to receive a single dose of CBD oil at a low-dose (2.5 or 5 mg/kg); a mid-dose (10 or 20 mg/kg); and a high-dose (40 or 80 mg/kg). Blood was collected at 0, 2, 4, 6, 12, and 24 hours for plasma CBD analysis. Between each dose, the cats underwent a two-week washout period. The mean half-life, time to maximal concentration, and mean retention time across all doses was 6.2, 4.1, and 9.6 hours respectively. The median maximal concentration for each dose (2.5, 5, 10, 20, and 40 mg/kg) was 16.1, 28.1, 73.5, 133.8, 87.0, 556.5 ng/mL respectively. The mean area under the curve for each dose was 134.5, 402.3, 483.4, 1,281.0, 724.1, 5,875.4 ng-h/mL respectively. There were no severe adverse effects noted in any of the cats. Several cats exhibited head shaking and excessive salivation at the time of CBD administration. Mild lethargy was noted in all cats at doses of 40 and 80 mg/kg. No clinically significant alterations in bloodwork (CBC and chemistry profile) were noted in any cat. A single oral dose of CBD up to 80 mg/kg was safe in this cohort of cats.

Abstract P05: Predicting Uropathogen Identity and Susceptibility Using Patient Signalment and Urine Biochemical Tests

William J. Love; Shelly Vaden; Mark Papich; Megan Jacob; Cristina Lanzas
College of Veterinary Medicine, North Carolina State University

Urinary tract infections (UTIs) are one of the most common indications for antimicrobial administration in dogs and cats. Despite the recent emphasis on antimicrobial stewardship, there has been little progress to improve the information available to clinicians for antimicrobial selection. Cumulative antibiogram tables are one of the few tools available, but the currently published tables are only divided by sample source if at all. More individualized antibiograms could provide more relevant susceptibility data and better stewardship. The objective of this study is to identify patient signalment and biochemical urinalysis (UA) data that can improve pathogen and susceptibility prediction. A retrospective study was performed using aerobic urine cultures and AST results (n=999) collected from dogs treated at the NC State Veterinary Hospital in Raleigh, NC between 2014 and 2018. Analysis was limited to the six most common pathogens: *Escherichia coli* (n=451), *Staphylococcus pseudintermedius* (n=101), *Enterococcus faecalis* (n=93), *Enteroc. faecium* (n=41), *Proteus mirabilis* (n=85) and *Klebsiella pneumoniae* (n=44). Susceptibility results were available for 11 antimicrobials: ampicillin, amoxicillin + clavulanic acid, cefazolin, cefovecin, cefpodoxime, chloramphenicol, doxycycline, enrofloxacin, marbofloxacin, gentamicin, and trimethoprim + sulfamethoxazole. Signalment data included patient sex, age, and body weight and UA data included urine protein, glucose, ketones, specific gravity, pH, and bacteriuria. The probability of isolate susceptibility given patient profile was estimated logistic regression models and a multinomial regression model was used to estimate probability of pathogen presence. Predictive performance was cross-validated using a 100 isolate subset. The logistic regression models identified several factors significantly associated with susceptibility probability for several antimicrobials. Fluroquinolone susceptibility was increased in alkaluric samples (enrofloxacin OR=1.6, $p=0.03$; marbofloxacin OR=1.7, $p=0.01$) and bacteriuric samples (OR=1.8, $p<0.01$ for both). Isolates from male dogs were more likely to be susceptible to marbofloxacin (OR=1.6, $p=0.03$) and gentamicin (OR=1.8, $p=0.01$). Doxycycline susceptibility was decreased in isolates from alkaluric samples (OR=0.5, $p<0.01$). Body weight and age were not found to significantly affect susceptibility. The multinomial regression found at least one signalment or UA covariate significantly associated with the probably of an organism compared to *E. coli* as the referent organism. The gram-positive species were less common in samples with significant bacteriuria (*E. faecalis* OR=0.47, *E. faecium* OR=0.25, *S. pseudintermedius* OR=0.34, $p<0.01$ for all results). Male dogs were more likely to culture *K. pneumoniae* (OR=2.8, $p=0.01$) and less likely to culture *P. mirabilis* (OR=0.35, $p<0.01$). Samples from alkaluric dogs were more likely to culture *S. mirabilis* (OR=6.1, $p<0.01$) or *S. pseudintermedius* (OR=3.1, $p<0.01$). Despite the significant results in the training models, neither the multinomial model ($\Delta G=6.8$, $p=0.81$) nor the neither the logistic models (best fit enrofloxacin $\Delta G=10.3$, $p=0.57$). This could possibly be improved with a larger sample size or alternative fitting approaches, e.g., machine learning techniques. Also, identification of uncomplicated versus complicated UTIs and underlying specific urogenital pathology, e.g., urolithiasis, anatomic abnormalities, and metabolic disorders, may also provide more valuable predictive information and will be explored in future studies. The current study demonstrates that patient profile and UA data may influence bacterial uropathogen identity and susceptibility but does not significantly improve prediction using these methods. These results indicate which patient data could be used to improve interactive dynamic cumulative antibiogram tables, either by weighting or stratifying historic susceptibility data, to be more relevant to individual patients.

Abstract P06: Efficacy of Long-acting Methadone Using a Combination of Oral and Injectable Formulations in Perioperative Dogs

Butch KuKanich; Kate KuKanich; Emily Klocke; Yoshihiro Azuma; Nora Springer; Gina Jensen; Micaela Freeman; Alyson Fitzgerald
College of Veterinary Medicine, Kansas State University

The purpose of the study was to evaluate the perioperative efficacy and tolerability of subcutaneous methadone/fluconazole followed by oral methadone/fluconazole/naltrexone (methadone) compared to intramuscular buprenorphine followed by oral codeine (buprenorphine). Enrollment included intact dogs (n=239 total; n=119 for methadone, n=120 for buprenorphine) for ovariohysterectomy or castration as part of a surgery class (IACUC approved). Dogs were later placed for adoption. One dog (methadone) was excluded from postoperative assessment due to an additional surgery to control hemorrhage. Methadone/fluconazole (0.5/2.5 mg/kg) was administered SC preoperatively and once postoperatively; oral methadone/fluconazole/naltrexone (0.5/2.5/0.125 mg/kg) q12h started ~24 h postoperatively. Buprenorphine was administered preoperatively (0.02 mg/kg IM) and postoperatively (0.01 mg/kg IM); oral codeine administration (1–2 mg/kg q8h) started ~8 h postoperatively. All dogs received NSAIDs daily starting ~24 h (methadone) or ~8 h (buprenorphine) postoperatively. All dogs received acepromazine (premedication), ketamine/midazolam (induction) and isoflurane anesthesia. Third year veterinary students performed anesthesia and surgery with faculty oversight. The Glasgow Composite Pain Scale-Short Form assessed postoperative pain; treatment failures included total pain scores exceeding 6 or need for additional postoperative analgesia/sedation. Enrollment included 47 castration, 72 ovariohysterectomy (methadone) and 44 castration, 76 ovariohysterectomy (buprenorphine) procedures. Propofol was additionally needed for anesthesia induction in 2/119 methadone and 8/120 buprenorphine dogs. Ephedrine was administered for intraoperative hypotension in 4/119 (methadone) and 8/120 (buprenorphine) dogs. One dog (methadone) received intraoperative atropine. Mean±SD postoperative rectal temperatures were 96.1±2.0°F (methadone) and 97.0±1.7°F (buprenorphine). Treatment failures included 1/118 (methadone) and 8/120 (buprenorphine) dogs. Methadone/fluconazole ± naltrexone was effective and well tolerated in perioperative dogs.

Abstract P07: Clinical Assessment of Transdermal Gabapentin in Cats

Jennifer E. Slovak[1]; Nicolas Villarino[2]
[1]Animal Medical Center-NYC; [2]Washington State University

The clinical use of gabapentin in veterinary medicine has recently surged and has provided clinicians an alternative to opioids and non-steroidal pharmaceuticals for painful patients, especially cats. However, when used chronically, the frequency, dose, and manner of administration can be challenging for some owners. Transdermal drug administration is often desirable for medicating cats and can help eliminate negative human-animal bond interactions, increase owner compliance, and minimize stress in a diseased patient population. There is evidence of transdermal *in vitro* gabapentin penetration in cats, but it is unclear what occurs *in vivo*. The use of transdermal delivery mechanisms is challenging because of species differences in skin structure, barrier properties of skin, ratio of the skin area required for optimal drug absorption and lipid solubility of the medications used. The purpose of our research was to investigate whether transdermal gabapentin used in a proprietary base, (Lipoderm®) at pre-determined concentrations would permeate feline skin *in vivo* in a healthy patient population. A second objective of this study was to determine if painful cats responded to transdermal gabapentin administration based on validated pain scores before, during, and after treatment. The initial pilot study (IACUC ASAF # 6379) consisted of 8 client-owned young (< 6 years) healthy cats assigned to one of the following groups: 5 mg/kg TID or 10 mg/kg TID of transdermal gabapentin applied to either the cervical region or ear pinna, with 4 groups in total. Two cats were randomly assigned to each group. Gabapentin was pre-formulated to 250 mg/ml (not exceeding 0.2 mls) volume of medication per dose in Lipoderm®. All cats had serum samples obtained pre-dose and at 1, and 5 days after starting the transdermal medication. Serum samples were immediately frozen at -80°C and sent for batch drug concentration analysis utilizing a validated HPLC mass-spectrometry method. Results showed detectable levels of gabapentin for all cats in all groups at day 1 (0.18–1.3 μg/mL) and day 5 (0.11–5.4 μl/mL). There was no difference when comparing gabapentin concentration (5 vs 10 mg/kg) and skin location (cervical versus ear pinna) for this group of cats ($p>0.05$). Descriptive statistics and an overall F-test were used to analyze the data. The clinical study (IACUC ASAF # F-8-19-19-G) consisted of 15 client-owned cats over 8 years of age. All cats were examined to be clinically stable and, on no medications despite underlying conditions (up to IRIS stage 2 kidney disease, osteoarthritis, and/or obesity). All cats were administered 10 mg/kg of transdermal gabapentin in a Lipoderm® base TID. Recorded pain scores using 2 validated feline pain assessment scales were recorded pre-dose and on days 1, 5 and 8 (gabapentin was discontinued after day 5). Quantification of serum gabapentin levels were performed pre-dose and day 1 and 5 of the study. Serum samples were immediately frozen at -80°C and sent to the same laboratory used in the pilot study for batch drug concentration analysis utilizing a validated HPLC mass-spectrometry method. Descriptive statistics and a repeated measures ANOVA were utilized for data analysis. Results showed detectable levels of gabapentin at day 1 (0.1–1.47 μg/mL) and day 5 (0.31–3.38 μg/mL) in all cats. The mean gabapentin level at day 1 was 0.6 μg /mL SD±0.51 and 1.32 μg/mL SD±0.99 at day 5. There was a significant difference in gabapentin levels from day 1 to day 5 ($p<0.02$). There was no significant statistical difference in pain scores between pre- dose and day 8 in any of the cats for either pain scale. There was a significant difference in pain scores of the cats for both pain scales when comparing pre-dose to day 5 ($p<0.05$) and when comparing day 1 to day 5 ($p<0.05$). The results of this study confirm that gabapentin can be absorbed trans-dermally in cats and that gabapentin levels increase from day 1 to day 5 after trans-dermal administration. Pain scores also improve from pre-dose to day 5 and day 1 to day 5 after transdermal application of gabapentin. By day 8 (3 days post discontinuation of transdermal gabapentin), pain scores returned to pre-dose values. Future studies evaluating the chronic use of gabapentin (>1 month) in cats and its use in cats with various stages of kidney disease should be evaluated.

Abstract P08: Pharmacokinetics, Sedative, and Physiological Effects of Trazadone in Cats Alone or in Combination with Gabapentin

Laura Tucker; Andrea Sanchez; Shauna Blois; Ron Johnson; Alexander Valverde
Ontario Veterinary College

The objective of this study was to determine pharmacokinetics, physiological, and sedative properties of trazadone alone or in combination with gabapentin in healthy cats. Six systemically healthy research cats were given trazodone (3 mg/kg) IV, trazodone (5 mg/kg) PO, or a combination of trazodone (5 mg/kg) and gabapentin (10 mg/kg) PO in a randomized, crossover design with a one-week washout period. Heart rate, respiratory rate, indirect blood pressure, and level of sedation were assessed for 24 hours. Venous blood samples were collected at 3, 5, 10, 20, 30, 45, 60 minutes and 2, 4, 6, 8, 10, 12, and 24 hours post-administration in the IV group, and at the same time points starting at 10 minutes in the PO groups. Analysis of plasma trazodone concentration was performed using liquid chromatography-tandem mass spectrometry (LC-MS/MS). Oral trazodone administration resulted in rapid absorption (Tmax=10 min). Addition of gabapentin resulted in lower bioavailability (29.7% vs 59.2%), shorter half-life (4 vs 5.2 hr), larger volume of distribution (7159 vs 4658 ml), and faster clearance (1234.7 vs 619.8 ml/hr) compared to trazodone alone. Sedation was significant in all groups from 1 to 8 hr post-trazodone; addition of gabapentin resulted in a similar degree of sedation. Physiological parameters were not significantly impacted by any treatment. This study concludes that oral trazodone at the dose tested is a safe and effective option for sedation in cats with some lag between peak plasma concentration and sedation effects. Combination with gabapentin did not result in significant clinical advantages.

Abstract P09: The Influence of Butorphanol on the Sedative and Cardioventilatory Effects When Co-administered Intramuscularly with Medetomidine-Vatinoxan

Heta Turunen
Vetcare Oy

Butorphanol can be used in combination with α_2-adrenoceptor agonists, such as medetomidine or its pharmacologically active enantiomer dexmedetomidine, to enhance sedation and muscle relaxation in dogs (Kuo & Keegan 2004). Vatinoxan (also known as MK-467 and L659'066) is a peripheral α_2-adrenoceptor antagonist that reduces the adverse cardiovascular effects of α_2-adrenoceptor agonists without substantially altering the level of sedation (Honkavaara *et al.* 2008; Rolfe *et al.* 2012). We aimed to study the influence of butorphanol on the intensity and duration of sedation as well as cardioventilatory effects in dogs treated with a medetomidine-vatinoxan combination drug. We hypothesized that the dose of the medetomidine-vatinoxan combination drug could be reduced with the addition of butorphanol to achieve a similar or improved sedative effect compared to that produced by a full dose of the medetomidine-vatinoxan combination drug, and that butorphanol would not cause detrimental cardioventilatory effects. This was a prospective, randomized, blinded, experimental cross-over study where eight healthy instrumented Beagle dogs were given intramuscularly a combination of medetomidine (1 mg/m^2) and vatinoxan (20 mg/m^2) or a combination of medetomidine (0.6 mg/m^2) and vatinoxan (12 mg/m^2) with butorphanol (0.1 mg/kg) mixed in the same syringe. Sedation scores were determined with a validated sedation scale (Wagner *et al.* 2017) by a blinded investigator at 3, 5, 10, 15, 20, 25, 30, 35, 40, 45, 50, 60, 70, 80, 90 and 120 minutes after drug administration. Cardioventilatory safety of the treatments was evaluated by continuous heart rate, electrocardiogram, arterial blood pressure and respiratory rate recordings. Differences between treatments were evaluated with a linear mixed model and Dunnett adjustment for multiple comparison ($p<0.05$). Sedation scores were significantly higher with treatment including butorphanol at 10 minutes and from 35 to 120 minutes after intramuscular drug administration. Coadministration of butorphanol with the medetomidine-vatinoxan combination drug provided faster onset of deeper sedation and also increased the duration of sedation with the doses used in the present study. No adverse effects or significant differences in heart rate, arterial blood pressure and respiratory rate between treatments were detected. To conclude, butorphanol can be used without disadvantageous cardiovascular or ventilatory effects to increase the depth and length of sedation induced by a medetomidine-vatinoxan combination drug when administered at a 40% reduced dose. **References**: Bennett R, Salla K, Raekallio M *et al.* (2016) Effects of MK-467 on the antinociceptive and sedative actions and pharmacokinetics of medetomidine in dogs. *Journal of Veterinary Pharmacology and Therapeutics* 39, 336–343. Honkavaara J, Raekallio M, Kuusela E *et al.* (2008) The effects of L-659,066, a peripheral α_2-adrenoceptor antagonist, on dexmedetomidine-induced sedation and bradycardia in dogs. *Veterinary Anaesthesia and Analgesia* 35, 409–413. Honkavaara J, Restitutti F, Raekallio M *et al.* (2012) Influence of MK-467, a peripherally acting α_2-adrenoceptor antagonist on the disposition of intravenous dexmedetomidine in dogs. *Drug Metabolism and Disposition* 40, 445–449. Kuo W and Keegan R (2004) Comparative cardiovascular, analgesic, and sedative effects of medetomidine, medetomidine-hydromorphone, and medetomidine-butorphanol in dogs. *American Journal of Veterinary Research* 65, 931–937. Rolfe N, Kerr C and Mcdonell W (2012)

Cardiopulmonary and sedative effects of the peripheral α_2-adrenoceptor antagonist MK 0467 administered intravenously or intramuscularly concurrently with medetomidine in dogs. *American Journal of Veterinary Research* 73, 587–594. Wagner M, Hecker K and Pang D (2017) Sedation levels in dogs: a validation study. *BMC Veterinary Research* 13, 1–8.

Abstract P10: Pharmacokinetics of Cannabinoids from Industrial Hemp Oil in Felines

Rebekah C. Strunk; Carolyn Warner; Robert Gillette; David Griffin
Travco Products, Inc

In recent years, interest in use of cannabinoids for medicinal purposes in humans has risen causing an increase in curiosity in the veterinary world. Currently, research is limited showing the clinical benefits of cannabidiol or other phytocannabinoids in dogs or cats. There are, however, multiple studies published showing bioavailability of cannabidiol in humans and dogs, but only one in cats currently exists. The findings of that study showed low bioavailability with oil in capsules for cats. The present study evaluated the pharmacokinetics (PK) of full spectrum hemp as an oil or gel formulation (containing NMXCB1220™) given orally in a research population of cats. Eight cats (n=8) were randomized to two treatment groups (oil or gel) and crossed-over after a 7-day washout. The plasma samples were collected at the following time points pre-treatment, 1, 2, 4, 8, and 24 hours. Calculated C_{max}, T_{max}, and area under the curve (AUC) parameters for plasma CBD concentrations were subject to logarithmic transformation to adjust for normality and analyzed using one-way ANOVA to compare treatment effects. Significant differences were noted between treatment groups for C_{max} ($p<0.001$) and AUC ($p<0.001$), with the gel showing six times higher mean peak concentrations (C_{max}) compared to the oil (relative bioavailability (F) =1.57). T_{max} was not significantly different between the two groups. This study shows that cannabidiol from hemp given orally to cats is greatly influenced by delivery method. The gel hemp formulation in the present study displayed greater bioavailability than the oil. Data suggests that an oil may not be the most efficient delivery of hemp phytocannabinoids in felines. Further research is needed to determine the specific effects cannabidiol may have in cats.

Abstract P11: Evaluation of Proton Pump Inhibitor Use in Canine Patients Hospitalized in a Tertiary Referral Hospital

Samantha Duxbury[1]; Emily Sorah[1]; Patti Andrews[1]; M. Katherine Tolbert[2]
[1]College of Veterinary Medicine, North Carolina State University; [2]Texas A&M College of Veterinary Medicine

Although proton pump inhibitors (PPIs) are commonly administered to hospitalized dogs, prescribing patterns and appropriateness of use has been inadequately characterized. The objective of this study is to describe the prescription patterns and appropriateness of use associated with PPIs in hospitalized dogs at a tertiary care facility. We hypothesized that the majority of prescriptions would not comply with current guidelines for the rational use of acid suppressants. A retrospective evaluation was performed of the medical records associated with a randomly selected population of hospitalized dogs receiving PPIs January 2013–December 2018. A total of 12,610 canine patients were admitted for a first-time hospitalization January 2013-December 2018. 40% of these patients (5062/12610) were prescribed an oral or intravenous PPI. An adequate indication for use was identified in 27% of patients (54/200). Of the patients surviving to discharge 54% (95/175) were discharged with a PPI and 61% (58/95) of those patients were prescribed an inappropriate dose. The majority of PPI prescriptions to hospitalized dogs at a tertiary care hospital lacked an appropriate indication. With the growing concern of adverse effects secondary to PPIs in human medicine, rationale use of these medications should be emphasized and reserved for patients with physical exam and clinicopathologic findings supportive of an adequate indication for use. Furthermore, analysis of the prescribing patterns of dispensed PPIs revealed a frequent occurrence of dosages considered inadequate for treatment of upper GI ulceration and bleeding, raising concern for ineffective therapy even with appropriate indications of use.

The Pharmacokinetics of Long-Term, Low-Dose Oral Rapamycin in Healthy, Middle-Aged, Medium-to-Large Breed Dogs

Jeremy B. Evans[1]; Ashley Morrison[1]; Martin Javors[2]; Marisa Lopez-Cruzan[2]; Matt Kaeberlein[3]; Daniel Promislow[3]; Kate Creevy[3]
[1]Texas A&M Veterinary Teaching Hospital; [2]UT Health San Antonio; [3]University of Washington; [4]Texas A&M University

Rapamycin is a macrolide antibiotic that targets mTORC1, an extensive regulatory metabolic pathway. While rapamycin is used as an immunosuppressant, numerous other potential uses are being investigated. Rapamycin has been studied extensively in humans and rodents, but few studies in canines exist and only utilize short courses of therapy. This study aimed to determine the pharmacokinetics of long-term (range 1–6 months; mean=3.5±1.8), low-dose (0.25 mg/kg) oral rapamycin in dogs. Seven healthy, middle-aged to senior dogs (three spayed females, four castrated males) between the ages of 6–10 (mean=9.4±1.1) years-old and weighing between 40–80 lbs were enrolled. All dogs had previously been recruited into a 12-month prospective, double-blinded, placebo-controlled study where they would receive rapamycin or a placebo 3 times a week (Monday, Wednesday, and Friday) for 6 months. Participating dogs underwent a fasted blood draw at timepoint 0, were given breakfast and their treatment, and underwent additional blood collection at 1, 2, 6, and 24 hours afterwards. Prior to treatment, no dog had detectable levels of rapamycin. In the 4 dogs that received rapamycin, drug concentration peaked 2 hours after ingestion, with the mean serum concentrations being 0.99±0.1 ng/ml at 1 hour, 1.49±0.6 at 2 hours, 0.96±0.7 at 6 hours, and 0.5±0.6 at 24 hours. Dog #4, the only dog who ate breakfast, had consistently but not significantly higher rapamycin concentrations at all time points, suggesting a possible influence of prior food consumption on rapamycin absorption. The results of this study indicate that the employed low-dose regimen of rapamycin results in low but clinically measurable serum concentrations of rapamycin which do not persist for 48 hours.

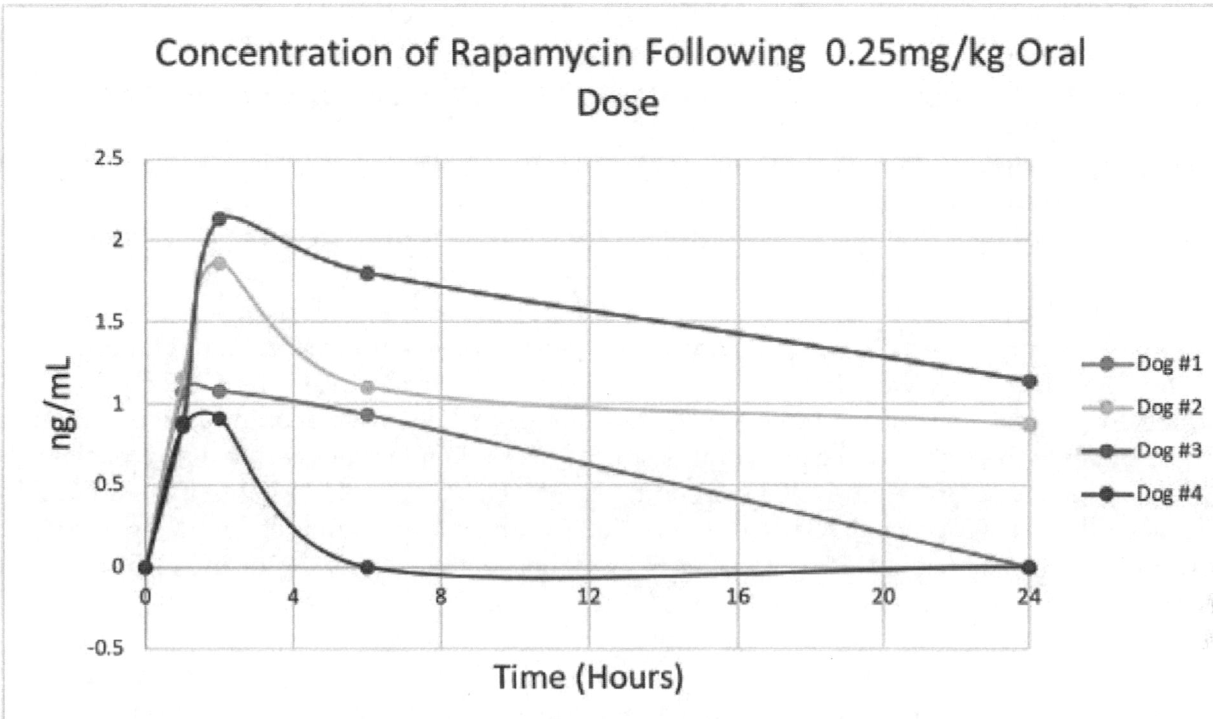
Concentration of Rapamycin Following 0.25mg/kg Oral Dose
2.5
2
1.5
1
0.5
0
-0.5
ng/mL
0
4
8
12
16
20
24
Time (Hours)
Dog #1
Dog #2
Dog #3
Dog #4

Abstract P13: Variability in Plasma Cannabidiol Concentrations in Dogs Receiving CBD-Containing Products

Jeremy A. George[1]; Benjamin Driggers[2]; Crisanta Cruz-Espindola[2]; Christina Hargis[2]; Harmon Roy[2]; Dawn Boothe[2]
[1]Auburn University; [2]College of Veterinary Medicine, Auburn University

Cannabidiol (CBD), one of two major phytocannabinoids in marijuana, is associated with evidence of therapeutic benefit as is demonstrated by its approval for treatment of drug refractory epilepsy in children. However, pet owners have been administering CBD-containing products for several years despite lack of scientific evidence. Legalization of industrial hemp is likely to increase this use. The purpose of this study was to demonstrate variability of plasma CBD in canine patients receiving CBD-containing products. Using UPLC-MS (assay validated for canine plasma), CBD and THC were quantitated in canine blood samples (n=338) received by Auburn University Clinical Pharmacology Lab's Therapeutic Drug Monitoring (TDM) program from February 2016 through October 2019. Samples were submitted specifically for CBD analysis (n=183) or were analyzed because the submission form for an alternate drug indicated the patient was also receiving CBD (n= 55). Data accompanying submission forms included patient signalment, diagnosis, concomitant drugs, the name and formulation of the CBD product, and dosing regimen. Not all information was provided for all samples. Of the 206 submission for which products manufacturer were identified, 44 different products were being used. The most common formulation was hemp oil (n=254) followed by capsule (n=38) and biscuit/treat/chewable/tincture (n=20) (31 unreported). The labeled active ingredient concentration ranged from 1 to 100 mg/mL for oils and 15 to 525 mg/capsule. The dosing regimens varied from 1 to 710 mg (median 14.97 mg, interquartile range (IQR) of 6-30.15 mg, n=182) or 0.056 to 35.5 mg/kg (median 0.7 mg/kg, IQR=0.28-1.56 mg/kg, n=178). CBD (ng/mL) ranged from nondetectable (ND) to >1000 ng/mL (median 13.7 ng/mL, IQR=2.01-55.95 ng/mL, n=187). THC ranged from ND to 87.43 ng/mL (median 0.6 ng/mL, IQR=0.01-1.85 ng/mL). When adjusted for dose, CBD concentrations in patient receiving oils ranged from 0.072 to 976 ng/mL (median 35.38 ng/mL, IQR 5.34-106.51 ng/mL, n=64) compared to non-oils of 0.052 to 425.74 ng/mL (median 10.92 ng/mL, IQR 1.68-200.48 ng/mL, n=15). This data demonstrates the marked variability in CBD concentrations achieved in patients receiving supplements, reflecting in part, variability in products, formulations, and dose. The need to provide standard recommendations is evident, particularly when implementing clinical trials, and monitoring may be a useful tool for standardizing product uses.

Abstract P14: Pharmacokinetics and Diuretic Effect of Intravenous, Tablet, Developed Oral Disintegrating Film of Furosemide in Dogs

Dong-hyuk Kwak[1]; Suk-kyu Koh[1]; Jong-woo Jeong[2]; Gwan-hyung Jo[3]; Tae-sung Koo[2]; Kyoung-won Seo[1]
[1]College of Veterinary Medicine, Chungnam National University; [2]Graduate School of New Drug Discovery and Development, Chungnam National University; [3]College of Pharmacy, Inje University

Furosemide, a diuretic acting on the Henle's loop, is commonly used to treat congestive heart failure in veterinary medicine. Some owners have difficulty administering oral tablet medications to animals - who are noncompliant and long-term administrations can also make it very challenging.

Oral disintegrating film (ODF) is a non-invasive application with rapid dissolution, which prevents the risk of swallowing. Objectives of this study were to research pharmacokinetic and pharmacodynamics profiles by comparing each routes; intravenous (IV), orally uncoated tablet (OUT) and newly developed ODF, following administration of furosemide in healthy beagle dogs. Five beagle dogs were administered a single dose (2 mg/kg) of furosemide loaded ODF, developed by the research team, using a cross-over design. The most suitable film forming agent was sodium alginate which was used to manufacture ODF. In aspect to size of the films, it was developed for ease of use in small animals. There were no significant differences in the pharmacokinetic profiles between OUT and ODF. Cmax was higher but T1/2 and Tmax were slightly shorter with ODF, when compared to OUT method. ODF resulted in a similar hourly urinary output, when compared with OUT, during the initial 2 hours after administration. These finding suggest that the pharmacokinetics and pharmacodynamics after administration of newly developed ODF are equivalent to that of OUT. Therefore, ODF formulations of furosemide have the benefit of easy and convenient administration, which would be helpful to patients and owners in veterinary medicine.

Table 1. Pharmacokinetic parameters of furosemide (2 mg/kg) by route of administration in beagle dogs

Pharmacokinetic parameters	IV	OUT	FS-ODF	P OUT vs. FS-ODF
C_{max}(μg/mL)	6.30±0.78 (6.46)	0.61±0.31 (0.69)	0.81±0.81 (0.58)	0.619
T_{max}(h)	0.083 (0.083)	1.05±0.45 (1)	0.65±0.29 (0.75)	0.130
$T_{1/2}$(h)	2.07±0.76 (1.63)	3.46±1.82 (2.86)	2.65±0.77 (2.46)	0.602
MRT (h)	0.90±0.25 (0.782)	4.76±1.59 (3.99)	3.36±0.77 (3.45)	0.128
AUC_{inf}(μg·h/mL)	2.74±0.25 (2.60)	1.62±0.77 (1.77)	1.28±0.70 (1.36)	0.493
AUC_{last}(μg·h/mL)	2.67±0.260 (2.56)	1.27±0.49 (1.53)	1.14±0.64 (1.11)	0.734
F (%)	100	59.06±27.95 (64.74)	46.91±25.49 (50.01)	0.493

C_{max}, maximum plasma concentration; T_{max}, time at the maximum concentration; $T_{1/2}$, elimination half-life, MRT, mean residence time; AUC_{inf}, area under the curve from time zero to time of infinity measurable concentration; AUC_{last}, area under the curve from time zero to time of last measurable concentration; F,fraction of oral dose absorbed(=bioavailability), expressed as a percentage; IV, intravenous; OUT, orally uncoated tablet; FS-ODF, furosemide-loaded orally disintegrating film. Results are presented as mean ±SD (median). (*), values are significantly different ($P < 0.05$); (**) values are significantly different ($P < 0.01$).

This page intentionally left blank.

Abstract R01: Use of an *In Vitro* Larval Motility Assay Evaluating Anthelmintic Efficacy against Canine and Feline Metastrongyloids

Daniela Bedenice[1]; Haifaa Mahjoub[2]; Henrik Stryhn[2]; Spencer Greenwood[2]; Gary Conboy[2]
[1]Tufts Cummings School of Veterinary Medicine; [2]Atlantic Veterinary College, University of Prince Edward Island

Metastrongyloids are important causative agents of canine and feline respiratory and cardiopulmonary disease. The purpose of this study was to determine the *in vitro* efficacy of 6 common anthelmintics (fenbendazole, eprinomectin, ivermectin, milbemycin oxime, moxidectin, selamectin) on the motility (viability) of infective third-stage larvae (L3) of *Angiostrongylus vasorum*, *Crenosoma vulpis*, and *Aelurostrongylus abstrusus*. First-stage larvae (L1) of *A. vasorum*, *C. vulpis* and *A. abstrusus* were isolated from infected lung tissue or feces (red fox, dog or cat origin) using a lung flush or Baermann technique. Laboratory-raised *Limax maximus* slugs were fed 1,500–3000 L1 and held at 16°C for 4 weeks to produce L3. Slugs were digested and 50–100 L3/well were incubated at 16°C for 3 days in culture media alone and media containing 4 different concentrations for each anthelmintic. Drug concentrations were chosen to bracket the expected *in vivo* drug plasma concentrations in anthelmintic-treated dogs and cats (based on published pharmacokinetic studies). Motile and non-motile L3 were manually counted using a dissecting microscope after each well was pulsed with warm digest solution to stimulate larval activity. L3 mortality was analyzed by multilevel logistic models, generating dose-response relationships. Drug concentration estimates corresponding to a 50% larval mortality (M-50) identified that *C. vulpis* was the most sensitive species to the anthelmintics tested. *A. abstrusus* was most susceptible to moxidectin (M-50=117 ng/ml, 95%CI: 34–399) and selamectin (M-50=1193 ng/ml, 95%CI: 694–2051), while *A. vasorum* was highly resistant to all anthelmintics tested, except for selamectin at high drug concentrations (M-50=14372 ng/ml, 95%CI: 6547–31547).

Abstract R02: Whole Genome Sequence Analysis of Yorkshire Terriers with Tracheal Collapse: Identification of an Associated Variant

Dylan J. DeProspero[1]; Eleanor Hawkins[1]; Steven Friedenberg[2]; Kathryn Meurs[1]
[1]NCSU College of Veterinary Medicine; [2]UMN College of Veterinary Medicine

Tracheal collapse, a progressive degenerative disease of the tracheal cartilage rings, results in dorsoventral flattening of the dorsal tracheal membrane and significant airway compromise. Increased prevalence in toy and small breed dogs, like the Yorkshire Terrier, suggests a genetic origin for this disease. We hypothesized that whole genome sequencing (WGS) could identify pathologic variant(s) that segregated with tracheal collapse. We aligned and compared WGS data from eight affected Yorkshire Terriers to the canine reference sequence and our WGS database of 174 control dogs (29 breeds) and 107 risk breed dogs (small breed dogs at increased risk of tracheal collapse). Variants in affected dogs were evaluated for pathogenic implications using the Standards and Guidelines for the Interpretation of Sequence Variants. Missense variants were evaluated using in-silico programs (Polyphen, Sift, and Provean) for deleterious effects. Gene function was analyzed for impact on tracheal cartilage. Alcian Blue and Safranin O staining was performed on archival tracheal tissue from affected dogs to assess the cartilage for levels of glycosaminoglycans as well as proteoglycans and cell chondrogenesis, respectively. One variant, a deleterious polymorphism at Chr3:26,679,306 (Endofin, ZFYVE16), was found in seven out of eight affected Yorkshire Terriers but not in any control dogs; the variant was also identified in 22 dogs from at-risk breeds. Additional Sanger sequencing of 41 affected Yorkshire Terriers and 59 unaffected risk-breed dogs showed that the Endofin variant was significantly associated with tracheal collapse in affected Yorkshire Terriers ($p=0.007$). Alcian Blue and Safranin O staining of tracheal tissue from affected dogs demonstrated a loss of chondroitin sulfate from the cartilage. Endofin functions to regulate membrane trafficking in the endosome and is involved in multiple signaling pathways, including the bone morphogenetic protein (BMP) pathway. Through the BMP pathway, Endofin regulates bone and cartilage formation and Endofin dysfunction could lead to decreased BMP signaling, resulting in the degenerative and dysfunctional cartilage seen in tracheal collapse dogs. Endofin's specific function in this pathway is to facilitate phosphorylation of the signal transducer complex SMAD1/5/8. The disruption of the BMP pathway by a non-functional Endofin protein could explain the decrease in chondroitin sulfate commonly seen in affected tracheal collapse samples. Attempts to identify the difference in levels of Endofin and SMAD1/5/8 phosphorylation in the tissues using IHC was unsuccessful due to a uniform loss of cartilage, likely due to processing issues. Future work could entail repeating staining on additional fresher samples and with a wider spectrum of antibodies, which could be beneficial to discern if there truly was a difference between affected and unaffected samples. We identify here a variant in Endofin that is strongly associated with tracheal collapse and could result in the degenerative and dysfunctional cartilage seen in tracheal collapse dogs. This information furthers our understanding of the development of tracheal collapse in the Yorkshire Terrier and may lead to improved diagnosis and treatment for affected dogs.

Abstract R03: Detection of Canine Nasal Disease Using Infrared Thermography

Tekla Lee-Fowler; Stuart Clark-Price; Kara Lascola
College of Veterinary Medicine, Auburn University

Infrared thermography detects variations in heat signature and has been utilized in other species to non-invasively identify respiratory disease. This pilot study aimed to determine if infrared thermography could be used to detect nasal disease in dogs. Dogs presenting for nasal disease were recruited for the study. Dorsal and rostral images were acquired using a Fluke TiX580 60 Hz thermal imaging camera. Images were analyzed using the accompanying software, and regions of interest were defined over the right and left nasal passages to determine the average temperatures. Temperature differences and imaging patterns were subjectively correlated with computed tomography (CT) and histopathology. Images acquired from 8 dogs (5 spayed females; 3 neutered males). Clinical sings included epistaxis (n=6), bilateral mucopurulent discharge (n=2), and sneezing (n=1). Image analysis revealed a median temperature difference of 2.7°F on rostral images and 1.5°F on dorsal images between nasal passages, with the affected nasal passage having a higher temperature. The nasal passage with higher temperature correlated with obstructive lesions on CT (neoplasia=6/8 dogs). Subjective analysis of images allowed correct identification of affected side in 7/8 dorsal and 6/8 rostral images. Minimal airflow obstruction (lymphoplasmacytic rhinitis=1, focal inflammation/granuloma=1) resulted in incorrect subjective assessment. Obstructive nasal disease results in a local temperature increase in the affected nasal passage that can be non-invasively detected by infrared thermography. Further investigation is warranted to determine if thermography can assist in differentiation of nasal diseases and/or used for monitoring of disease progression and treatment.

Abstract R04: Documenting Bacterial Infection in Canine Aspiration Pneumonia

Jennifer Howard[1]; Carol Reinero[2]; Gregory Almond[1]; Aida Vientós-Plotts[2]; Leah Cohn[2]; Megan Grobman[1]
[1]College of Veterinary Medicine, Auburn University; [2]College of Veterinary Medicine, University of Missouri;

The background is aspiration pneumonia (AP) may occur without secondary bacterial infection. Widespread antimicrobial use is in part driven by lack of study documenting prevalence of bacterial infection in AP (b-AP).

Our hypothesis and objectives are to determine canine b-AP prevalence using bronchoalveolar lavage (BAL) cytology and culture, and identify risk factors and non-invasive criteria for predicting b-AP. We hypothesized <50% of dogs with AP would have b-AP (positive cytology and culture), and non-invasive markers would poorly predict b-AP.

There are 21 client-owned dogs.

This is a retrospective study at the University of Missouri and Auburn University (01/01/09-10/01/19). Inclusion criteria were radiographic diagnosis of AP and ≥1 risk factor, CBC, thoracic radiographs, and BAL cytology/culture collected within 48hrs of presentation. Dogs receiving >1 dose of antimicrobials within 7 days of BAL were excluded. Thirteen predictors of b-AP were assessed including predisposing conditions and radiographic score. Cohn's kappa coefficient determined agreement between cytology and culture. Mann-Whitney-Rank Sum test identified differences between groups (significance, $p<0.05$).

Only 21/426 dogs qualified for study enrollment; the most common reason for exclusion was lack of BAL collection (n=378 dogs). Nine dogs had concordant BAL cytology and culture (both positive, n=7; both negative n=2) and 12 dogs had discordant results. No agreement was found between culture & cytology (K:0.0). Bacterial-AP was identified more frequently with megaesophagus ($p=0.03$). No other markers were predictive of b-AP ($p>0.05$).

The conclusion are bronchoscopy and BAL are underutilized in AP. Diagnosis of b-AP is challenging with only one-third of dogs having clear-cut b-AP.

Abstract R05: Six Minute Walk Test in West Highland White Terriers with Pulmonary Fibrosis

Elizabeth Rozanski[1]; Luis Dos Santos[1]; Kay McGuire[2]; Lindsay Merkel[3]; John Rush[1]
[1]Tufts University; [2]Westie Foundation of America; [3]University of Minnesota

West Highland White Terriers (WHWT) are recognized to have a fibrosing pulmonary disease. In people with pulmonary fibrosis (PF), the 6-minute walk test (6MWT) is widely used to monitor progression of disease. The goal is this study was to evaluate the 6 MWT in healthy WHWT and in WHWT with PF. Healthy dogs were recruited at the National Specialty in Kimberton, PA in October 2019. Following owner consent, dogs were walked for 6 minutes outdoors and the distance recorded. The ambient temperature was 66 ° F and overcast. WHWT with PF were recruited from the hospital population of affected dogs, and from a Facebook Support Group (Westie Lung Disease-IPF in USA). Affected dogs were walked either outside or inside (e.g., during inclement weather). A diagnosis of PF was made based upon clinical examination and radiographs and/or CT scan. Distances between healthy dogs and affected dogs were compared. Additionally, as affected dogs are commonly older, the subgroup of healthy older WHWT (10 years and older) was separately compared to the affected dogs. Comparisons were made using a T-test and commercially available software. A p value of <0.05 was considered significant. Thirty-three healthy WHWT and 12 affected WHWT were enrolled. The mean distance walked by healthy dogs was 486±27 meters while the affected dogs walked 221±83 meters ($p<0.001$). Older healthy WHWT (n=8) walked 480±32 meters, which was also significantly ($p<0.001$) farther than affected WHWT. There was no different between young and older healthy WHWTs.

Abstract R06: Laryngotracheobronchoscopy Via Laryngeal Mask Airway in Cats and Dogs: A 16-Year Experience

Yukihito Shiroshita[1]; Teppei Suganuma[1]; Tomohiro Sakurai[2]; Kazuya Mamada[1]; Kenichi Inaba[1]; Sho Kusaba[1]
[1]VeRMS Study Group; [2]VeRMS

Tracheobronchoscopy in cats and small dogs is conventionally performed without intubation, with a complication rate of 30–40% and a mortality rate of 5.9%. Laryngotracheobronchoscopy via laryngeal mask airway (LTBS-LMA) using a flexible bronchoscope provides a constant airway and oxygen supply during examinations from the larynx to the lungs in cats and small dogs and is easy to perform with the animals in supine position. However long-term clinical studies are lacking. This study aimed to assess the practicality and safety of LTBS-LMA in cats and small dogs. The endobronchial anatomy of dogs in supine position and bronchial distribution on chest X-ray (CXR) were illustrated using four bronchoscopies, a 3D print, and a celluloid cast of the tracheobronchial tree of healthy dogs. Subsequently, the medical records of dogs and cats that underwent LTBS-LMA between 2002 and 2019 were reviewed. Animals were categorized into three groups: cats (F), small dogs (S, <5 kg), and medium-to-large dogs (ML, >5 kg). Indications and LTBS-LMA procedure outcomes were compared between groups using descriptive statistics and the chi-squared test. Altogether, 820 LTBS-LMAs in 572 cases (436 dogs, 136 cats) were included. The most common indications were abnormal CXR findings, chronic cough, and chronic dyspnea in F; and abnormal CXR findings, stridor, and chronic cough in the S and ML groups. LTBS-LMAs were performed in 212, 325, and 283 times in the F, S, and ML groups, respectively, with 90% of the animals in supine position. In the F, S, and ML groups, brushing was performed in 153, 234, and 130; biopsy (laryngeal, n=58; tracheal, n=33; and bronchial, n=39) in 55, 34, and 28; transbronchial lung biopsy in 23, 2, and 11; and bronchoalveolar lavage in 86, 139, and 127 LTBS-LMAs, respectively. Endoscopic interventions included stenting in 51, argon plasma coagulation in 46, debulking in 21, balloon dilation in 9, foreign body removal in 6, and snare resection in 6 LTBS-LMAs. The complication rates were 20.8%, 7.7%, and 4.6%; the mortality rates were 3.8%, 0.9%, and 1.1%, in the F, S, and ML groups, respectively. The mortality rate in cats decreased significantly ($p<0.01$) from 25.0% (3/12) in the initial 2-year period to 2.5% (5/200) in the 14-year period after defining a candidate criterion of requiring a partial pressure of oxygen in arterial blood >60 mm Hg in room air. LTBS-LMA provided endoscopic examinations and interventions from the larynx to the lungs in cats and small dogs with comparable practicality and safety to those in medium and large dogs.

Abstract R07: Quantitative Proteomics of Bronchoalveolar Lavage in West Highland White Terriers with Canine Idiopathic Pulmonary Fibrosis

Henna P. Laurila[1]; Merita Määttä[2]; Rosemary Maher[3]; Rob Beynon[3]; Paul McNamara[4]; Minna Rajamäki[5]
[1]Department of Equine and Small Animal Medicine, Faculty of Veterinary Medicine, University of Helsinki, Finland; [2]Faculty of Veterinary Medicine, University of Helsinki, Finland; [3]Institute of Integrative Biology, University of Liverpool; [4]Institute of Translational Medicine, University of Liverpool; [5]Faculty of Veterinary Medicine, University of Helsinki, Finland

Canine idiopathic pulmonary fibrosis (CIPF) is a fibrotic lung disease of West Highland white terriers (WHWTs) with poorly understood etiology. By measuring bile acids from bronchoalveolar lavage fluid (BALF), we showed that reflux microaspiration (MA) of gastrointestinal contents is associated with CIPF. Our aim was to identify BALF proteome differences between CIPF and healthy WHWTs by quantitative proteomics to find evidence of MA. BALF from 10 CIPF and 10 healthy WHWTs, as well as vomit and gastric juice from one dog, were digested by either trypsin or endopeptidase GluC. Resulting peptides were analysed by shotgun proteomic analysis using liquid chromatography tandem-mass spectrometry. Relative protein abundances for BALF from CIPF and healthy WHWTs were compared using principal component analysis (PCA) and ANOVA testing with adjustments for false discovery rate. BALF was additionally searched for pepsin and other gastric proteins. BALF total protein concentration was higher in CIPF than in healthy. Macrophage-secreted proteins cathelicidin, protein S100 and transforming growth factor beta-induced protein had a higher fold change in CIPF (11.06, 11.77 and 2.69, respectively) and carboxylic ester hydrolase had a lower fold in CIPF than in healthy (-3.67), q-value <0.05. Only minor differences were detected in gastric protein abundance between CIPF and healthy with trypsin digestion and none with GluC. No pepsin peptides were identified. CIPF BALF proteome differs from healthy WHWTs in abundance of proteins related to immune defence and inflammation. Using proteome analysis, we showed no overt reflux aspiration in CIPF WHWTs when compared to healthy WHWTs.

Abstract R08: Impact of Spacer Design on Drug Delivery and Potential Drug Cost Implications

Elizabeth Rozanski[1]; Mark Nagel[2]; Shae Sartori[2]
[1]Tufts University; [2]Trudell Medical International

The use of valved holding chambers (VHCs), also referred to as 'spacers', enables inhaled medications to be delivered to animals. Inhaled steroids are effective therapeutics for the management of respiratory diseases and are an alternative to systemic steroids which may have short- and long-term side effects. However, the delivery process is a critical component of aerosol drug delivery and must be considered when assessing drug performance. VHCs are designed to decelerate and capture the aerosol plume from a Metered Dose Inhaler (MDI) in order to alleviate coordination issues to give animals time to inhale the medication properly. Many animals have small tidal volumes that will require multiple breaths to empty the chamber of medication. Therefore, chamber design elements such as volume and length, the use of electrostatic-dissipative materials, inhalation valve function, and the design of the mask interface can influence delivery performance, which impacts pulmonary deposition and desired therapeutic effect. The purpose of this *in vitro* study was to assess drug availability in a range of chambers in a clinically relevant setting to provide guidance to veterinarians when recommending a VHC. This study evaluated the impact of drug availability when using fluticasone (Flovent HFA 110 µg) MDI with a range of different VHCs available on the market: AeroKat* Chamber (Trudell Medical International), Vet one 2in1 Feline Chamber (MWI), and four devices found on Amazon.com; Canack Aerosol Chamber, AeroFlow FAC, PawGreet FAC, and Inhaler Spacer for Cat. Each chamber was evaluated by breathing simulator that was programmed to simulate the tidal breathing pattern of a cat (tidal volume 50 mL, Inspiratory/expiratory ratio =1:2, 20 breaths per minute). Coordination delays between MDI actuation and inhalation by tidal breathing were imposed to simulate real-world use conditions. Delays of 2, 5, and 10 seconds were investigated to determine the impact on delivered drug mass. The breathing simulator was coupled to a filter which in turn was attached to the mouthpiece of the device on test. The mass of drug collected on the filter is indicative of total drug availability to the animal. Following activation of the MDI and appropriate delay interval, the filter was subsequently assayed via high performance liquid chromatography and the mass of fluticasone propionate determined. VHC 2s Delay 5s Delay 10s Delay AeroKat* Chamber 22.4±5.2 18.5±0.5 17.7±0.9 Vet one 2in1 Feline Chamber 12.8±2.4 9.0±1.6 7.3±1.4 Canack Aerosol Chamber 9.9±0.3 7.1±0.4 5.3±1.1 AeroFlow FAC 9.3±0.4 7.7±2.1 3.2±0.5 PawGreet FAC 8.2±0.6 6.0±3.4 2.1±0.8 Inhaler Spacer for Cat 0.2±0.1 0.2±0.1 0.2±0.0 Availability of medication has implications for disease control and pet owner expense. Animals not well controlled may need higher doses of medication to control the animal's disease. For example, using the 2s Delay results and AeroKat* Chamber as a reference, owners delivering a Flovent* 110 inhaler (2 puffs per day, 60/month, $280 per MDI) could experience increased drug costs of between $1,246–$3,774 annually. With new products entering the veterinary sphere and increasing access to online selling marketplaces, these devices need to be evaluated in order to validate drug delivery performance. Without this information, devices that appear similar but perform differently could mean that some patients will not receive the intended dose and as a result be at heightened risk of poor disease control and potential for respiratory exacerbation.

Using Antibiograms and Culture and Susceptibility Reports in Clinical Decision-Making

Erin Frey, DVM, MPH, DACVPM
Raleigh, NC, USA

Introduction

Antimicrobial-resistant infections in companion animals compromise the health and welfare of patients and caregivers and complicate empirical antibiotic decision-making. To address concerns of increasing antimicrobial resistance, professional organizations have outlined core principles of antimicrobial stewardship and defined the veterinarian's role as a steward to "preserve the effectiveness and availability of antimicrobial drugs."[2,3,23] Consensus groups, such as the Task Force for Antimicrobial Stewardship in Companion Practice and the Animal International Society for Companion Animal Infectious Diseases have published guidelines to educate veterinarians and provide tools that support antimicrobial stewardship principles.[4,13,16,24] Antimicrobial susceptibility testing (AST) and local or regional cumulative antibiograms can inform clinical decision-making and the judicious use of antimicrobial drugs (AMD).[4,6,12,23] When constructed using consensus-driven guidelines, antibiograms can guide empirical choice of antibiotics in the time between recognition of a patient's bacterial infection and when an AST report identifies the bacterial species and susceptibility pattern. In the face of antimicrobial resistance, bacterial infections, and limited prospects of new therapeutic options, veterinarians are urged to use these tools to make the best use out of the AMD we have for companion animals.[10]

Antibiograms

The Clinical and Laboratory Standards Institute (CLSI) defines an antibiogram as, "the report generated by analysis of results on isolates from a particular institution(s) in a defined period of time that reflects the percentage of first isolates (per patient) of a given species that is susceptible to each of the antimicrobial agents routinely tested."[7] Currently some academic veterinary specialty hospitals make use of their in-house microbiology laboratories and infectious disease expertise to build and use custom antibiograms from their patients.

Antibiograms can also help overcome the cost barrier related to AST testing. Benchmark and fee studies from 2017 and 2018 have indicated a median cost to the client of $174 and $125 (IQR: $93.00–167.50), respectively, and a 2015 survey of AVMA members noted that 84% of respondents felt that the cost of culture and AST prevented them from recommending them to clients.[1,11,21] Despite this cost barrier, ongoing AST testing is needed to ensure antibiograms remain current.

With supportive history and clinical signs, veterinarians can use point-of-care diagnostic testing (such as dry-mount urine sediment staining, lesional cytology of skin, and rapid urine tests) to support the presence of a bacterial infection.[14,15,17-20,22,25] Antibiograms can then be used to identify a subset of antimicrobial drugs with historically high patterns of susceptibility and in conjunction with consensus guidelines for selection of first-line therapies.[13,16,24] Once available, the AST report can guide practitioners through an "antibiotic time-out" to review their initial empiric choice and decide whether to continue, stop, or switch the previously prescribed antibiotic.[5]

Antimicrobial Susceptibility Testing (AST)

Antimicrobial susceptibility testing (AST) is a crucial component to managing bacterial disease in animals, and it is important that veterinarians understand how testing is performed and how to interpret their results when they are making decisions about patients.[9] Veterinarians should be aware that antimicrobial drugs may be listed in alphabetical order or by drug class on their AST reports, and neither of these conventions indicates a hierarchy of recommended use. Similarly, AMD may not be reported on the AST report due to tiered reporting in accordance with antimicrobial stewardship plans, because a drug is considered a drug of last resort in people, or when intrinsic resistance to a particular agent is expected.[9] Practitioners should consult with their laboratory to determine what type of tiered reporting or suppression of AMD is being used and to establish how frequently the laboratory updates their breakpoints to reflect current standards.

Species: Canine			
Source: Skin			
Organism ID: *Staphylococcus pseudintermedius*			
Antimicrobial agent	**Interpretation**	**MIC**	**Test range or comments**
Amoxicillin-clavulanate	S	≤0.25	Canine-specific breakpoint

Submitting Samples for AST

When considering when to send samples for AST testing, clinicians should decide whether the bacterial infection is clinically significant. Sampling of a contaminated site, such as the mouth, may yield commensal bacteria, and environmental sampling may yield bacterial growth unrelated to a patient's clinical condition.[9] Clinicians may anticipate a decreased response to AMD in the face of disease effects like chronic disease processes (e.g., necrosis), pharmacokinetic effects like decreased tissue blood flow (e.g., abscess), or local effects (e.g., foreign bodies).[9] Although MIC interpretations on isolates may suggest susceptibility, this may not be the case in these patients.

Interpreting AST Reports

In addition to determining whether the lab uses standardized AST methodologies, clinicians need to appreciate whether breakpoints are veterinary-specific or an extrapolation of human breakpoints. While some human breakpoints can be extrapolated to dogs, this is not universally true, and care should be taken to understand which breakpoints are specifically for canine patients, for example.

Critically, the breakpoints listed are matched to specific drug regimens (i.e., dose, route, frequency, and duration of use). In a clinical setting, drugs must be used at the same dosage regimen that was used to make the breakpoint, since any variance from this regimen will prevent the application of the associated MIC and interpretation (S,I,R).[8,9] Breakpoints are made with an understanding of the time- or concentration-dependent levels related to the MIC and the amount of active, non-protein-bound portion of a drug available, amongst other considerations.

Any extrapolation to other bacteria, other body sites, or other drugs should be made using consensus-driven reports or in consultation with a veterinary microbiologist or pharmacologist. Because breakpoints are based on systemic administration, they cannot be applied to topical uses (for example, ocular, otic, or dermal application).[9] In general, skin and soft tissue breakpoints can be used for lungs, bones, synovial fluid, and internal organs but not for protected sites in the body (i.e., the brain, eye, and intracellular sites) where tissue barriers prevent drug diffusion.[9] Additionally, Enterobacteriaceae susceptibility patterns from fecal sample isolates cannot be used to predict a clinical response in enteric infections.[9]

While some laboratories choose to report both the MIC and interpretation (S,I,R), others will only provide the interpretation of the MIC on the AST report. When choosing an AMD for a bacterial infection in a patient, clinicians should **not** compare the MIC between two AMD on their AST report. In the example below, although the MIC is different for the two drugs listed, the interpretation of susceptible applies to both cephalexin and trimethoprim-sulfa.

Patient species: Canine			
Source: Urine, cystocentesis			
Organism ID: *Escherichia coli*			
Antimicrobial agent	**Interpretation**	**MIC**	**Test range or comments**
Cephalexin	S	8	Canine-specific breakpoint
Trimethoprim-sulfa	S	2	Human breakpoint

The use of antibiograms and consensus-driven prescribing guidelines should direct the choice of appropriate drug from amongst those susceptible.[13,16,23] Where possible and available, an antimicrobial drug with a species-specific breakpoint should be chosen since it has been derived from studies of a particular dosage regimen in that same species (i.e., canine breakpoint for a canine patient).

REFERENCES

1. American Animal Hospital Association. *The Veterinary Fee Reference: Vital Statistics for Your Veterinary Practice.* Lakewood, CO: American Animal Hospital Association Press; 2018.
2. American Association of Feline Practitioners/American Animal Hospital Association. AAFP/AAHA basic guidelines of judicious therapeutic use of antimicrobials. Available at: www.aaha.org/globalassets/02-guidelines/infection-control/aafp_aaha_antimicrobialguidelines.pdf. Accessed February 14, 2020.
3. American Veterinary Medical Association. Antimicrobial stewardship definition and core principles. Available at: www.avma.org/resources-tools/avma-policies/antimicrobial-stewardship-definition-and-core-principles. Accessed February 14, 2020.
4. American Veterinary Medical Association. Task force on antimicrobial stewardship in companion animal practice: activities August 2013–December 2015. Available at: www.avma.org/sites/default/files/resources/TFASCAP_Report.pdf. Accessed February 14, 2020.
5. American Veterinary Medical Association. Veterinary checklist for antimicrobial stewardship. Available at: www.avma.org/sites/default/files/2019-11/Veterinary-Checklist-Antimicrobial-Stewardship.pdf. Accessed February 14, 2020.
6. Bender JB, *et al.* Antimicrobial stewardship in companion animal practice. *J Am Vet Med Assoc.* 2015;246:287–288.
7. CLSI. *Analysis and Presentation of Cumulative Antimicrobial Susceptibility Test Data; Approved Guideline.* 4th ed. CLSI document M39-A4. Wayne, PA: Clinical and Laboratory Standards Institute; 2014.
8. CLSI. *Performance Standards for Antimicrobial Disk and Dilution Susceptibility Tests for Bacteria Isolated from Animals.* 4th ed. CLSI supplement VET08. Wayne, PA: Clinical and Laboratory Standards Institute; 2018.
9. CLSI. *Understanding Susceptibility Test Data as a Component of Antimicrobial Stewardship in Veterinary Settings.* 1st ed. CLSI report VET09. Wayne, PA: Clinical and Laboratory Standards Institute; 2019.
10. Frey E. The role of companion animal veterinarians in one-health efforts to combat antimicrobial resistance. *J Am Vet Med Assoc.* 2018;253:11.
11. Grayzel SE, *et al.* Understanding companion animal practitioners' attitudes toward issues of antimicrobial stewardship. *J Am Vet Med Assoc.* 2015;247:883–884.
12. Guardabassi L, *et al.* Antimicrobial stewardship in small animal veterinary practice: from theory to practice. *Vet Clin North Am Small Anim Pract.* 2015;45:361–376.

13. Hillier A, *et al*. Guidelines for the diagnosis and antimicrobial therapy of canine superficial bacterial folliculitis (Antimicrobial Guidelines Working Group of the International Society for Companion Animal Infectious Diseases). *Vet Dermatol.* 2014;25:163–e43.
14. Jacob M, *et al*. Diagnostic accuracy of a rapid immunoassay for point-of-care detection of urinary tract infection in dogs. *Am J Vet Res.* 2016;77(2):162–166.
15. Kvitko-White H, *et al*. Evaluation of a catalase-based urine test for the detection of urinary tract infection in dogs and cats. *J Vet Intern Med.* 2013;27(6):1379–1384.
16. Lappin MR, *et al*. Antimicrobial use guidelines for treatment of respiratory tract disease in dogs and cats: Antimicrobial Guidelines Working Group of the International Society for Companion Animal Infectious Diseases. *J Vet Intern Med.* 2017;31:279–294.
17. Mendelsohn C, *et al*. Practical cytology for inflammatory skin diseases. *Clin Tech Small Anim Pract.* 2006;21(3):117–127.
18. O'Neill E, Horney B, Burton S, *et al*. Comparison of wet-mount, Wright-Giemsa and Gram-stained urine sediment for predicting bacteriuria in dogs and cats. *Can Vet J.* 2013;54:1061–1066.
19. Swenson CL, Boisvert AM, Kruger JM, *et al*. Evaluation of modified Wright-staining of urine sediment as a method for accurate detection of bacteriuria in dogs. *J Am Vet Med Assoc.* 2004;224:1282–1289.
20. Swenson C, Boisvert A, Gibbons-Burgener S, *et al*. Evaluation of modified Wright-staining of dried urinary sediment as a method for accurate detection of bacteriuria in cats. *Vet Clin Pathol.* 2011;40(2):256–264.
21. Tumblin D, *et al*. *The Well-Managed Practice Benchmarks Study 2017.* Columbus, OH: WMPB, LLC; 2017.
22. Way LI, Sullivan LA, Johnson V, *et al*. Comparison of routine urinalysis and urine Gram stain for detection of bacteriuria in dogs. *J Vet Emerg Crit Care.* 2013;23:23–28.
23. Weese JS, *et al*. ACVIM consensus statement on therapeutic antimicrobial use in animals and antimicrobial resistance. *J Vet Intern Med.* 2015;29(2):487–498.
24. Weese J, *et al*. International Society for Companion Animal Infectious Diseases (ISCAID) guidelines for the diagnosis and management of bacterial urinary tract infections in dogs and cats. *Vet J.* 2019;247:8–25.
25. Ybarra WL, *et al*. Performance of a veterinary urine dipstick paddle system for diagnosis and identification of urinary tract infections in dogs and cats. *J Am Vet Med Assoc.* 2014;244:814–819.

Calcium Oxalate Urolithiasis: New Insights into Mechanisms and Prevention

Eva Furrow, VMD, PhD, DACVIM (Small Animal Internal Medicine)
St. Paul, MN, USA

Introduction

Calcium oxalate (CaOx) is one of the most common stone types in dogs and yet one of the least well understood. Most patients are grouped into the category of "idiopathic CaOx stone formers," meaning that they do not have a systemic condition (such as hypercalcemia) underlying their risk.[1] In idiopathic stone formers, the etiology of CaOx urolithiasis is assumed to be complex, with risk determined by the combined effects of fixed (e.g., genetics, sex, size) and modifiable (e.g., diet, water intake, voiding behavior) factors. The contributions of these factors vary between individuals, and those with strong contributions from fixed factors are likely at greatest risk for stones. For example, the Miniature Schnauzer breed is consistently reported to have one of the highest odds ratios of CaOx uroliths and is at increased risk for recurrence relative to other breeds.[2-4]

Metabolic Derangements

Hypercalciuria is the most common abnormality associated with CaOx urolithiasis in dogs.[1,5,6] Hypercalciuria can be identified by measurement of urine calcium-to-creatinine ratios; ratios ≥0.06 mg/kg are typical for dogs with current or historic CaOx urolithiasis.[1,5,6] Hyperoxaluria is rare.[1,6-9] Deficiencies in crystallization inhibitors are also reported in dogs with CaOx urolithiasis and other urolith types, including struvite and urate.[10-12]

Greater than 90% of dogs with CaOx urolithiasis are normocalcemic, and the hypercalciuria is therefore deemed idiopathic.[1] Idiopathic hypercalciuria is categorized into three groups based on the underlying pathophysiologic mechanism: 1) absorptive, caused by hyperabsorption of calcium by the gut; 2) resorptive, caused by increased bone turnover; and 3) renal leak, caused by decreased renal resorption of calcium. In humans, idiopathic CaOx stone formers often have a combination of these derangements contributing to their hypercalciuria.[13] In dogs with idiopathic CaOx uroliths, the data supports absorptive as the most common or predominant mechanism of hypercalciuria (**Table 1**).

Table 1. Mechanisms of idiopathic hypercalciuria in dogs with CaOx urolithiasis

Category	Evidence
Absorptive (gut)	Likely the most common/predominant mechanism • On average, dogs with CaOx uroliths have higher blood ionized calcium than controls.[1] • Serum β-CrossLaps concentrations are lower in dogs with CaOx uroliths than controls, supporting decreased bone resorption as a compensatory response.[14] • A third of dogs with CaOx uroliths have increased ratios of serum 25(OH)D/24,25(OH)$_2$D, consistent with decreased inactivation of vitamin D metabolites, which could drive calcium absorption.[15]
Resorptive (bone)	Not yet documented
Renal leak	Primary renal leak hypercalciuria is present in a subset of stone formers affected by a hereditary disorder (**see Table 2**).

Genetic Disorders

In humans, up to one-third of young patients (≤25 years old) with nephroliths or nephrocalcinosis are affected by a monogenic (single gene) disorder.[16] In order to identify dogs that might have monogenic disorders, we reviewed over 140,000 CaOx stone submissions to the Minnesota Urolith Center. We extracted data from juvenile dogs (≤2 years old at diagnosis) and compared it to mature dogs (7–9 years at diagnosis). Two breeds stood out as being significantly overrepresented in the juvenile group compared to the mature group: the English and French Bulldog.[17] We performed whole genome sequencing of English Bulldogs with CaOx urolithiasis and were able to filter down the variant list from several million to a single variant that met the criteria of being present in all affected English Bulldogs sequenced, absent from other dogs (including bulldogs without CaOx stones) and predicted to impact protein function. We refer to this as the Hereditary Calcium Oxalate Urolithiasis, Type 1 (CaOx1) variant. The Hereditary CaOx1 variant resides in a protein that is exclusively expressed in the kidney and is released in the urine. The protein functions to regulate urinary calcium resorption and inhibit urinary crystallization, as well as contribute to blood pressure control and immunity against urinary tract infections.

Hereditary CaOx1 is an autosomal recessive disorder; only dogs with two copies of the variant are at increased risk for CaOx uroliths. The disorder is present in 60–65% of English Bulldogs with CaOx uroliths compared to <1% of the general English Bulldog population, resulting in an odds ratio of >100. Male dogs with CaOx1, regardless of reproductive status, are more likely to develop uroliths, and the median age at first stone diagnosis is 2.5 years (range 1–8 years). The Hereditary CaOx1 variant also exists in several non-bulldog breeds (**Table 2**). Of note, Hereditary CaOx1 has not been identified in French Bulldogs, supporting the presence of a distinct genetic disorder causing stone risk in this breed. Hereditary CaOx1 has also not been detected in the breeds with greatest overall risk for CaOx urolithiasis, such as the Miniature Schnauzer, Bichon Frise, Lhasa Apso, or Shih Tzu.[2,3] Genetic testing for Hereditary CaOx1 is available through the University of Minnesota Canine Genetics Laboratory: https://vetmed.umn.edu/research/labs/canine-genetics-lab.

Table 2. Breeds affected by Hereditary CaOx1

American Staffordshire Terrier	Border Collie	Boston Terrier
Bullmastiff	English Bulldog	Havanese
"Pit Bull"	Rottweiler	Staffordshire Bull Terrier

Another monogenic CaOx disorder reported in dogs is primary hyperoxaluria. A causal variant in a hepatic enzyme that metabolizes glyoxylate, *AGXT* was discovered as the cause of primary hyperoxaluria type 1 in Coton de Tulear dogs with a severe pediatric-onset crystal nephropathy.[9]

Moving Toward Precision Medicine

Historically, the recommendations for prevention of CaOx stone recurrence have been applied to all CaOx stone formers, i.e., a "one-size-fits-all" approach. The efficacy of this general management strategy is low, with recurrence rates as high as 50% in <2 years for some breeds.[4] The application of research findings to clinical care is critical to moving toward precision medicine for CaOx urolithiasis. Tailoring therapy to the underlying metabolic or genetic disorder in an individual patient could improve success rates. A list of therapies proposed for prevention of CaOx urolithiasis is provided in **Table 3** with the rationale for their use and whether or not the evidence to date supports it. Additional recommendations for CaOx urolith prevention can be found at the Minnesota Urolith Center's website: https://urolithcenter.org.

Table 3. Rationale versus evidence for the use of preventative therapies for CaOx urolithiasis in dogs

Therapy	Rationale	Evidence
Potassium citrate*	Citrate reduces CaOx crystallization by binding calcium, forming soluble complexes with calcium ions, and increasing the activity of other crystallization inhibitors.	Hypocitraturia has not been reported in dogs, but deficiencies in crystallization inhibitors are common in dogs with idiopathic CaOx stones and are a major feature of Hereditary CaOx1.
Hydrochlorothiazide	Thiazide diuretics promote calcium resorption in the proximal tubule and thereby decrease hypercalciuria.	Hereditary CaOx1 causes hypercalciuria due to decreased resorption of calcium in the distal convoluted tubule. Thiazide diuretics are a specific therapy for this disorder. Idiopathic CaOx stone formers also often have hypercalciuria, but it is unlikely due to renal leak. While hydrochlorothiazide could still reduce calciuresis in these patients, blood ionized calcium concentration should be measured prior to and after starting this medication.
Vitamin B6	A derivative of vitamin B6 is a co-factor for *AGX1*, a hepatic enzyme that metabolizes glyoxylate.	Commercial dog foods should be formulated with sufficient B6; supplementation is recommended for dogs on homemade diets. Vitamin B6 is recommended for patients with primary hyperoxaluria type 1, regardless of diet, but this condition is extremely rare.
Bisphosphonates	Bisphosphonates inhibit osteoclast function and decrease resorptive hypercalciuria.	Resorptive hypercalciuria has not been documented in dogs with CaOx urolithiasis. Rather, there is data supporting reduced bone resorption, likely as a compensatory mechanism to absorptive hypercalciuria. The use of these drugs in patients that already have reduced bone resorption might increase risk for adverse effects.
Sodium	Dietary sodium promotes water intake and thereby dilutes urinary calcium and oxalate.	Sodium decreases urinary calcium resorption by the kidneys, resulting in increased calcium excretion. In short-term trials in dogs, this effect appears to be canceled out by the increased urine volume,[18] but it is unknown how dogs with idiopathic hypercalciuria respond to sodium supplementation, especially over time. In rodent models, a high sodium diet markedly increases CaOx stone risk, and a low sodium diet decreases risk, despite increased diuresis on the higher sodium diet.[19,20] In people, a high sodium diet is considered a risk factor for CaOx stones, and sodium restriction decreases recurrence rates.[21]

*Avoid supplements containing cranberry extract, which increases urinary oxalate excretion in people.[22]

References

1. Furrow E, *et al.* Fasting urinary calcium-to-creatinine and oxalate-to-creatinine ratios in dogs with calcium oxalate urolithiasis and breed-matched controls. *J Vet Intern Med.* 2015;29:113.
2. Low WW, *et al.* Evaluation of trends in urolith composition and characteristics of dogs with urolithiasis: 25,399 cases (1985–2006). *J Am Vet Med Assoc.* 2010;236(2):193.

3. Hunprasit V, *et al.* Epidemiologic evaluation of calcium oxalate urolithiasis in dogs in the United States: 2010–2015. *J Vet Intern Med.* 2019;33:2090.
4. Allen HS, *et al.* Associations of diet and breed with recurrence of calcium oxalate cystic calculi in dogs. *J Am Vet Med Assoc.* 2015;246:1098.
5. Carr SV, *et al.* Measurement of preprandial and postprandial urine calcium to creatinine ratios in male miniature schnauzers with and without urolithiasis. *J Vet Intern Med.* 2020. doi: 10.1111/jvim.15690. [Epub ahead of print]
6. Dijcker JC, *et al.* Urinary oxalate and calcium excretion in dogs and cats diagnosed with calcium oxalate urolithiasis. *Vet Rec.* 2012;171:646.
7. Stevenson AE, *et al.* Risk factor analysis and relative supersaturation as tools for identifying calcium oxalate stone-forming dogs. *J Small Anim Pract.* 2003;44:491.
8. Danpure CJ, *et al.* Enzymological characterization of a putative canine analogue of primary hyperoxaluria type 1. *Biochim Biophys Acta.* 1991;1096:134.
9. Vidgren G, *et al.* Primary hyperoxaluria in Coton de Tulear. *Anim Genet.* 2012;43:356.
10. Carvalho M, *et al.* Role of urinary inhibitors of crystallization in uric acid nephrolithiasis: dalmatian dog model. *Urology.* 2003;62:566.
11. Raila J, *et al.* Levels of retinol and retinyl esters in plasma and urine of dogs with urolithiasis. *J Vet Med A Physiol Pathol Clin Med.* 2003;50:380.
12. Carvalho M, *et al.* Defective urinary crystallization inhibition and urinary stone formation. *Int Braz J Urol.* 2006;32:342.
13. Worcester EM, *et al.* New insights into the pathogenesis of idiopathic hypercalciuria. *Semin Nephrol.* 2008;28:120.
14. Luskin AC, *et al.* Bone resorption in dogs with calcium oxalate urolithiasis and idiopathic hypercalciuria. *Res Vet Sci.* 2019;123:129.
15. Groth EM, *et al.* Vitamin D metabolism in dogs with and without hypercalciuric calcium oxalate urolithiasis. *J Vet Intern Med.* 2019;33:758.
16. Daga A, *et al.* Whole exome sequencing frequently detects a monogenic cause in early onset nephrolithiasis and nephrocalcinosis. *Kidney Int.* 2018;93:204.
17. Saver A, *et al.* Calcium oxalate urolithiasis in juvenile dogs. *J Vet Intern Med.* 2018;32:2278.
18. Stevenson AE, *et al.* Effect of dietary moisture and sodium content on urine composition and calcium oxalate relative supersaturation in healthy miniature schnauzers and Labrador retrievers. *Res Vet Sci.* 2003;74:145.
19. Huang HS, *et al.* High sodium-induced oxidative stress and poor anticrystallization defense aggravate calcium oxalate crystal formation in rat hyperoxaluric kidneys. *PLoS One.* 2015;10:e0134764.
20. Krieger NS, *et al.* Low sodium diet decreases stone formation in genetic hypercalciuric stone-forming rats. *Nephron.* 2019;142:147.
21. Qaseem A, *et al.* Dietary and pharmacologic management to prevent recurrent nephrolithiasis in adults: a clinical practice guideline from the American College of Physicians. *Ann Intern Med.* 2014;161:659.
22. Terris MK, *et al.* Dietary supplementation with cranberry concentrate tablets may increase the risk of nephrolithiasis. *Urology.* 2001;57:26.

New Insights in Biomarkers of Urinary Diseases

Gilad Segev, DVM, DECVIM-CA
Jerusalem, Israel

INTRODUCTION

For decades, serum creatinine (sCr), urea and urine concentrating ability were the only markers for kidney function, despite their low sensitivity and specificity. The development of novel kidney biomarkers allows identification of kidney injury even when the latter is not associated with any decrease in kidney function, therefore, the growing availability of these markers might change some of the common paradigms in veterinary nephrology in the next years.

Limitations of Serum Creatinine Concentration

Despite the diagnostic advancements made in other medical fields (e.g., the use of biomarkers in cardiology), sCr is still being used as the marker for kidney function, despite its multiple shortcomings, including: 1) High variability among dog breeds, resulting in a very wide reference range. Consequently, sCr is not expected to rise above the reference range in most dog breeds until ~75% of nephrons become non-functional and in many cases early decrease in kidney function results in changes within the reference range. 2) sCr is affected by extra-renal factors, particularly muscle mass and hydration status, therefore lacks specificity. 3) sCr is a functional marker thus it is blinded to kidney injury that is not accompanied by decreased kidney function. 4) sCr does not represent the severity of the dysfunction until a steady-state has been reached. Consequently, substantial changes in GFR at the early stages of kidney injury are associated with relatively small changes in sCr.

The above limitations of sCr are being reflected in the findings of several studies indicating that a small, and even transient, increase in sCr in human patients is detrimental. In one study, as little as 0.5 mg/dL increase in sCr was associated with increased in-hospital mortality.[1] In another study, a transient increase in sCr (for 1–3 days) was also associated with increased odds ratio for in-hospital mortality.[2] Finally, even a small and transient increase in sCr in patients that were discharged from the hospital, was associated with the need for chronic dialysis over the ensuing three years.[3] The aforementioned limitations and studies imply that relying on sCr as the only marker of kidney function does not provide all the information needed to accurately assess kidney function, to promote timely manner intervention, and to determine the prognosis. Symmetric dimethylated arginine (SDMA) overcomes some of the aforementioned limitations, as this marker is not affected by muscle mass and has a narrower reference range. Indeed, there is growing evidence regarding the utility of SDMA as an early marker of kidney dysfunction, and recent evidence suggests that SDMA is an earlier marker of CKD both in dogs and cats compared with sCr.[4,5] Yet, SDMA is a functional marker as well and like sCr does not detect injury that is not associated with a decrease in kidney function.

POTENTIAL UTILITIES OF RENAL BIOMARKERS

Characteristics

In recent years, many researches in nephrology have focused on identifying sensitive and specific biomarkers. An ideal biomarker should meet several requirements: 1) be detectable in urine and/or plasma, 2) assays should be readily available and cost effective, 3) the marker should high prediction of kidney injury (namely high sensitivity and specificity), 4) provide information regarding the etiology, 5) provide information regarding the location of the injury (i.e., glomerulus, tubules), 6) reflect the severity of the injury, 7) indicate both kidney injury and repair processes and, 8) predict the likelihood of recovery and the outcome.

Early Diagnosis of Kidney Injury

Renal biomarkers have several potential advantages, of which early diagnosis is considered a major one.[4] It has been shown that kidney injury can be identified using sensitive biomarkers, days before any increase in sCr is documented,[6] thus measurement of biomarkers may alert the clinician's attention to ongoing kidney damage, before there is a measurable reduction in kidney function. For example, it would be more reasonable to monitor a patient receiving a nephrotoxic drug using biomarkers with the capability to indicate injury, rather than using markers that can only indicate presence of decreased kidney function. Such practice is expected to be associated with better outcome, since once kidney function decreases, recovery is expected to be prolonged and the patient might die before the kidney had the opportunity to recover (even if the latter is possible).

Screening Patients at Risk for AKI

Biomarkers indicating active kidney injury are likely more sensitive compared with markers of GFR (e.g., SDMA, sCr). It has been shown that the prevalence of AKI in hospitalized patients is relatively high when evaluating serial changes in sCr or urine production, yet these reported data likely under represent the true prevalence of AKI, as these data rely on functional markers and not on injury markers. Early recognition of kidney injury in hospitalized patients will allow timely identification, and potentially elimination of the cause, and might also facilitate therapy initiation before the injury progresses to failure.

Differentiating Upper and Lower Urinary Tract Infection

A recent study of neutrophil gelatinase-associated lipocalin (NGAL) demonstrated increased urinary NGAL concentration in a subset of dogs with apparently lower urinary tract disease, potentially indicating upper urinary tract involvement.[7] Yet, any increase in urinary NGAL concentration might be the result of local inflammation within the lower urinary system, and not necessarily due to kidney damage, since NGAL originates also from neutrophils that are being recruited as part of the local inflammatory process.[8] Nonetheless, it is possible that increase in urinary NGAL concentration actually indicated kidney damage (e.g., pyelonephritis) in some of these animals. Development of markers that are specific to the kidney (i.e., kidney-specific biomarkers) would likely aid in detecting upper tract infection when UTI is documented. Moreover, sequential changes of these biomarkers could guide treatment.

Markers of Chronic Kidney Disease Progression

The traditional conception is that AKI and chronic kidney disease (CKD) represent two distinct processes of kidney damage. AKI represents rapidly progressing active damage whereas CKD represent slowly progressive damage. Lately, it has been suggested that AKI and CKD might not be separate entities, as they are influenced by various similar conditions, share common risk factors and ultimately impact each other (AKI is a risk factor for CKD and vice versa).[9] In both dogs and cats with CKD, the etiology is often unidentified, but regardless of the etiology, there is gradual replacement of the normal kidney parenchyma with inflammation and scar tissue. Results of recent studies indicate that active kidney damage is present in a subset of dogs with apparently stable (based on functional markers) CKD. This finding might account, at least in part, for the wide variation in progression rate of CKD that occurs among dogs and cats.[10,11] Presence of active kidney injury, demonstrated by increased concentrations of biomarkers, likely predicts risk for rapid progression of the disease, whereas absence of active kidney injury, predicts a more slowly progressive kidney disease.

Identification of markers of kidney damage in dogs with CKD might also facilitate the diagnosis of IRIS Stage I CKD, which is currently challenging using the available tools. Early identification of the disease will allow early intervention, aimed to preserve kidney function. Moreover, assessment of biomarkers indicating active kidney damage in animals with CKD will facilitate the investigation of novel therapeutic interventions. Active injury biomarkers will be monitored sequentially after the application of these therapies, at the same manner as alanine aminotransferase is being used to assess the efficacy of different interventions during the management of liver diseases. The currently available markers of kidney damage preclude such an assessment, as short-term interventions are unlikely to alter kidney function, despite either benefit or harmful effects.

Active Research

In recent years, plethora of studies have been published exploring the utility of different biomarkers in different clinical settings. These include N-acetyl-β-glucosaminidase, Kidney Injury Molecule-1, Retinol Binding Protein, NGAL, Interleukins and others. The limitation of many of these markers is that these markers are either functional markers or markers that are not specific to the urinary system. The latter is an important limitation especially in AKI patients or in patients with systemic diseases, in which multiple organs may be involved in the disease, and thus biomarkers originating from other organs can reach the urine via glomerular filtration, regardless of kidney damage. Development of kidney specific biomarkers overcomes this major limitation, as their presence in the urine indicate kidney damage.
Recent unpublished data suggest that kidney specific serum inosine, urinary cystatin B and urinary clusterin are some of the promising biomarkers. These markers are alluring, as kidney specific assays have been developed and encompass the potential to predict accurately active and ongoing injury processes. These markers thus have the potential to identify Grade I AKI, Stage 1 CKD, upper urinary tract involvement in UTI, and be used as surrogate markers for active ongoing damage and progression rate in animals with CKD.

References

1. Chertow GM, *et al.* Acute kidney injury, mortality, length of stay, and costs in hospitalized patients. *J Am Soc Nephrol.* 2005;16:3365.
2. Uchino S, *et al.* Transient azotaemia is associated with a high risk of death in hospitalized patients. *Nephrol Dial Transplant.* 2010;25:1833.
3. Wald R, *et al.* Chronic dialysis and death among survivors of acute kidney injury requiring dialysis. *JAMA.* 2009;302:1179.
4. Yerramilli M, *et al.* Kidney disease and the nexus of chronic kidney disease and acute kidney injury: the role of novel biomarkers as early and accurate diagnostics. *Vet Clin North Am Small Anim Pract.* 2016;46:961.
5. Nabity M, *et al.* Symmetric dimethylarginine assay validation, stability, and evaluation as a marker for the early detection of chronic kidney disease in dogs. *J Vet Intern Med.* 2015;29:1036.
6. Palm CA, *et al.* Urinary neutrophil gelatinase-associated lipocalin as a marker for identification of acute kidney injury and recovery in dogs with gentamicin-induced nephrotoxicity. *J Vet Intern Med.* 2016;30:200.
7. Segev G, *et al.* Evaluation of neutrophil gelatinase-associated lipocalin as a marker of kidney injury in dogs. *J Vet Intern Med.* 2013;27:1362.
8. Decavele, *et al.* Increased urinary neutrophil gelatinase associated lipocalin in urinary tract infections and leukocyturia. *Clin Chem Lab Med.* 2011;49:999.
9. Cowgill LD, *et al.* Is progressive chronic kidney disease a slow acute kidney injury? In: *Vet Clin North Am Small Anim Pract.* United States: Elsevier Inc; 2016:995.
10. Jacob F, *et al.* Clinical evaluation of dietary modification for treatment of spontaneous chronic renal failure in dogs. *J Am Vet Med Assoc.* 2002;220:1163.
11. Ross SJ, *et al.* Clinical evaluation of dietary modification for treatment of spontaneous chronic kidney disease in cats. *J Am Vet Med Assoc.* 2006;229:949.

Role of Dietary Fiber in Canine and Feline Gastrointestinal Disease

Aarti Kathrani, BVetMed (Hons), PhD, DACVIM (SAIM), DACVN, FHEA, MRCVS
London, UK

DEFINITION OF FIBER

Fiber is defined as mainly carbohydrate polymers, which are present in edible plants and are not hydrolysed by mammalian digestive enzymes present in the small intestine nor absorbed in the small intestine[1]. Dietary fiber includes celluloses, hemicelluloses, lignins, oligosaccharides, pectins, gums, waxes, resistant starches and associated compounds such as polyphenols.

CLASSIFICATION OF FIBER

Fiber can be classified based on its solubility in water, its viscosity and its fermentation. The action of fiber depends on these three properties. Solubility refers to the ability of the fiber to be fully dispersed when mixed with water. Soluble fibers include pectins, gums and mucilages, whereas cellulose and lignin are examples of insoluble fiber[2]. Fibers with increased viscosity can reduce the rate of passage of digesta through the gastrointestinal tract, slow down the rate of digestion of nutrients and can increase satiety leading to decreased food intake[3]. Fiber can also be described as rapidly to slowly fermentable. Fibers that are more rapidly fermented result in more short-chain fatty acid production in a shorter period of time versus fiber sources that ferment more slowly. The proportions of the different short-chain fatty acids (butyrate, propionate and acetate) produced are determined by the bacterial species that are fermenting the fiber, as well as the specific type of dietary fiber being fermented[4]. Short-chain fatty acids promote proliferation of the intestinal mucosal epithelium and villus height[3]. They also help to preserve the mucosal barrier by stimulating differentiation of goblet cells and enhancing mucus production[3]. Short-chain fatty acids promote colonic regulatory T cells to help prevent excessive inflammation[3]. In addition to short-chain fatty acid production, soluble fiber decreases fecal bulk, increases gastric retention by increasing viscosity of gastric contents, lowers the pH of the intestinal lumen and alters bile acid profiles[3].

Table 1. Properties, examples and functional description of different dietary fibers[5]

Properties	Examples	Functional description
Soluble, non-viscous, readily fermentable	Inulin Wheat dextrin	Readily dissolve in water and are rapidly and completely fermented
Soluble, viscous, readily fermentable fibers	Gums Pectins	Form a gel-like consistency in water; however, these properties are then lost following fermentation
Soluble, viscous, slowly fermentable fibers	Psyllium	Forms a gel-like consistency in water, which does not undergo extensive fermentation; hence, its viscosity if preserved throughout the colon, which helps to soften feces in constipation and add form to loose feces
Insoluble fibers	Wheat bran Cellulose Lignin	Exert a laxative effect via stimulation and irritation of the intestinal mucosa to increase secretion and peristalsis

DETERMINATION OF DIETARY FIBER

The regulations in the USA require that the maximum amount of crude fiber be listed on pet food labels. Crude fiber measures most of the insoluble fibers (cellulose, lignin and some hemicelluloses that are slowly fermentable) but none of the soluble fibers[6]. Therefore, the crude fiber measurement does not accurately depict the total dietary fiber. Indeed, the crude fiber method accounts for only 5 to 20% of total dietary fiber in a food[7]. Conversely, total dietary fiber measures soluble fibers (fructans, galactans, mannoses, mucilages, pectins), as well as the insoluble fibers, but does not measure lower molecular weight saccharides such as fructooligosaccharides[6,7]. However, total dietary fiber is able to separate out the soluble fiber and insoluble fiber fractions. The product guides of various therapeutic diets in the USA commonly list the total dietary fiber, as well as the insoluble and soluble fiber content.

FUNCTION OF FIBER IN THE GASTROINTESTINAL TRACT

Fiber has many effects on the gastrointestinal tract. Fiber can help to increase fecal bulk, as well as moisture content and help regulate bowel motility. Fermentation of dietary fiber by bacteria mostly results in the production of short-chain fatty acids, such as butyrate[8]. Short-chain fatty acids are important promoters of homeostasis of the colon, as they control colonic motility and blood flow and intestinal pH[8,9]. Short-chain fatty acids are weak anions and therefore may exert osmotic pressure within the colon and increase the water content of the stool. However, their osmotic and reabsorptive properties appear to be dose dependent.

Butyrate is a major fuel source for colonocytes[10] and helps to keep the intestine anaerobic, thereby inhibiting the growth of pathogenic bacteria. Butyrate also has direct effects on intestinal barrier function, the intestinal immune system by its direct effects on intestinal epithelial cells, dendritic cells, macrophages and mast cells and by regulating transcriptional activity[11]. Specifically, butyrate has been shown to exert anti-inflammatory and anti-carcinogenic effects via epigenetic modifications via the inhibition of histone deacetylases[4]. Butyrate also increases the expression of mucins and antimicrobial peptides in intestinal epithelial cells[12], thereby preserving intestinal barrier function[13]. Butyrate's effects on the immune system include: 1) it can induce the expression of PPAR-gamma, which is able to exert anti-inflammatory effects via antagonism of NF-kappa B[14], and 2) it is able to enhance the stability and activity of T regulatory cells[15,16].

In addition, certain dietary fibers can also affect the intestinal immune system through binding to toll-like receptors and carbohydrate binding domains[11]. Dietary fibers are also able to support colonization of beneficial microbiota and block adhesion of pathogenic microbiota by providing alternative adhesion sites[3]. In a pig microbiome study, high-fiber/low-fat diet was associated with higher concentrations of *bifidobacteria, lactobacilli,* and *faecalibacterium prausnitzii,* which have a protective role in intestinal inflammation, as well as increased short-chain fatty acid production[17,18]. A recent systematic review found that fiber intervention particularly with fructans and galacto-oligosaccharides in healthy adults leads to increased abundance of both *Bifidobacterium* and *Lactobacillus* species[19]. Human IBD patients are known to have dysbiosis with decreased numbers of short-chain fatty acid producing bacteria and reduced butyrate concentrations[13]. Interestingly, the richness of the donor microbiota with butyrate-producing bacteria is correlated with success of fecal microbiota transplantation in humans with IBD[20]. Poorly fermentable fibers, such as cellulose, cause a stool bulking effect that causes shorter intestinal transit times and increases stool mass[7].

Certain fermentable fibers act as prebiotics, which are defined as "selectively fermented non-digestible food ingredients or substances that specifically support the growth and or activity of health promoting bacteria that colonize the gastrointestinal tract"[21].

Advantages of Fiber

Although dietary fiber is not considered essential for dogs and cats, they are important for the maintenance of colonic health. Total deficiency of fiber in typical pet foods is not a practical problem, as many ingredients used in pet foods contain some fiber[6]. The addition of fermentable fiber to the diet can provide many benefits, such as regular bowel movements, competition of healthy bacteria against potentially harmful bacteria and generation of short-chain fatty acids. Also, the addition of dietary fiber prevents the fermentation of protein, which therefore avoids the generation of possible toxic and cancer-promoting metabolites[2]. The proposed mechanisms for the protective effect of fiber on colorectal cancer include dilution of carcinogens in the feces, quicker intestinal transit times, short-chain fatty acid production and binding of carcinogenic bile acids[22,23]. Although short-chain fatty acids provide less than 5% of the energy needs of dogs and cats[6], generation of the short-chain fatty acid butyrate is not only an important energy source for colonocytes, but also has many anti-inflammatory properties by downregulating the immune response via inhibiting the transcription of proinflammatory cytokines and promoting the differentiation of T-regulatory cells[15,24].

Dogs fed a blend of fermentable fiber had longer jejunal villi compared to a diet with poorly fermentable fiber[25]. Dogs fed the insoluble fiber cellulose had the highest incidence of colonic crypt mucus distension and cryptitis compared to those fed the soluble fibers beet pulp or pectin and gum arabic[7]. Dogs fed fructooligosaccharides were shown to have greater blood flow via the left colic artery than dogs fed cellulose or beet pulp[7]. However, studies using porcine and poultry have shown that the beneficial effects of dietary fiber on gut health depends not just on the type and form and the physiochemical characteristics of the dietary fiber, but also on the matrix in which it lies[26,27].

Potential Disadvantages of Fiber

The amount of dietary fiber and utilization of nutrients are inversely proportional to each other[3]. Therefore, as the amount of dietary fiber increases, the utilization of nutrients decreases. Dietary fiber can also bind to ions such as calcium, zinc and copper, which can be released in the distal gut as the fiber is fermented, where they can then exert effects such as local antimicrobial action[28]. However, some fiber types reduce and others enhance mineral absorption[6]. Increased amounts of dietary fiber may also dilute energy and nutrient density to such an extent the animal may not be able to consume enough to meet their needs, and this may negatively impact on their growth rate. Dietary fibers that are viscous may form gels in the gastrointestinal tract and therefore act as steric hindrance to enzymatic hydrolysis[7], which can reduce nutrient absorption. The inclusion of dietary fiber in pet foods may reduce food intake. Possible causes for the decreased intake include reduced palatability of the diet, gastrointestinal upset (for example, flatulence and diarrhoea) and mechanical satiation. Inclusion of increased amounts of fermentable fiber can result in increased production of short-chain fatty acids, which may result in flatulence, abdominal distention, diarrhoea and increased fecal frequency. Similarly, the abrupt addition of or change in fiber intake can also lead to bloating, abdominal cramps and increased flatulence[29]. Also, soluble and viscous fibers can delay gastric emptying and therefore may worsen signs of dyspepsia. This was demonstrated in a small study in people where transient lower oesophageal sphincter relaxations were observed with fiber supplementation[30]. Due to concerns regarding the negative effects of certain dietary fiber, safe upper limits are in place (for example, the safe upper limit for beet pulp in adult dog diets is set at 7.5% DM basis)[7].

Veterinary Studies Assessing Fiber in Gastrointestinal Disease

There have been limited studies assessing the efficacy of dietary fiber in canine and feline chronic enteropathy. One retrospective study showed that treatment with a commercial therapeutic, highly digestible diet supplemented with soluble fiber resulted in a very good to excellent response in most dogs with chronic idiopathic large bowel diarrhoea[31]. In another study, 19 dogs with chronic colitis that had failed a low-fat diet responded to a fiber-supplemented diet, and maintenance could be achieved without the concurrent need for other medications[32]. In cats with colitis, one study showed response to diets high in fiber or supplemented with fiber[33]. One uncontrolled trial in cats with constipation showed that a commercial psyllium-enriched dry diet was able to improve stool quality and clinical outcome in over 80% of cats within a week of starting the diet.[34]

Fiber can also serve as a prebiotic. Healthy dogs fed oligofructose had fewer *Enterobacteriaceae* and *clostridia* and greater numbers of *lactobacilli*[35]. Similarly, healthy cats fed diets containing oligofructose had significantly increased fecal concentrations of *lactobacilli* and decreased concentrations of *C. perfringens* and *E. coli* compared to controls[36]. Studies have also demonstrated positive effects on the systemic immune response following oral supplementation of mannan oligosaccharides in dogs[7,37]. One study assessing the effects of chondroitin sulfate and several prebiotics including resistant starch, beta-glucan and mannan oligosaccharides in 27 dogs with IBD showed that the histologic score of dogs that received the supplement together with a hydrolyzed diet decreased by 1.53 fold[38]. Another study demonstrated that feeding 1% fructooligosaccharide decreased the number of aerobic and facultative anaerobic bacteria in the small intestine of German Shepherd dogs susceptible to small intestinal bacterial overgrowth[39].

The response to increased dietary fiber intake varies from individual to individual due to differences in the composition of the microbiota, which can be influenced by diet[5]. Of the *Bacteroidetes*, a *Prevotella*-enriched microbiome is associated with higher concentrations of short-chain fatty acids because of specific enzymes for polysaccharide breakdown[40]. Therefore, personalized nutrition is important in relation to fiber intake, since there is variability in the microbiome, which will lead to significant variability in response to dietary fiber intake. Therefore, the effects of dietary fiber in individual cases still remain as trial and error. In addition, recent studies would suggest that reintroduction of specific fiber-sensitive organisms may be required in combination with a high-fiber diet[5]. Microbiome read-outs may help to determine those individuals that are likely to benefit the most from dietary fiber and may therefore help to inform personalized dietary recommendations with regard to specific fiber types.

References

1. Jones JM. CODEX-aligned dietary fiber definitions help to bridge the 'fiber gap'. *Nutr J.* 2014;13:34.
2. Williams BA, Mikkelsen D, Flanagan BM, *et al.* "Dietary fibre": moving beyond the "soluble/insoluble" classification for monogastric nutrition, with an emphasis on humans and pigs. *J Anim Sci Biotechnol.* 2019;10:45.
3. Jha R, Fouhse JM, Tiwari UP, *et al.* Dietary fiber and intestinal health of monogastric animals. *Front Vet Sci.* 2019;6:48.
4. Wong C, Harris PJ, Ferguson LR. Potential benefits of dietary fibre intervention in inflammatory bowel disease. *Int J Mol Sci.* 2016;17.
5. O'Grady J, O'Connor EM, Shanahan F. Review article: dietary fibre in the era of microbiome science. *Aliment Pharmacol Ther.* 2019;49:506–515.
6. Hand MS, Thatcher CD, Remillard RL, *et al.* *Small Animal Clinical Nutrition*, 5th ed. Mark Morris Institute, Topeka, KS; 2010.
7. NRC. *Nutrient Requirements of Dogs and Cats.* Washington DC: The National Academy Press; 2006.
8. Tan J, McKenzie C, Potamitis M, *et al.* The role of short-chain fatty acids in health and disease. *Adv Immunol.* 2014;121:91–119.

9. Topping DL, Clifton PM. Short-chain fatty acids and human colonic function: roles of resistant starch and nonstarch polysaccharides. *Physiol Rev*. 2001;81:1031–1064.
10. Robles Alonso V, Guarner F. Linking the gut microbiota to human health. *Br J Nutr*. 2013;109 Suppl 2:S21–26.
11. Cai Y, Folkerts J, Folkerts G, *et al*. Microbiota-dependent and -independent effects of dietary fibre on human health. *Br J Pharmacol*. 2019.
12. Raqib R, Sarker P, Bergman P, *et al*. Improved outcome in shigellosis associated with butyrate induction of an endogenous peptide antibiotic. *Proc Natl Acad Sci U S A*. 2006;103:9178–9183.
13. Goncalves P, Araujo JR, Di Santo JP. A cross-talk between microbiota-derived short-chain fatty acids and the host mucosal immune system regulates intestinal homeostasis and inflammatory bowel disease. *Inflamm Bowel Dis*. 2018;24:558–572.
14. Kinoshita M, Suzuki Y, Saito Y. Butyrate reduces colonic paracellular permeability by enhancing PPARgamma activation. *Biochem Biophys Res Commun*. 2002;293:827–831.
15. Furusawa Y, Obata Y, Fukuda S, *et al*. Commensal microbe-derived butyrate induces the differentiation of colonic regulatory T cells. *Nature*. 2013;504:446–450.
16. Arpaia N, Campbell C, Fan X, *et al*. Metabolites produced by commensal bacteria promote peripheral regulatory T-cell generation. *Nature*. 2013;504:451–455.
17. Miquel S, Martin R, Rossi O, *et al*. *Faecalibacterium prausnitzii* and human intestinal health. *Curr Opin Microbiol*. 2013;16:255–261.
18. Heinritz SN, Weiss E, Eklund M, *et al*. Intestinal microbiota and microbial metabolites are changed in a pig model fed a high-fat/low-fiber or a low-fat/high-fiber diet. *PLoS One*. 2016;11:e0154329.
19. So D, Whelan K, Rossi M, *et al*. Dietary fiber intervention on gut microbiota composition in healthy adults: a systematic review and meta-analysis. *Am J Clin Nutr*. 2018;107:965–983.
20. Vermeire S, Joossens M, Verbeke K, *et al*. Donor species richness determines faecal microbiota transplantation success in inflammatory bowel disease. *J Crohns Colitis*. 2016;10:387–394.
21. Gibson GR, Willems A, Reading S, *et al*. Fermentation of non-digestible oligosaccharides by human colonic bacteria. *Proc Nutr Soc*. 1996;55:899–912.
22. Park Y, Hunter DJ, Spiegelman D, *et al*. Dietary fiber intake and risk of colorectal cancer: a pooled analysis of prospective cohort studies. *J Am Med Assoc*. 2005;294:2849–2857.
23. Lipkin M, Reddy B, Newmark H, *et al*. Dietary factors in human colorectal cancer. *Annu Rev Nutr*. 1999;19:545–586.
24. Wang H, Shi P, Zuo L, *et al*. Dietary non-digestible polysaccharides ameliorate intestinal epithelial barrier dysfunction in IL-10 knockout mice. *J Crohns Colitis*. 2016;10:1076–1086.
25. Massimino SP, McBurney MI, Field CJ, *et al*. Fermentable dietary fiber increases GLP-1 secretion and improves glucose homeostasis despite increased intestinal glucose transport capacity in healthy dogs. *J Nutr*. 1998;128:1786–1793.
26. Jha R, Berrocoso JD. Review: dietary fiber utilization and its effects on physiological functions and gut health of swine. *Animal*. 2015;9:1441–1452.
27. Jha R, Woyengo TA, Li J, *et al*. Enzymes enhance degradation of the fiber-starch-protein matrix of distillers dried grains with solubles as revealed by a porcine *in vitro* fermentation model and microscopy. *J Anim Sci*. 2015;93:1039–1051.
28. Makki K, Deehan EC, Walter J, *et al*. The impact of dietary fiber on gut microbiota in host health and disease. *Cell Host Microbe*. 2018;23:705–715.
29. Eswaran S, Muir J, Chey WD. Fiber and functional gastrointestinal disorders. *Am J Gastroenterol*. 2013;108:718–727.
30. Piche T, des Varannes SB, Sacher-Huvelin S, *et al*. Colonic fermentation influences lower esophageal sphincter function in gastroesophageal reflux disease. *Gastroenterol*. 2003;124:894–902.
31. Leib MS. Treatment of chronic idiopathic large-bowel diarrhea in dogs with a highly digestible diet and soluble fiber: a retrospective review of 37 cases. *J Vet Intern Med*. 2000;14:27–32.
32. Lecoindre P, Gaschen FP. Chronic idiopathic large bowel diarrhea in the dog. *Vet Clin North Am Small Anim Pract*. 2011;41:447–456.
33. Dennis JS, Kruger JM, Mullaney TP. Lymphocytic/plasmacytic colitis in cats: 14 cases (1985–1990). *J Am Vet Med Assoc*. 1993;202:313–318.
34. Freiche V, Houston D, Weese H, *et al*. Uncontrolled study assessing the impact of a psyllium-enriched extruded dry diet on faecal consistency in cats with constipation. *J Feline Med Surg*. 2011;13:903–911.
35. Buddington RK, Buddington KK, Sunvold GD. Influence of fermentable fiber on small intestinal dimensions and transport of glucose and proline in dogs. *Am J Vet Res*. 1999;60:354–358.

36. Sparkes AH, Papasouliotis K, Sunvold G, *et al.* Effect of dietary supplementation with fructo-oligosaccharides on fecal flora of healthy cats. *Am J Vet Res.* 1998;59:436–440.
37. Swanson KS, Grieshop CM, Flickinger EA, *et al.* Supplemental fructooligosaccharides and mannanoligosaccharides influence immune function, ileal and total tract nutrient digestibilities, microbial populations and concentrations of protein catabolites in the large bowel of dogs. *J Nutr.* 2002;132:980–989.
38. Segarra S, Martinez-Subiela S, Cerda-Cuellar M, *et al.* Oral chondroitin sulfate and prebiotics for the treatment of canine inflammatory bowel disease: a randomized, controlled clinical trial. *BMC Vet Res.* 2016;12:49.
39. Willard MD, Simpson RB, Delles EK, *et al.* Effects of dietary supplementation of fructo-oligosaccharides on small intestinal bacterial overgrowth in dogs. *Am J Vet Res.* 1994;55:654–659.
40. Ou J, Carbonero F, Zoetendal EG, *et al.* Diet, microbiota, and microbial metabolites in colon cancer risk in rural Africans and African Americans. *Am J Clin Nutr.* 2013;98:111–120.

Role of Coagulation Assessment in the Management of Hepatobiliary Disease

Cynthia R.L. Webster, DVM, DACVIM (Small Animal Internal Medicine)
Grafton, MA, USA

Key Points

1. Coagulation testing is used diagnostically, prognostically and to assess bleeding and thrombotic risk.
2. Hypocoagulable, hypercoagulable and hyperfibrinolytic states exist in hepatobiliary disease.
3. Minimum bleeding risk assessment should include platelet count, PT, aPTT, hematocrit and fibrinogen.
4. There is growing evidence that thromboelastography may have value in predicting bleeding and thrombotic tendencies.
5. Evaluation of bleeding and thrombotic risk should include an individualized assessment of risk factors in each patient.
6. Treatment is aimed at perceived bleeding and thrombotic risks and not simply at results of individual coagulation tests.

Coagulation Parameters and Diagnosis of Hepatobiliary Disease

In animals with an acute onset of liver disease and the presence of hyperbilirubinemia, a >1.5 times prolongation in prothrombin time (PT) sub-classifies acute liver disease (ALD) as acute liver failure (ALF). This is an important differentiation as ALF carries a much worse prognosis and in human medicine is a trigger for referral to a liver transplant center.[1-7] In this setting PT is being used to more as an indication of a profound loss of hepatic synthetic capacity rather than a harbinger of bleeding tendencies.

Dogs with congenital portosystemic shunts (CPSS) consistently have low protein C activity and this can help differentiate the presence of these macroscopic shunts from the microscopic abnormalities that are seen in primary hypoplasia of the portal vein (microvascular dysplasia).[4,5] A cautionary note however, low protein C activity will also accompany multiple acquired portosystemic shunts secondary to portal hypertension so low values do not have discriminatory value in differentiating CPSS from other hepatic parenchymal disease.

Coagulation Parameters and Prognosis in Hepatobiliary Disease

In dogs and cats with ALD/ALF due to hepatotoxic drugs, prolongations in PT and aPTT, as well as low protein C and AT activity predict non-survival. In dogs with chronic hepatitis (CH) prolongations in PT or aPTT predict shorter survival times.[7,8] When prolongation in PT is incorporated into a clinical scoring system in Labradors with CH, the score correlates with the stage of fibrosis on biopsy and survival time. The finding of hyperfibrinolysis in dogs with liver failure has also associated with non-survival.[9]

Assessing Bleeding and/or Thrombotic Risk

Establishing the coagulation status in a patient with hepatobiliary disease is challenging due to the wide range of coagulation alterations that can occur with hepatobiliary dysfunction. The majority of factors that regulate pro-coagulation, anti-coagulation and fibrinolysis are produced and/or activated in the liver. Fibrinogen, prothrombin, FV, FVII, FIX, FX, FXI, FXII, and FXIII are all synthesized at least to some extent in the liver. Hepatocytes are also the site of vitamin-K-dependent carboxylation of prothrombin, FVII, FIX, FX, protein C, and protein S. Many anticoagulants including AT, protein C and protein S, are synthesized in the liver. In addition, hepatocytes synthesize factors that control fibrinolysis including plasminogen, PAI-1, and alpha-2-antiplasmin. Lastly, the liver is responsible for the clearance of activated pro and anticoagulant factors and their by-products (e.g., FDPs and D-dimers). The current thinking in hepatology is that a precarious rebalanced state of hemostasis exists in liver disease. Coagulation testing in patients with liver disease reveal parameters consistent with hypercoagulable, hypocoagulable and/or hyperfibrinolytic states. Alternatively, patients can have normal coagulation assessment but be easily tipped into bleeding or thrombosis by risk factors such as concurrent presence of uremia, SIRS, inflammation in the splanchnic bed, portal hypertension, hypervolemia, sepsis, anemia, ascites, or drugs which affect coagulation such as corticosteroids or NSAIDs.

The ideal way to assess bleeding risk in dogs and cats with liver disease is not known and has long been a subject of controversy in human patients as well. The current consensus in human medicine is that it should consist of packed cell volume, platelet count, PT, aPTT and fibrinogen.[10,11] It is reasonable to assume this same evaluation would have value in dogs and cats. The parameter values that likely delineate a high risk for bleeding are listed in Table 1.

Table 1. Bleeding risk assessment

Bleeding risk assessment test	**High risk**
Platelet count	<50,000/µl
PT, aPTT	Either >1.5xULN
Fibrinogen	<100 mg/dl
vWF activity (in predisposed breeds)	<50%

Although additional coagulation testing in human patients with liver disease can demonstrate deficiencies in antithrombin, protein C and increases in vWF, Factor VIII and D-dimers, the value of these parameters in predicting bleeding/thrombotic risk have not proven. Similar changes in these coagulation parameters occur in dogs and cats with liver disease.[3,4,6,12-14]

The application of more global coagulation tests, such as thromboelastography (TEG), may ultimately provide a better understanding of coagulopathy and liver disease.[15-20] TEG follows clot formation as a dynamic process, measuring clot time and strength, as well as the kinetics of clot formation. For many years, TEG has been used to predict coagulation status and plan for interventions in liver transplant and cardiac surgery patients. In the last few years there is a growing body of evidence that TEG is more effective than traditional coagulation parameters in predicting bleeding tendencies.

Several studies in dogs and cats with liver disease have described TEG results and while none have been conducted to look at its ability to detect bleeding tendencies, several functional states have been described.[3,4,12] Dogs with ALF are more likely hypocoagulable and hyperfibrinolytic than dogs with ALD.[3] The hypocoagulable state (low MA/G) is associated with low fibrinogen and protein C activity while hyperfibrinolysis (high LY 30) correlates with low protein C activity and an elevated white blood cell count. Hyperfibrinolysis appears to be primary as it is not accompanied by evidence of secondary hyperfibrinolysis such as increases in D-dimers, RBC fragmentation or end organ damage from microthrombi. The pathogenesis of this hyperfibrinolytic state is unknown. In people with liver disease it is due to increased endothelial cell production and decreased hepatic clearance of TPA and PAI as well as decreased hepatic synthesis of TAFI and anti-plasmin.[13] Whether or not this hyperfibrinolytic state leads to an increased tendency to bleed in dog as does in humans[21,22] has not been investigated. In a recent retrospective study of ALD in dogs, however, spontaneous bleeding, particularly from the gastrointestinal tract, did occur,[2] although none of the dogs had TEG analysis to check for hyperfibrinolysis. In the same study, thrombosis of the splanchnic vasculature was also found in several dogs at necropsy suggesting that under the right condition dogs with ALD can be hypercoagulable as well.

In a recent report of TEG analysis in dogs with CH, overall coagulation status was variable.[14] Dogs with higher clinical scores, lower fibrinogen and those with complications associated with end stage disease (portal hypertension) tended to be hypocoagulable. Hyperfibrinolysis occurred in 23% of dog and was associated with the severity of injury as assessed by serum transaminases. Some dogs were hypercoagulable and this fits with the observation that dogs with CH are predisposed to portal vein thrombosis.[23]

On TEG analysis dogs with CPSS tend to have mild hypercoagulability (as evaluated by increased G).[24] Dogs with high G value are more likely to have hepatic encephalopathy and high fibrinogen suggesting that the pro-inflammatory state of hepatic encephalopathy may contribute to the thrombotic tendency. Portal vein thrombosis has been reported as a complication of CPSS.[25]

Two studies looking at obstructive jaundice in dogs have found hypercoagulable parameters (increased fibrinogen, D-dimer and TEG G value).[26,27] We recently completed a study of coagulation parameters in dogs with gallbladder mucoceles. This study demonstrated that most dogs exhibited hypercoagulable tendencies with high TEG G values, hyperfibrinogenemia, thrombocytosis and increased D-dimers. These hypercoagulable tendencies are in line with a recent study that found pathologic conditions associated with thrombosis accounted for up to 10% of the post cholecystectomy mortality. The role that hypercoagulability and associated occult thrombosis might play in the reported morbidity and mortality associated with GB mucoceles treatment needs further study.

Our unpublished results on TEG analysis in cats with cholestatic disease shows a mixture of both potential bleeding and thrombotic tendencies. Cats with inflammatory disease tended to have hypercoagulable TEGs while cats with lipidosis were hypocoagulable. TEG indices of coagulation state (G value) and hyperfibrinolysis were significantly negatively correlated with PT, aPTT and serum alkaline phosphatase suggesting that grade of disease is an important determining factor. Most cats did not have provocative procedures so no correlation with bleeding could be established but the finding of a hyperfibrinolytic state in some cats suggests that delayed bleeding may be an issue.

TREATMENT OPTIONS FOR COAGULOPATHIES

Treatment should be aimed at reversal of clinically relevant problems (bleeding, thrombosis) and not at correcting abnormal laboratory parameters which may or may not provide an accurate assessment of bleeding/thrombotic risk.

Management of nonbleeding patients. In the human literature, multiple studies have demonstrated that attempted correction of mild prolongations in PT and aPTT or thrombocytopenia with FFP administration or platelet transfusions, respectively, does not always have an effect on the incidence of post-procedural bleeding. These observations suggest that the reason for the underlying bleeding tendency may not be due to defects in coagulation factors or platelet function, but instead be secondary to some component of the coagulation cascade that is not impacted by replenishment of soluble factors or platelets, such as vascular endothelial dysfunction or premature clot lysis.[1,11,29]

Despite the inherent limitations of bleeding risk assessment and the evidence for lack of efficacy in interventional management, there are nonetheless some guidelines that can be applied to the management of coagulation status in hepatobiliary disease. All patients with prolonged PT should be treated with subcutaneous vitamin K1 to ensure that this vitamin deficiency is not contributing to the coagulopathy. Considering the poor sensitivity of PT in detecting VK deficiency however, it may be prudent to prophylactically treat all cholestatic (hyperbilirubinemic) patients with VK.

In patients with prolonged clotting times, but no evidence of spontaneous bleeding in which no provocative procedures are planned, there is no need to administer blood products solely for the sake of improving bleeding times. Even if provocative procedures are necessary there is growing feeling in human hepatology that these patients should not receive prophylactic therapy, but instead be provided with rescue therapy, if necessary. Prior to performing invasive procedures, the clinician should be prepared to address any post procedure hemorrhage by ensuring that the patient is blood typed and that blood products are available. In addition, patients should be monitored in the hospital closely.[17-20]

Other factors that figure into the decision to give prophylactic blood product therapy are a history of bleeding disorder, the perceived bleeding risk associated with procedure (aspirate vs. biopsy vs. surgery), likelihood of rescue therapy being successful after the proposed procedure, the cost of the intervention, and ease of managing potential complications or side effects of prophylactic treatment. Any benefit of prophylactic therapy needs to be weighed against potential side effects of blood product support which include transmission of infectious agents, occurrence of transfusion related lung injury(TRALI) or other immune-related transfusion reaction and volume overload. Indeed the latter may be particularly deleterious as there is literature in the human field to support the observation that volume overload potentiates portal hypertension which actually promotes greater bleeding.

Cats with obstructive jaundice present a unique challenge.[30,31] Multiple studies have shown that these cats have significant bleeding associated with surgery and often require post-operative blood product support despite the presence of normal pre-operative PT and aPPT and adequate VK supplementation. It is interesting to speculate that perhaps some of these cats may be hypocoagulable due to heparin like effects secondary to endothelial dysfunction or from hyperfibrinolytic. Our hospital has noted that the routine practice of intra-operative transfusion in the cats with a PCV less than 30% anecdotally improves outcome in our hospital.

Management of Patients with Active Hemorrhage

Fresh frozen plasma (FFP), it is the first line treatment for spontaneous hemorrhage or iatrogenic bleeding as it supplies the widest range of coagulation factors and is readily available. The side effects of FFP in veterinary patients include volume overload (with exacerbation of portal hypertension), infections, transfusion reactions, and acute lung injury.[32-34] The recommended dose range is 5–15 ml/kg as a starting point. Given the previously discussed poor correlation of clotting times to bleeding tendency, the authors recommend using discontinuation of bleeding as an endpoint for FFP transfusions and not correction of PT and aPTT. FFP transfusion reactions were reported in 14% of cats in one study.[34] In 2 studies in dogs, 46% and 80% of the FFP transfusions given to dogs were for coagulation tests abnormalities.[32,33] These studies in dogs and cats do document improvements in PT and aPTT with transfusion in many cases, although data specifically on those animals with hepatic disease is not given.

Cryoprecipitate is indicated for hypofibrinogenemia (fibrinogen <100 mg/dl), and can also be used to supply FVIII, vWF, and FXIII. The advantages of cryoprecipitate over FFP include decreased volume of administration, and possibly a decreased tendency for transfusion reactions.

Antifibrinolytic therapy would be indicated if hyperfibrinolysis is documented with TEG or suspected based on delayed bleeding from mucosal surfaces after invasive procedures. The most frequently used therapies in human patients are aminocaproic acid and tranexamic acid, which inhibit plasmin. In humans undergoing liver transplantation, these therapies have been shown to reverse the evidence of hyperfibrinolysis seen on TEG, and case series suggest they can be effective in controlling hemorrhage in patients with cirrhosis. Veterinary studies evaluating these anti-fibrinolytic agents, and their possible use in patients with liver disease are needed.

REFERENCES

1. Kavanagh C, *et al. J Vet Emerg Crit Care.* 2011;21:589–604.
2. Lester C, *et al. J Vet Emerg Crit Care.* 2016;26:559–567.
3. Kelley D, *et al. J Vet Intern Med.* 2015;29:1053–1062.
4. Toulza O, *et al. J Am Vet Med Assoc.* 2006;229:1761–1771.
5. Prins M, *et al. Vet J.* 2010;185:163–168.
6. Dereszynski DM, *et al. J Am Vet Med Assoc.* 2008;232:1329–1337.
7. Shih JL, *et al. J Vet Intern Med.* 2007;21:33–39.
8. Strombeck DR, *et al. J Am Vet Med Assoc.* 1988;193:1109–1113.
9. Sigrist NE, *et al. Vet J.* 2018;242:67–73.
10. DeAngelis GA, *et al. J Vasc Interv Radiol.* 2016;27:1665–1674.
11. Patel IJ, *et al. J Vasc Interv Radiol.* 2012;23:727–773.
12. Poldervaart JH, *et al. J Vet Intern Med.* 2009;23:72–80.
13. Dereszynski DM, *et al. J Am Vet Med Assoc.* 2008;232:1329–1337.
14. Fry W, *et al. J Vet Intern Med.* 2017;31:419–426.
15. De Pietri L, *et al. Hepatology.* 2016;63:566–573.
16. Davis JPE, *et al. Ann Hepatol.* 2018;17:205–213.
17. Hum J, *et al. J Clin Gastroenterol.* 2019. doi: 10.1097/MCG.0000000000001285.
18. Kohli R, *et al. Dig Dis Sci.* 2019;64:2661–2670.
19. Blasi A, *et al. Hepatology.* 2018;68:2325–2337.
20. Debernardi Venon W, *et al. Eur J Gastroenterol Hepatol.* 2015;27:1313–1319.
21. Birkbeck R, *et al. J Small Anim Pract.* 2019;60:641–655.
22. Leebeek FW, *et al. Semin Thromb Hemost.* 2015;41:474–480.
23. Respess M, *et al. J Vet Intern Med.* 2012;26:230–237.
24. Kelley D, *et al. J Vet Intern Med.* 2013;27:1262–1267.
25. Rogers CL, *et al. J Vet Intern Med.* 2008;22:282–287.
26. Mayhew PD, *et al. J Am Vet Med Assoc.* 2013;242:778–785.
27. Richardson R, *et al. J Vet Emerg Crit Care.* 2018;28:551–558.
28. Jaffey JA, *et al. Vet J.* 2019;251:105350.
29. Webster CR. *Vet Clin North Am Small Anim Pract.* 2017;47:601–615.
30. Boute NJ, *et al. J Am Vet Med Assoc.* 2006;9:1376–1382.
31. Mayhew PD, *et al. J Small Anim Pract.* 2002;43:247–253.
32. Logan JC, *et al. J Am Vet Med Assoc.* 2001;218:1449–1455.
33. Snow SJ, *et al. J Vet Emerg Crit Care.* 2010;20:441–445.
34. Mansi ET, *et al. J Feline Med Surg.* 2019:1098612X19876728.

Techniques to Search for Articles When Making Clinical Decisions

Heather K. Moberly, AHIP, FHEA, PgCert (Veterinary Education)
College Station, TX, USA

Introduction

Evidence-based Veterinary Medicine is a methodology that balances the evidence, the clinician, and the client to determine the best option for each individual clinical scenario. These elements are not represented equally; it's a balance. It's the opposite of "cookbook medicine."1

It uses a five step process:

1. Ask an answerable question
2. Acquire the published evidence
3. Appraise the relevance and quality of the evidence
4. Apply evidence to the clinical context
5. Assess the impact

This session supports the "Acquire" step (step 2).

The First Five Considerations

Clinical searching competence needs to function in the unconscious competence or unconscious supercompetence2 realm to fit into a manageable workflow.

The first consideration of searching, perhaps basic and overlooked, is a conscious thought, however brief, about the search. This exists in the space between identifying an information need and beginning the search. The ready ease and availability of an Internet connection and a search capable device supercharges this leap from "need" to "search"; however takes time to identify:

1. What information do you need?
2. Is it likely to have been published?
3. By whom?
4. In what publication(s) or type of publication?
5. Where is/are the publication(s) available or indexed?

This leads into where and how to search effectively and efficiently.

Inventory Your Access

Keep the list of resources available to you in mind when constructing your search. This session focuses on freely available resources. This does not imply they are the sole or best resources for any or all clinical veterinary situations; it recognizes searching techniques share commonalities and may be adapted to a variety of resources. Focusing on freely available resources avoids the appearance of favoritism or bias.

After you assemble a list of your available resources it is simpler to match your need to them. For those at universities and other well-resourced establishments, check in periodically with the librarians for updates and changes.

It may be of benefit to intentionally think through what resources you have immediate access to and those which you could access if needed. Include journal and database subscriptions and freely available resources you know about. Not all information needs are immediate and you may have deeper reserves than you realize. Personal, practice, and institutional subscriptions are the clearest and most obvious resources. Memberships in professional organizations can also include either access to resources or a discount for access. Additionally, if a resource is of interest, ask these organizations if they can add either access or a discount as a future benefit.

Structure Your Question

Place your need in a framework. This assures you include all the necessary components and identify optional components. It also assists identification of relevant articles in search results. The most common structure in human and veterinary medicine is PICO - Patient Intervention Comparison Outcome. When searching veterinary questions, the patient portion includes relevant species and perhaps breeds. A best practice is to identify **assessable** outcomes rather than generalizations. Additionally, it is notable that the outcomes are often used during selection of relevant results not as search terms.

Choose Search and Combine Terms

The key to efficiently retrieving relevant results hinges on two factors: choosing terms and combining them. The bottom line message is searching is iterative; there is no one perfect search. The best process is to search, review results, and refine and rerun the search until satisfied.

Choosing Terms

Create a list of synonyms for each item of your PICO. List synonyms, singular and plural, regional variations, acronyms, and variant spellings. These terms that you choose, independent of where you will search for articles are considered free-text terms. The next step is to add controlled terms which will be specific to each database. Controlled terms, often called subject headings although there are a variety of possibilities, are assigned by the resource and gather together like items regardless of the free-text terms the authors used. It is important to include both free-text and controlled terms if you are conducting a comprehensive search. In PubMed, for example, search for your free-text terms in the MeSH index; MeSH stands for **Me**dical **S**ubject **H**eadings. Refer to the help documentation for the database to ascertain how to label the controlled vocabulary.

Depending on the database, each term may be broadened using truncation or stemming. This is most often terminal truncation accomplished by determining the root of a term and adding a symbol to indicate the search should include variant endings. An example is arthrit* which may retrieve both arthritis and arthritic. Some databases allow internal truncation which is useful for retrieving American and British spellings, for example, catheteri*e may return both catheterize and catheterise. Rarely can leading truncation be used, for example *estrous to return both estrous and oestrous, instead listing these variants as synonyms and combining using a Boolean OR is appropriate. Exercise caution and consult database help documents because truncation may change other search functionality.

Similarly, forcing terms into a phrase is database specific. Most frequently words are forced together into a phrase by enclosing the terms in inverted brackets or quotation marks, for example " ". As in the case of truncation, forcing phrases can affect search functionality and consulting help documents is advised. If phrase forcing is used it may be best to also include the terms separately in addition to the quoted phrase and combine these using a Boolean OR.

A best practice is identifying items you may search frequently, for example a species, and constructing a robust set of search terms to represent it which is then saved and reused. Depending on the resource being searched you may save the search within it, for example for PubMed searches use the My NCBI account, or simply as a document on your computer or a cloud drive.

Choose the Scope of Your Results

Having created a list of free-text and controlled terms, the next choice before combining them to create the search is the scope of the results. This is described as specificity and sensitivity;

however, because those terms have more than one context in medicine and statistics the terms precision and recall can be substituted.[3] A high recall search returns a large set of results, which, presumably, include a higher proportion of relevant items. The disadvantage is the overall large set means reading through more irrelevant items to locate the relevant items. A high precision search returns a smaller set of results and, perhaps, a smaller number of relevant items. The advantage is reading through a smaller set to locate the relevant items. Neither is always better. Neither is always correct. It depends on the individual search need.

Combine Terms

Having determined search terms, combine these using the Boolean connectors AND, OR, and NOT.[4] This is the power behind online searching. Combine synonyms for each single concept using OR to encompass the breadth of descriptions for the concept. Combine the terms representing each separate concept using AND to require that each concept is present in each result. Use NOT with caution because it will exclude all the articles that contain the term you're excluding, even if the rest of the article is spot on. If you need to reduce the number of search results, instead of using NOT, consider using filters. In each database, results filters and sorting options help to focus results without adding more search terms. Consider filtering by publication date, language, or publication type. In each database peruse the results filter options to know what may be useful.

Increase Your Relevant Results: Cited and Citing References

A time-honored method for increasing relevant results is locating one, or a few, articles and examining the reference list for additional relevant articles. These cited articles were necessarily chronologically older than the starting point article. Additionally, for those with access to Science Citation Index (SCI), results could be expanded by to chronologically newer articles by identifying articles in which the starting point occurred in the reference list— citing articles. Currently, citing article citations are available through Google Scholar (GS). The most efficient way to locate these is searching GS for the title of the starting point article. If the starting point article is found in GS, read the last line of the citation for the notation "# citing articles" and click that link. It is important to note that SCI citing articles represent citations from a specific list of journals. GS citing articles represent whatever GS could locate online.

Self-Paced Searching Tutorials

Name	URL
EBVM Learning	www.ebvmlearning.org
PubMed for Veterinarians	https://tinyurl.com/tw35a3s
PubMed (general)	www.nlm.nih.gov/bsd/disted/pubmedtutorial

Freely Available Resources. Examples

Articles

Name	URL	Producer	Subject coverage
PubMed	https://pubmed.gov	National Library of Medicine	Biomedical sciences, includes veterinary medicine
PICO.vet	https://pico.vet	Dr. Stuart Turner	Tool to construct a PICO based veterinary PubMed search

Global Index Medicus	www.globalindexmedicus.net	World Health Organization	Biomedical sciences, includes veterinary medicine
VetSRev	http://webapps.nottingham.ac.uk/refbase (VIN editor: link could not be accessed on 5-22-20)	University of Nottingham	Veterinary, systematic reviews
Google Scholar	https://scholar.google.com	Google	All

Research Syntheses

Name	URL	Notes
Best Bets for Vets	https://bestbetsforvets.org	
Knowledge Summaries	www.veterinaryevidence.org	Published in Veterinary Evidence journal

REFERENCES

1. Moberly HK. Researching that difficult case – Where to find information for free. VetsNorth and VetsSouth, Improve International; 2019.
2. Howell WS. *The Empathic Communicator*. Belmont, CA: Wadsworth; 1982.
3. Moberly HK. Searching for veterinary information: balancing precision and recall. *Texas Veterinarian*. 2018;80(2):57. https://oaktrust.library.tamu.edu/handle/1969.1/166423.
4. Keefe L, Moberly HK. PubMed for Veterinarians: PubMed and Boolean Logic (Handout). Texas A&M University. 2015. https://oaktrust.library.tamu.edu/handle/1969.1/158260.
5. Keefe L, Moberly HK. PubMed for Veterinarians: Using PubMed Filters (Handout). Texas A&M University. 2015. https://oaktrust.library.tamu.edu/handle/1969.1/158259.

Tools for Finding Free Legal Copies of Articles

Heather K. Moberly, AHIP, FHEA, PgCert (Veterinary Education)
College Station, TX, USA

Introduction

Access to full-text articles and other data can be frustrating. Locating free and legal access can be even more so. Fortunately, there are an increasing number of tools and techniques to assist and keep you on the right side of the law. This session is about free resources—nothing is free—it is about acquiring copies of articles that have no cost to you to acquire and use and with that there is always a Caveat Emptor. Products cost money and time to create. If you are not paying for the product you are acquiring in currency, who is? Perhaps you are the product? Is the producer watching your behavior? Are they tracking the resources that you choose? Is the cost targeted advertising? The bottom line will always be that you need to take responsibility to investigate the options and make the choices that you are comfortable with.

This session focuses on locating Open Access (OA) articles and data[1] although it includes other strategies as well. OA refers to this information being available to the consumer at the point of need without cost or other barrier. Information specialists, publishers, and others often describe restricted access through metaphors of walls—paywalls, datawalls, and geographic borders.[2] Paywalls require money as the toll to pass through and receive content. Datawalls generally require registration and tracking of some sort, so the toll is data about you and your habits. Geographic borders as a barrier refers to the current situation in which an article that is open access in one country may not be in another, so where you are matters. Think data collection is not a big deal and is not lucrative? ResearchGate received a US patent in February 2020 for "Enhanced Online User-Interaction Tracking and Document Rendition," for a summary use the link to InfoDocket at the reference.[3]

Article Versions

A consideration of seeking free access to articles is the article version. Many publishers allow free distribution of the pre-print or the post-print of articles and restrict the distribution of the final formatted publisher version. Pre-print generally refers to the pre-review manuscript (e.g., the author submitted version). Post-print generally refers to the post-review manuscript. A best practice for authors is, when possible, to consider distribution when choosing the journal. Further, authors should read publisher copyright agreements closely, negotiate to retain the most generous distribution rights, and use those rights to post the appropriate version of their articles online.

Locating Articles

OA Journals

The first thought when seeking free full-text is whether the journal is, itself, OA. It is a best practice to be aware of OA journals in your field and changes to these journals. An example near and dear to you as ACVIM members is the *Journal of Veterinary Internal Medicine* (*JVIM*) move to an OA model. In some cases of a subscription journal moving to OA, only the new articles will be OA; in the case of *JVIM* all articles past and present are freely available. A quite recent flip from subscription to OA is *Bovine Practitioner* from the American Association of Bovine Practitioners.

If an entire journal is not OA, it may be that individual articles are OA. Individual articles may be OA because of a variety of reasons including a funding mandate, an author ownership (copyright) agreement, or a hybrid publishing model. SHERPA/RoMEO, http://sherpa.ac.uk/romeo, is a free online tool with information about individual journals and definitions.

Search Engine Integration

A straightforward method to locate freely available article copies can be through Google Scholar (GS), https://scholar.google.com. GS identifies freely available articles and provides links in the column to the right of the citation result. Paste the title of the sought article into the search box and search. The full text links are displayed in an order of preference and more than one link may be identified; in the last line of the citation, choose # versions to display all of the choices.

Specific Tools

The following tools are offered with a caveat emptor warning. You need to be comfortable with the terms and conditions of loading any of these executable programs and the data tracking or other hidden cost they may incur. Remember the introduction, nothing is free and in this case you are gaining easy and convenient access identification.

Online articles are often assigned a unique Digital Object Identifier, or DOI, number. This allows search engines and other tools to locate the article even if the URL changes. Internet Browser Extensions, also called Browser Buttons or Browser Plug-Ins, are small free executable programs available for in the online stores for browsers such as Google Chrome and Firefox. Unpaywall, https://unpaywall.org/, and Open Access Button, https://openaccessbutton.org/, are non-profit projects that identify and link to free, legal copies of articles. Unpaywall is a Browser Extension only. Open Access Button is both a Browser Extension and a website with a search function. If you are affiliated with a university or large company, ask your information specialists about additional options including LibKey Nomad and Kopernio to link to OA and your institutional subscriptions.

Online Collections

Institutional and national repositories often host freely available publications not found elsewhere. Obvious examples include institution-related documents such as theses and dissertations, white papers, and teaching materials. Less obvious are commercially published articles for which the publisher has granted the author permission to share the finally formatted pdf on a non-commercial or institutional website. In these instances the publisher has not granted permission for a commercial site like ResearchGate or Academia.edu, more about these below. One strategy is to use a tool like Unpaywall or Open Access button which harvest data from institutional repositories. Another is to visit the institutional library website and search the repository for the paper or author; a strategy that can yield treasure.

Other hosted open access life science and biomedical science repositories include PMC, formerly called PubMed Central, and Europe PMC, formerly called UK PubMed Central. These may be searched for a topic to locate a body of results or for a specific article.

Collections that are more difficult to categorize but worth a look to see if they would be of use to a specific searcher or topic include the Directory of Open Access Journals (DOAJ), https://doaj.org/, The Internet Archive, https://archive.org/, and the Hathi Trust, www.hathitrust.org. Internet Archive and Hathi Trust are book focused.

Social Media Sites

Following authors, institutions, and journals on Twitter has several advantages. One is early announcements of publications. Another is linked items are sometimes freely available from the Tweet for a while. ResearchGate and Academia.edu are researcher social media sites that often include full-text article access. These articles are also often linked from GS. ResearchGate and Academia.edu are for profit, venture capital funded sites. Academia.edu is not part of the US education system; it received their URL before 2001 when eligibility requirements were changed. The .edu domain is managed by Educause and the current eligibility requirements, https://net.educause.edu/eligibility.htm, are quite clear.

Added Benefit of Membership or Status

Always check the resources you purchase and your memberships, sometimes they poorly advertise subscriptions, full-text content, or a discount for you to purchase content. Ask the veterinary library from your university and your veterinary associations about benefits. Depending where you are in the world there are some great benefits to take advantage of.

Libraries

Libraries can be a fantastic source of low or no cost access. Check with your local public library to ascertain both the content they can provide and whether they can borrow from another library for you (Interlibrary loan). Some services are not free; however, the nominal cost is lower than purchasing access from the publisher as an individual. Many US states provide the public library system with consortial access to online journals. As a library card holder you have access in person and online to materials and really, even if this increases your access by a journal or two that's a journal or two you did not know you had free access previously.

Proceedings and Grey Literature

The peer-reviewed journal article is a gold standard for scientific information sharing but you may have access to other valuable publications. Conference proceedings can be valuable and in some specialties may contain reports and data not published elsewhere. Check association websites for their proceedings; many are open access, others are limited to members or attendees. IVIS, the International Veterinary Information Service, is free in the sense that no currency changes hands; however, it does require a login. IVIS has a variety of content including full-text proceedings.

References

1. Swan A. Policy guidelines for the development and promotion of open access. UNESCO. 2012. https://unesdoc.unesco.org/ark:/48223/pf0000215863
2. Hinchcliffe LJ. From Paywall to Datawall. Scholarly Kitchen, Official Blog of the Society of Scholarly Publishing. 11 October 2018. https://scholarlykitchen.sspnet.org/2018/10/11/from-paywall-to-datawall/
3. Price G. ResearchGate Receives U.S. Patent For "Enhanced Online User-Interaction Tracking and Document Rendition". InfoDocket, 11 February 2020. www.infodocket.com/2020/02/11/researchgate-receives-u-s-patent-for-enhanced-online-user-interaction-tracking-and-document-rendition/

Near Infrared Spectroscopic Analysis of Biofluids for Rapid Detection of *Mannheimia haemolytica* Infection in Cattle

Amelia Woolums; Mariana Santos-Rivera; Merrilee Thoresen; Victoria Jefferson; Florencia Meyer; Carrie Vance
Mississippi State University

Mannheimia haemolytica (MH) is a leading cause of bovine respiratory disease (BRD). Diagnosis of BRD traditionally relies on visual assessment, which can be untimely, insensitive, and nonspecific; more rapid and accurate diagnostic tests are needed. Near Infrared Spectroscopy (NIRS) allows fast, noninvasive assessment of biofluids, whereby the chemical profile is described by the intensity and structure of sample reflectance or transmittance spectra in the NIR band (780 to 2500 nm). We hypothesized that NIRS evaluation of breath condensate, nasal secretions, saliva, plasma, and/or serum (biofluids) could profile signatures differentiating cattle before and after MH infection. Biofluids were collected from five 6-month-old Holstein steers over four days prior to and 23 days after intrabronchial challenge with MH. Clinical signs (rectal temperature, heart and respiratory rate, and respiratory score) and complete blood counts (CBC) were assessed pre- and post-challenge. By Day 2 post-challenge calves had signs of acute MH infection with significantly increased ($p<0.05$) rectal temperatures, heart rates, and respiratory rates. CBCs revealed significantly increased ($p<0.05$) leukocyte and neutrophil counts. Due to severe disease calves were treated with a 5-day course of antimicrobial (AM, ceftiofur hydrochloride). Biofluid samples were grouped into four categories: a) baseline, b) infected, pre-AM, c) infected, post-AM, and d) recovered. Spectral signatures of biofluids (n=3800) collected via ASD FieldSpec®3 portable spectrometer were analyzed with multivariate statistics using Unscrambler®X v. 10.5. Chemometrics were applied to the wavelength range 1300–1600 nm using a balanced randomly selected subset of samples (n=150) with an 80/20 Calibration/Validation distribution for Linear Discriminant Analysis (LDA-PCA). Analysis discriminated samples according to category; LDA-PCA showed correct classification between 89.6–100% and 81.7–100% for the calibration and validation process, respectively. NIRS with multivariable analysis rapidly discriminated MH infection states in cattle, establishing a basis for further development of NIRS as a BRD diagnostic tool.

Gastrointestinal Pharmacokinetics and Phenotypic Persistence of Antimicrobial Resistance Over Time after Florfenicol Administration in Steers

Jennifer L. Halleran; Hannah Sylvester; Ryker Minch; Claire Bublitz; Derek Foster
North Carolina State University

Antimicrobial resistance (AMR) is a global threat to both human and veterinary medicine leading to recommendations to use lower tier antimicrobials such as florfenicol, a common antibiotic used in cattle. Our objective was to quantify co-selection of enteric AMR bacteria in cattle treated with two labeled dosing regimens of florfenicol. In addition, we wanted to gain insight on the gastrointestinal pharmacokinetics of florfenicol and how it correlates with AMR. We hypothesized the prevalence of AMR isolates would be higher and persist longer in the steers administered the repeated, lower dose of florfenicol. We also hypothesized that the gastrointestinal pharmacokinetics of florfenicol for the lower, repeated dose would be below a therapeutic level for a longer period of time.

Twelve 6 mo old steers underwent gastrointestinal surgery to facilitate placement of an ultrafiltration probe in the ileum and colon. An interstitial probe was placed in the region of the withers. Jugular vein catheterization allowed for blood to be collected. Ultrafiltrate, interstitial fluid and blood were collected for the pharmacokinetic analysis. Twenty-four hours after surgery, the steers were either administered 20 mg/kg florfenicol intramuscularly (IM) (n=6) 48 hr apart or 40 mg/kg florfenicol subcutaneously (SC) (n=6) once. Feces was collected manually until day 38. To determine the pharmacokinetic parameters of florfenicol administration, HPLC with UV detection will be utilized. To assess the phenotypic resistance changes over time, two different methods were used. Feces was serially diluted and plated on selective media to ensure growth of *E. coli* and *Enterococcus*. The pure colonies were grown on blood agar plates and used on gram positive and gram negative NARMs plates to assess a wide variety of resistance to antibiotics. In addition, the serially diluted feces were plated on either plain MacConkey, plain *Enterococcus* or MacConkey or *Enterococcus* plates infused with either tetracycline (16 ug/mL), ampicillin (32 ug/mL, *E. coli*; 16 ug/mL, *Enterococcus*), or ceftiofur (8 ug/mL) in triplicate. After incubation, plates were counted and averaged to determine the concentration (CFU/g) of either *E. coli* or *Enterococcus* at each time point. To determine the prevalence AMR isolates, the log CFU/g of the resistant isolates was divided by the log CFU/g of wild type isolates at each time point. These were averaged across steers at each time to compare over time within the dosing group and between the dosing groups using student T-tests with appropriate Bonferroni correction to account for multiple testing. Pharmacokinetic analysis will be completed using Phoenix to determine pharmacokinetic parameters.

There was no significant difference from baseline log growth for *E. coli* or *Enterococcus* in either dosing group. Based upon the antibiotic infused media, there was a significant increase in log growth of tetracycline resistant *E. coli* compared to baseline in the IM group (p=0.007). No other comparisons were significant. We anticipate that these findings will provide guidance to veterinarians on the most appropriate dosing regimen of florfenicol in cattle in order to minimize selection of AMR enteric bacteria.

Effect of Dilution on Specificity of Saline Agglutination Tests for Diagnosis of Immune-Mediated Hemolytic Anemia

Unity Jeffery[1]; Prudence Sun[2]
[1]College of Veterinary Medicine, Texas A&M University; [2]Texas A&M University

Saline agglutination tests (SATs) are commonly used in the diagnosis of canine immune-mediated hemolytic anemia (IMHA), but frequent false positive results have been reported using a 1:1 saline to blood ratio. To determine if increased saline to blood ratios improve specificity, SATs were performed at a saline to blood ratio of 1:1, 4:1, 9:1 and 49:1 for 104 dogs with hematocrits ≤30%. SATs were considered positive if at least 1 of 5 randomly selected x40 objective fields contained ≥1 aggregate of ≥4 erythrocytes or ≥2 aggregates of 3 erythrocytes. Of the 104 dogs, 5 were classified as affected by IMHA based on a positive direct antiglobulin gel test and clinical evidence of hemolysis. These criteria were not met in 96/104 dogs classified as unaffected by IMHA and 3/104 were excluded because IMHA status was unclear. Specificity for IMHA was 32% (95% confidence interval 22–43) for 1:1, 56% (45–66) for 4:1, 70% (60–79) for 9:1 and 95% (88–98%) for 49:1. Sensitivity was 100% (40–100) for 1:1 and 4:1 and 80% (30–100) for 9:1 and 49:1. Positive predictive value increased from 7% (6–8) for 1:1 to 44% (23–68) for 49:1. Negative predictive value was 100% at 1:1 and 4:1 and 99% at 9:1 and 49:1 (92–99). Diagnostic accuracy was 35% (25–46) for 1:1, 58% (47–67) for 4:1, 70% (60–79) for 9:1 and 94% (88–98) for 49:1. Based on these results, SATs should be performed at a dilution of at least 49:1 to improve specificity for diagnosis of IMHA.

Development and Implementation of a Hemovigilance Program at a University Veterinary Teaching Hospital

Jillian Haines[1]; Tandi Ngwenyama[2]; Linda Martin[1]; Jane Wardrop[1]
[1]Washington State University; [2]Oregon State University

Patient blood management (PBM) is an evidence based multidisciplinary approach to optimizing the care of patients who may require a transfusion. One aspect of PBM is hemovigilance, which focuses on recognizing and reporting transfusion reactions. The potential for fatal consequences make the prevention, rapid recognition, and treatment of transfusion reactions a critical requirement for the safety of veterinary patients receiving blood products. The objective of this study was to describe the development and implementation of a small animal hemovigilance program at a university veterinary teaching hospital. A hemovigilance working group composed of veterinary specialists in clinical pathology, internal medicine, and emergency and critical care was established. This group developed evidence based definitions of transfusion reactions, reaction classification systems, and a transfusion reaction reporting form. The reporting form contained sections for patient information, transfusion information, administration details, and reaction details. Reaction events were classified by reaction type, severity grade, and imputability to transfusion. Following implementation of the hemovigilance program, transfusion reaction data was collected and examined for the period spanning October 2014 and March 2019. During the study period 718 canine transfusions [4 whole blood, 400 packed red blood cell (pRBC), 300 fresh frozen plasma (FFP), 7 platelet rich plasma, 7 cryoprecipitate] and 124 feline transfusions [5 whole blood, 95 pRBC, 24 FFP] were administered. There were 32 total reactions (27 canine and 5 feline), with the most common reaction being febrile non-hemolytic transfusion reactions (19/32, 59%). The incidence rate of transfusion reactions was found to be 3.8% in dogs and 4.0% in cats. For the confirmed reactions, classification criteria for case definition, reaction severity grade, and imputability were able to be determined and recorded. This allowed targeted interventions to be implemented in order to potentially reduce future reactions. This report shows that a hemovigilance program can be instituted successfully in a veterinary hospital setting. Once developed, standardized reporting tools could be utilized by multiple hospitals and provide the basis for more widespread reaction reporting in veterinary medicine.

Feline Pneumonia: A 10-Year Retrospective

Jonathan D. Dear; Johnson Eric; William Vernau; Sean Hulsebosch; Lynelle Johnson
School of Veterinary Medicine, University of California, Davis

Pneumonia is a poorly described disease in cats. Objectives of this study were to describe clinicopathologic, radiographic and microbiologic features of affected cats. Medical records at the UC Davis Veterinary Medical Teaching Hospital were searched to identify cats with a clinical diagnosis of bronchopneumonia (BP) or aspiration pneumonia (AP). Results extracted from the record included signalment, exam findings, pertinent historical details, and presence of conditions suspected to be risk factors for aspiration. Diagnostics were collated including CBC, BALF analysis and microbial culture. Radiographs were scored by a single, masked radiologist using a published rubric. Sixty cats with identified including 32 with AP and 28 with BP. Airway samples were obtained in 32 cats with BP and 6 with AP. 44 cats with inflammatory airway disease (IAD) were used as controls. Duration and prevalence of cough were significantly less in cats with AP (13 days, 8/32) compared to cats with BP (240 days, 27/28; $p=0.02$, $p<0.0001$ respectively) and respiratory rate was higher (43 versus 26, $p=0.007$). Gastrointestinal disease was the most common risk factor associated with AP. Radiographically, 57 cats had evidence of disease. Cats with AP were more likely to have an alveolar pattern and higher total score than cats with BP or IAD and less likely to have a bronchial pattern. *Mycoplasma* (12) spp. were the most frequently cultured organisms from BALF in cats with BP. These results suggest that bacterial pneumonia should be suspected in cats with cough and severe radiographic infiltrates.

Canine Influenza Virus: Dog and Person Factors Likely Influencing Transmission in Canine Show Settings

Jason W. Stull[1]; J.S. Weese[2]
[1]Atlantic Veterinary College; [2]University of Guelph

Canine influenza is an important disease affecting dogs, especially in situations where many dogs come together (e.g., boarding, shows). Recent outbreaks of canine influenza have involved dog shows in the United States with anecdotally high levels of dog-dog transmission of canine influenza virus (CIV), resulting in dog illness and death, and wide-spread transmission into the general dog community. Despite the importance of this disease, little is known of modifiable factors linked to CIV spread in dogs participating in shows. Study objectives were to describe the knowledge, attitudes, practices (KAP) and observations related to CIV transmission and disease prevention by individuals participating in dog shows in the United States and Canada and pilot a voluntary surveillance tool to serve as an early warning mechanism to identify disease events linked to dog shows. Two questionnaires were developed, widely distributed through national and regional kennel club organizations and administered (May–Nov 2019) on-line to a large convenience sample of individuals that had ever shown at least one dog in the past, with a focus on those showing in Canada and the USA. The KAP questionnaire was completed by 2,301 respondents, while the illness surveillance questionnaire was completed by 449 respondents (had participated in a show within the past 14 days). Overall, respondents were very/somewhat concerned (64%) that one of the dogs they show will become infected with CIV at a show. Despite this concern, respondents' knowledge of CIV was lacking in several key areas (e.g., routes of transmission, infectious period) and high-risk practices frequently reported (e.g., rarely/never immediately disinfect high contact items brought back from a show). Although a high proportion (>93%) of respondents reported all of their show dogs had received core vaccines (or had titers performed, as appropriate) in the past three years, a minority had received 'lifestyle' vaccines in the past year [CIV (51%), Bordetella (60%), leptospirosis (40%), Lyme (20%)]. Of the illness surveillance respondents, a minority reported that since the event one or more of the dogs they showed had developed cough (2%), diarrhea (7%) or vomiting (2%). Of these 44 ill dogs, 8 (18%) required veterinary care at which time *Bordetella* (n=4) and CIV (n=1) were diagnosed. A high proportion (39%) of the respondents that reported ill dogs shortly after their most recent event, stated that since the dog became ill additional dogs at their facility not at the event had also become ill, suggesting further transmission. This study suggests several concerning CIV knowledge gaps and high-risk practices among the dog showing community, but also a high level of motivation to make changes in practices. Although not definitive evidence for show-associated transmission, these findings are suggestive of such an occurrence and highlight the potential utility of this surveillance tool. Together, these findings can be used to inform veterinary recommendations for CIV prevention.

Canine Leptospirosis in Canada, Test-Positive Prevalence and Risk Factors (2009–2018)

Michelle Evason[1,2]; Jason Stull[3]; Donald Szlosek[4]; Scott Weese[5]
[1]University of Prince Edward Island; [2]University of Guelph; [3]Department of Health Management, Atlantic Veterinary College, University of Prince Edward Island; [4]IDEXX Laboratories; [5]Ontario Veterinary College, University of Guelph

In dogs, the incidence and seroprevalence of *Leptospira* spp. appears to be increasing, particularly in North America where it is considered a re-emerging disease. Research on leptospirosis regional distribution and dog-related risk factors is limited in Canada and does not include currently used clinical testing methodology such as PCR. Study objectives were: 1) Evaluate the temporal and spatial distribution of canine leptospirosis in Canada, and 2) Describe dog, geographic, and temporal risk factors for leptospirosis. This repeat cross-sectional study evaluated canine leptospirosis PCR testing (n=10,437 dogs) submitted by Canadian veterinarians between July 2009 and May 2018. Logistic regression was used to identify significant risk factors for test-positive dogs. One or more positive leptospirosis PCR test results were reported from 880 dogs (8.4%). Number and proportion positive varied by year; number of dogs positive ranged from 12 (2010) to 367 (2017), while proportion positive ranged from 4.8% (2011) to 14.2% (2017). There was regional variation with the proportion positive ranging from 1.0% (3/307; Alberta/Saskatchewan/Manitoba) to 18.5% (104/562; Nova Scotia). The final multivariable model identified male (OR=1.3), young (≤1.0 year; OR=2.1), toy breed (OR=3.3), urban (OR=1.3) dogs to be at significantly increased odds for leptospirosis. Significant 2-way interactions of season with both province and year highlighted the complex spatial and temporal influences on leptospirosis occurrence. This research builds on prior studies on canine leptospirosis and provides important new disease distribution and risk factor information for Canada. Our findings will inform veterinary recommendations for leptospirosis diagnosis and prevention in Canada.

Evaluation of Antimicrobial Prescriptions in Dogs with Suspected Bacterial Urinary Tract Disease

J. Scott Weese[1]; Dennis Ballance[2]; Philip Bergman[2]; Talon McKee[2]; Jinelle Webb[3]
[1]University of Guelph; [2]VCA Clinical Studies; [3]VCA Mississauga-Oakville Veterinary Emergency and Specialty Hospital

Antimicrobial stewardship is getting increasing attention in veterinary medicine and assessment of antimicrobial use practices is a core component of stewardship activities. A retrospective review of antimicrobial prescriptions for dogs diagnosed with infectious urinary tract disease in veterinary practices in the United States and Canada was performed to evaluate antimicrobial prescribing. A total of 7803 antimicrobial prescriptions from 7387 dogs were evaluated; 7127 dogs from 673 clinics from the US and 260 dogs from 50 clinics in Canada. The main clinical concerns were acute cystitis (n=6582), recurrent cystitis (n=428) and pyelonephritis (n=326). Amoxicillin/clavulanic acid (2702, 41%), cefpodoxime (1024, 16%) and amoxicillin (874, 13%) were the most commonly prescribed antimicrobials for acute cystitis. The median duration was 12 d (range 3–60 days, interquartile range (IQR) 4 days). There was a significant difference in duration between 2016 and 2018, and 2016 and 2018 (both $p<0.001$), with lower duration in 2018. 55% (3634) dogs were treated with a recommended first line drug, rising from 52% (736/1410) in 2016, 53% (1348/2536) in 2017 and 59% (1550/2636) in 2018 ($p<0.0001$). There were significant differences in prescription patterns and overall use of 'highest priority critically important antimicrobials' between Canada and the US, and between different US regions. Amoxicillin/clavulanic acid (146, 33%), marbofloxacin (95, 21%) and cefpodoxime (65, 14%) were most commonly used for recurrent cystitis, with a median duration of 14 d (range 3–77, IQR 10.5). Amoxicillin/clavulanic acid (86, 26%), marbofloxacin (56, 17%) and enrofloxacin (36, 11%) were most commonly prescribed for pyelonephritis; however, 91 (28%) dogs received two antimicrobials and two (0.6%) three antimicrobials. The median duration of treatment was 14 d (range 3–77 d, IQR 11 d). Fluoroquinolones were prescribed significantly more often ($p<0.0001$) to dogs with pyelonephritis compared to dogs with acute cystitis, while first ($p=0.018$) and third generation cephalosporins ($p=0.009$) and penicillins ($p=0.0007$) were more commonly prescribed to acute cystitis cases Decreases in duration and increased use of recommended first line antimicrobials over the study period were encouraging but room for interventions to optimize antimicrobial use remains.

Carriage of Extended Spectrum Betalactamase-Producing *E. coli* and *K. pneumoniae* in Healthy Dogs and Cats

Cooper Brookshire; Kwang Won Seo; Heather Daw; Peter McGinn; Joo Youn Park; Sunghyun Yoon; Keun Seok Seo
Mississippi State University

Antimicrobial resistance is a widely recognized concern in companion animal medicine. Fecal, oral and aural culture swabs were obtained from 77 dogs and 32 cats at the time of routine spay/neuter surgery. Each swab was processed utilizing sequential selective media broths (MacConkey + cephalothin followed by MacConkey + cephalothin + ceftiofur) followed by MacConkey agar plating. Colonies phenotypically and morphologically consistent with *E. coli* or *K. pneumoniae* were positively identified to species level with PCR, and a multiplex PCR protocol was utilized for identification of extended spectrum betalactamase (ESBL) genes. Numerous ESBL genes, including those from the ACC, CIT, CTM, EBC, FOX, OXA, TEM and VIM families, were detected. Kirby Bauer antimicrobial susceptibility testing was performed on all isolates. Multidrug resistance was common in *E. coli* isolates as indicated by high resistance rates for tetracycline (76.8%), trimethoprim sulfamethoxazole (78.6%), enrofloxacin (37.5%), gentamicin (30.4%), and chloramphenicol (32.1%). No isolates were resistant to imipenem. Resistance to cefpodoxime was detected in 100% of *E. coli* and *K. pneumoniae* isolates. *Escherichia coli* and *K. pneumoniae* with confirmed ESBL characteristics were detected in 39% and 11.7% of dogs, and in 28.1% and 25% of cats, respectively. In total, an ESBL-producing *E. coli* and/or *K. pneumoniae* was detected in 53.2% of dogs and cats presenting for routine surgery. This alarmingly high prevalence of ESBL-producing *E. coli* and *K. pneumoniae* in apparently healthy dogs and cats suggests that there is an urgent need for improved antibiotic stewardship guidelines and compliance in the veterinary community.

Pharmacokinetics of Gabapentin in Cats with Chronic Kidney Disease

Jessica M. Quimby[1]; Sarah Lorbach[2]; Ashlie Saffire[3]; Amanda Kennedy[3]; Luke Wittenburg[4]; Sarah Jones[2]; Rene Paschall[2]; Karen van Haaften[5]
[1]College of Veterinary Medicine, The Ohio State University; [2]The Ohio State University; [3]Cats Only Veterinary Clinic; [4]University of California-Davis; [5]BC SPCA

Gabapentin (20 mg/kg orally once 1–2 hours prior to visit) has become popular as a medication prescribed to reduce stress and improve compliance in feline patients during veterinary visits. However this dose may be inappropriate for elderly cats, specifically those with chronic kidney disease (CKD) as gabapentin is cleared only by renal excretion. Therefore the purpose of this study was to assess the pharmacokinetics of gabapentin in cats with CKD in comparison to clinically healthy cats using a limited-sampling approach. Apparently healthy cats and cats with stable IRIS Stage 2 and 3 CKD were enrolled. Diagnostic tests required before enrollment included serum biochemistry profile, complete blood count, urinalysis, blood pressure, and serum total thyroxine measurement. Blood was drawn 3 and 8 hours after gabapentin administration (10 mg/kg capsule orally once followed by 3 ml water). Gabapentin serum concentration was measured using liquid chromatography coupled to tandem mass spectrometry. Half-life was determined by calculating the rate of gabapentin elimination between the 3 hour and 8 hour time points. Comparison of half-life between the CKD and clinically healthy cats was performed with Mann-Whitney test and correlation between serum creatinine and gabapentin half-life was performed with Spearman Rank. Five apparently healthy and nine CKD cats (4 IRIS Stage 2, 5 IRIS Stage 3) completed the study. CKD cats had a median age of 15 years (range 5–18) and a median serum creatinine 2.6 mg/dL (2.4–3.3). Apparently healthy cats had a median age of 1.6 years (range 1–7) and a median serum creatinine 0.9 mg/dL (0.9–1.3). After administration of a single oral 10 mg/kg dose of gabapentin (median 10, range 9.6–10.3 mg/kg), CKD cats had a significantly longer median calculated half-life (13.1 hours, range 7.3–19.9) in comparison to normal cats (4.3 hours, range 3.9–6.9)($p=0.001$). Serum creatinine was significantly correlated with calculated half-life ($p=0.0002$, $r=0.86$). CKD significantly delays the clearance of gabapentin and this information should be taken into account when prescribing to these patients.

Role of Angiotensin II and Prostaglandins in the Renal Function Following an Oral Protein Load

Marileda B. Carvalho[1]; Gabriela Covizzi[2]; Maricy Appricio[3]; Aracelle Alves[4]; Wilter Vicente[1]
[1]Universidade Estadual Paulista - UNESP; [2]Centro Universitário - UNIFAS; [3]UNIFRAN; [4]Universidade Federal de Uberlândia

Glomerular proteinuria can be a primary manifestation of kidney disease as well as an important factor involved in the secondary progressive parenchyma damage and functional impairment. Although there is relevant evidence for understanding the pathophysiology and therapeutic intervention, some questions are still waiting for answers. The objective of this study was to evaluate the effects of oral protein load on the glomerular and tubulointerstitial functions, considering the roles of renal prostaglandin and angiotensin II (AII), by means of pharmacological inhibition of the mechanisms involved. Seven intact healthy dogs (3 males and 4 females) were enrolled. They were medium-sized, adult, multi-mixed-breed dogs, with age (years) and body weight (kg), expressed as median (range), of 5 (3–6) and 13 (10–22), respectively. All of them were adapted to the experimental management. Three consecutive pairs of experiments were conducted at 7-day intervals. The first pair included a baseline assessment (control) and oral protein load (OPL) consisting of powdered egg white at 4.7 g/kg of body weight (partially rehydrated for consumption). For the second pair, the dogs were treated with flunixin meglumine, a nonsteroidal anti-inflammatory drug (NSAID), at 1.1 mg/kg, q 12 h, for two days, and OPL on the second day (NSAID + OPL). For the third pair of experiments, the dogs were treated with enalapril maleate, an angiotensin-converting-enzyme inhibitor (ACEi), at 0.5 mg/kg, q 12 h and OPL on the second day (ACEi + OPL). Endogenous creatinine clearance (eCcr), urine protein-to-creatinine ratio (UP/C), urine output (UOP), sodium fractional excretion (NaFE), and serum and urine osmolaliy (sOsm; uOsm) were obtained during a 4 hours period of urine collection, after overnight 12 hours fasting followed by OPL when indicated. Water *ad libitum* was available during fasting and evaluation. Oral protein load increased (P<.05) glomerular filtration rate (GFR), even when subjects were under the effect of NSAID or ACEi, although, at a lower rate (P<.05). Urine output and NaFE increased (P<.05) as well, even under the ACEi effect. With NSAID the GFR did not change significantly, suggesting that prostanoids would not participate in the regulatory mechanism of glomerular filtration. However, renal hydrosaline regulation was modified, although OPL addition was effective in inducing a different response. The increase of GFR following ACEi inhibition evidenced a possible role of AII in maintaining a low GFR during fasting as observed in the baseline control. The association with the OPL, unexpectedly, resulted in a statistically non-significant increase in UP/C (range 0.1 to 0.4), that reached a pathological level, although the GFR increase was lower (P=.015) than obtained previously (only ACEi). The study presented, despite having a small n, raises a possible basis for further investigations that may explain and confirm the benefits of protein restrictions in CKD.

Table 1. Variables of renal function from seven healthy dogs under evaluated by Wilcoxon matched pairs signed rank test (basal *vs* each treated).

Basal control data	treatments

	median (range)	OPL	NSAID	NSAID+OPL	ACEi	ACEi+OPL
eCcr (mL/min/kg)	2.0 (1.2–2.8)	↑	ns	ns	↑	↑
UOP (μL/min/kg)	6.8 (3.1–21)	↑	↓	ns	↑	↑
NaFE (%)	0.16 (0.06–1.0)	↑	↓	↑	ns	ns
UP/C	0.12 (0.06–0.2)	ns	ns	ns	ns	ns
uOsm/sOsm (%)	362 (180–627)	ns	↑	↑	ns	ns
sOsm (mEq/L)	301 (285–304)	ns	↑	ns	ns	↑

↑: significantly increased (P<.05); ↓: significantly decreased (P<.05)
Abbreviations: OPL, oral protein load; NSAID, non-steroidal anti-inflammatory drug; ACEi, angiotensin-converting-enzyme inhibitor, eCcr, endogenous creatinine clearance; UOP, urine output; FENa, fractional excretion of sodium; UP/C, urine protein-to-creatinine ratio; sCr, serum creatinine; uOsm, urine osmolality, sOsm, serum osmolality.

Cats with Early Stage Chronic Kidney Disease May Benefit from Moderate Restriction in Dietary Phosphorus

Sofia Schauf Bailon[1]; Esther Bijsmans[2]; Jujhar Atwal[1]; Jen Coltherd[1]; Matthew Gilham[1]; Vincent Biourge[2]; Jonathan Elliott[3]; Denise Elliott[2]; Phillip Watson[1]; Anne Marie Bakke[1]
[1]Waltham Petcare Science Institute/Mars Global Petcare; [2]Royal Canin; [3]The Royal Veterinary College

Dietary phosphorus (P) restriction for the long term management of feline chronic kidney disease (CKD) significantly improves clinical parameters and prolongs life expectancy of cats.[1] Here we describe responses in a cohort of cats diagnosed with early-stage CKD to different dietary regimens and levels of phosphorus restriction. The study was designed as a 45-month observational study of a cohort initially consisting of 19 adult domestic shorthair cats, age range 1.8–9.25 yrs, 10 males and 9 females, all neutered. CKD and IRIS stage was based on creatinine, urine specific gravity, and ultrasound structural changes to kidney tissue and renolith detection. CKD progression was monitored by approximately yearly ultrasounds and 6-monthly blood and urine analyses. Cats were initially fed a renal dry diet (P 0.96 g/1000 kcal; Ca:P 1.75; protein 60 g/1000 kcal) for 17 months. Subsequently, they were transitioned onto a mix of a renal wet (P 0.79 g/1000 kcal; Ca:P 1.59; protein 69 g/1000 kcal) with the dry diet for 4 months, and finally a further transition to more moderately restricted wet (P 1.60 g/1000 kcal; Ca:P 1.32; protein 101 g/1000 kcal) and dry (P 1.35 g/1000 kcal; Ca:P 1.55; protein 78 g/1000 kcal) diets for 24 months. Nine and ten cats were diagnosed with CKD IRIS stage 1 and non-azotaemic stage 2, respectively. Four cats were euthanized, one each during months 12, 15, 20 and 40. Serum urea, creatinine, phosphorus, vitamin D and parathyroid hormone (PTH) did not significantly change over time while on the renal diet, while after 6 months fibroblast growth factor 23 (FGF-23) and blood total and ionized calcium had started to increase. By month 17, FGF-23 above 700 pg/mL was observed in three cats (17.6%, all IRIS stage 2) and 88% had developed total and/or ionised hypercalcaemia (15/17 iCa>1.35; 5/17 tCa>2.95 mmol/L). An increase in renolith and urolith prevalence from 1/17 to 7/17 cats was also observed. Upon transition to the moderately restricted phosphate diets, FGF23, total and ionised calcium decreased, with one cat remaining hypercalcaemic (6%), and prevalence of renoliths and uroliths decreasing as the end of the study period approached. Cats with early CKD fed a dry therapeutic renal diet with P 0.96 g/1000 kcal for 17 months developed hypercalcaemia, some also renoliths and uroliths, which were largely resolved following transition onto a diet regime with P content of 1.35–1.60 g/1000 kcal. Other parameters of kidney function remained stable for the 24 months on the moderately restricted phosphorus diet. While the pathophysiology of hypercalcaemia associated with feeding renal diets remains to be determined, these results suggest that this may occur more frequently in cats with early stage, non-azotaemic CKD. In these cases, feeding a less phosphate restricted diet may be a more appropriate long term management strategy.

1. Elliott J, Rawlings JM, Markwell PJ, *et al*. Survival of cats with naturally occurring chronic renal failure: effect of dietary management. *Journal of Small Animal Practice*. 2000;41:235–242.

Measuring Endogenous ACTH: How, Why, and When?

Patty Lathan, VMD, MS, DACVIM (SAIM)
Starkville, MS, USA

Background

Adrenocorticotropic hormone (ACTH) is a 39-amino acid peptide hormone synthesized by the pituitary gland. The first 24 amino acids are responsible for its biological activity and the structure is highly conserved between species. ACTH has a plasma half-life of approximately 10 minutes. Following stimulation by corticotropin releasing hormone (CRH) from the hypothalamus, ACTH is secreted from the pituitary gland and travels through circulation to bind to receptors in the adrenal cortex, where it stimulates the synthesis of cortisol and other adrenocortical hormones. In healthy dogs and dogs with pituitary-dependent hyperadrenocorticism (PDH), secretion of ACTH is episodic and pulsatile. Unlike in people, a circadian pattern has not been definitively identified.[1,2]

ACTH is derived from a large precursor protein, pro-opiomelanocortin (POMC). POMC is then broken into multiple biologically active fragments: N-terminal fragment, ACTH, and β-lipotropin . Metabolism of ACTH results in the formation of α-MSH and corticotropin-like intermediate lobe peptide (CLIP), neither of which is secreted separately in people.[1,2]

ACTH in Dogs with Hyperadrenocorticism

Approximately 85% of all naturally occurring cases of canine HAC are pituitary-dependent. The remaining 15% are caused by adrenal tumors (ADH). Dogs with PDH secrete a much larger amount of ACTH than normal dogs. This excess stimulation from ACTH then causes over-secretion of cortisol from **both** adrenal glands and results in bilateral adrenomegaly.

Dogs with AT usually secrete excessive cortisol from only one adrenal gland. Since the hypothalamus and pituitary gland detect high levels of cortisol in the blood, CRH and ACTH secretion are dramatically decreased.

Measurement of ACTH concentrations can help differentiate dogs with PDH from dogs with ADH. Intuitively, high ACTH levels are consistent with PDH, and very low levels are consistent with ADH. Unfortunately, since ACTH is secreted episodically by the pituitary, it is possible for a dog with PDH to have a low ACTH value. So, although a high value is diagnostic for PDH, a low value does **not** rule it out.

ACTH in Dogs with Hypoadrenocorticism

Due to the time and expense associated with the ACTH stimulation test, investigators have evaluated alternative methods to diagnose hypoadrenocorticism. One strategy uses the feedback principles of the hypothalamic-pituitary-adrenal axis. ACTH stimulates the release of cortisol in normal dogs, and cortisol feeds back to the pituitary to inhibit further ACTH secretion. In hypoadrenocorticism, lack of negative feedback of cortisol on the pituitary gland results in an increased cortisol-to-ACTH ratio (CAR).

Javadi *et al.* (*J Vet Intern Med*, 2006) were the first group to evaluate the CAR in dogs. They compared normal dogs with Addisonian dogs, and found that while there was overlap between cortisol and ACTH concentrations, there was no overlap between CAR.[3] Two later studies compared CAR in normal dogs, dogs with non-adrenal illness (or diseases mimicking Addison's), and dogs with Addison's.[4,5] There was no overlap between the CAR in dogs with Addison's vs. healthy dogs and dogs with non-adrenal illness in the Lathan study, but there was overlap between two dogs with Addison's and dogs in the non-adrenal illness category in the larger Boretti study.[4,5] No dogs with secondary hypoadrenocorticism were included in the former study, and only one dog in the latter. Thus, no conclusions could be made regarding the CAR in dogs with secondary Addison's. Importantly, even though there was minimal overlap between the CARs of Addisonians and non-Addisonians in all three studies, the ranges for the CARs for each group differed between studies.

Measurement of ACTH

Currently, the two assays used for measurement of ACTH are the radioimmunoassay and the chemiluminescent assay. Whereas the radioimmunoassay requires the use of radioisotopes, the chemiluminescent assay does not. Unfortunately, ACTH is prone to rapid degradation, so samples must arrive at the laboratory frozen (on dry ice), or collected into a vial with the protease inhibitor, aprotinin. Unfortunately, samples collected with aprotinin have significantly lower ACTH concentrations than those without aprotinin when the chemiluminescent assay is used.[2] Thus, laboratories recommend that these samples be shipped on dry ice, which leads to significant logistical issues for practitioners without access to dry ice. Thus, an accurate point-of-care (POC) assay for the measurement of ACTH is desirable. There is currently a POC ACTH assay under development; however, as of the writing of this manuscript, its accuracy and feasibility for use in clinics without reference lab expertise has not been evaluated independently.[6]

References

1. Behrend EN. Canine hyperadrenocorticism. In: Feldman EC, Nelson RW, Reusch CE, Scott-Moncrieff JCR, eds. *Feline and Canine Endocrinology*. 4th ed. Philadelphia, PA: Saunders; 2015:377–417.
2. Scott-Moncrieff JC, Koshko MA, Brown JA, Hill K, Refsal KR. Validation of a chemiluminescent enzyme immunometric assay for plasma adrenocorticotropic hormone in the dog. *Vet Clin Pathol*. 2003;32:180–187.
3. Javadi S, Galc S, Boer P, Joris H, Tesk E, Kooistra HS. Aldosterone-renin and cortisol-to-adrenocorticotropic hormone ratios in healthy dogs and dogs with primary hypoadrenocorticism. *J Vet Intern Med*. 2006;20:556–561.
4. Lathan P, Scott-Moncrieff JC, Wills RW. Use of the cortisol-to-ACTH ratio for diagnosis of primary hypoadrenocorticism in dogs. *J Vet Intern Med*. 2014;28:1546–1550.
5. Boretti FS, Meyer F, Burkhardt WA, Riond B, Hofmann-Lehmann R, Reusch CE, Sieber-Ruckstuhl NS. Evaluation of the cortisol-ACTH ratio in dogs with hypoadrenocorticism, dogs with diseases mimicking hypoadrenocorticism and in healthy dogs. *J Vet Intern Med*. 2015;29:1335–1341.
6. Zomedica Product Brief. Bulk acoustic wave sensor technology in veterinary medicine. Received 2020.

Foodborne Infectious Diseases

J. Scott Weese, DVM, DVSc, DACVIM
Guelph, ON, Canada

As something that animals (and people) are exposed to on a daily basis, pet food and treats represent important potential sources of illness. Foodborne disease has been poorly quantified in dogs and cats, but it clearly occurs. Further, human infections can result from direct exposure to contaminated pet foods or from exposure to pathogens subsequently shed in feces. Because of the potential implications of contamination with pathogens or toxins, or health problems because of excessive or inadequate levels of nutrients, vitamins and other components, rigorous quality control programs are indicated in pet food and treat manufacturing. However, the pet food and treat manufacturing industry is highly variable. It includes large multinational corporations, small single-site manufacturers, small manufacturers that contract production to other companies and small-scale (including in-home) commercial processing. Further, some pets receive food or treats prepared in the household. The foodborne pathogen challenges can vary greatly between these types, as do preventive measures.

Approaches to regulation of pet food vary, and the range of national regulations is beyond the scope of this review. As an example, in the U.S., the Food and Drug Administration (FDA) requires that food be 'safe to eat, produced under sanitary conditions, free of harmful substances and truthfully labelled'. Manufacturing facilities must have a food safety plan that assesses potential food safety hazards and establishes good manufacturing practices for manufacturing, processing, packing and storage.

Pathogens of Concern

Most of the attention has focused on *Salmonella,* but a variety of pathogens are of concern, such as *Listeria* and Shiga toxin-producing *E. coli* (e.g., *E. coli* O157), with the latter of greater health concern in humans exposed to the food. *Campylobacter* is an important pathogen but of somewhat lesser concern because of its relatively poor tolerance to environmental stressors such as freezing. *Brucella suis* infection has also been identified.[1] Additional bacteria, ranging from minimally relevant spoilage organisms to serious concerns such as *Clostridium botulinum,* can thrive in inadequately processed food (e.g., improper canning). Some pathogens are of broad concern, involving many different food types and production systems, while others are very narrow, such as infection by the bacterium *Neorickettsia helminthoeca* within the parasite *Nanophyetus salmincola* in raw salmonids.[2] Parasites, most notably *Toxoplasma gondii,* are also of potential concern. Disease concerns include both animals and humans, since pet food-associated human infections with *Salmonella* and *E. coli* O157 have been identified[3,4] (www.fda.gov/animal-veterinary/outbreaks-and-advisories/fda-investigates-outbreak-salmonella-infections-linked-raws-paws-ground-turkey-food-pets; www.health.state.mn.us/news/pressrel/2019/salmonella012819.html).

Incidence and Impact of Foodborne Disease

The incidence of foodborne disease in dogs and cats is unclear. Outbreaks receive most of the attention and are undeniably important. However, they may not represent the main burden of disease, particularly associated with raw diet or boutique manufacturers, where small batch contamination is less likely to be identified because of the lower likelihood of large numbers of cases. Sporadic disease can go unidentified but represents the main disease burden. The true impact of foodborne disease on animal and human is almost impossible to accurately identify. Investigation is most often to occur when humans develop a reportable disease (e.g., salmonellosis, *E. coli* O157 infection).

Potential Sources of Contamination

Contamination can occur at many steps in the food preparation process. This includes raw ingredients, palatability enhancers, manufacturing equipment, post-manufacturing storage, contamination during retail storage, and contamination in the household after containers are opened. A range of preventive measures at different levels may be required to identify and eliminate these risks.

Preventive Measures

Prevention of contamination and monitoring are a complex process that can involve multiple steps, each of which is designed to achieve a particular goal. Examples are discussed briefly below.

Heat

Heat is one of the most effective and reliable pathogen elimination methods and is a key component of foodborne pathogen elimination. Canned (wet) food undergoes high temperatures during the cooking and canning process, with temperatures that are adequate to render the food sterile. Pet foods are considered to be low-acid foods, which carry an increased risk of *Clostridium botulinum* growth if improperly cooked and canned, so canned food must follow low-acid canned food regulations that dictate time and temperature, along with the type of monitoring and monitoring equipment that are required. Dry foods undergo cooking, with heat and desiccation eliminating many pathogens. The extrusion process also results in generation of heat and pressure that can inactivate many pathogens.

Irradiation

Irradiation is a highly effective means of eliminating pathogens, particularly from items that are not amenable to heat treatment, but cost and consumer perceptions limit widespread use. Irradiation of cat (but not dog) food was banned in Australia after an outbreak of neurological disease in cats fed a specific imported diet that had been irradiated,[5] potentially because of irradiation-induced reduction in vitamin A.[6] While this phenomenon has not been reported elsewhere, it is an important consideration if whole diets are being irradiated, as opposed to treats.

High-Pressure Pasteurization

This process involves pressurizing foods to 100–800 MPa, with only a slight (approximately 3°C/100 MPa) increase in temperature (http://www.fda.gov/Food/FoodScienceResearch/SafePracticesforFoodProcesses/ucm101456.htm). Some raw food manufacturers use this process, largely to reduce *Salmonella* contamination. The bacterium, food matrix and pressurization conditions will impact efficacy of high-pressure pasteurization. Published data regarding the efficacy of methods used on pet foods are lacking, and recalls of treated diets have occurred. Companies using this process should ensure that proper testing is done with their raw food, batch size and pressurization conditions (ideally with experimentally contaminated food), and make that information available publicly.

Preservatives
A variety of preservatives may be added to pet food. Regardless of whether they are 'natural' or 'artificial' (a marketing designation rather than a microbiologically relevant factor), preservatives are not typically intended to kill contaminants. They are mainly added to prevent oxidation of the food and associated discolouration or spoilage. Bacterial inhibition by commonly used preservatives is not adequate to be relied on as a pathogen control method.

Freezing
While not a method used in commercial food preparation, freezing will eliminate some pathogens and have little effect on others. Freezing is most often used to eliminate parasites. For example, freezing (e.g., 2 days at -20°C) can effectively eliminate *Toxoplasma*.[7] *Campylobacter* spp. are also relatively intolerant of freezing, although freezing is not a tool that can be used with confidence to eliminate the risk of campylobacteriosis.

Freeze-drying
Freeze-drying does not eliminate most pathogens. In fact, freeze-drying is a common and effective tool for long-term preservation of bacteria. The freezing component of the process may eliminate pathogens such as parasites or *Campylobacter* that are susceptible to freezing.

Cleaning and Disinfection
Persistence of *Salmonella* in processing facilities has been implicated in large *Salmonella* outbreaks involving commercial dry food, from combination of poor equipment, poor maintenance and inadequate cleaning and disinfection. Detailed protocols and practices are used in well-run manufacturing facilities, but deficiencies can be present and can lead to contamination. Key aspects of cleaning and disinfection are the use of proper cleaners and disinfectants, the presence of surfaces amenable to cleaning and disinfection, provision of adequate time, and, perhaps most importantly, actually performing the task thoroughly.

Pathogen Testing
Testing can allow for detection of contamination before food reaches the consumer, as well as provide an indication that something may have gone wrong in the food preparation process (with subsequent investigation). However, testing does not cover all potential pathogens. Sporadic testing may not be adequate to determine uncommon (but still relevant) contamination. The quality of results also depends strongly on the methods (i.e., if you don't really want to find *Salmonella*, you can use a protocol that is poorly sensitive). Specific pathogen testing practices and thresholds are typically proprietary. However, the FDA has a zero-tolerance policy for *Salmonella* in pet food, so any positive *Salmonella* test would result in a recall.

Hazard Analysis and Response
This is a standard food safety procedure that identifies and evaluates potential (foreseeable) hazards encountered during the manufacture, processing, packing and storage of food. The concept of hazard and critical control point (HACCP) analysis is well established and an expectation of any manufacturer. This approach is designed to identify key control points in the production system and to employ preventive and/or monitoring measures at those points to prevent and detect contamination. Thorough description of HACCP and individual control points is beyond the scope of these proceedings.

REFERENCES

1. van Dijk MAM. *Emerg Infect Dis.* 2018;24(6):1127–1129.
2. Headley SA. *Vet J Lond Engl 1997.* 2011;187(2):165–173.

3. Imanishi M. *J Am Vet Med Assoc.* 2014;244(5):545–553.
4. Public Health England. London, UK; 2018.
5. Child G. *Aust Vet J.* 2009;87(9):349–351.
6. Caulfield CD. *Vet Pathol.* 2009;46(6):1258–1269.
7. Kijlstra A. *Int J Parasitol.* 2008;38(12):1359–1370.

PETS TOO: Emergence of Carbapenem-Resistant *Enterobacteriaceae* in Companion Animals

Shelley C. Rankin, PhD; Stephen D. Cole, VMD, MS, DACVM (Bacteriology/Mycology, Immunology); Donna Oakley, CVT, VTS (ECC)
Philadelphia, PA, USA

SESSION OBJECTIVES

1. Participants will be able to recognize CRE on a laboratory report.
2. Participants will learn how to determine if there are treatment options for CRE.
3. Participants will determine if it is a requirement to report a case/organism to a local or state department of health.
4. Participants will use the principles of infection control to identify clinic reservoirs of CRE and how to clean and disinfect the hospital environment.

BACKGROUND

The emergence of carbapenem-resistant *Enterobacteriaceae* (CRE) in companion animal veterinary medicine was inevitable—and yet we are completely unprepared. In April 2019, passive surveillance identified the blaNDM-5 gene in a carbapenem-resistant *E. coli* isolated from a dog at a veterinary teaching hospital (VTH) in Philadelphia, USA. In May 2019, the microbiology lab became aware that, in April 2018, the Philadelphia Department of Public Health (PDPH) issued a Health Advisory declaring carbapenem-resistant *Enterobacteriaceae* as "reportable" organisms. The microbiology lab director contacted PDPH.

A retrospective review of hospital records showed that from July 2018 through June 2019, CR-*E. coli* was isolated from 15 animals (14 dogs, 1 cat). Six patients were identified as part of a cluster associated with the ICU [Cole *et al.* 2020]. The review also showed that over the course of the next year, CRE was isolated from patients from an additional 9 animals and a variety of clinical specimens (N=15 total).

This session (and these proceedings) highlight "lessons learned" from this outbreak.

A PATIENT TO HANDLE WITH CARE

Samantha, a 13 yo FS Great Dane with a history of megaesophagus was admitted to the ICU following admission to the emergency service the previous evening. On presentation, Samantha was lethargic with obvious increased respiratory rate and effort. Physical exam revealed a rectal temperature of 104.2°F and crackles on auscultation of the lungs. Radiographs suggested aspiration pneumonia. Samantha underwent anesthesia for placement of a gastrotomy tube and an endotracheal wash was performed for aerobic culture. Two days later the microbiology lab reported pure growth of a beta-hemolytic *Streptococcus* that was susceptible to ampicillin. Samantha was being treated with ampicillin and enrofloxacin. Lack of clinical improvement and history of anesthesia led clinicians to question if Samantha had re-aspirated and a different population of bacteria was causing the observed clinical signs. Samantha was re-anesthetized and a second endotracheal wash was performed. While awaiting culture results, Samantha was switched to clindamycin/ceftazidime combination therapy and clinical improvement (improved appetite, resolution for fever and improvement of radiographic signs) was documented by the primary clinician. Three days later the microbiology lab reported a carbapenem-resistant isolate of *E. coli* (Figure 1).

Figure 1

Sensitivity Analysis	Escherichia coli	
Amikacin	<=2	S
Amoxicillin/K Clavulanate	>=32	R
Ampicillin	>=32	R
Cefalexin	>=8	R
Chloramphenicol	>=64	R
Enrofloxacin	>=4	R
Gentamicin	8	R
Imipenem	4	R
Marbofloxacin	>=4	R
Penicillin		
Piperacillin	>=128	R
Tetracycline	>=16	R
Tobramycin	>=16	R
Trimethoprim/Sulfamethoxazole	>=320	R
Vancomycin		

In light of Samantha's culture result, she was re-evaluated and the clinical relevance of the CRE isolate was questioned given her improvement. Bruno, a 10 yo MC Rottweiler was admitted to the ICU on the same day as Samantha for suspected aspiration following adrenalectomy. Persistent fever and dyspnea led to lung lobectomy and aerobic culture of the lung tissue. Three days later the microbiology lab reported a carbapenem-resistant *E. coli* with the same antimicrobial susceptibility profile as Samantha's isolate.

In the Cardiology service, Della, a 9 yo mixed-breed dog, who was previously housed in the ICU during pacemaker placement, presented for a recheck. During her procedure Della suffered an AKI secondary to prolonged anesthesia (suspected) and at recheck bacterial rods were observed on urinalysis. Aerobic culture was added and two days later the microbiology lab reported >100,000 CFU/mL of carbapenem-resistant *E. coli*, with the same susceptibility profile as above.

The internal medicine service also had a recheck of a dog that had been housed in the ICU the week prior with suspect aspiration pneumonia (laryngeal paralysis). The dog was clinically static to mildly improved, but radiographically the lungs appeared more affected. An endotracheal wash was performed and submitted for aerobic bacterial culture. Two days later the microbiology lab reported a carbapenem-resistant *E. coli*, with the same susceptibility profile as above.

Recognizing CRE on Your Lab Report

The most likely scenario is that you will identify CRE in a patient by routine culture and sensitivity for what you suspect is a bacterial infection. If your microbiology lab reports a Gram-negative organism such as *E. coli*, *Klebsiella* spp. or *Enterobacter* spp. that is resistant to a carbapenem (imipenem or meropenem with a MIC ≥4 µg/mL), then you should contact your lab to confirm the result. Ask the lab to determine if the organism produces a carbapenemase or send the isolate to a reference lab. Not all labs routinely test carbapenem antibiotics and you should request additional testing if organisms are resistant to all or most other antimicrobial classes, as shown in Figure. 1. If you suspect CRE due to therapeutic failure of a carbapenem, then you should request additional susceptibility testing.

Carbapenem-resistant *Enterobacteriaceae* are an emerging global healthcare threat that has rarely been reported in companion animals. CRE have acquired genes for production of carbapenemase enzymes that hydrolyze carbapenems. Major classes of carbapenemases include KPC, NDM, IMP, VIM and OXA-48. These genes are often found on plasmids and can be transferred between bacterial strains and species. Containment of cases will be critical to prevent the spread of these organisms to other animals, and most importantly, to humans. Every case of CRE in the USA should be considered significant. These organisms spread easily in healthcare settings due to the susceptible nature of the patient population and the role of "silent carriers" who are colonized in the GI tract. Infections with CRE are opportunistic and can include wound infections, aspiration pneumonia, implant infections and UTI. Blood stream infections with CRE in human infection are associated with significant mortality (>60%), prolonged hospitalization and increased costs.

Antibiotic Use and CRE Infections

It is important to consider the role of CRE in clinical disease before initiating antimicrobial therapy. For example, if CRE is isolated from the urine of a patient without clinical signs of UTI, then therapy may not be warranted. Additionally, patients who have clinical signs that are improving, in spite of inappropriate therapy, may not require antimicrobial therapy. In patients where CRE is thought to contribute significantly to disease, treatment with a topical antibiotic (i.e., wound infections) is preferable to systemic therapy. Systemic antimicrobial therapy for CRE infections should be guided by antimicrobial susceptibility testing results and the advice of experts.

There is no evidence that antimicrobials can be used to decolonize the gastrointestinal tract; therefore, it should not be attempted. In patients that are colonized, it is important to avoid systemic antimicrobials in order to avoid giving CRE a selective advantage over other organisms. It is not yet known, but antibiotics may prolong shedding and increase the number of CRE in gastrointestinal tract because they may suppress other bacteria.

It is also important to consider use of antimicrobials in non-colonized patients during an outbreak. Antimicrobial use may affect the normal bacteria of the GI tract and may make animals more susceptible to colonization.

Infection Prevention for CRE

Outbreaks of CRE in a veterinary facility dictate the need for a multifaceted approach to review and improve infection prevention policies and procedures. It is critical to highlight to all involved that this is best practice to prevent spread to animals and people in your facility. During the outbreak described here, key areas of improvement to infection prevention included evaluation of medical device reprocessing, promotion of hand hygiene, identification and cohorting of colonized patients, detection of environmental reservoirs and improvement of cleaning practices.

During an outbreak of CRE all infection prevention policies should be reviewed and, if possible, audited. In particular, there should be focus on reprocessing of critical (i.e., surgical instruments and other devices that enter sterile sites) and semi-critical (i.e., devices that have contact with mucous membranes such as endotracheal tubes and endoscopes) medical device reprocessing. In several human healthcare outbreaks, duodenoscopes have been identified as sources of colonization. A review of the reprocessing of endotracheal tubes at the VTH revealed use of an inappropriate disinfectant and prompted a switch to use of new endotracheal tubes for all patients.

Hand hygiene is one of the most important parts of an infection prevention program to combat CRE. Increasing access to alcohol-based hand sanitizer and hand hygiene training with products such as "glo-germ" has been shown to have positive effects on compliance with hand hygiene policies.

It is critical to "flag" all CRE-positive patients in the medical record. CRE-positive animals that revisit a facility must be promptly identified before readmission to the hospital. Positive patients should be cohorted in the same area of the hospital and contact with caregivers should be limited to as few people as possible. Full contact precautions should be implemented when direct contact with the animal is anticipated. CRE-positive animals should not be permitted to move throughout the hospital and eliminations should be contained to their housing space. Targeted environmental surveillance should be performed if on-going transmission within a veterinary facility is suspected. We have developed a Swiffer-based method for the targeted detection of CRE in the environment. Important areas to consider sampling include floor drains, sinks, high-touch areas and indoor or outdoor elimination areas. The outdoor elimination area with synthetic turf is thought to be a major environmental source during the prolonged transmission period. In addition, sites that have housed colonized, or infected, animals should be cleaned thoroughly and then disinfected. These environmental sites should be tested and cleared as "CRE negative" prior to housing a new patient in these spaces. Most commercial disinfectants kill CRE if the manufacturer's instructions for use and contact time are followed. We found value in employing a UV-based disinfection strategy during the outbreak, but this approach was adjunctive and not in any way a replacement for the cleaning and disinfection protocols.

REPORTING CRE

Currently, there are no specific federal regulations that require reporting of CRE from animals. Requirements for reporting will vary from state to state, or even local public health jurisdiction to jurisdiction. Should you report a CRE case? The answer is currently not clear. For example, CRE are reportable in the city of Philadelphia but not in the state of Pennsylvania. **Even if CRE are not reportable in your region it is important to note that many jurisdictions require veterinarians to report unusual clusters of disease or pathogens of high consequence or concern**. If you do not know your state or local requirements, you should always contact your state or local public health department for clarification. Some public health departments may make recommendations regarding fecal screening of owners and hospital staff to colonized humans.

CONCLUSIONS

The emergence of CRE in companion animal veterinary medicine will likely become a major challenge to our community. Veterinary medicine is a complex healthcare network that sees animals moving between hospitals for many different reasons, as do humans associated with the profession. It is likely that animals infected, or colonized, with CRE will be found across the country.

We highly recommend that you evaluate the risk CRE poses to your practice and develop a plan to respond to the identification of CRE from your patients.

A variety of useful (human) guidelines and important information can be found at:
www.cdc.gov/hai/organisms/cre/index.html

REFERENCES

1. Cole SD, Peak L, Tyson GH, Reimschuessel R, Ceric O, Rankin SC. New Delhi metallo-β-lactamase-5-producing *Escherichia coli* in companion animals, United States. *Emerg Infect Dis.* 2020;26(2):381–383. https://dx.doi.org/10.3201/eid2602.191221
2. Tyson GH, Li C, Ceric O, Reimschuessel R, Cole S, Peak L, *et al.* Complete genome sequence of a carbapenem-resistant canine *Escherichia coli* isolate with bla_{NDM-5} from a dog in the United States. *Microbiol Resour Announc.* 2019;8:e00872-19. https://doi.org/10.1128/MRA.00872-19
3. Gupta N, Limbago BM, Patel JB, Kallen AJ. Carbapenem-resistant *Enterobacteriaceae*: epidemiology and prevention. *Clin Infect Dis.* 2011;53:60–7. https://doi.org/10.1093/cid/cir202
4. Köck R, Daniels-Haardt I, Becker K, Mellmann A, Friedrich AW, Mevius D, *et al.* Carbapenem-resistant *Enterobacteriaceae* in wildlife, food-producing, and companion animals: a systematic review. *Clin Microbiol Infect.* 2018;24:1241–50. https://doi.org/10.1016/j.cmi.2018.04.004
5. Grönthal T, Österblad M, Eklund M, Jalava J, Nykäsenoja S, Pekkanen K, *et al.* Sharing more than friendship—transmission of NDM-5 ST167 and CTX-M-9 ST69 *Escherichia coli* between dogs and humans in a family, Finland, 2015. *Euro Surveill.* 2018;23. https://doi.org/10.2807/1560-7917.ES.2018.23.27.1700497.
6. Hong JS, Song W, Park HM, Oh JY, Chae JC, Han JI, *et al.* First detection of New Delhi metallo-β-lactamase-5-producing *Escherichia coli* from companion animals in Korea. *Microb Drug Resist.* 2019;25:344–9. https://doi.org/10.1089/mdr.2018.0237
7. Yousfi M, Mairi A, Bakour S, Touati A, Hassissen L, Hadjadj L, *et al.* First report of NDM-5-producing *Escherichia coli* ST1284 isolated from dog in Bejaia, Algeria. *New Microbes New Infect.* 2015;8:17–8. https://doi.org/10.1016/j.nmni.2015.09.002
8. Hornsey M, Phee L, Wareham DW. A novel variant, NDM-5, of the New Delhi metallo-β-lactamase in a multidrug-resistant *Escherichia coli* ST648 isolate recovered from a patient in the United Kingdom. *Antimicrob Agents Chemother.* 2011;55:5952–4. https://doi.org/10.1128/AAC.05108-11
9. Iregui A, Ha K, Meleney K, Landman D, Quale J. Carbapenemases in New York City: the continued decline of KPC-producing *Klebsiella pneumoniae,* but a new threat emerges. *J Antimicrob Chemother.* 2018;73:2997–3000. https://doi.org/10.1093/jac/dky322
10. Giufrè M, Errico G, Accogli M, Monaco M, Villa L, Distasi MA, *et al.* Emergence of NDM-5-producing *Escherichia coli* sequence type 167 clone in Italy. *Int J Antimicrob Agents.* 2018;52:76–81. https://doi.org/10.1016/j.ijantimicag.2018.02.020
11. Zong Z, Fenn S, Connor C, Feng Y, McNally A. Complete genomic characterization of two *Escherichia coli* lineages responsible for a cluster of carbapenem-resistant infections in a Chinese hospital. *J Antimicrob Chemother.* 2018;73:2340–6. https://doi.org/10.1093/jac/dky210

Phosphate and Bone Mineral Disturbances in Early Stage Feline Chronic Kidney Disease

Jonathan Elliott, MA, VetMB, PhD, Cert, SAC, DECVPT, FHEA, MRCVS
London, UK

Introduction

Chronic kidney disease (CKD) is the most common cause of death in cats surviving to 5 years of age (O'Neill *et al.* 2015). It is a heterogeneous syndrome, which progresses at variable rates. Progression is driven by both extrinsic and intrinsic factors. Plasma phosphate at diagnosis is a factor that is associated with progression of cats with azotemic CKD (Chakrabarti *et al.* 2012). Accumulating evidence suggests that feeding healthy cats diets that are high in phosphate, particularly where highly bioavailable inorganic phosphate has been included, is associated with kidney damage (Alexander *et al.* 2019). Thus, phosphate homeostasis in general and bone mineral disturbance (formerly secondary renal hyperparathyroidism) in particular are important factors in the progression and potentially the initiation of CKD.

Major Players in Phosphate Homeostasis

Until the year 2000, the major hormones thought to influence phosphate homeostasis were those primarily concerned with calcium balance, namely parathyroid hormone (PTH) and calcitriol. PTH increases phosphate excretion via the kidney but, at the same time, brings more phosphate into the system by increasing calcium and phosphate release from bone and increasing calcitriol formation. Calcitriol acts on the intestine to increase both calcium and phosphate absorption and is synergistic with PTH in its actions on the kidney and bone. It was postulated a hormone existed that increased renal excretion of phosphate without releasing phosphate from bone or increasing phosphate absorption from the gastrointestinal tract. That proved to be the case with the discovery of FGF23, a phosphotonin that inhibits (a) the expression of the proximal tubular phosphate transporters (thus increasing renal excretion of phosphate), (b) PTH secretion and (c) the conversion of calcidiol to calcitriol by the kidney (Geddes *et al.* 2013a). As with many hormones, its discovery came about because of diseases recognised where excess hormone occurred, such as X-linked hypophosphatemic rickets (Gohil, Imel 2019).

FGF23 is produced by osteocytes and osteoblasts. The exact stimulus for its secretion remains to be determined. Acute ingestion of phosphate in cats does not lead to increased FGF23 secretion (Coltherd *et al.* 2018) but chronic ingestion of increased amounts of phosphate does lead to elevated circulating levels of FGF23 (Alexander *et al.* 2019). There is still much to understand about phosphate homeostasis in the cat. The pathways for intestinal absorption and factors influencing these remain completely unknown for example. These vary between species and so characterisation of the feline phosphate transporter systems seems to be an important goal for future research.

Disturbances of Phosphate Homeostasis in Feline CKD

Research in the 1980s demonstrated that cats fed high levels of phosphate when renal mass had been surgically reduced developed renal lesions in the remnant kidney (fibrosis and mineralisation in particular) (Ross *et al.* 1982). Naturally occurring CKD in cats leads to increased circulating levels of PTH (Barber, Elliott 1998), which reduce in response to feeding of a phosphate restricted diet (Barber *et al.* 1999). Cats with IRIS stage 2 and 3 CKD fed a commercial renal diet with the aim of controlling PTH survive for longer (Elliott *et al.* 2000) and have less uremic crises (Ross *et al.* 2006). These early studies showed that secondary renal hyperparathyroidism is common in the later stages of CKD and that dietary phosphate restriction (+ or - intestinal phosphate binders) reduced serum PTH and phosphate and improved survival.

Phosphate, however, is predominantly an intracellular anion. It seems likely that measurement of serum phosphorus is like looking at the tip of the iceberg and that accumulation of phosphate occurs long before plasma phosphate and PTH increase. Some evidence that this was the case came from unpublished observations that whilst plasma phosphate reduced and stabilised in cats with CKD within the first 2 to 3 months of feeding a clinical renal diet whereas plasma PTH could continue to decrease for up to 12 months. Conversely, if the owner decided to feed a standard maintenance diet again, interestingly, PTH increased before plasma phosphate did.

Phosphate Homeostasis in Early CKD

The discovery of FGF23 gave us the tool to assess disturbed phosphate homeostasis long before plasma phosphate or PTH increase. As glomerular filtration rate (GFR) falls, if dietary phosphate intake remains the same, the rate of phosphate excretion reduces (as phosphate excretion is dependent on GFR). Thus, if plasma phosphate does not increase, phosphate in another body compartment (such as intracellular fluid or bound stores) must increase. Studies in the last decade at the Royal Veterinary College have suggested this to be the case.

Finch *et al.* (2013) demonstrated that plasma concentrations of FGF23 predicted incident azotemia in healthy geriatric cats. Healthy cats over the age of 9 years with plasma creatinine concentrations within the reference range were recruited to a prospective study and followed for 12 months. After 12 months, they were divided into three groups based on their kidney function at that time. Group 1 had normal kidney function, group 2 had plasma creatinine concentrations that were borderline (but within the reference range) associated with urine specific gravity below 1.035 and group 3 were azotemic. Figure 1 shows box and whisker plots of the plasma FGF23 concentrations of these three groups of cats at the initial screening time points.

Figure 1

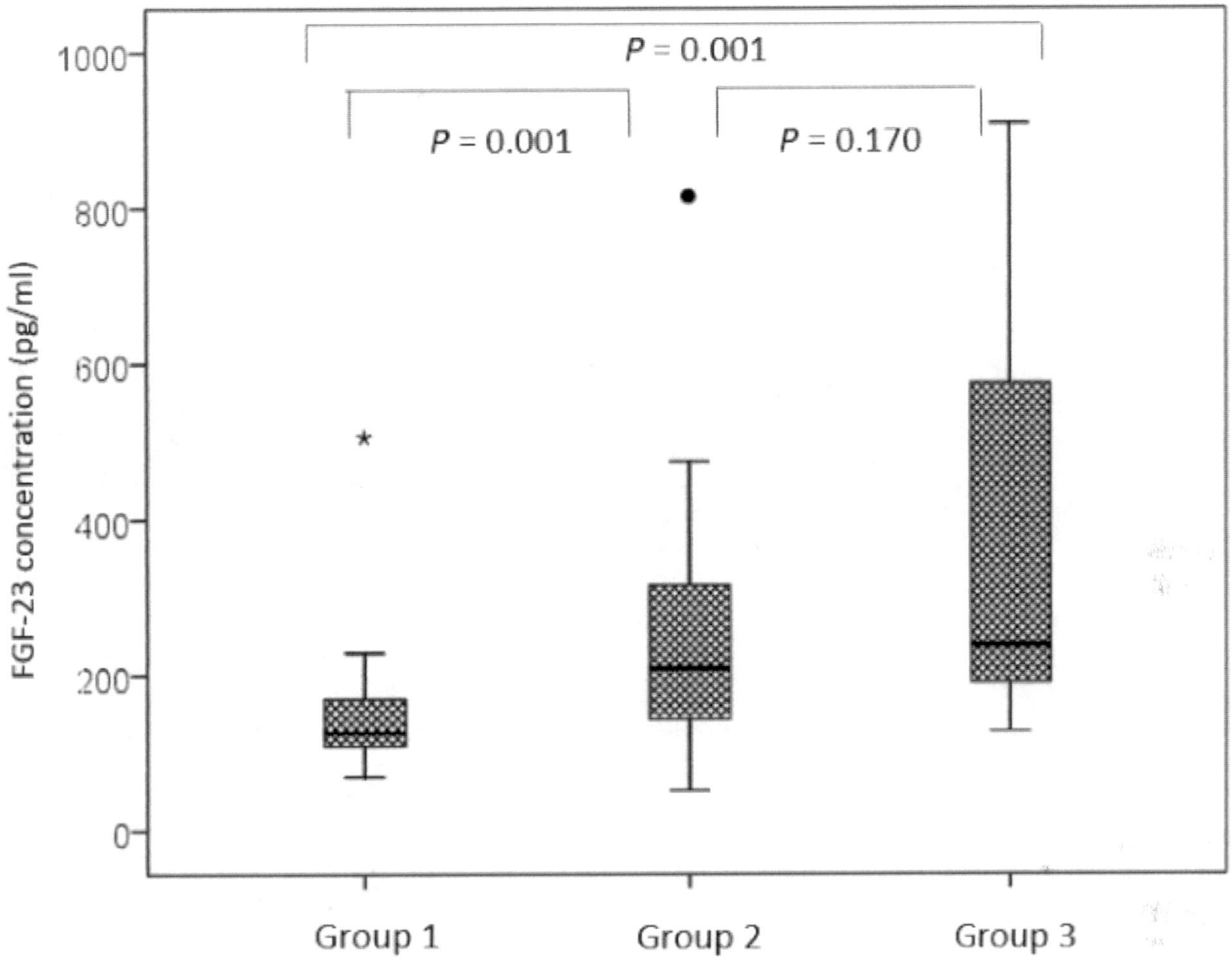

Reproduced with permission from Finch, *et al. J Vet Intern Med.* 2013;27(2):227.

Clearly, cats whose kidney function remained normal over the 12 month period (group 1) had significantly lower plasma FGF23 concentrations than those developing azotemic CKD (group 3) at the initial screening period.

This supports the view that bone mineral disturbances occur early in the course of CKD and provides the rational for the concept that feeding a phosphate restricted diet in early CKD would be beneficial. However, prospective intervention studies are required to address this question. To date, the evidence is equivocal (Geddes *et al.* 2016). Clinical use of FGF23 to determine which cats would benefit from dietary phosphate restriction most is hampered by the lack of a commercial test for this hormone and so its study is confined to the research setting. As a circulating marker it shows great promise as a prognostic indicator predicting all-cause mortality and progression (Geddes *et al.* 2015) and as a means of monitoring the need for and response to treatment in cats whose plasma phosphate concentrations are within the IRIS treatment target ranges for their stage (Geddes *et al.* 2013b).

IS DIETARY PHOSPHATE RESTRICTION APPROPRIATE FOR ALL CATS WITH CKD?

From the above discussion, it would seem that the application of dietary phosphate restriction would be appropriate for all stages of CKD. However, it is well recognised that some cats with

CKD develop hypercalcemia (van den Broek *et al.* 2017) and sometimes this appears to be triggered by feeding a clinical renal diet (Barber *et al.* 1999). Phosphate restricted clinical renal diets tend to have higher Ca:P ratios than standard grocery diets (as a result of phosphate restriction). In some cases, reducing the degree of phosphate restriction leads to a reduction in serum calcium without compromising the control of bone mineral disturbance. Prospective studies are required to determine how cats that become hypercalcemic when fed a clinical renal diet can best be identified and to understand the optimum dietary management of these cats.

Why is hyperphosphatemia detrimental?

In human medicine, the most common cause of death in CKD patients that are on dialysis is cardiovascular disease, which is thought to occur secondary to vascular mineralisation (Hill *et al.* 2016). Radiographic evidence shows that this occurs in cats with IRIS CKD stage 4. *In vitro* data demonstrate that exposure of vascular smooth muscle cells to media containing phosphate of 2 mmol/l (6.2 mg/dl) changes the phenotype of these cells so that they secrete extracellular matrix which then becomes mineralised (Louvet *et al.* 2013). Magnesium is an antagonist of this process *in vitro*. Low serum magnesium concentrations in cats with CKD are associated with increased FGF23 concentrations (independent of plasma phosphate) (van den Broek *et al.* 2018). There is an interaction between serum phosphate and magnesium in predicting progression of CKD. Whether correcting the low serum magnesium improves the outcome of hypomagnesemic cats with CKD remains to be determined. Cats with serum magnesium below 0.8 mmol/l (1.9 mg/dl) should definitely be monitored closely and attention should be paid to controlling plasma phosphate to mitigate the bone mineral disturbances as good control of this complication of CKD is associated with favourable outcomes (Elliott *et al.* 2000).

Conclusions

Bone mineral disturbance is an important and early consequence of feline CKD. A greater understanding of the physiology of phosphate homeostasis in the cat and the factors influencing this in individual cats will, in the future, assist in veterinarians tailoring the management of CKD to the individual patient at all stages.

References

1. Alexander J, *et al.* Effects of the long-term feeding of diets enriched with inorganic phosphorus on the adult feline kidney and phosphorus metabolism. *Br J Nutr.* 2019;121(3):249.
2. Barber PJ, Elliott J. Feline chronic renal failure: calcium homeostasis in 80 cases diagnosed between 1992 and 1995. *J Small Anim Pract.* 1998;39(3):108.
3. Barber PJ, *et al.* Effect of dietary phosphate restriction on renal secondary hyperparathyroidism in the cat. *J Small Anim Pract.* 1999;40(2):62.
4. Chakrabarti S, Syme HM, Elliott J. Clinicopathological variables predicting progression of azotemia in cats with chronic kidney disease. *J Vet Intern Med.* 2012;26(2):275.
5. Coltherd JC, *et al.* Not all forms of dietary phosphorus are equal: an evaluation of postprandial phosphorus concentrations in the plasma of the cat. *Br J Nutr.* 2019;121(3):270.
6. Elliott J, *et al.* Survival of cats with naturally occurring chronic renal failure: effect of dietary management. *J Small Anim Pract.* 2000;41(6):235.
7. Finch NC, *et al.* Fibroblast growth factor 23 (FGF-23) concentrations in cats with early nonazotemic chronic kidney disease (CKD) and in healthy geriatric cats. *J Vet Intern Med.* 2013;27(2):227.
8. Geddes RF, *et al.* The role of phosphorus in the pathophysiology of chronic kidney disease. *J Vet Emerg Crit Care (San Antonio).* 2013a;23(2):122–133.
9. Geddes RF, *et al.* The effect of feeding a renal diet on plasma fibroblast growth factor 23 concentrations in cats with stable azotemic chronic kidney disease. *J Vet Intern Med.* 2013b;27(6):1354.
10. Geddes RF, *et al.* Relationship between plasma fibroblast growth factor-23 concentration and survival time in cats with chronic kidney disease. *J Vet Intern Med.* 2015;29(6):1494.

11. Geddes RF, *et al*. The effect of moderate dietary protein and phosphate restriction on calcium-phosphate homeostasis in healthy older cats. *J Vet Intern Med*. 2016;30(5):1690.
12. Gohil A, Imel EA. FGF23 and associated disorders of phosphate wasting. *Pediatr Endocrinol Rev*. 2019;17(1):17.
13. Hill NR, *et al*. Global prevalence of chronic kidney disease - a systematic review and meta-analysis. *PLoS One*. 2016;11(7):e0158765.
14. Louvet L, *et al*. Magnesium prevents phosphate-induced calcification in human aortic vascular smooth muscle cells. *Nephrol Dial Transplant*. 2013;28(4):869.
15. O'Neill DG, *et al*. Longevity and mortality of cats attending primary care veterinary practices in England. *J Feline Med Surg*. 2015;17(2):125.
16. Ross LA, *et al*. Effect of dietary phosphorus restriction on the kidneys of cats with reduced renal mass. *Am J Vet Res*. 1982;43(6):1023.
17. Ross SJ, *et al*. Clinical evaluation of dietary modification for treatment of spontaneous chronic kidney disease in cats. *J Am Vet Med Assoc*. 2006;229(6):949.
18. van den Broek DH, *et al*. Chronic kidney disease in cats and the risk of total hypercalcemia. *J Vet Intern Med*. 2017;31(2):465.
19. van den Broek DHN, *et al*. Prognostic importance of plasma total magnesium in a cohort of cats with azotemic chronic kidney disease. *J Vet Intern Med*. 2018;32(4):1359.

The Case for Treating Anemia Earlier in Feline Chronic Kidney Disease (CKD)

Jonathan Elliott, MA, VetMB, PhD, Cert, SAC, DECVPT, FHEA, MRCVS
London, UK

Introduction

Chronic kidney disease (CKD) is a heterogeneous syndrome afflicting an increasing number of cats as they age. It is the most common cause of death in cats over the age of 5 years (O'Neill *et al.* 2015). The kidney is the organ that senses the oxygen carrying capacity of the blood and communicates with the bone marrow via the hormone erythropoietin to adjust the circulating red cell mass according to needs. Thus, it is not surprising that with loss of functional kidney mass, red cell mass reduces and non-regenerative anemia contributes to the clinical signs of late stage CKD (lethargy and depression). However, epidemiological evidence suggests that within reference range reductions in packed cell volume (PCV) are predictive of more progressive CKD and all cause mortality, leading to the hypothesis that small reductions in red cell mass may contribute to progressive renal injury.

Current recommendations from the International Renal Interest Society (IRIS) are to provide erythropoietin stimulating agents (ESAs) when the PCV has reduced to below 20% (IRIS 2019) which is when the anemia is thought to be affecting the quality of life of the cat. This is not normally encountered until IRIS stage 4 CKD (serum creatinine >5 mg/dl; serum SDMA >38 µg/dl) and the current treatment recommendations state human recombinant erythropoietin is the most effective treatment but is not approved for veterinary use: darbepoetin is preferable as it is less antigenic than epoetin alfa. Part of the reason for waiting until this stage is that in some cases, these treatments can be self-limiting due to antibody production leading to pure red cell aplasia (Chalhoub *et al.* 2011) and so the risk benefit of using human recombinant products has to be carefully considered. Other complications of use of these products (such as hypertension, iron deficiency) can be managed, so if products were available which did not carry the risk of pure red cell aplasia, their use earlier in the CKD feline patient might be considered. The case for doing so is discussed below.

Association Between Low Red Cell Mass and Progression and Survival in Feline CKD

The prevalence of reduced red cell mass associated with CKD increases with the stage of CKD (Elliott, Barber 1998; Lulich *et al.* 1992; DiBartola *et al.* 1987). The IRIS staging system is helpful in standardising the categorisation of CKD although all the descriptive papers that quantify the prevalence of anemia in CKD feline patients were published before the IRIS CKD staging system was introduced.

Most of the cases included in the studies published by Lulich *et al.* (1992) and DiBartola *et al.* (1987), showed many clinical signs and these cats had relatively short survival times. In the study reported by Elliott & Barber (1998), cats diagnosed with CKD were categorised according to their clinical presentation into three groups namely compensated, uremic and end stage CKD patients, which approximate to IRIS stages 2, 3 and 4 respectively. Compensated showed no clinical signs and were identified on routine screening, uremic showed variable signs compatible with CKD but were eating and relatively stable and end stage required hospitalisation and fluid therapy but had a history compatible with chronic disease. The prevalence of anemia (defined as a PCV <27%) in each of these groups was 0, 32 and 65%

respectively. From these data, it is clear that anemia is associated with more severe clinical presentation in cats with CKD. Indeed, the prevalence of anemia in the end stage CKD group may have been under-estimated as a number of cats within this group were dehydrated at the time of diagnosis, leading to an increase in their PCV.

Few studies have examined the influence of red cell mass on survival in feline CKD although it is recognised that low red cell mass is associated with a poor prognosis regardless of the stage of CKD the cat presents. Elliott & Barber (1998) looked at the survival of cats in the uremic group of their study (32% of which were anemic). There were 33 cats in this group, 12 of which had PCVs below 27%. The median survival of the group as a whole (excluding 5 cats whose owners decided to euthanase their cats on diagnosis of CKD) was 233 days, whereas the median survival time of the anemic cats in this group was 84 days. This finding supports the clinical impression that low red cell mass is associated with poor outcome in the CKD feline patient. Unpublished data from our research group shows that PCV at initial diagnosis of CKD is associated with survival. The population of cats (n=139) was divided into quartiles based on their PCV at diagnosis and Kaplan-Meier survival curves constructed using death (by natural causes or euthanasia as the event). Cats that were still alive at the time of analysis were censored. The median survival times and PCVs of the four groups are presented in Table 1. Log rank analysis demonstrated a significant difference between quartiles (p<0.0001). Multivariable Cox proportional hazards analysis showed PCV to be an independent risk factor (independent of age, sex and plasma creatinine) with the lowest quartile having significantly greater risk of death (shorter survival) than quartiles 3 and 4.

Table 1. Effect of PCV at diagnosis on survival time (all-cause mortality) in cats with CKD

Group	Median (days)	95% CL	Hazard ratio (95% CL)*	P value
1st quartile (PCV <29)	189	111-267	Referent group	NA
2nd quartile (PCV 29-33)	264	97-431	0.61 (0.33, 1.11)	0.105
3rd quartile (PCV 34-37)	586	343-883	0.32 (0.17, 0.61)	<0.0001
4th quartile (PCV >37)	794	497-1091	0.15 (0.07, 0.32)	<0.0001

* Multivariable Cox regression analysis; CL = confidence limit

Chakrabarti *et al.* (2012) studied factors associated with progression of CKD in cats. Progression was defined as a 25% increase in plasma creatinine compared to the value at diagnosis within 12 months of initial diagnosis. This definition was empirical, and this study was biased towards studying cats with relatively long follow-up. Those which progressed rapidly following an initial diagnosis at a relatively advanced stage of CKD often had not shown the requisite 25% increase in plasma creatinine concentration prior to a decision being made to euthanise the cat. Two hundred and thirteen cats met the inclusion criteria for this study and 47% (101 cats) of them met the definition of progression. In the univariable analysis of risk factors for progression, PCV was significantly lower in the progressive cats when compared to the non-progressive cats (univariate binary logistic regression analysis). The median PCV (interquartile range) of the progressive group was 31 (25 to 35%) compared to 35 (32 to 38%) for the non-progressive (stable) group. Figure 1 presents these data as a scatter plot to provide an illustration of the spread of the PCV values for the two groups.

Figure 1. Scatter plot of PCV data from cats with CKD that progress (progressive) within 12 months of diagnosis vs. those that do not (stable)

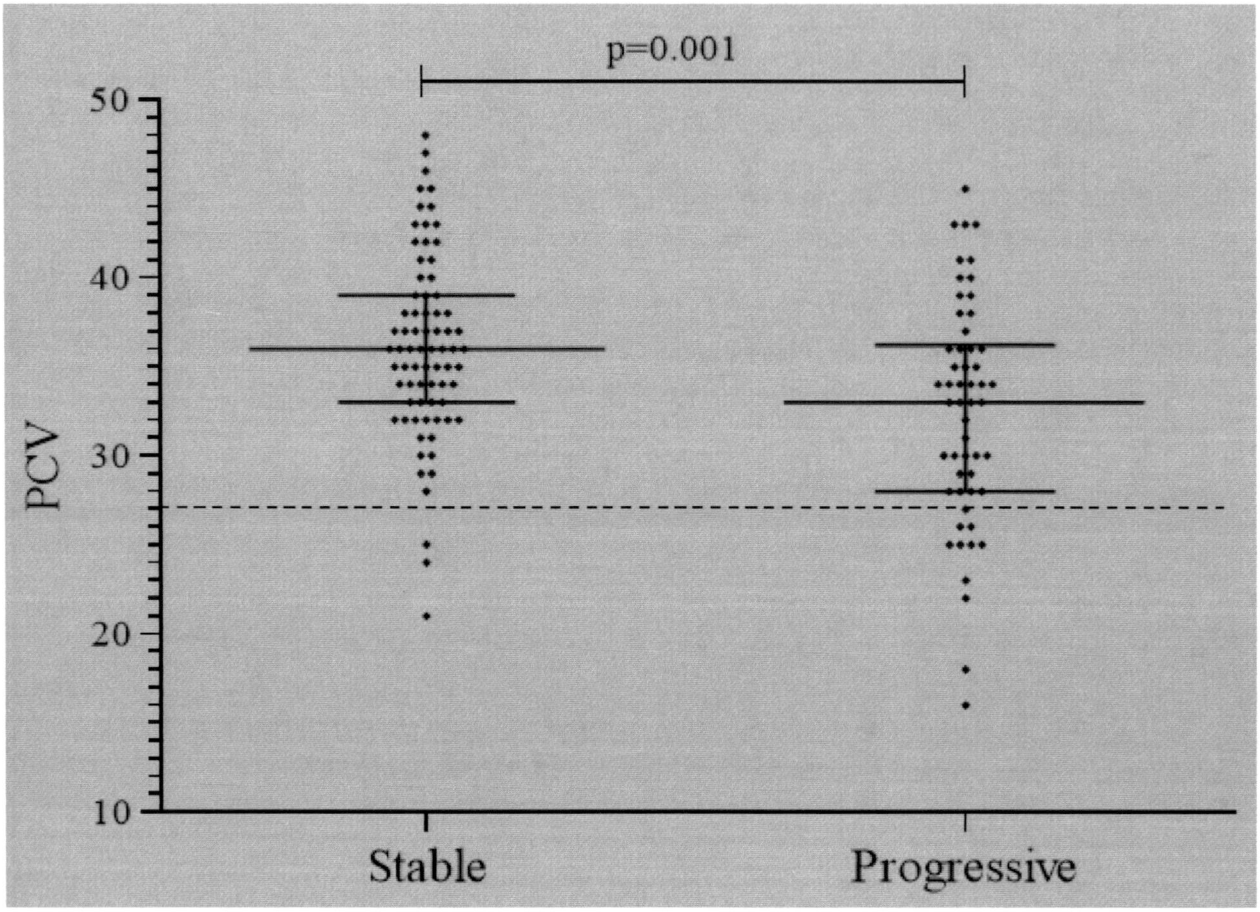

Data from Chakrabarti *et al.* 2012

This study demonstrates that a relatively small reduction in PCV (within the lower end of the reference range) is a factor associated with (possibly contributing to) an increased risk of progression. Multivariate analysis demonstrated that PCV was an independent risk factor associated with progression in cats in IRIS CKD stage 2 (serum creatinine concentration below 2.9 mg/dl). As it is an epidemiological study, the association does not necessarily mean that the reduced PCV is contributing to progression; it could merely be an indirect association. Nevertheless, there are theoretical reasons discussed below why even marginally reduced PCV could exacerbate continuing renal pathology and therefore why treatments that increase red cell mass would be renoprotective (see below).

Determination of Threshold of Hematocrit at Which to Initiate Treatment

The relationship between hematocrit and tissue oxygen delivery is a bell-shaped curve with a steep decline in oxygen delivery where reduced hematocrit limits tissue oxygenation (below 30%) and when blood viscosity limits microcirculatory flow (above 60%) (Fan *et al.* 1980). These experiments were conducted in dogs. This is an important concept in a highly metabolically active tissue, such as the kidney, and in a sedentary species such as the cat where normal PCVs often run below 35%. Furthermore, the proximal tubule of the kidney is highly metabolically active transporting ions and as numbers of nephrons are reduced in kidney disease, adaptations occur within the remaining functioning nephrons, which hypertrophy and increase their work rate (Brown, Brown 1995). Growth of the peritubular vascular network to supply increased oxygen demand is not likely to keep pace with this tubular hypertrophy process, the fibrosis increases the diffusion distance for oxygen from the peritubular capillary and the tubule and thus the red cell mass delivering oxygen to proximal tubules that are hypermetabolic will be critical. In theory, this could explain the strong relationship between reduced hematocrit and progression of CKD in the cat and, in turn, with the reduced survival.

Tubular hypoxia is a strong candidate for driving nephron loss and progressive renal injury in the cat and other species. Data published by Schmiedt *et al.* (2016) showed that a single bout of prolonged ischemia induces chronic structural changes within the feline kidney that mirror findings in feline CKD, specifically tubular atrophy, interstitial fibrosis and mononuclear inflammation. This model is of particular interest, given the suggestion that CKD might be a result of repetitive bouts of acute ischemic injury (Brown *et al.* 2015) and clearly is of relevance to the discussion here that PCV values at the lower end of the reference range and below could increase the susceptibility of the CKD patient to further bouts of ischemia and therefore progressive kidney damage.

Conclusions

Clearly, the above hypothesis remains to be proven through well designed clinical trials. The data presented are observational and the associations found do not necessarily imply cause and effect. There are too few published prospective randomised clinical trials involving cats with CKD to provide the evidence on which we select appropriate treatments for our patients. This is particularly true of ESAs, where the only published evidence involves uncontrolled studies of series of clinical cases (Chalhoub *et al.* 2012). With the current interest in developing novel ESA products for the cat, the hypothesis presented above should be addressed in future clinical trials.

References

1. Brown SA, Brown CA. Single-nephron adaptations to partial renal ablation in cats. *Am J Physiol.* 1995;269(5 Pt 2):R1002–R1008.
2. Brown CA, *et al.* Chronic kidney disease in aged cats: clinical features, morphology, and proposed pathogeneses. *Vet Pathol.* 2015;53(2):309–326.
3. Chakrabarti S, Syme HM, Elliott J. Clinicopathological variables predicting progression of azotemia in cats with chronic kidney disease. *J Vet Intern Med.* 2012;26(2):275.
4. Chalhoub S, *et al.* Anemia of renal disease: what it is, what to do and what's new. *J Feline Med Surg.* 2011;13(9):629.
5. DiBartola SP, *et al.* Clinicopathologic findings associated with chronic renal disease in cats: 74 cases (1973–1984). *J Am Vet Med Assoc.* 1987;190(9):1196.
6. Elliott J, Barber PJ. Feline chronic renal failure: clinical findings in 80 cases diagnosed between 1992 and 1995. *J Small Anim Pract.* 1998;39(2):78.
7. Fan FC, *et al.* Effects of hematocrit variations on regional hemodynamics and oxygen transport in the dog. *Am J Physiol.* 1980;238(4):H545.
8. International Renal Interest Society (IRIS) treatment recommendations. http://www.iris-kidney.com/pdf/IRIS_CAT_Treatment_Recommendations_2019.pdf. 2019.
9. Lulich JP, *et al.* Feline chronic renal failure: questions, answers, questions. *Compend Contin Educ Vet.* 1992;14:127–152.
10. O'Neill DG, *et al.* Longevity and mortality of cats attending primary care veterinary practices in England. *J Feline Med Surg.* 2015;17(2):125.
11. Schmiedt CW, *et al.* Unilateral renal ischemia as a model of acute kidney injury and renal fibrosis in cats. *Vet Pathol.* 2016;53(1):87–101.

Weight and Muscle Loss in Chronic Disease

Lisa M. Freeman, DVM, PhD, DACVN
North Grafton, MA, USA

Cachexia and Sarcopenia

Cachexia, a loss of muscle, has been described since the time of Hippocrates.[1,2] It is estimated to affect 6–12 million people in the United States, Europe, and Japan.[3] Cachexia occurs in a variety of chronic diseases in people, including congestive heart failure (CHF; cardiac cachexia), cancer (cancer cachexia), and chronic kidney disease (CKD; renal cachexia), as well as in people with acute illnesses and injuries. The syndrome of cachexia also is common in dogs and cats with these conditions.[1,2] Like cachexia, sarcopenia is characterized by muscle loss, but sarcopenia occurs during aging even in the absence of disease. Cachexia and sarcopenia can occur concurrently since many chronic diseases occur in the aging population.

The weight loss in cachexia is unlike that seen in a healthy dog, cat, or person that loses weight. In a healthy animal that is receiving insufficient calories to meet requirements, metabolic adaptations allow fat to be used as the primary fuel source, thus preserving muscle. In animals with illnesses or injury, the primary fuel source is amino acids from muscle so these animals quickly catabolize muscle and lean body mass. Although there are some differences among forms of cachexia in different diseases (e.g., cardiac cachexia versus cancer cachexia), a quantitative and qualitative loss of muscle is a hallmark of cachexia in dogs and cats. Fat and bone also are lost to a lesser degree, although in advanced cases of cachexia, all body compartments are depleted. The loss of muscle has direct and deleterious effects on strength, immune function, and survival.

In humans, the definition of cachexia continues to be debated but emphasizes unintended weight loss (≥5% in 12 months or less), along with some combination of decreased muscle strength, fatigue, anorexia, low fat-free mass index, and abnormal biochemistry.[4] Weight loss is an insensitive measure of muscle loss, so using it as a diagnostic criterion delays identification of cachexia until its more advanced stages and a resulting underdiagnosis of the prevalence of this syndrome. In addition, there are certain types of cachexia (e.g., cardiac cachexia with ascites) in which the weight loss is masked by accumulation of fat or water. Therefore, waiting until weight loss occurs often prevents an early diagnosis and misses the underlying issue of muscle loss and inflammation. Another reason for using factors other than total weight loss for a diagnosis of cachexia is that cachexia is a process—i.e., muscle loss—and not necessarily an end-stage syndrome. Muscle loss occurs before significant weight loss can be detected. If a loss of muscle were used as a criterion for the diagnosis of cachexia in humans, the prevalence of this syndrome in chronic diseases would be even higher.

In addition to the insensitivity of weight loss as the criterion for cachexia, it is not only the quantitative loss of muscle that results in deleterious effects.[1,2] There also are qualitative changes in function of the remaining muscle including fat infiltration, altered mitochondrial function, and a shift in fibers from type I to type IIb fibers. This shift accelerates muscle loss as type II fibers are more susceptible to atrophy. In addition, recent studies have shown that it is the muscle loss that is associated with functional deficits, not the weight loss. Therefore, in veterinary medicine, the diagnosis of cachexia and sarcopenia is based on muscle loss, rather than weight loss in an effort to diagnose these conditions at an earlier stage.[1,2]

COMMON FORMS OF CACHEXIA

Despite many common mechanisms that are consistent across all types of cachexia, there also are some details specific to individual types of cachexia.

Cardiac Cachexia

Cardiac cachexia is common in dogs and cats with CHF, with 48–69% of dogs[5-7] and 42–48% of cats[8,9] with CHF affected by some degree of cachexia. It is important to note that this is despite the fact that very few dogs (4.5%) or cats (12%) were thin (i.e., <4/9 BCS). Therefore, clinicians must not wait until animals are thin to diagnose cardiac cachexia. Cardiac cachexia typically is recognized only after CHF has developed (although can develop when CHF is imminent), and animals with right-sided CHF have more muscle loss compared to those with left-sided CHF. Anorexia, hyporexia, or dysrexia are extremely common in dogs and cats with CHF which limits the substrates available for maintaining weight and muscle.[1,2] However, inflammatory cytokines, such as tumor necrosis factor (TNF) and interleukin-1 (IL-1), are primary mediators of cachexia. These inflammatory cytokines directly cause reduced food intake, increase energy requirements, and increase muscle catabolism. Cardiac cachexia causes morbidity but also impacts mortality. Cardiac cachexia was recently shown to be associated with shorter survival times in dogs and cats with CHF.[7,9]

Cancer Cachexia

In dogs and cats with cancer, only between 4–5.5% were assessed as being thin (i.e., having a low BCS) at the time of diagnosis, with a large percentage (36–55%) being overweight.[10-12] However, this reflects the importance of looking beyond body condition score (BCS), which primarily assesses fat stores: one study reported that 69% of dogs had experienced weight loss before diagnosis (31% had <5% weight loss, 14% had lost 5-10%, and 23% had weight loss of >10%).[10] In addition, although the prevalence of thin animals was low, 35% of dogs had mild to severe muscle loss. A study in cats with cancer identified muscle loss in 91% of affected cats.[13] Muscle loss was common (72%) even in cats that were overweight. Studies in cats and dogs have shown that thin animals had a significantly shorter survival time compared to those that were normal or overweight.[12,13] This underscores the importance of assessing not only BCS, but also muscle condition score (MCS; which specifically assesses muscle) and changes in body weight in order to detect cancer cachexia.

Renal Cachexia

Although the prevalence of cachexia in dogs and cats with CKD has not specifically been measured, studies noting body condition of cats with CKD have reported that most are thin and have lost weight. Cats with CKD that are thin have a shorter survival time compared to cats that are normal or overweight.[14,15] Similar to cats with CHF, there appears to be an obesity paradox for cats with CKD, with a U-shaped curve: the lowest survival is for cats with a low body weight, followed by cats that are very overweight, with cats in the middle weight having the longest survival times.[14] The obesity paradox also appears to exist in dogs with CKD, with thin dogs having significantly shorter survival times compared to normal weight or even overweight dogs.[16]

DIAGNOSIS OF CACHEXIA AND SARCOPENIA

Moderate to severe cachexia is not difficult to identify in a dog with CHF or a cat with advanced CKD. However, identification of cachexia is more difficult when it is more subtle. However, detecting cachexia and sarcopenia when they are mild offers more opportunities for successful intervention. Therefore, developing clinically applicable techniques are necessary to identify muscle loss at an early stage. Some of the currently used research tools, such as dual-energy x-ray absorptiometry, have limitations for measurement of muscle due to inherent assumptions. Computed tomography is an excellent method for quantifying muscle but requires general anesthesia or sedation. However, clinically-applicable methods of assessing muscle, such as the MCS,[17] can and should be evaluated in every patient at every visit, in addition to body weight, BCS, and diet history. Body condition score and MCS are not directly related because an animal can be obese but still have substantial muscle loss ("sarcopenic obesity," which may confer an even worse outcome); conversely, animals can be thin but have a normal MCS. Ultrasound is an excellent method for quantification of muscle loss and techniques specific to dogs and cats have been developed.[18,19] Our group has validated two ultrasound techniques, the vertebral epaxial muscle score (VEMS) and the forelimb epaxial muscle score (FLEMS), to more accurately quantify muscle in dogs and cats with cachexia and sarcopenia.[18,19] These methods can be used for earlier detection of cachexia and sarcopenia in dogs, as well as for quantitative outcome measures in studies of potential treatments for these conditions in dogs and cats.

PRACTICAL IMPLICATIONS

Always make a specific recommendation for diet—this may be only to say that the diet the animal is currently eating is a good choice and that the owner is feeding an appropriate amount to keep the dog or cat in ideal body condition. It is important to reinforce good nutritional choices that owners are making, since there are so many myths about pet nutrition. It also is important to make recommendations specific for an individual animal based not only on the stage of disease, but also on the animal's nutritional assessment, physical examination, laboratory results, individual food preferences, and concurrent diseases. Ensuring adequate intake of protein and other nutrients is critical to maintain lean body mass. For patients with CHF or cancer or in aging animals, avoid renal diets unless advanced CKD or proteinuria are present. Even certain maintenance or senior diets with protein levels at or just above the Association of American Feed Control Officials minimums may not have enough protein for animals with heart disease, especially if they have cachexia or sarcopenia. In complicated cases, especially those with concurrent diseases, consider referring the patient to a board-certified veterinary nutritionist (https://acvn.org/ or https://www.esvcn.eu/college).

Omega-3 fatty acid supplementation may provide benefits for animals with cachexia or sarcopenia. The omega-3 fatty acids, eicosapentaenoic acid (EPA) and docosahexaenoic acid (DHA), are normally are found in low concentrations in cell membranes compared to the omega-6 fatty acids, but they can be increased by a food or supplement enriched in omega-3 fatty acids. Omega-3 fatty acids reduce production of the more inflammatory eicosanoids and of inflammatory cytokines, such as tumor necrosis factor and interleukin-1. Omega-3 fatty acids may also have other potential benefits in aging animals or those with certain chronic diseases associated with cachexia due to their anti-inflammatory, antiarrhythmic, and antithrombotic effects.

NEW INTERVENTIONS

Because of the important implications of cachexia and sarcopenia on morbidity and mortality in people, there is extensive research into the prevention, diagnosis, and treatment these common syndromes. There are exciting opportunities for new and effective targets in cachexia to reduce energy requirements, enhance energy intake, improve nutrient absorption, and modify metabolic alterations to prevent and even reverse muscle loss as the mechanisms of these syndromes are becoming better understood (e.g., ghrelin receptor agonists, myostatin antagonists). To be most effective, treatment for cachexia should be multi-modal and include an anti-catabolic effect to decrease muscle loss (e.g., anti-inflammatory agents) and an anabolic effect to enhance protein synthesis (e.g., resistance exercise), but also requires adequate substrate (i.e., calories, protein, and other nutrients). Therefore, single-agent treatments are less likely to be completely effective. It is important to recognize that this is above and beyond optimal treatment for the underlying disease.

Conclusions

Cachexia and sarcopenia are important to recognize, prevent, and treat in veterinary practice because of their high prevalence and deleterious effects, and a better understanding of these syndromes is critical to optimize patient care. New drugs, diets, and other treatments to specifically target cachexia and sarcopenia are being developed and are likely to benefit canine and feline, as well as human patients.

References

1. Freeman LM. Cachexia and sarcopenia. *J Vet Intern Med*. 2012;26:3.
2. Freeman LM. Cachexia and sarcopenia in companion animals. *J Cachexia Sarcopenia Muscle*. 2018;1:1.
3. von Haehling S, *et al*. Prevalence and clinical impact of cachexia in chronic illness in Europe, USA, and Japan: facts and numbers update *J Cachexia Sarcopenia Muscle*. 2016;7:507.
4. Evans WJ, *et al*. Cachexia: a new definition. *Clin Nutr*. 2008;27:793.
5. Freeman LM, *et al*. Nutritional alterations and the effect of fish oil supplementation in dogs with heart failure. *J Vet Intern Med*. 1998;12:440.
6. Beaumier A, *et al*. Clinical findings and survival time in dogs with advanced heart failure. *J Vet Intern Med*. 2018;32:944.
7. Ineson D, *et al*. Clinical and laboratory findings and survival time associated with cardiac cachexia in dogs with congestive heart failure. *J Vet Intern Med*. 2019;33:1902.
8. Pierce K, *et al*. Association between survival time and changes in NT-proBNP in cats treated for congestive heart failure. *J Vet Intern Med*. 2017;31(3):678.
9. Santiago SL, *et al*. Cardiac cachexia in cats with congestive heart failure: prevalence and clinical, laboratory, and survival findings. *J Vet Intern Med*. 2020;34:35.
10. Michel KE, *et al*. Evaluation of body condition and weight loss in dogs presented to a veterinary oncology service. *J Vet Intern Med*. 2004;18:692.
11. Weeth LP, *et al*. Prevalence of obese dogs in a population of dogs with cancer. *Am J Vet Res*. 2007;68:389.
12. Romano FR, *et al*. Association between body condition score and cancer prognosis in dogs with lymphoma and osteosarcoma. *J Vet Intern Med*. 2016:30:1179.
13. Baez JL, *et al*. A prospective investigation of the prevalence and prognostic significance of weight loss and changes in body condition in feline cancer patients. *J Feline Med Surg*. 2007;9:411.
14. Freeman LM, *et al*. Evaluation of weight loss over time in cats with chronic kidney disease. *J Vet Intern Med*. 2016;30:1661.
15. King JN, *et al*. Prognostic factors in cats with chronic kidney disease. *J Vet Intern Med*. 2007;21:906.
16. Parker VJ, *et al*. Association between body condition and survival in dogs with acquired chronic kidney disease. *J Vet Intern Med*. 2011;25:1306.
17. World Small Veterinary Association Global Nutrition Committee. Muscle condition score for dogs. https://wsava.org/global-guidelines/global-nutrition-guidelines/. Accessed February 9, 2020.
18. Freeman LM, *et al*. Validation of muscle condition score and ultrasound for assessment of muscle mass in dogs with cachexia and sarcopenia. *Am J Vet Res*. 2019;80:595.
19. Freeman LM, *et al*. Evaluation of the use of muscle condition score and ultrasonographic measurements for assessment of muscle mass in cats with cachexia and sarcopenia. *Am J Vet Res*. In Press.

Current Literature Review of Cannabinoids in Veterinary Species

Stephen Cital, RVT, SRA, RLAT, VCC, CVPP, VTS-LAM
Portland, ME, USA

The endocannabinoid system (ECS) has been evolving since the beginning of vertebrate species, but why did veterinary practitioners never learn about it in school? Just like our human counterparts' veterinary practitioners have no or only a cursory understanding of the ECS and cannabis therapy. Veterinary practitioners have also been echoing that there "is no evidence" in animals. To the contrary, we have numerous studies utilizing cannabinoids and other cannabis molecules for therapeutic relief and translational studies that could be considered for any vertebrate creature. Here we review the current and pertinent literature in utilizing cannabis derivatives in animals and discuss the future forecast of cannabis in veterinary medicine. Veterinary medicine has historically lagged behind human medicine since its inception. Veterinarians and veterinary technicians are cautious creatures that often have not only a human as the owner to deal with but a patient that cannot speak and tell you how treatment is or isn't working. We also see less uniformity in regulations and practice acts in veterinary medicine compared to our human counterparts. What veterinary medicine and human medicine do have in common is the lack of education on the topic of cannabis as a legitimate medical therapy in veterinary school. We also see less interest in pharmaceutical companies wanting to invest in clinical studies for companion animals (cats and dogs). Moreover, we see more conservative organized veterinary medicine, like the American Veterinary Medical Association, still holding on to outdated information and prohibition era beliefs surrounding cannabis, perpetuating false assertions that there is no evidence for safety or efficacy of cannabis products in animals.[1] To the contrary, we have hundreds, if not thousands, of articles in laboratory animal species, which technically includes dogs and cats, verifying the safety of cannabinoids at extremely high doses and therapeutic potential for numerous conditions. Yet, the bias persists. However, the use and demand for cannabis products in veterinary medicine is growing rapidly, mainly by pet owner demand.[2,3] Unfortunately, it is growing faster than most practitioners have the time to educate themselves about it. Another confounding factor feeding a negative bias in veterinary medicine is the all too often tetrahydrocannabinol (THC) toxicity in companion animals. Since the legalization of medical marijuana in the United States began, the Animal Poison Control and Pet Poison Hotline has seen a 330% increase in THC toxicities.[4] There is no question to the risk and occurrence of THC toxicity in animals, yet it should be noted there are no reported deaths that can be definitively attributed to THC or other phytocannabinoids without other factoring chemicals also present in the system. The suspected lethal dose of THC in dogs is >9 g/kg, a nearly impossible dose to achieve. The most common route of exposure to THC in companion animals is via ingestion. Approximately 66% of exposures involve pets ingesting homemade or commercial edible goods. The second most common source of cannabis exposures involves ingestion of plant material, followed by cannabis oils or tinctures. Symptoms of THC toxicity include lethargy, central nervous system depression, ataxia, vomiting, urinary incontinence or dribbling, increased sensitivity to motion or sound, dilated pupils, hypersalivation, and bradycardia. Less common symptoms include aggression, agitation, low blood pressure, low respiratory rates, elevated heart rates, and **nystagmus** (continuous abnormal movements of the eyes). Rare signs include seizures or comatose conditions.[5] A 2018 study investigating the susceptibility of cannabis-induced convulsions in rats and dogs, reported no seizures in dogs. But central nervous system signs including ataxia, tremors, and hypoactivity were observed when dogs were given chronic daily oral doses of cannabis extracts containing concentrations as high as 27 mg/kg THC combined with 25 mg/kg cannabidiol (CBD) (1.08:1 ratio of THC to CBD) for 56 weeks.[6] Interestingly, GW also did a long-term study in dogs for 39 weeks with a CBD isolate at 100 mg/mg with some minor concern of liver distress. While 100 mg/kg is an unrealistic dose we'd ever give our patients, it does highlight the potential of adverse effects.

More recently, we have a 12-week study published showing the safety of using a hemp CBD product from ElleVet Sciences in healthy dogs and cats. The study concluded that at a dose of 2 mg/kg, PO, BID animals can safely tolerate longer term dosing with minimal to no side-effects.

There are some fundamental distinctions one must make on the topic of medical cannabis, and even specific terminology used when approaching medical cannabis as a valid medical therapy. The first distinction is between a "marijuana" plant versus a "hemp" plant. The hemp plant has much lower levels of THC (less than 0.3% by dry weight) and has found favor among veterinary professionals since there is a reduced risk of THC toxicity.[7] This distinction is particularly important for recommendations made by veterinary professionals. At the time this article was written there were no states that allow for medical marijuana prescriptions for veterinary use, with many states also denying veterinarians the ability to even "recommend" an over the counter hemp-based product. California was the first state to pass legislation at the end of 2018 with AB2215 that allows veterinarians to discuss cannabis (marijuana specifically) as a therapeutic option, but the legislation still prohibits veterinarians from prescribing, dispensing, or recommending marijuana to animals. Currently hemp remains a legal OTC product for consumers to purchase without the recommendation of a DVM in a majority of states.

There are numerous agonists or chemical compounds extracted from marijuana and hemp plants, that act on the endocannabinoid system (ECS) in animals, just like in humans. The **endocannabinoid system** is an extensive group of endogenous cannabinoid receptors found in nearly every vertebrate, both centrally (in the brain) and peripherally. In fact, we see the evolution of ECS receptors in lesser species, such as invertebrates through the millennia via ancient genetic markers. This ECS is responsible for a variety of the same homeostatic processes seen in humans, with more physiological and genetic effects being discovered every year, often in research animal models. Some of these therapeutic effects, when proven in animal models, do not translate well into human models. This is likely because of the difference in the distribution and density of ECS receptors in animals versus humans.[8]

Veterinary Cannabis Studies

To date, we have an ever-growing list of relevant studies for practical use of cannabis in companion animals. Most notably, we now have the results from four studies, two conducted at Colorado State University (CSU) and two from Cornell University, to help shed light on effective and safe dosing of CBD dominant cannabis products in dogs. In the Colorado State University study, conducted by Dr. Stephanie McGrath, we see dogs given three different dosing strategies. A group of 30 healthy beagle dogs were randomly assigned to receive a cannabidiol dominant product in the form of a capsule, oil, and CBD transdermal cream at a dose of 10 mg/kg/day or 20 mg/kg/day for 6 weeks. In the study, the dogs had complete blood counts, chemistry panels, urinalysis, and bile acids performed at 0, 2, 4, and 6 weeks. The most notable effect was elevations in serum alkaline phosphatase (ALP) that occurred in some dogs. All of the dogs in the study also experienced diarrhea, while the dogs that received the transdermal formula had reddened skin after application that was not of clinical concern.

Because the products used in the study were plant-based, the variability between batches were measured. Variability was <10% for the CBD-infused transdermal cream and CBD-infused oil. There was considerable variation, 28–31%, between the CBD concentration in the capsules and the amount stated on the label. Higher systemic exposures were observed with the oral CBD-infused oil formulation, and the half-life after a 75 mg and 150 mg dose was 199.7±55.9 min and 127.5±32.2 min, respectively. Exposure was dose-proportional, and the oral CBD-infused oil provided the most favorable pharmacokinetic profile. While the study mentions that the diarrhea was not related to the formula, it should be noted that this assertion cannot be made with certainty. The study concluded that this particular CBD dominant product, with limited terpenes, appeared to be well tolerated in dogs.[9]

Also, at CSU there are two continuing studies: one on osteoarthritis (OA) and the other with a longer-term study focused on the utility for canine epilepsy. The pilot study for epilepsy published in the summer of 2019 used a 5 mg/kg daily dose for 6 weeks. More subjective clinical differences were seen between the placebo group and treatment group using a product from Applied Basic Sciences. However, these differences were not of statistical difference from the placebo group in the conclusion. The author of these notes suspects the dose in this study is too low compared to human extrapolation of dosing. A long-term study over three years will follow with higher dosing. A similar study is also being conducted at the University of Florida with ElleVet Sciences product.[10]

In a canine study conducted at Cornell University under the direction of Dr. Joseph Wakshlag, we see similar, yet more favorable results with no diarrhea, utilizing a product made by ElleVet Sciences. A single dose pharmacokinetic study was performed using two different doses of CBD enriched oil. The industrial hemp used in this study has ~10 mg/mL CBD and an equal mix of ~10 mg/mL cannabidiolic acid (CBDA), 0.24 mg/mL THC, 0.27 mg/mL cannabichromene (CBC), and 0.11 mg/mL cannabigerol (CBG). All other cannabinoids were less than 0.01 mg/mL with a robust terpene profile. The initial investigation into single-dose oral pharmacokinetics was performed with four beagles. Each dog received a 2 mg/kg and an 8 mg/kg oral dosage of CBD oil. The dogs were fed 2 h after dosing. Blood was collected at 0, 0.5, 1, 2, 4, 8, 12, and 24 h after oil administration. Pharmacokinetics demonstrated that CBD half-life of elimination median was 4.2 h (3.8–6.8 h) for the 2 mg/kg dose, and 4.2 h (3.8–4.8 h) for the 8 mg/kg dose. These results led to dosing during the clinical trial at 2 mg/kg body weight every 12 h. For the clinical efficacy study, which assessed the use for dogs with radiographically confirmed OA, a randomized placebo-controlled, veterinarian and owner blinded, cross-over study was used. Dogs received CBD oil (2 mg/kg) or placebo oil every 12 h. Hematology, serum chemistry, and physical examinations were performed on every visit. A canine brief pain inventory and Hudson activity scores showed a significant decrease in pain and an increase in activity with CBD oil. Veterinary assessment showed decreased pain during CBD treatment. Owners reported no adverse side effects; however, serum chemistry showed an increase in alkaline phosphatase (ALP) similarly to the CSU study during CBD treatment which normalized over time. Conclusions of the clinical study suggest that 2 mg/kg of ElleVet Sciences CBD product twice daily can help increase comfort and activity in dogs with OA. It should also be noted that some dogs in the study were also on traditional nonsteroidal anti-inflammatory drugs with no adverse effects.[11]

Data has shown in both studies that the other nonpsychotropic cannabinoids, primarily CBD, has a wide safety margin with only minimal side effects. In both studies, the elevated ALP was notable. Interestingly, the increase in liver values was not associated with any other elevated

liver values (gamma-glutamyl transferase, bile acids, or alanine aminotransferase) and may be a response to cannabinoid metabolism through the cytochrome P450 (CYP450) pathway.[12,13]
Animal models of CBD utility for anxiety or panic attacks are supported by studies placing a prey species in front of a predator species and conditioned escape responses in mice and rats. According to these studies, anxiety or panic attacks would be related to the flight and freezing defensive responses elicited by threats which expression was decreased in both models.[14]
Canopy Growth has announced the completion of their anxiety study in companion animals that are currently awaiting publication. The author is aware of other anxiety-related research being conducted in dogs, cats and birds with hopeful publication in the next year. Levels of stress markers such as oxytocin, vasopressin and cortisol will be measured alongside a behavioral survey.
Besides the alleviation of disease processes, veterinary scientists are exploring hemp for its anxiety-reducing effects and nutritional content increases in production animals. Stress on food-producing animals is directly correlated to poorer production of eggs, milk, down, wool, or muscle growth for meat producing animals. By exploring the use of naturally occurring cannabinoids in biomass from hemp production for other uses, we are beginning to see the utility for this welfare and economic challenge. The nutritional content of cannabis biomass is also being studied as feed for production animals, since seeds are particularly full of beneficial fatty acids.[15-18]
Aside from pharmacokinetic, pain, seizure, and anxiety studies, we have seen scientific articles looking at cannabinoid receptor proliferation in feline and canine epidermal tissues suggesting the efficacy of topical and systemic applications for atopic dermatitis in dogs and hypersensitivity dermatitis in cats.[19,20]
Other studies looking at ECS distribution in various tissues are those interested in gastrointestinal function. One study published by Glaiazzo and colleagues looked at CB1, CB2, GPR55, and PPARa in canine gastrointestinal tissue, giving us deeper insight to the anatomical basis of supporting therapeutic cannabis in relieving motility disorders and visceral hypersensitivity in canine acute or chronic enteropathies.[21] We have also seen studies looking at protective effects, specifically for gastrointestinal mucosal lesions secondary to acute pancreatitis in rat models. This is of particular interest in companion animals because pancreatitis is a common occurrence.[22]
Anticancer effects are one of the more common interests with owners. The scientific literature (in rodent models) is promising in several different cancer types, and certainly for the alleviation of symptoms related to chemotherapy or radiation therapy.[23,24] A canine cancer study at the University of Florida has some exciting preliminary *in vitro* results for three different types of canine cancer cells. Hopefully, we will see similar results in the on-going *in vivo* study.

CONCLUSION

While the legal status of cannabis and hemp products continues to play out, it is critical that we continue to push for quality scientific data to support therapeutic evidence. Just like in human medical cannabis circles, the veterinary side of things will continue to evolve, looking for specific cannabinoid and terpene profiles for various ailments or ECS support. As scientists, consumers, and animal lovers, we must pressure cannabis manufacturers to produce products following good manufacturing guidelines, use safe ingredients for animals, and be transparent with what is in their products. To that end, manufacturers should suggest dosing regiments based on science instead of anecdotes. It is critical to note dose extrapolation from one tested product to the next would not necessarily provide the same efficacy or have the same safety profile as mentioned in the conclusions of a specific study based on the wide array of cannabinoid and terpene profiles on the market. We must also pressure local governments, mainly state veterinary and pharmacy boards, to adopt legislative language to allow veterinary professionals to discuss, recommend, and, in some cases, prescribe cannabis product for our pets. Lastly, we must encourage the veterinary profession to educate themselves on this topic. A veterinary specific textbook is currently underway, with contributions from over 30 veterinary specialists. The textbook is expected to hit the shelves on 2020 and is being published by Springer Nature.

REFERENCES

1. Rumple S. *Trends Magazine*. American Animal Hospital Association; 2018:29–34.
2. Landa L, Sulcova A, Gbelec P. *Veterinarni Medicina*. 2016;61(3):111–122.
3. Kogan LR, Hellyer PW, Robinson NG. *Journal of the American Holistic Veterinary Medical Association*. 2016;42(Spring):40–48.
4. Meola SD, Tearney CC, Haas SA, *et al. Journal of Veterinary Emergency and Critical Care*. 2012;22(6):690–696.
5. Brutlag A, Hommerding H. Toxicology of marijuana, synthetic cannabinoids, and cannabidiol in dogs and cats. *The Veterinary Clinics of North America - Small Animal Practice*. 2018;8:1087–1102.
6. Whalley BJ, Lin H, Bell L, *et al. British Journal of Pharmacology*. 2018. doi:10.1111/bph.14165.
7. Greb A, Puschner B. *Toxicology Communications*. 2018;2:10–4.
8. Marzo VD, Bifulco M, Petrocellis LD. *Nature Reviews. Drug Discovery*. 2004;3;771–784.
9. Bartner LR, McGrath S, Rao S, *et al. The Canadian Journal of Veterinary Research*. 2018;82:178–183.
10. Coloradostate.edu. Preliminary data from CBD clinical trials 'promising.' https://cvmbs.source.colostate.edu/preliminary-data-from-cbd-clinical-trials-promising/. Accessed 2/2019.
11. Gamble LJ, Boesch JM, Frye CW, *et al. Frontiers in Veterinary Science, Pharmacokinetics, Safety, and Clinical Efficacy of Cannabidiol Treatment in Osteoarthritic Dogs*. 2018. doi: 10.3389/fvets.2018.00165.
12. Harvey DJ, Samara E, Mechoulam R. *Pharmacology, Biochemistry, and Behavior*. 1991;40:523–532.
13. Narimatsu S, Watanabe K, Matsunaga T, *et al. Chemical & Pharmaceutical Bulletin*. 1990;38:1365–1368.
14. Soared VP, Campos AC. *Current Neuropharmacology*. 2017;15:291–299.
15. Kriese U, Schumann E, Weber WE, *et al. Euphytica*. 2004;137:339–351.
16. Neijat M, Gakhar N, Neufeld J, House JD. *Poultry Science*. 2014;93:2827–2840.
17. Neijat M, Suh M, Neufeld J, House JD. *Lipids*. 2006;51:601–614.
18. Stastnik O, Karasek F, Stenclova H, *et al*. The effect of hemp by-products feeding on gut microbiota and growth of broiler chickens. In: *Proceedings from the International PhD Students Conference MendelNet*. Brno, Czech Republic: 2016;23.
19. Campora L, Miragliotta V, Ricci E, *et al. American Journal of Veterinary Research*. 2012;73(7):988–995.
20. Miragliotta V, Ricci PL, Albanese F, *et al. Veterinary Dermatology*. 2018. doi: 10.1111/vde.12658.
21. Galiazzo G, Giancola F, Stanzani A, *et al. Histochemistry and Cell Biology*. 2018. https://doi.org/10.1007/s00418-018-1684-7.
22. Cao M, Li Y, Xu J, *et al. PLoS One*. 2012;12(7):e52921.
23. Cancer.gov. Antitumor effects. https://www.cancer.gov/about-cancer/treatment/cam/hp/cannabis-pdq#_7. Accessed 2/2019.
24. Honarmand M, Namazi F, Mohammadi A, Nazifi S. *Comparative clinical pathology*. 2018. https://doi.org/10.1007/s00580-018-2810-6.

The Variable Role of the Credentialed Veterinary Technician

Ed Carlson, CVT, VTS (Nutrition)
Woburn, MA, USA

History

Veterinary technicians, as a profession, have a very interesting history. The U.S. Air Force developed an animal technician training program for enlisted Air Force members in 1951. The first civilian educational program was established by the State University of New York at Delhi in 1961. The first veterinary technician educational program began around 1968. In 1972, the first accreditation procedures for animal technician programs were established by the American Veterinary Medical Association (AVMA) Council of Education. In 1973, Michigan State University and Nebraska Technical College were the first animal technician educational programs to receive accreditation by the AVMA. In 1981, the North American Veterinary Technician Association was formed to represent all veterinary technicians. The name has since changed to the National Association of Veterinary Technicians in America (NAVTA), an organization that works closely with AVMA to protect, support, and promote the profession of veterinary technology. In 1989, the AVMA officially approved the term "veterinary technician."

Ongoing Issues in the Profession

Obviously, issues such as lack of job satisfaction, low wages, and underutilization of veterinary technicians are not new problems. Multiple surveys that have been conducted over the past 27 years have yielded the same responses.

According to the responses to the 1992 NAVTA survey, published as "Increase Your Salary by Utilization" in *Vet Technician* (May 1993), the four greatest concerns of veterinary technicians include: low salary (88% of survey respondents agreed with the statement "veterinary technicians are so underpaid that the feasibility of staying in the profession is declining"); lack of career advancement; lack of employment benefits; lack of professional recognition, noting that "the variety of terminology used across the country to identify veterinary technicians contributes to this."

R. David Brunner, DVM, wrote in the 1994 article, "How to Leverage Your Staff for Proficiency and Profitability": "Inadequate compensation and ineffective utilization begins at the top when the DVMs perform both their duties and those better delegated to the technicians. The cycle continues with the technician performing technical tasks and those of the assistant." The cover of the June 1995 *NAVTA News* headline was "Profession in Crisis, Veterinary Technicians Caught in a Vicious Circle."

The 2016 NAVTA Demographic Survey reported that "the top 6 most significant problems that face individuals as credentialed VTs include low income, burnout, lack of recognition and career advancement, the underutilization of skills, and the competition with on the job trained technicians."

However, why are these issues continuing? And, why is our profession not progressing? These are two critical questions that need answers.

The Current State of Credentialing and Title Protection

Answers may lie in reviewing current professional credentialing, a widely accepted job description, and title protection nationwide. Lack of a standardized credential, defined scope of practice, and title protection across the country are major contributing factors to the lack of professional recognition and respect, low wages, and underutilization of veterinary technicians/veterinary nurses.

In terms of credentialing, the Veterinary Technician National Exam (VTNE) is a requirement to become credentialed in 48 states. Graduation from an AVMA-accredited veterinary technology or veterinary nursing program is a requirement to obtain a credential in 42 states. Nine (9) states require certification (CVT), while voluntary certification (CVT) is offered in 10 states through the state's veterinary technician association. Licensing (LVT) is required in 14 states. Registration (RVT) is required in 15 states. And, one state credentials veterinarian technicians as Licensed Veterinary Medical Technicians (LVMT), with one state having no credential options at all.

The next issue of title protection is nonexistent in some states. In other states, the credentialing requirement **only** offers title protection. For example, in Texas, only those who are legally licensed in the state of Texas are allowed to use the term Veterinary Technician or Licensed Veterinary Technician (LVT). However, under Texas law, an LVT **and** a non-licensed employee of a licensed veterinarian may perform all the same tasks and duties. The same is true in Hawaii where registration is required to use the title Veterinary Technician, Registered Veterinary Technician, or RVT. However, non-credentialed employees may perform all the same skills, tasks, and procedures as an RVT. In the state of New Hampshire, "any person, not a licensed veterinarian, who assists a veterinarian in the care and treatment of animals shall be the professional responsibility of the veterinarian being assisted, who shall personally train and supervise the assistant/technician. The assistant's/technician's acts and conduct shall be deemed at all times to be the acts and conduct of the licensed veterinarian employing her or him. For the purpose of this rule, assistant and technician shall be deemed equivalent."

Allowing non-credentialed individuals to perform medical procedures, technical skills, and nursing duties on veterinary patients is a significant contributing factor to credentialed veterinary technicians/veterinary nurses leaving the profession, being undervalued, underpaid, and underutilized.

BARRIERS TO UTILIZATION

Lack of knowledge, liability, negative past experiences, and lack of trust can be considered barriers to the utilization or underutilization of veterinary technicians. Many veterinarians appear to be unaware of the education, knowledge, and technical skills necessary to earn a degree in veterinary technology. Many are also unaware of credentialing requirements, including continuing education requirements, for credentialed veterinary technicians or the process for obtaining and maintaining a veterinary technician specialty (VTS).

Some veterinarians express concern for their own liability should an error or accident occur while a veterinary nurse/technician is performing a procedure. Some may say they can perform the procedure more quickly themselves rather than delegating this to a veterinary nurse/technician, not taking into consideration the potential long-term financial benefit to the practice and themselves.

Past negative experiences should not be a barrier to utilization. However, negative experiences often are roadblocks to implementation. Talking about what went wrong, how it could have been avoided, and training and practice under direct supervision can build trust and allow for proper utilization.

Perhaps the most significant barrier to utilization is a lack of trust. Veterinarians who do not trust a veterinary technician are unlikely to delegate to that individual. Technicians who do not trust a veterinarian are unlikely to ask for more responsibility.

BENEFITS OF PROPER UTILIZATION

Optimal utilization allows veterinarians to see more patients and enables them to focus their time on the tasks that require a veterinary license, such as diagnosing, prescribing, and performing surgical procedures.

Additionally, optimum patient care and client service are the primary benefits of the proper utilization of veterinary technicians/nurses! Veterinary technicians/nurses often work closely with patients in the veterinary hospital and may be the first to notice changes in patient status. Those who are encouraged to make informed decisions within the parameters of the doctors' orders when caring for their patients are more fulfilled in their jobs and are strong patient advocates.

Two studies have shown substantial economic benefit when veterinary practices utilized credentialed veterinary technicians. A study by Fanning and Shepherd (2010) notes that the average veterinary practice generated $93,311 more gross revenue for each credentialed veterinary technician per veterinarian. The other study, by the Ontario Association of Veterinary Technicians (OAVT) in January 2019, found increased gross revenue of over $78,000 per veterinarian per each credentialed veterinary technician.

It has been theorized (and the author believes) that when veterinary nurses/technicians are encouraged to continue to learn and grow professionally and to use the technical skills they are trained to perform, they are generally happier and more likely to stay with the practice than those who are only allowed to do less interesting tasks (Felsted 2019).

Individuals who are satisfied in their role, feel respected, and are valued as team members are more likely to hold themselves and others accountable. Such teams often develop effective communication and problem-solving skills, which promote a well-run veterinary hospital, optimal patient care, and exceptional client service.

PROMOTING UTILIZATION

There are easy ways to start the conversation about effective, optimal use of veterinary technicians/nurses at your hospital by involving the whole team! Do not fall into the trap of, "We have always done it this way!" Discuss what tasks currently being performed by doctors could be delegated to technicians and which tasks that are being performed by technicians could be delegated to assistants.

Evaluate one procedure at a time. Ask, "Are tasks falling to the lowest-paid team member who is legally allowed to perform the job?" Optimal technician utilization includes effectively using veterinary assistants, which allows technicians to focus on tasks that require additional education and advanced skill. Stratification of labor also makes good economic sense for the practice!

Next, ask if there is a written protocol or standard operating procedure (SOP) describing how to correctly and safely perform the task. Have team members been trained to perform the task following the written guidelines? Veterinary professionals often enjoy teaching others. Helping others to learn can be very rewarding for the instructor. Team members who enjoy teaching could be encouraged to give in-hospital lectures and hands-on labs.

Consider what tools or support you need to advance your or your teams' knowledge, skills (including soft skills), and abilities. Ask the leadership of your organization how you can obtain what you need to become more effective. Better yet, create a proposal of what it might take to meet your goals and how it will benefit the hospital or organization.

When all members of the veterinary healthcare team are utilized to their fullest potential, everyone wins. Team members feel appreciated and empowered; patients receive excellent medical care; and client service and compliance improve. Start the conversation at your hospital today!

References

1. ACER Consulting. Exploring the value that registered veterinary technicians bring to Ontario companion animal practices. Ontario Association of Veterinary Technicians. 2019.
2. Chadderdon L, Lloyd JW, Pazak HE. New directions for veterinary technology. *J Vet Med Educ.* 2014;41(1):96–101. doi: 10.3138/jvme.0713-102R.
3. Fanning J, Shepherd AJ. Contribution of veterinary technicians to veterinary business revenue, 2007. *J Am Vet Med Assoc.* 2010;236(8):846. doi: 10.2460/javma.236.8.846.
4. Felsted KE. VHMA technician utilization survey results. Veterinary Hospital Managers Association Insider's Insight Monthly Report. 2019.
5. NAVTA 2016 demographic survey results. https://cdn.ymaws.com/www.navta.net/resource/resmgr/docs/2016_demographic_results.pdf
6. Salois M. AVMA: More veterinary technicians, higher practice efficiency. 2019. https://dvm360.com/avma-more-veterinary-technicians-higher-practice-efficiency

How to Make a Definitive Diagnosis of Poisonous Plants Toxicosis in Food Animals

Steve Ensley, DVM, PhD
Manhattan, KS, USA

Definitive diagnosis on any case is challenging but to definitely diagnose a poisonous plant toxicosis is high on the list of things not often done or maybe not often done well. Just defining the term, sudden death in animals can mean many things. Sometimes sudden death means I just suddenly found an animal dead, or I just observed this animal suddenly die or somewhere between both of these. If more than 1 animal dies it is logical to think that all deaths occurring during this time are related or all most certainly caused by the same agent. This is often not the case.

Poisonous plant intoxication is one of the differentials on the list for a variety of livestock diseases and unexplained food animal deaths. Toxic plants commonly poison livestock but obtaining a definitive diagnosis is often challenging. The objectives of this paper are to provide a logical thought process to go through to help veterinarians and diagnosticians arrive at the most accurate and definitive diagnosis possible.

Too successfully conduct a poisonous plant investigation I think it is a must that you visit the site of the toxicosis. As a diagnostician I handle the majority of my cases from my office and lab. For most cases I can obtain samples from a submitter and reach a definitive diagnosis. For a successful poisonous plant investigation this is not the case. I practiced 14 years before becoming a diagnostician and I learned that there are many pieces of the puzzle that you won't identify unless you do a field investigation.

Developing a Differential Diagnosis

Great observations skills and a repeatable thought process are key to be able to make a challenging diagnosis. Great observations skills are built over time mostly from cases where you didn't have great observations skills. These skills are honed over time by being honest with yourself with what you didn't do a great job with and remembering what you did right. Clinical signs associated with poisonous plants can be subtle with changes in animal movement, behavior or change in appetite around herd mates. We are taught very well in the science of veterinary medicine. The art of veterinary medicine is just as important but is a skill set that can take experience and time to develop.

To begin with, to diagnose a difficult case you need an accurate timeline. If the animal caretaker can keep a herd book, health records, voice recording or an electronic calendar with notes this can be vital to getting the information you need to solve the case. Being able to document when and where these clinical signs occurred, observed poisonous plant exposures, abundance and ID of plants; and the ages, body condition and history of the affected animals. Knowing the environmental conditions, health status of animals across the fence line, feeding practices and drinking water supplies is also useful.

If you can obtain whole blood, serum or urine from a live animal exhibiting clinical signs typical for the toxicosis you are seeing, that is the best fluid sample. Obtaining blood or serum after death is not a sample that you would want to use for clinical pathology. Post mortem blood has undergone too many changes to make it useful for clinical pathology, CBC or serum chemistry. Urine collected after death is a good, reliable sample in contrast to a postmortem blood/serum sample. I always want to get fixed and fresh tissues from every major organ including the brain. I want ocular fluid and urine also. I also want to get these samples into the lab as quickly as possible in a preserved condition. I don't have to do testing with every tissue or fluid I receive but I want to be able to have access to them if they are needed. I don't want to have to go back to a dead animal and recover samples days after death just because I didn't collect it immediately. Many of these cases will be done after hours, in the dark, with or without a flashlight and with more calls to go on after this one is processed.

For every infectious disease there is a toxin that will produce similar clinical signs. Working through a poisonous plant case will involve getting help with plant ID from either a local plant expert like an extension person, a botanist at a university, diagnosticians at labs like the Poisonous Plant lab in Logan UT, local veterinarians and diagnosticians at the state veterinary diagnostic labs. Collection and preservation of the plant is critical. Plants will degrade quickly when put into plastic bags and left in the heat and humidity. Most toxins are stable but excessive heat could possibly degrade even these compounds. A plant press is the ideal method of preserving plants. If plants cannot be pressed then using paper bags to hold them is the next best alternative. Using individual bags for each plant specimen will make sure plants don't get mixed up either inside the bag or outside. Pressing plants can be done many ways with simple supplies. The main goal with pressing is to remove the moisture and maintain the anatomy of the plant. This is done by drying the plant in a paper bag in an oven at 150°F (65°C) for 10–12 hours. Do not microwave plant samples. Refrigerated samples, even in a paper bag quickly deteriorate.

Table 1. Samples to collect for potential poisonous plant poisoning

Analysis	Sample	Size	Shipping
CBC	Purple top blood tube (any anticoagulant)	3–5 ml	Chilled shipped on ice in a styrofoam cooler
Serum chemistries	Red top blood tube, remove serum from red cells as soon as possible	5–10 ml	Chilled on ice, if frozen serum should be separated from the red cells
Microscopic evaluation of tissues	Tissues from all major organs. Any tissues with lesions.	1x1x2 cm sections	Formalin fixation
Gross necropsy	Fresh dead animal	Intact whole animal	Fresh
Chemical analysis of ocular fluid or urine	Ocular fluid and urine can be collected post mortem.	20 ml	Stored in plastic tubes and shipped on ice or frozen a styrofoam cooler
Chemical analysis of serum, blood or milk	These samples ideally are from antemortem collection.	Minimum of 10 ml	Stored in plastic tubes and shipped on ice or frozen in a styrofoam cooler
Chemical evaluation of tissue	Fresh liver and kidney	4x4x4 cm sections	Stored in plastic tubes and shipped on ice or frozen in a styrofoam cooler
Chemical evaluation of feces or gastrointestinal contents	Feces, stomach/rumen content. Intestinal samples.	50–100 ml	Stored in plastic bags and shipped frozen

Plant identification	Entire plant	Whole plant including flowers, pods, leaves, stems and roots	Pressed and dried
Plant chemical analysis	Entire plant	5 or 6 whole plants	In a paper bag, fresh or frozen
Baled forage for weed contamination and weed identification	Stored baled forage	Random samples of suspect forage, bale probe samples	Dry
Silage or green chopped feed	Feeds	Representative feed samples - 10–20 g	Frozen

Contact the laboratory prior to collection of samples as they often require specific sampling, sample preparation and shipping. Label all materials with permanent markers; provide date, owner, location and contact information. Postmortem blood is of limited value. It is usually more helpful to analyze tissue, GI content, urine or ocular fluid than post mortem blood. In acute toxicosis most plants and toxins will be in the rumen or upper small intestine. There are a few labs that will provide a feed/ingest microscopy service.

The USDA/ARS Poisonous Plant Research Laboratory (PPRL) in Logan, UT is the best lab in the US for help with chemical identification of many plants. They are a research lab primarily but also assist with diagnostic cases. Some plant compounds like dehydropyrrolizidine alkaloids (DHPAs) have been characterized by this lab. The pyrrolizidine alkaloids are a common plant causing chronic hepatic damage. We have worked with the PPRL in an investigation of pine needle abortion in 2015. A toxin-induced abortion outbreak in Nebraska cattle was confirmed to be caused by pine needle consumption following detection of tetrahydroagathic acid in fetal fluids using gas chromatography-mass spectroscopy. *Journal of Veterinary Diagnostic Investigation*. 27(1):74–79. UC Davis veterinary diagnostic lab can analyze feed samples and rumen content for pyrrolizidine alkaloids, strophanthidin (cardiotoxins), oxalates and oleander glycosides. They also analyze urine for pyrogallol and rumen content for gallic acid.

SUMMARY

With all the diagnostic tools available today we have a better chance to make a definitive diagnosis of a poisonous plant toxicosis. A toxicology involved diagnosis is still a challenge and aligning all the data points can be difficult. I try to make the most probable diagnosis based on the facts. All diagnosis cannot be made on data alone. You still have to use the art of veterinary medicine and previous experience to make a diagnosis.

I try to make a rapid diagnosis especially in the case of continuing losses. If there is high death loss I will run all available analytical analysis that I can as rapidly as possible. An approach like this is expensive up front but may be more cost effective in the long run. In most cases I do the analysis that I think will return the answers I need and then work through the case in a slower progression but stop doing analysis when I have an answer.

Adipose Tissue as an Integrator of Metabolic and Inflammatory Signals in Periparturient Cows

G. Andres Contreras, DVM, MS, PhD
East Lansing, MI, USA

The adipose tissue (AT) is a multisite organ that plays significant roles in the homeorhetic adaptation to lactation in periparturient dairy cows. First, AT releases free fatty acids from its triglycerides reserves through lipolysis that are used as a substrate for lactogenesis and bodily functions during periparturient negative energy balance. Second, AT secretes numerous adipokines that directly and indirectly adapt other organs to use free fatty acids (FFA) as energy substrates.[1] Third, AT goes through a remodeling process that includes infiltration of anti-inflammatory macrophages that promote the differentiation of new adipocytes that are capable of buffering FFA excess accumulated during the first week of lactation.[2] A successful transition from gestation to lactation requires that the AT efficiently integrates systemic metabolic and inflammatory signals. This manuscript summarizes our current understanding of AT biology in the periparturient period of dairy cows, focusing on essential integrative functions that support lactational homeorhesis and metabolic and immune responses during disease.

Lipid Mobilization

FFA are constantly being esterified (i.e., lipogenesis) and hydrolyzed (i.e., lipolysis) to and from triglyceride molecules within adipocytes. This process is known as lipid mobilization. During negative energy balance, the rate of lipolysis surpasses that of lipogenesis, and thus FFA are released into circulation. Lipolysis can be broadly divided into two categories: basal and demand lipolysis.[3] The rate of basal lipolysis is strongly associated with adipocyte size and increases steadily throughout lactation; thus, over-conditioned cows release more FFA from adipocytes at basal conditions than lean cows.[4] In periparturient dairy cows, demand lipolysis is regulated hormonally and is activated by catecholamines, growth hormone, angiopoietin-like 4, and prolactin (Figure 1). As demand lipolysis is activated, lipogenesis is drastically reduced by different factors, including low plasma insulin, insulin resistance in adipocytes, and inflammatory responses within the AT. All these factors inhibit the transcription of gene networks regulating *de novo* lipogenesis and triglyceride assembly. In healthy cows, as lactation progresses, energy balance becomes positive, plasma insulin returns to pre-calving levels, and AT lipolysis and inflammation are reduced. However, in cows that do not transition adequately into lactation, AT exhibits an impaired response to the anti-lipolytic effects of insulin driven by chronic AT inflammation, leading to lipolysis dysregulation.[5] These animals are more susceptible to metabolic and inflammatory diseases, and their lactation and reproductive performance are negatively impacted.[6]

Figure 1

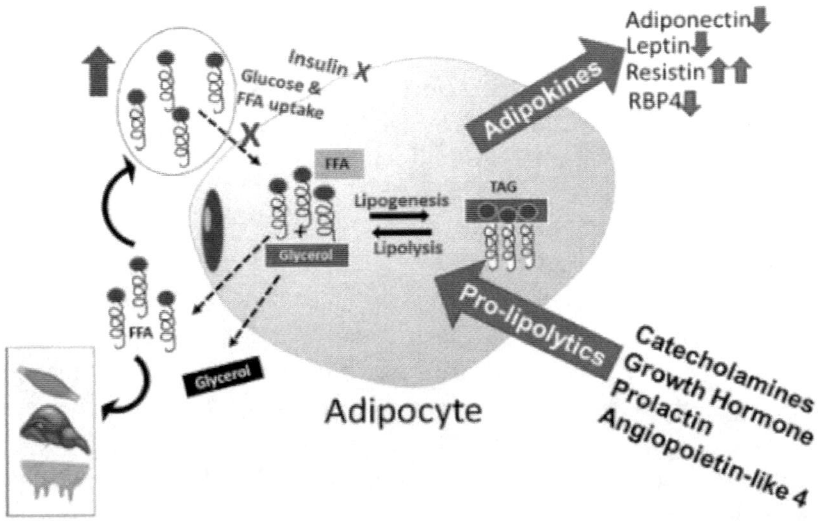

Lipid mobilization includes lipogenesis and lipolysis. Lipolysis during the periparturient period is induced by pro-lipolytic agents and releases free fatty acids (FFA) and glycerol from triglycerides (TAG). Adipocytes secrete adipokines that modulate metabolism systemically prioritizing the use of FFA for lactogenesis and thus supporting lactational homeorhesis.

ADIPOKINES, SYNCHRONIZERS OF PERIPARTURIENT METABOLIC FUNCTION

The AT regulates metabolism by controlling the availability of the energy-dense FFA and by secreting adipokines that have endo-, para-, and autocrine functions. These peptides are produced by the adipocytes and by cells of the stromal vascular fraction (SVF) that include immune and vascular cells. There are over 200 adipokines described in the literature, but only a few have been studied in dairy cattle. Importantly, the secretion patterns of the adipokines described in periparturient cows support lactational homeorhesis by redirecting glucose to the mammary gland, increasing FFA flow to the liver and epithelial cells in the mammary gland, and modulating energy intake[7] (Figure 1). For example, plasma concentrations of adiponectin, which enhances systemic insulin sensitivity and reduces lipolysis, drop from 35 μg/mL during the dry period to <20 μg/mL immediately postpartum.[8] Also, the expression of its receptors is reduced during the first three weeks postpartum.[9] Reflecting its antilipolytic effects, plasma adiponectin is negatively associated with circulating FFA.[10] Similar to adiponectin, plasma leptin peaks during the dry period (>6 ng/mL), and its nadir (<4 ng/mL) is reached the first week after calving.[11] Adiposity influences leptin concentration, with over-conditioned cows exhibiting higher plasma leptin pre-calving compared to lean animals.[12] This difference explains, in part, the more dramatic drop in dry matter intake and higher rates of lipolysis observed in over-conditioned cows. Since leptin is a potent anorexigenic peptide, its low levels postpartum promote the return of dry matter intake to pre-calving levels.

In contrast to adiponectin and leptin, the synthesis of resistin increases during the periparturient period. Resistin is a lipolytic adipokine that is secreted by adipocytes and macrophages. Resistin promotes the hydrolysis of triglycerides by inhibiting insulin signaling and promoting inflammatory responses within the AT.[13] In dairy cows, plasma resistin concentrations rise from 45 ng/mL during the dry period to values above 75 ng/mL postpartum.[14] The degree of adiposity (e.g., body condition score) is positively associated with the production of resistin by AT macrophages.[14] Therefore, over-conditioned cows will have higher circulating resistin compared to lean cows, making them more susceptible to excessive lipolysis and AT inflammation. Retinol-binding protein 4 is a potent inhibitor of adipocyte glucose uptake that also impairs the differentiation of preadipocytes into adipocytes.[15] Plasma levels of this adipokine fall from >50 mg/mL one week before parturition to <30 mg/mL immediately after calving.[16] By inhibiting AT glucose utilization, retinol-binding protein 4 ensures energy prioritization to the mammary gland; however, impairing adipogenesis reduces the capacity of AT to buffer FFA excess predisposing to lipolysis dysregulation.

Adipose Tissue Remodeling

Excessive lipolysis in AT induces a remodeling process within the organ that is characterized by a moderate inflammatory response with infiltration of immune cells, the proliferation of the cellular components of the stromal vascular fraction (i.e., non-adipocyte cells), and changes in its extracellular matrix.[5,17] Macrophages are the primary immune cell infiltrating AT during lipolysis. The specific inflammatory phenotype of these mononuclear cells has been broadly classified in classical (M1), which have active proinflammatory responses, and alternative phenotype (M2), which promote inflammation resolution. The central role of M2 macrophages in AT is to remove products of lipolysis that can be toxic to the cell such as FFA, neutral lipids (triglycerides).[18] For this reason, a moderate infiltration of M2 macrophages into AT is beneficial for periparturient cows.

During lipolysis dysregulation, that is when lipolysis is protracted and intense, macrophages aggregate forming crown-like structures and polarize toward the M1 phenotype (Figure 2).[19] These macrophages secrete potent cytokines such as TNFa, which impair insulin signaling, leading to a vicious circle where AT inflammation exacerbates lipolysis, which in turn can aggravate AT inflammation.

Figure 2. Adipose tissue remodeling during the periparturient period of dairy cows

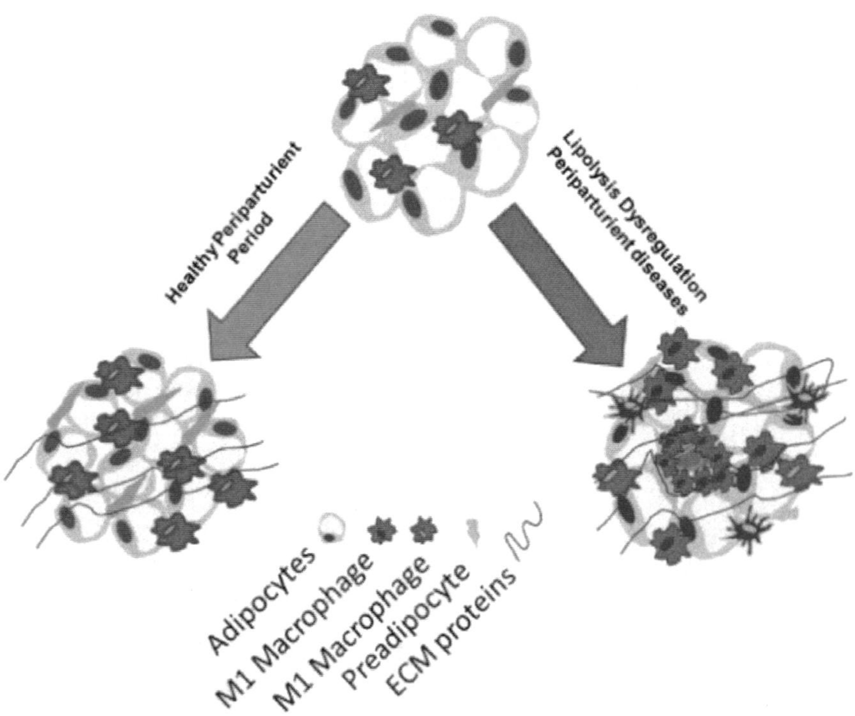

In healthy cows, lipolysis induces infiltration of M2 macrophages and minimal changes in the structure and composition of the extracellular matrix (ECM). In cows that succumb to periparturient diseases, M1 macrophages infiltrate AT forming aggregates (crown-like structures), leading to excessive lipolysis and inflammation.

AT mass is reduced by 25–35%[20] during the first 40 days after calving. Although not demonstrated in dairy cattle, rapid bodyweight loss induced by extended caloric restriction causes adipocyte death and the release of lipid remnants.[18] As a response, preadipocytes proliferate to generate new adipocytes that will replenish fat cell populations in a process termed adipogenesis. An adequate adaptation to periparturient NEB requires adipogenesis.

The third component of AT remodeling that remains largely unstudied is the changes in the ECM structure. Initial transcriptomic profiling of AT indicates that the expression of collagen 1 and 3A1, and thrombospondin 1 increase in ruminants immediately after calving.[21,22] High accumulation of these proteins in AT has been associated with fibrosis and insulin resistance in humans.[17] Although conjecture, AT fibrosis may predispose cows to excessive lipolytic responses, especially older animals.

Adipose Tissue (Dys)Function During Periparturient Diseases

The periparturient period is the lactation stage with the highest risk for metabolic and infectious diseases in dairy cows. Periparturient health events pose severe welfare issues[2] and also result in significant economic losses associated with milk yield loss, cost of treatment, and culling.[3] To make things more complicated, periparturient illnesses often are presented as complexes of metabolic and inflammatory/infectious diseases.[6] Two significant risk factors for increased disease susceptibility around parturition are lipolysis dysregulation (described above) and the dramatic increase in circulating endotoxins (e.g., lipopolysaccharide [LPS], lipoteichoic acids). Remarkably, common periparturient diseases such as mastitis, metritis, pneumonia, and metabolic events such as ruminal acidosis, heat stress, and parturition often result in high circulating levels of LPS.[23] In humans and rodent models of disease, the inflammatory response to endotoxins, especially LPS, impairs the metabolic function of AT.[24] In periparturient cows, experimental LPS exposure was associated with a higher incidence of displaced abomasum and placental retention and changes in metabolic parameters, including low plasma cholesterol, and high β-hydroxybutyrate and FFA.[25] The profile of these parameters indicates that endotoxemia possibly induces the development of lipolysis dysregulation in bovine AT.

The possible mechanisms by which the endotoxemia associated with multiple periparturient diseases triggers AT dysfunction are twofold. First, endotoxins activate lipolysis by the classic and inflammatory lipolytic pathways. In the classic pathway, LPS binds to its receptor TLR4 and increases the levels of intracellular cAMP through a calcium-dependent pathway,[26,27] leading to the activation of hormone-sensitive lipase. LPS activation of lipolysis by the inflammatory pathway involves two major signaling cascades:

1. NF-κB: TLR4 binding with LPS stimulates the phosphorylation and degradation of IκB proteins, leading to the translocation of NF-κB that triggers the synthesis of proinflammatory cytokines including TNFα.[28] The latter promotes lipolysis by impairing the expression/function of perilipin, causing the thinning of the protein envelope of the lipid droplet and making it more susceptible to the action of hormone-sensitive lipase.[29]
2. The activation of the mitogen-activated protein kinase/extracellular signal-regulated kinase (MAPK and ERK1/2). This pathway activates beta-adrenergic receptors that ultimately trigger the lipolytic activity of hormone-sensitive lipase.[30,31] In rodent adipocytes, LPS activates lipolysis preferentially by ERK1/2, as these species are resistant to NF-κB-triggered lipolysis.[31-33] Currently, it is unknown if LPS directly triggers lipolysis in bovine AT and which lipolytic pathway(s) are activated.

Initial experiments in our laboratory demonstrated that LPS induces lipolysis in AT from periparturient dairy cows. Subcutaneous AT (SCAT) explants were collected from Holstein dairy cows at -14 d prepartum and at +6 d and +12 d after calving. Next, explants were incubated in the presence of the adrenergic agonist isoproterenol (ISO=1 μM; BAS=0 μM), LPS (20 μg/ml), and ISO plus LPS (LPSISO). Lipolytic responses were quantified by the release of glycerol in the incubation media. LPS increased glycerol release from SCAT by 73±18% across all time points ($p<0.001$). Importantly, LPSISO increased the lipolytic response of SCAT explants by 40±17% compared with ISO ($p<0.05$) and by 255±37% compared with BAS ($p<0.001$). These data indicate that lipolytic responses in SCAT are increased upon LPS exposure and strongly suggest that LPS potentiates AT responses to common stimulants of postpartum lipolysis such as catecholamines.

The second mechanism by which endotoxins may induce AT dysfunction is by altering the inflammatory phenotype of AT macrophages. Endotoxins promote macrophage M1 polarization; therefore, exposure to these bacterial byproducts early in the periparturient period may predispose cows to excessive lipolytic response during NEB postpartum. However, endotoxin-driven M1 polarization in AT may be affected by the degree of adiposity and by the development of endotoxin tolerance.[34] Therefore, future studies need to evaluate the effects of endotoxins on the phenotype of AT immune cells and its impact on metabolic function during the periparturient period.

Conclusion

Our knowledge of AT biology in periparturient dairy cows has advanced dramatically in the last two decades. However, there are still gaps in our understanding of AT remodeling and its role in homeorhetic adaptation to lactation, including the responses of AT to infectious and inflammatory diseases and the anatomical differences in the immunobiology of AT depots. Filling these gaps will support the development of preventive and corrective nutritional or pharmacological interventions to maintain periparturient AT function.

References

1. Contreras GA, Strieder-Barboza C, Raphael W. Adipose tissue lipolysis and remodeling during the transition period of dairy cows. *J Anim Sci Biotechnol.* 2017;8:41.
2. Contreras GA, Strieder-Barboza C, De Koster J. Symposium review: modulating adipose tissue lipolysis and remodeling to improve immune function during the transition period and early lactation of dairy cows. *J Dairy Sci.* 2018;101:2737–2752.
3. Lafontan M, Langin D. Lipolysis and lipid mobilization in human adipose tissue. *Prog Lipid Res.* 2009;48:275–297.
4. De Koster J, Van den Broeck W, Hulpio L, *et al.* Influence of adipocyte size and adipose depot on the *in vitro* lipolytic activity and insulin sensitivity of adipose tissue in dairy cows at the end of the dry period. *J Dairy Sci.* 2016;99:2319–2328.
5. Contreras GA, Kabara E, Brester J, *et al.* Macrophage infiltration in the omental and subcutaneous adipose tissues of dairy cows with displaced abomasum. *J Dairy Sci.* 2015;98:6176–6187.
6. Ospina PA, McArt JA, Overton TR, *et al.* Using nonesterified fatty acids and β-hydroxybutyrate concentrations during the transition period for herd-level monitoring of increased risk of disease and decreased reproductive and milking performance. *Vet Clin North Am Food Anim Pract.* 2013;29:387–412.
7. Giesy SL, Yoon B, Currie WB, *et al.* Adiponectin deficit during the precarious glucose economy of early lactation in dairy cows. *Endocrinology.* 2012;153:5834–5844.
8. Singh SP, Haussler S, Gross JJ, *et al.* Short communication: circulating and milk adiponectin change differently during energy deficiency at different stages of lactation in dairy cows. *J Dairy Sci.* 2014;97:1535–1542.
9. Saremi B, Winand S, Friedrichs P, *et al.* Longitudinal profiling of the tissue-specific expression of genes related with insulin sensitivity in dairy cows during lactation focusing on different fat depots. *PloS ONE.* 2014;9:e86211–e86211.
10. Kabara E, Sordillo LM, Holcombe S, *et al.* Adiponectin links adipose tissue function and monocyte inflammatory responses during bovine metabolic stress. *Comp Immunol Microbiol Infect Dis.* 2014;37:49–58.
11. Holtenius K, Agenas S, Delavaud C, *et al.* Effects of feeding intensity during the dry period. 2. Metabolic and hormonal responses. *J Dairy Sci.* 2003;86:883–891.

12. Leon HV, Hernandez-Ceron J, Keislert DH, *et al*. Plasma concentrations of leptin, insulin-like growth factor-I, and insulin in relation to changes in body condition score in heifers. *J Anim Sci*. 2004;82:445–451.
13. Park HK, Kwak MK, Kim HJ, *et al*. Linking resistin, inflammation, and cardiometabolic diseases. *Korean J Intern Med*. 2017;32:239–247.
14. Reverchon M, Ramé C, Cognié J, *et al*. Resistin in dairy cows: plasma concentrations during early lactation, expression and potential role in adipose tissue. *PloS ONE*. 2014;9:e93198–e93198.
15. Romacho T, Elsen M, Rohrborn D, *et al*. Adipose tissue and its role in organ crosstalk. *Acta Physiol (Oxf)*. 2014;210:733–753.
16. Abd Eldaim MA, Kamikawa A, Soliman MM, *et al*. Retinol binding protein 4 in dairy cows: its presence in colostrum and alteration in plasma during fasting, inflammation, and the peripartum period. *J Dairy Res*. 2010;77:27–32.
17. Rutkowski JM, Stern JH, Scherer PE. The cell biology of fat expansion. *J Cell Biol*. 2015;208:501–512.
18. Kosteli A, Sugaru E, Haemmerle G, *et al*. Weight loss and lipolysis promote a dynamic immune response in murine adipose tissue. *J Clin Invest*. 2010;120:3466–3479.
19. Newman AW, Miller A, Leal Yepes FA, *et al*. The effect of the transition period and postpartum body weight loss on macrophage infiltrates in bovine subcutaneous adipose tissue. *J Dairy Sci*. 2019;102:1693–1701.
20. Akter SH, Haussler S, Danicke S, *et al*. Physiological and conjugated linoleic acid-induced changes of adipocyte size in different fat depots of dairy cows during early lactation. *J Dairy Sci*. 2011;94:2871–2882.
21. Faulconnier Y, Chilliard Y, Torbati MB, *et al*. The transcriptomic profiles of adipose tissues are modified by feed deprivation in lactating goats. *Comp Biochem Physiol Part D Genomics Proteomics*. 2011;6:139–149.
22. Sumner-Thomson JM, Vierck JL, McNamara JP. Differential expression of genes in adipose tissue of first-lactation dairy cattle. *J Dairy Sci*. 2011;94:361–369.
23. Dickson MJ, Kvidera SK, Horst EA, *et al*. Impacts of chronic and increasing lipopolysaccharide exposure on production and reproductive parameters in lactating Holstein dairy cows. *J Dairy Sci*. 2019;102:3569–3583.
24. Hersoug L-G, Møller P, Loft S. Role of microbiota-derived lipopolysaccharide in adipose tissue inflammation, adipocyte size and pyroptosis during obesity. *Nutr Res Rev*. 2018;31:153–163.
25. Zebeli Q, Sivaraman S, Dunn SM, *et al*. Intermittent parenteral administration of endotoxin triggers metabolic and immunological alterations typically associated with displaced abomasum and retained placenta in periparturient dairy cows. *J Dairy Sci*. 2011;94:4968–4983.
26. Song J, Duncan MJ, Li G, *et al*. A novel TLR4-mediated signaling pathway leading to IL-6 responses in human bladder epithelial cells. *PLoS Pathog*. 2007;3:e60.
27. Moon EY, Lee YS, Choi WS, *et al*. Toll-like receptor 4-mediated cAMP production up-regulates B-cell activating factor expression in Raw264.7 macrophages. *Exp Cell Res*. 2011;317:2447–2455.
28. Lu Y-C, Yeh W-C, Ohashi PS. LPS/TLR4 signal transduction pathway. *Cytokine*. 2008;42:145–151.
29. Laurencikiene J, van Harmelen V, Arvidsson Nordstrom E, *et al*. NF-kappaB is important for TNF-alpha-induced lipolysis in human adipocytes. *J Lipid Res*. 2007;48:1069–1077.
30. Hong S, Song W, Zushin P-JH, *et al*. Phosphorylation of Beta-3 adrenergic receptor at serine 247 by ERK MAP kinase drives lipolysis in obese adipocytes. *Mol Metab*. 2018;12:25–38.
31. Zu L, He J, Jiang H, *et al*. Bacterial endotoxin stimulates adipose lipolysis via toll-like receptor 4 and extracellular signal-regulated kinase pathway. *J Biol Chem*. 2009;284:5915–5926.
32. Bergan HE, Kittilson JD, Sheridan MA. PKC and ERK mediate GH-stimulated lipolysis. *J Mol Endocrinol*. 2013;51:213–224.
33. Chi W, Dao D, Lau TC, *et al*. Bacterial peptidoglycan stimulates adipocyte lipolysis via NOD1. *PLoS ONE*. 2014;9:e97675.
34. Komegae EN, Fonseca MT, da Silveira Cruz-Machado S, *et al*. Site-specific reprogramming of macrophage responsiveness to bacterial lipopolysaccharide in obesity. *Front Immunol*. 2019;10:1496.

Evaluating and Modulating Adipose Tissue Function in Periparturient Cows

G. Andres Contreras, DVM, MS, PhD
East Lansing, MI, USA

Introduction

During the transition period, dairy cows mobilize triglyceride reserves stored in adipose tissues (AT) to offset the negative energy balance (NEB) induced by fetal requirements, the onset of lactation, and limited dry matter intake. Lipolysis hydrolyzes triglycerides and releases free fatty acids (NEFA) to be used as an energy substrate in different organs. In modern dairy cows, the lipolytic process is exacerbated because of their homeorhetic drive to meet the genetic potential for high lactation efficiency.[1] Excessive rates of lipolysis, usually measured as plasma NEFA and BHB, are a significant risk factor for the development of periparturient diseases and reduced lactation performance.[2] Periparturient health events have a high economic impact on dairy farms; for example, in the United States, costs associated with treating clinical cases of metritis, displaced abomasum (DA), and ketosis range between $145 and $340 per case. For these reasons, adequate management of periparturient cows is necessary to maintain the health and productivity of dairy herds. When designing and implementing periparturient cow programs, it is critical to monitor adipose tissue function and to consider tools that minimize lipolysis and maximize lipogenesis during the weeks following calving. This manuscript summarizes recent advances in clinical, pharmacological, and nutritional tools for the evaluation and modulation of adipose tissue function in periparturient cows.

Evaluating Adipose Tissue Function During the Periparturient Period

Periparturient cow management programs often include routine measures of clinical and production parameters that can directly or indirectly assess AT function. Body condition score (BCS) is a good measure of subcutaneous adiposity and the dynamics of BCS changes around parturition subjectively describe lipolysis rates.[3-5] Although BCS assessment is practical, there are issues with its reproducibility and sensitivity, as it is a subjective measure that varies depending on the scorer. To effectively use BCS scoring as a monitor tool, it is necessary to standardize its measurement, taking into account scorer training, the location where scoring is performed, etc. If using BCS as part of the periparturient cows' management program, cows should approach calving with a BCS of 3.25–3.5, as excessively thin or over-conditioned cows are more susceptible to disease.

Another alternative to measuring adiposity dynamics (i.e., AT lipolysis) is the use of image biomarkers obtained during AT ultrasound examination.[6] Images are captured in real-time, are easy to obtain, and avoid the variability associated with subjective visual measurements. Regression models are available to predict the AT mass of subcutaneous, retroperitoneal, omental, and mesenteric depots.[7] Subcutaneous AT depth is strongly correlated with BCS evaluation when measured by trained personnel and is highly sensitive and specific in predicting plasma NEFA values at close-up and calving in dairy cattle.[8]

Bodyweight (BW) loss during the first month of lactation is primarily explained by reductions in AT mass that usually average more than 25%[9]; for this reason, BW is a good proxy for AT lipolysis intensity and function. On average, heifers lose 6–7% of their BW in the first 30 days of lactation, while cows lose 8–9%.[10] Cows losing more than 10% of their BW at calving during the first weeks postpartum are at a higher risk for diseases and poor lactation performance.[11]

Currently, the most common direct measure of lipolysis is plasma NEFA. In preventive herd medicine, pre- and post-calving plasma NEFA values are used as early lactation disease predictors (Table 1). Similar to plasma NEFA, postpartum plasma BHB indicate NEB and predict disease risk in early lactation.[2] Lipolysis can also be evaluated at the group or individual animal level using the milk fat to milk protein percentage ratio (Table 1). Milk fat increases as plasma NEFA rise. Cows with milk fat to milk protein ratio values higher than 2 during the first week after calving are at a higher risk for developing retained fetal membranes, DA, clinical endometritis, and being culled before the end of lactation.[12]

Table 1. Selected biomarkers of periparturient cow adipose tissue function

Biomarker	Reference values
NEFA[2]	<0.50 mEq/L
BHB[26]	12.5 mg/dL subclinical ketosis; 14.5 mg/dL clinical ketosis
Cholesterol[49]	1.7–4.3 mmol/L prepartum; 2.7–5.3 mmol/L postpartum
Triglycerides[50]	0.12–0.65 mmol/L
GOT/AST[50]	46.5–103 IU/L
GGT[50]	21–37 IU/L
Acetoacetate/acetone[51]	>10 mg/dL of acetoacetate and acetone
Albumin	3.2–3.7 g/dL blood
Milk fat:protein ratio	1.00–1.25
Fetuin-A[13]	1 mg/mL plasma prepartum 0.8 mg/mL plasma postpartum
Isoprostanes[15]	0.12 nM plasma periparturient period <0.05 nM plasma mid lactation

NEFA=nonesterified fatty acids; BHB=beta-hydroxybutyrate; GOT/AST=glutamic-oxaloacetic transaminase/aspartate aminotransferase; GGT=gamma-glutamyl transferase

Single biomarkers do not provide enough information to support management decisions during the periparturient period. However, when multiple biomarkers are analyzed together and combined with health, production, nutritional, and environmental data, biomarkers become essential for identifying metabolic problems related to extended periods of intense lipolysis. Novel biomarkers that are strongly associated with lipolysis intensity and AT inflammation include the acute-phase proteins and fatty acid transporters fetuin-A[13] and albumin,[14] and the oxidative stress markers isoprostanes[15]. Oxidized linoleic acid metabolites, including 9-, 12-, and 13-hydroxy-octadecadienoic acids (HODE), are also indicators of lipolysis intensity and AT inflammation.[16] These HODEs are generated during lipolysis by cyclooxygenases, lipoxygenases, and non-enzymatic oxidation by reactive oxygen species. In healthy periparturient cows, enzymatic reactions predominantly produce 9- and 13-HODEs. In contrast, 12-HODE is synthesized exclusively by non-enzymatic oxidation, and increased levels reflect oxidative stress and inflammation within AT. Reference values for plasma concentrations of these recently described biomarkers of AT function are provided in Table 1. However, to validate the sensitivity and specificity of these markers to assess AT function, extensive epidemiological studies are required.

Modulating Adipose Tissue Function in Periparturient Cows

A key goal of any periparturient cow management program should be to limit BW loss and excessive AT lipolysis.[17] Some of the more commonly used nutritional, pharmacological, and management tools that could be used to prevent excessive lipolysis are listed below.

Optimizing Dry Matter Intake

Severe reductions in DMI during the final weeks of the dry period trigger BW loss and have a marked impact on postpartum DMI.[18] For this reason, cow housing must be optimal and must prioritize feed bunk access in the weeks before parturition.[19] The currently recommended strategy to feed cows during the dry period in view of limiting lipolysis is to provide high levels of fiber.[20] In the postpartum, limiting the amount of highly fermentable starch is advisable, as it increases the oxidation of energetic compounds in the liver that activate satiety and thus depresses DMI.[21] Providing moderately fermentable starch sources (e.g., dry ground corn) will maximize postpartum DMI while maintaining energy balance.[20] Importantly, adequate levels of DMI decrease AT lipolysis and could boost plasma insulin concentrations.[22]

Niacin

Niacin or vitamin B_3 in the form of nicotinic acid has a robust antilipolytic effect in AT by reducing the activation of hormone-sensitive lipase through inhibition of adenylate cyclase.[23] Although niacin inhibits lipolysis in a dose-dependent manner,[24] the results of field studies have been inconsistent. Some report minimal lipolysis inhibition[25,26] while others[27,28] demonstrated strong NEFA-lowering effects when fed pre- and postpartum. The reasons behind this variability can be related to the timing of supplementation, as experiments that fed niacin before and after calving[27,28] were more effective in limiting lipolysis compared to those that only supplemented niacin after parturition.[29] Cows with high rates of lipolysis prepartum develop inflammatory responses within AT that activate lipolysis through toll-like receptor signaling that is independent of adenylate cyclase and therefore not blocked by niacin.

Methyl Donors

Choline and methionine supplementation reduces lipolysis in AT when fed during the periparturient period.[30,31] Methionine appears to be critical in the control of AT lipolysis, at least in rodents, as its deficiency dramatically increases triglyceride breakdown and blunts fatty acid reesterification.[32] Since both methyl donors reduce hepatic triglyceride accumulation and promote NEFA oxidation capacity and lipid-exporting efficiency of hepatocytes,[33] including them in periparturient diets can be beneficial. However, since both are degraded in the rumen, it is necessary to supplement rumen-protected forms.

Monensin

When fed during the close-up and early lactation, monensin decreases plasma NEFA post-calving.[34] There are no reports of a direct effect of monensin on AT; however, by increasing the production of propionic acid, it reduces the rate of lipolysis in adipocytes. In rat fat cells, propionic acid blunts the activation of adenylyl cyclase and thus hormone-sensitive lipase phosphorylation.[35]

Chromium

Unlike niacin, methyl donors, and monensin, chromium does not have an effect on lipolysis. Instead, chromium enhances lipogenesis in adipocytes. The mechanism of action is by binding to chromodulin and then enhancing the lipogenic responses to the activation of the insulin receptor.[36,37] Chromium has been supplemented as propionate, methionine, and picolinate.[38-40] Similar to niacin, the effects of chromium have been inconsistent, and more extensive feeding trials are needed to determine the best source and the ideal timing for supplementation.

Glucogenic Precursors

Propylene glycol (PG), glycerol, and propionate have been used as antilipolytic agents in periparturient cows.[41] In dairy farms, PG is routinely used to treat and prevent clinical hyperketonemia. The antilipolytic mechanism of PG is related to the increased production of ruminal propionic acid. As described above, propionic acid is a potent inhibitor of lipolysis in adipocytes. When PG is orally administered for 3 to 5 days (310 g), it limits adipose lipolysis effectively and promotes a rapid return to a positive energy balance. It is important to note that PG has antilipolytic properties only when administered as an oral bolus. When PG is included in the TMR, it does not affect AT lipolytic responses.[42]

Glucose

Glucose is often administered as an IV bolus for the treatment of clinical hyperketonemia. This infusion rapidly decreases the rate of lipolysis in AT and reduces circulating FFA levels. When administered alone as a bolus, glucose has a <3-hour antilipolytic effect.[43] Glucose alone is not an effective long-term antilipolytic therapy and has to be paired with oral administration of glucogenic precursors.[44]

Saturated and Monounsaturated Fatty Acids

Feeding supplemental fat in the immediate postpartum increases energy intake, possibly reducing negative energy balance. However, not all fatty acids have the same effects on adipocyte lipolytic and lipogenic rates. For example, when fed alone, palmitic acid promotes lipolysis, and thus BW loss, by inhibiting insulin secretion and signaling.[45] When cows are fed a combination of palmitic and oleic acid, there is a pro-lipogenic effect that minimizes BW loss.[46] This effect is probably related to the capacity of oleic acid to promote triglyceride synthesis and lipid droplet formation. Also, oleic acid improves insulin receptor signaling and increases the oxidative capacity of cells, therefore reducing lipolytic activity.[47,48] In periparturient cows, feeding a combination of palmitic and oleic acids maintains energy availability while enhancing insulin signaling and promoting fatty acid esterification and triglyceride formation in AT.

Conclusions and Future Directions

Effective periparturient cow management ensures productive lactations and minimal incidence of diseases during early lactation. Continuous monitoring of AT lipolysis intensity and inflammatory responses provides data to support management decisions, including dietary or pharmacological interventions. The current pool of biomarkers of AT function is limited in number, specificity, and sensitivity; therefore, further research is needed to identify new biomarkers that facilitate management decisions. The majority of tools available to modulate AT function during the periparturient period reduce lipolysis; however, it is necessary to develop interventions that promote lipogenesis harnessing the capacity of AT to buffer energy excess.

References

1. Contreras GA, Strieder-Barboza C, Raphael W. Adipose tissue lipolysis and remodeling during the transition period of dairy cows. *J Anim Sci Biotechnol.* 2017;8:41.
2. Ospina PA, McArt JA, Overton TR, *et al.* Using nonesterified fatty acids and β-hydroxybutyrate concentrations during the transition period for herd-level monitoring of increased risk of disease and decreased reproductive and milking performance. *Vet Clin North Am: Food Anim Pract.* 2013;29:387–412.
3. Pires JAA, Delavaud C, Faulconnier Y, *et al.* Effects of body condition score at calving on indicators of fat and protein mobilization of periparturient Holstein-Friesian cows. *J Dairy Sci.* 2013;96:6423–6439.
4. Roche JR, Kay JK, Friggens NC, *et al.* Assessing and managing body condition score for the prevention of metabolic disease in dairy cows. *Vet Clin North Am: Food Anim Pract.* 2013;29:323–336.

5. Schulz K, Frahm J, Meyer U, *et al.* Effects of prepartal body condition score and peripartal energy supply of dairy cows on postpartal lipolysis, energy balance and ketogenesis: an animal model to investigate subclinical ketosis. *J Dairy Res.* 2014;81:257–266.
6. Bell MJ, Maak M, Sorley M, *et al.* Comparison of methods for monitoring the body condition of dairy cows. *Front Sustain Food Syst.* 2018;2.
7. Raschka C, Ruda L, Wenning P, *et al. In vivo* determination of subcutaneous and abdominal adipose tissue depots in German Holstein dairy cattle. *J Anim Sci.* 2016;94:2821–2834.
8. Strieder-Barboza C, Zondlak A, Kayitsinga J, *et al.* Lipid mobilization assessment in transition dairy cattle using ultrasound image biomarkers. *Livestock Sci.* 2015;177:159–164.
9. Akter SH, Haussler S, Danicke S, *et al.* Physiological and conjugated linoleic acid-induced changes of adipocyte size in different fat depots of dairy cows during early lactation. *J Dairy Sci.* 2011;94:2871–2882.
10. van Straten M, Shpigel NY, Friger M. Analysis of daily body weight of high-producing dairy cows in the first one hundred twenty days of lactation and associations with ovarian inactivity. *J Dairy Sci.* 2008;91:3353–3362.
11. Zachut M, Moallem U. Consistent magnitude of postpartum body weight loss within cows across lactations and the relation to reproductive performance. *J Dairy Sci.* 2017;100:3143–3154.
12. Toni F, Vincenti L, Grigoletto L, *et al.* Early lactation ratio of fat and protein percentage in milk is associated with health, milk production, and survival. *J Dairy Sci.* 2011;94:1772–1783.
13. Strieder-Barboza C, Contreras GA. Fetuin-A modulates lipid mobilization in bovine adipose tissue by enhancing lipogenic activity of adipocytes. *J Dairy Sci.* 2019;102:4628–4638.
14. Ruprechter G, Adrien MdL, Larriestra A, *et al.* Metabolic predictors of peri-partum diseases and their association with parity in dairy cows. *Res Vet Sci.* 2018;118:191–198.
15. Kuhn MJ, Mavangira V, Gandy JC, *et al.* Production of 15-F_{2t}-isoprostane as an assessment of oxidative stress in dairy cows at different stages of lactation. *J Dairy Sci.* 2018;101:9287–9295.
16. Andres Contreras G, De Koster J, de Souza J, *et al.* Lipolysis modulates the biosynthesis of inflammatory lipid mediators derived from linoleic acid in adipose tissue of periparturient dairy cows. *J Dairy Sci.* 2020;103:1944–1955.
17. Contreras GA, Strieder-Barboza C, De Koster J. Symposium review: modulating adipose tissue lipolysis and remodeling to improve immune function during the transition period and early lactation of dairy cows. *J Dairy Sci.* 2018;101:2737–2752.
18. Grummer RR, Mashek DG, Hayirli A. Dry matter intake and energy balance in the transition period. *Vet Clin North Am: Food Anim Pract.* 2004;20:447–470.
19. Cook NB, Nordlund KV. Behavioral needs of the transition cow and considerations for special needs facility design. *Vet Clin North Am: Food Anim Pract.* 2004;20:495–520.
20. Allen MS, Piantoni P. Carbohydrate nutrition: managing energy intake and partitioning through lactation. *Vet Clin North Am: Food Anim Pract.* 2014;30:577–597.
21. Allen MS, Bradford BJ, Oba M. Board-invited review: the hepatic oxidation theory of the control of feed intake and its application to ruminants. *J Anim Sci.* 2009;87:3317–3334.
22. Bradford BJ, Allen MS. Depression in feed intake by a highly fermentable diet is related to plasma insulin concentration and insulin response to glucose infusion. *J Dairy Sci.* 2007;90:3838–3845.
23. Kenez A, Locher L, Rehage J, *et al.* Agonists of the G protein-coupled receptor 109A-mediated pathway promote antilipolysis by reducing serine residue 563 phosphorylation of hormone-sensitive lipase in bovine adipose tissue explants. *J Dairy Sci.* 2014;97:3626–3634.
24. Pescara JB, Pires JAA, Grummer RR. Antilipolytic and lipolytic effects of administering free or ruminally protected nicotinic acid to feed-restricted Holstein cows. *J Dairy Sci.* 2010;93:5385–5396.
25. Kinoshita A, Kenez A, Locher L, *et al.* Insulin signaling in liver and adipose tissues in periparturient dairy cows supplemented with dietary nicotinic acid. *PLoS ONE.* 2016;11.
26. Tienken R, Kersten S, Frahm J, *et al.* Effects of an energy-dense diet and nicotinic acid supplementation on production and metabolic variables of primiparous or multiparous cows in periparturient period. *Arch Anim Nutr.* 2015;69:319–339.
27. Morey SD, Mamedova LK, Anderson DE, *et al.* Effects of encapsulated niacin on metabolism and production of periparturient dairy cows. *J Dairy Sci.* 2011;94:5090–5104.
28. Yuan K, Shaver RD, Bertics SJ, *et al.* Effect of rumen-protected niacin on lipid metabolism, oxidative stress, and performance of transition dairy cows. *J Dairy Sci.* 2012;95:2673–2679.
29. Havlin J, Robinson P, Garrett J. Niacin feeding to fresh dairy cows: immediate effects on health and milk production. *Anim Prod Sci.* 2017;57:1069–1078.
30. Cooke RF, Del Río NS, Caraviello DZ, *et al.* Supplemental choline for prevention and alleviation of fatty liver in dairy cattle. *J Dairy Sci.* 2007;90:2413–2418.

31. Sun F, Cao Y, Cai C, *et al*. Regulation of nutritional metabolism in transition dairy cows: energy homeostasis and health in response to post-ruminal choline and methionine. *PLoS ONE*. 2016;11:e0160659.
32. Perrone CE, Mattocks DAL, Hristopoulos G, *et al*. Methionine restriction effects on 11β-HSD1 activity and lipogenic/lipolytic balance in F344 rat adipose tissue. *J Lipid Res*. 2008;49:12–23.
33. Sales J, Homolka P, Koukolová V. Effect of dietary rumen-protected choline on milk production of dairy cows: a meta-analysis. *J Dairy Sci*. 2010;93:3746–3754.
34. Duffield TF, Rabiee AR, Lean IJ. A meta-analysis of the impact of monensin in lactating dairy cattle. Part 1. Metabolic effects. *J Dairy Sci*. 2008;91:1334–1346.
35. Heimann E, Nyman M, Degerman E. Propionic acid and butyric acid inhibit lipolysis and de novo lipogenesis and increase insulin-stimulated glucose uptake in primary rat adipocytes. *Adipocyte*. 2014;4:81–88.
36. Hua Y, Clark S, Ren J, *et al*. Molecular mechanisms of chromium in alleviating insulin resistance. *J Nutr Biochem*. 2012;23:313–319.
37. Davis CM, Vincent JB. Chromium in carbohydrate and lipid metabolism. *J Biol Inorg Chem*. 1997;2:675–679.
38. Nikkhah A, Mirzaei M, Khorvash M, *et al*. Chromium improves production and alters metabolism of early lactation cows in summer. *J Anim Physiol Anim Nutr*. 2011;95(1):81–89.
39. Hayirli A, Bremmer DR, Bertics SJ, *et al*. Effect of chromium supplementation on production and metabolic parameters in periparturient dairy cows. *J Dairy Sci*. 2001;84:1218–1230.
40. Yasui T, McCarthy MM, Ryan CM, *et al*. Effects of monensin and starch level in early lactation diets on indices of immune function in dairy cows. *J Dairy Sci*. 2016;99:1351–1363.
41. Piantoni P, Allen MS. Evaluation of propylene glycol and glycerol infusions as treatments for ketosis in dairy cows. *J Dairy Sci*. 2015;98:5429–5439.
42. Bjerre-Harpoth V, Storm AC, Eslamizad M, *et al*. Effect of propylene glycol on adipose tissue mobilization in postpartum over-conditioned Holstein cows. *J Dairy Sci*. 2015;98:8581–8596.
43. Mann S, Yepes FAL, Duplessis M, *et al*. Dry period plane of energy: effects on glucose tolerance in transition dairy cows. *J Dairy Sci*. 2016;99:701–717.
44. Mann S, Yepes FL, Behling-Kelly E, *et al*. The effect of different treatments for early-lactation hyperketonemia on blood β-hydroxybutyrate, plasma nonesterified fatty acids, glucose, insulin, and glucagon in dairy cattle. *J Dairy Sci*. 2017;100(8):6470–6482.
45. de Souza J, Strieder-Barboza C, Contreras GA, *et al*. Effects of timing of palmitic acid supplementation during early lactation on nutrient digestibility, energy balance, and metabolism of dairy cows. *J Dairy Sci*. 2019;102:274–287.
46. de Souza J, Preseault CL, Lock AL. Altering the ratio of dietary palmitic, stearic, and oleic acids in diets with or without whole cottonseed affects nutrient digestibility, energy partitioning, and production responses of dairy cows. *J Dairy Sci*. 2018;101:172–185.
47. Tsuchiya A, Nagaya H, Kanno T, *et al*. Oleic acid stimulates glucose uptake into adipocytes by enhancing insulin receptor signaling. *J Pharmacol Sci*. 2014;126:337–343.
48. Najt CP, Khan SA, Heden TD, *et al*. Lipid droplet-derived monounsaturated fatty acids traffic via PLIN5 to allosterically activate SIRT1. *Mol Cell*. 2020;77(4):810–824.
49. Kaneene JB, Miller R, Herdt TH, *et al*. The association of serum nonesterified fatty acids and cholesterol, management and feeding practices with peripartum disease in dairy cows. *Prev Vet Med*. 1997;31:59–72.
50. Bertoni G, Trevisi E. Use of the liver activity index and other metabolic variables in the assessment of metabolic health in dairy herds. *Vet Clin North Am: Food Anim Pract*. 2013;29:413–431.
51. Krogh MA, Toft N, Enevoldsen C. Latent class evaluation of a milk test, a urine test, and the fat-to-protein percentage ratio in milk to diagnose ketosis in dairy cows. *J Dairy Sci*. 2011;94:2360–2367.

SOTA Metabolic Disease: Postpartum Metabolic Disease and Immune Dysfunction—What Have We Learned?

Sabine Mann, DVM, PhD, DECBHM, DACVPM (Epidemiology)
Ithaca, NY, USA

Introduction

Advances in the management of periparturient dairy cows have been significant. Yet morbidity and mortality are still increased dramatically in the first 30 days after calving when cows experience systemic inflammation, greater susceptibility to infection, and decreased ability for efficient pathogen clearance.[1,2] This coincides with the nadir in energy and protein balance of dairy cows. Prevention of disease is essential, but to date, we lack understanding of the link between macronutrient metabolism, metabolic disease, and regulation of immune response and inflammatory balance. To address this knowledge gap, research targeting the integration of immune function and metabolic disease, as well as the understanding of the underlying mechanisms, is needed. This presentation will give an overview of the latest advances in our understanding of the link between metabolic disease and immune dysfunction of the postpartum dairy cow.

Importance of Inflammation During Early Lactation and the Interplay with Metabolic Disease

Controlled inflammation is crucial for physiological functions associated with parturition and expulsion of the placenta, and a systemic inflammatory state is present in cows postpartum even when no disease is apparent.[3] Inflammation that is excessive in magnitude, duration, or both leads to detrimental host tissue injury, cell death, and inadequate pathogen clearance.[4] The immune system of the postpartum cow in a state of negative nutrient balance reacts to bacterial stimuli with excessive inflammatory cytokine production compared with cows in late lactation or the dry period when inflammation is quickly controlled.[5,6] An example of this dysregulation is the cow's response to a coliform mastitis, which differs in magnitude and host-damaging character between the postpartum period and other stages of the production cycle.

The persistence and magnitude of inflammation differ between individual cows, rise sharply at parturition, and are associated with increased disease risk and loss of milk production throughout the lactation.[7,8] The excessive nutrient deficit is further complicated by the intake-suppressing effect of proinflammatory cytokines as experimentally demonstrated for TNF-α in fresh cows.[9] Cows experiencing an excessive lack of nutrients are not only predisposed for reproductive diseases such as metritis, retained placenta, and endometritis,[10-12] but also mastitis early postpartum[13]. Each of these diseases has an estimated cost of $262 to $444,[14,15] and all health disorders contribute to reduced reproductive success, milk production, and a greater risk for early culling[16,17]. Economic loss due to early lactation involuntary culling alone is estimated between $500 and $1,000 per cow.[18] Inflammatory disease also adds to the risk for death and euthanasia, which is greatest in early lactation.[19] The cost of transition cow inflammatory dysfunction therefore consists of the combined losses due to increased morbidity and mortality, reduced milk production, lowered efficiency of production capacity, and decreased reproductive success—all of which further contribute to culling of cows from the herd.

Epidemiological evidence has established the association between metabolism and innate immune function and the greater disease susceptibility in metabolically challenged animals.[20-22] However, our understanding of the causative mechanisms is still developing. Efforts to explain how metabolic imbalance affects the immune system of the postpartum cow span the last decades. Early evidence pointed to a direct contribution of increased ketone concentrations and low glucose concentration in the development of immune dysfunction and lowered immunocompetence,[23-25] yet mechanisms underlying the observed immune dysfunction are still poorly understood today. The direct link between the demands of lactation and immune dysfunction was demonstrated by the blunting or absence of loss of function in cows that had undergone mastectomy prior to parturition.[26,27] The improvement in analytical technologies and advances in understanding of the interplay between metabolism and immune function in other species have enabled us to revisit possible mechanistic pathways to explain the physiology, a necessary progress that allows for targeted interventions in the future.

Nutrient Use and Nutrient Sensing by Immune Cells

Immune cells have different preferences for energy substrate use at rest and during activation. Glucose, glutamine, and fatty acids are among the most commonly used fuels.[28,29] Glucose is stored in immune cells in the form of glycogen and is available for rapid mobilization. Glucose and glycogen availability are decreased in the postpartum cow.[30,31] However, immune cells are able to sense a greater variety of nutrients and signals than these typical fuels, and they integrate this information to program immune outputs. Among these signals are growth factors and insulin, concentration of amino acids, cellular energy balance, and oxidative stress. Concentrations of insulin and amino acids are decreased in the immediate postpartum period,[32,33] and a decrease in the cellular energy balance is directly sensed by immune cells,[31] and it has been recognized that oxidative stress contributed to the metabolic disease processes of early lactation[34]. An example of a well-characterized pathway that integrates metabolic cues and immune function is the mammalian target of rapamycin (mTOR).

More than a decade of research in human medicine showed that the regulation of mTOR in innate immune cells has important consequences for the immunological response and orchestrates the direction and severity of inflammation. Recent evidence points to a role of the mTOR pathway as a possible link between metabolic disease and nutrient deficit on the one hand, and changes in immune function on the other hand.[31] Understanding how this and other potentially applicable pathways regulate innate and adaptive immune cells of cattle during periods of nutrient stress will provide important insights into ways to improve host resilience and possibly direct inflammatory response during this critical period.

References

1. Alvasen K, *et al. Prev Vet Med.* 2014;117(1):110–120.
2. Shahid MQ, *et al. J Dairy Sci.* 2015;98(7):4401–4413.
3. Bionaz M, *et al. J Dairy Sci.* 2007;90(4):1740–1750.
4. Bauerfeld CP, *et al. J Immunol.* 2012;188(6):2847–2857.
5. Sordillo LM, Pighetti GM, Davis MR. *Vet Immunol Immunopathol.* 1995;49(3):263–270.
6. Waldron MR, *et al. J Dairy Sci.* 2006;89(2):596–610.
7. Bradford BJ, *et al. J Dairy Sci.* 2015;98(10):6631–6650.
8. Jahan N, Minuti A, Trevisi E. *Vet Immunol Immunopathol.* 2015;165(3–4):119–126.
9. Yuan K, *et al. PLoS One.* 2013;8(11):e80316.
10. Yasui T, *et al. J Dairy Sci.* 2014;97(5):2763–2770.
11. Wagener K, Gabler C, Drillich M. *Theriogenology.* 2017;94:21–30.
12. Heppelmann M, *et al. Theriogenology.* 2016;85(7):1348–1356.
13. Ospina PA, *et al. Vet Clin North Am Food Anim Pract.* 2013;29(2):387–412.
14. Rollin E, *et al. Prev Vet Med.* 2015;122(3):257–264.
15. Liang D, *et al. J Dairy Sci.* 2017;100(2):1472–1486.

16. Ribeiro ES, *et al. J Dairy Sci*. 2016;99(3):2201–2220.
17. Gohary K, LeBlanc SJ. *J Am Vet Med Assoc*. 2018;252(12):1485–1489.
18. De Vries A. Cow Longevity Conference 2013. Tumba, Sweden: DeLaval International.
19. NAHMS, United States Department of Agriculture; Fort Collins, CO.
20. Schukken YH, Erb HN, Scarlett JM. *Cornell Vet*. 1989;79(4):319–326.
21. Esposito G, *et al. Anim Reprod Sci*. 2014;144(3–4):60–71.
22. Janosi S, *et al. Acta Vet Hung*. 2003;51(3):409–424.
23. Nonnecke BJ, Franklin ST, Young JW. *J Dairy Sci*. 1992;75(4):982–990.
24. Suriyasathaporn W, *et al. Vet Res*. 2000;31(4):397–412.
25. Zerbe H, *et al. Theriogenology*. 2000;54(5):771–786.
26. Nonnecke BJ, *et al. J Dairy Sci*. 2003;86(7):2359–2368.
27. Kimura K, *et al. J Dairy Sci*. 2002;85(6):1437–1444.
28. Pearce EL, Pearce EJ. *Immunity*. 2013;38(4):633–643.
29. Wang A, *et al. Science*. 2019;363(6423):eaar3932.
30. Galvao KN, *et al. J Dairy Sci*. 2010;93(7):2926–2937.
31. Mann S, *et al. J Dairy Sci*. 2018:101.
32. Meijer GA, *et al. J Dairy Sci*. 1995;78(5):1131–1141.
33. Mann S, *et al. J Dairy Sci*. 2016;99(1):701–717.
34. Mavangira V, Sordillo LM. *Res Vet Sci*. 2018;116:4–14.

Financial Literacy for Residents and Veterinary Professionals

Pedro Lopes Bento, DVM, DACVIM (SAIM)
Tinton Falls, NJ, USA

Financial Literacy - Why?

We've all spent countless years in clinical training. The goal is to obtain as much knowledge as possible and become the best doctor possible, regardless of if you're aiming for board certification or not. That's exactly what veterinary school, internships, and residencies are aimed at. For the vast majority of us, financial education wasn't a part of our training, other than the occasional lunch and learn from the financial advisor/planner that will work with you for "free." And while we continue to know only the bare minimum (or less) about finances, the student loan burden and debt-to-income ratio continue to increase. In most cases, your debt is greater than $200,000, so your net worth at graduation is negative. Let that sink in for a minute. You simply cannot take a backseat on this.

Taking care of your finances is your second job, and you should learn the very basics, even if you don't want to manage your own money and want to hire a financial professional to help you. But how will you know if they are truly helping you if you have no basic financial literacy? Most of the financial industry wants to make money off of you. They purposely target doctors and other high-income professionals! High income and minimal financial knowledge lead to high profits!

The goal of this is designed to spark your interest and make you want to learn more, not to provide an all-encompassing financial bible. Remember, what we're discussing today are the absolute basics of what you should know, but there's so much more out there. Plus, personal finance is personal. You can't take anything from these proceedings and directly apply it to your situation without putting some thought into evaluating your own financial circumstances.

Financial Advisors/Planners

Most of the financial industry is designed to offer services that are more beneficial to them than to you. In reality, the amount of money spent in sales training vastly outweighs what is spent in financial training for these professionals. Most training is done in-house, and many courses that offer designations can be obtained with very little training. This is a very foreign concept to us. In the US, you need at least 8 years of training to become a DVM. I encourage you to do some Internet searching on the requirements for many financial advisor positions. You'll find that in most cases, advanced degrees in finance and other significant certifications are not even a requirement! You'll also find that experience in sales is quite common and sometimes more important than the former. Are they trying to solidify your finances or trying to sell products to make a nice commission? Don't get me wrong, there are excellent financial professionals out there that will provide solid advice and truly help you increase your net worth. It's your job to learn how to differentiate between the two.

In order to identify potential conflicts of interest and to determine if you're going to receive the kind of help you want, use the following questions to guide you when meeting with a potential financial advisor:

- Fiduciary – if a financial professional is a fiduciary, they have an ethical and legal duty to act solely in your best interest. You must ask if they are fiduciaries in every aspect of their work. For example, they offer financial planning and are fiduciaries, but also sell insurance. Unless you ask, they might not disclose that they are not fiduciaries on the insurance side of things!

- Advanced finance degrees and/or certifications—some weekend courses or courses with minimal coursework with a multiple-choice exam at the end can offer letters at the end of someone's name. Look for the following instead:
    - Certified financial planner (CFP)
    - Chartered financial analyst (CFA)
    - Certified public accountant (CPA)
    - Chartered financial consultant (ChFC)
    - Chartered life underwriter (CLU)
    - Juris doctorate (JD)
    - Master's in business administration (MBA)

    No one will have all of the above, but they might have a combination. Keep in mind that the aforementioned professionals will focus on different things. For example, you don't want a JD if you're looking for financial planning or tax preparation. You'd want to look for a CFP, or CFA and CPA, respectively. This is no different from the way our medical system is organized; if you have an arrhythmia, you go see a cardiologist, not a dermatologist.
- Learn how they get paid—bottom line, you want an advisor that is **fee-only**. There are different types of compensation in this model: hourly, annual retainer, assets under management (AUM). Remember that AUM fees can amount to a large amount depending on the size of your portfolio. On the other hand, fee-based advisors charge a fee **and** receive commissions. And then there are commission-only professionals. As mentioned above, they only get paid when you buy/invest in certain products. There is a very clear-cut conflict of interest in trying to sell products that carry the highest commissions. Those commissions are paid by you! Unfortunately, products with the highest commissions are not the ones you're interested in. Remember the financial advisors that wanted to work with you for free in vet school or residency? The "free" lunch/dinner is never really free.
- Do not mix investing and insurance professionals (unless they're fiduciary).
- Run if they are not willing to, or hesitate to, disclose how they get paid, potential conflicts of interest, fiduciary duties, etc. Always ask questions!

Retirement Accounts

It is never too early to start saving for retirement. The interest compounding formula works both ways. The longer your money has to compound, the more you'll have. The longer it takes you to pay any type of debt (including student loans), the more you'll pay overall. And while the amount of money you'll be able to set aside for retirement during internship or residency won't make a huge difference (compared to what you'll be able to contribute when you're done), it's still extremely important to get in the habit of working toward saving. Plus, retirement plans in academic institutions (and in some private practices) will match some of the money you put into these accounts. This is effectively part of your salary, and it would be unwise to leave money on the table.

Account Types

To keep it simple, there are two main types of accounts: pre- and post-tax. They all have different rules and contribution limits. Depending on your employer, you will not necessarily have access to all of these, and it is extremely important to be familiar with the benefit options you have available (you can always reach out to the HR department with questions). Keep in mind that not all are employer dependent, and you can have access to them regardless (traditional IRA, Roth IRA, taxable account). The most common pre-tax accounts include 401(k), 403(b), 403(a), health savings account (HSA), 457(b), traditional individual retirement accounts (IRA), and SIMPLE IRAs. In all of these, employees can contribute to a retirement plan using income that has not been subject to income taxes. Taxes are paid when the money is withdrawn in retirement. HSA accounts are exceptions, as long as the funds are withdrawn to pay for qualified medical expenses. Post-tax accounts include Roth IRA, Roth 401(k) and taxable accounts. You make contributions to these using income that has been subjected to taxes, and will be able to withdraw funds without paying additional taxes, regardless of your tax bracket in the future. The exception will be a taxable account. These are simple investing accounts you can open at various entities (e.g., Vanguard, Fidelity, Schwab, etc.). Although you won't pay taxes on withdrawal of the money you used to invest in the first place, taxes will be due on the gains (capital gains and on interest income, such as dividends). This tax rate is lower than your federal tax bracket.

What to Invest In

Now that you're more familiar with the various types of accounts you can invest in, let's discuss in general terms what you should consider investing. First, there will need to be some thought put into your asset allocation. This is the percentage of your portfolio you'll put in different types of investments (or asset classes): stocks, bonds, real estate, precious metals, commodities. Within these, there are additional considerations, such as large cap stocks, value stocks, international stocks, and many more. Also important, is the concept of index funds (passively managed) and actively managed funds (where someone is actively buying and selling stocks). The former have lower expense ratios (what you pay to be invested in the fund), and data suggests better results in the long term with using these types of funds. Actively managed funds, despite the human factor, do not guarantee higher returns on your money. For example, if you're using an index fund and it returns 5.5% but has an expense ratio of 0.09%, you're making 5.41%. If you're using an actively managed fund and the expense ratio is 1.5% (annually) and the fund returned 6% on your money, you're effectively only making 4.5%. It's easy to forget this.

When thinking about portfolio diversification (i.e., investing in multiple stocks/bonds, depending on the fund), both options will do well. Diversification is key to reduce risk. Imagine putting all your money into two companies (regardless of how strong they appear to be at this time), and that in 20 years they go bankrupt. All your money just disappeared. On the other hand, imagine investing in an index fund composed of hundreds or thousands of companies. If a few of those companies in your fund go bankrupt, you probably won't notice as much since other companies in the fund will buffer those changes. Overall, when comparing passive index funds and actively managed funds, historical financial data suggests investment in low-cost (in terms of expense ratio), diversified, and passively managed funds to be more beneficial.

INSURANCE

It is crucial to understand the difference between insurance and investing. They should never be mixed and you should probably avoid professionals that deal with both (unless they are fiduciary for both). The focus of this lecture is on disability and term-life insurance. By obtaining both of these, you're insuring against financial catastrophe if something happens to you and you can no longer work. Other types of insurance to cover financial catastrophe include liability (personal and professional), health, and homeowner/renter's insurance.

Disability Insurance

What if, out of nowhere, you could not work as a veterinarian anymore? You'd lose your most valuable asset: the ability to work and make the income you anticipate. That's where disability insurance comes into play. There is both short- and long-term disability insurance. Here, I'm focusing solely on long-term disability since we're focusing on catastrophic insurance coverage.

Long-term disability is defined as ≥90 days, although depending on the policy, the waiting period (time period between the date you are disabled and when you start receiving payments) could be shorter or longer than that. Ideally, you want it to be 90 days. If you are disabled for life, the policy will typically pay until you're 65 years old. Since women are more likely to make a claim, policies are more expensive for this gender. This is in part due to the risks of pregnancy.

The first thing to do is to find an independent disability agent; because they're independent, they can sell you policies from different companies. If you shop directly with a company, you'll only be offered policies from the company that that agent works for, and likely won't get the best product. You'll need to provide not only financial information, but also medical information. If you have a medical condition (or multiple), you'll want your agent to informally check in with various insurance companies about how those might affect your application, **before** the application actually occurs. If you get denied disability insurance due to a medical condition, you are unlikely to get a policy with **any** other insurance company. Don't try to hide anything; the last thing you want is to be paying premiums only to have your benefit denied when you need it.

You'll want to obtain a policy that is "own occupation." If you are board-certified, you want to specify your specialty. You **do not** want "any occupation" listed. This means that if you cannot work as a veterinarian, but can perform any other job, a benefit will not be paid, independently of income differences. Also, make sure to get a non-cancellable policy. This means that as long as you pay your premium, the insurance company cannot cancel or increase the policy rates. Generally, you'll be able to replace 60% of your income with an individual policy. This benefit is paid without taxes being applied. Make sure to calculate your expenses and purchase a policy that allows you to cover them. The cost of individual disability policies is high, but when you get it, it's yours until you're 65 years old (assuming you continue to pay the premium and continue to be disabled). The policy will only pay 100% benefit if you are 100% disabled, with many variations depending on the disability. This is different from term-life insurance, which will pay 100% upon your death.

There are also group long-term disability policies. You can obtain these through your employer (if they offer this benefit). These are cheaper and easier to qualify for. Unfortunately, they are not as good as individual policies, and will lapse when you leave your current employer. They might start as own-occupation for a period of time and then become any-occupation. The benefit paid is subject to taxes, which is not the case for individual policies. Lastly, with certain employers, you cannot opt-out of the benefit. Therefore, when applying for an individual policy, they will not allow you to cover as much of your salary as listed above. If you don't qualify or can't afford an individual policy, make sure to obtain this benefit from your employer.

Life Insurance

The purpose of life insurance is to ensure that your loved ones are not financially burdened by your death. The amount of insurance to obtain is highly dependent on your financial and family situation. If you have four children, you'll need a higher amount than if it's only you and your partner. You'll want to add up all of your current expenses, loans, possible future expenses (i.e., college costs) and evaluate your current savings. You should calculate your current monthly expenses to an annual basis and then figure out how many years you'll need that amount for. The insurance you want is "term life insurance." You purchase life insurance that will pay if you die within that term. In other words, if you purchase a 30-year term policy today, the insurance company will pay the benefit if you die within the next 30 years. The premium will

remain the same for the duration of the policy. At a certain point in your career, hopefully you'll have accumulated enough wealth that you'll likely be able to self-insure.
Life insurance is a common employer benefit. Keep in mind that this benefit is not enough to cover your family in the event of your death, as it usually only covers up to one time your annual salary (**always** read your benefits carefully). For term-life insurance you're likely talking about obtaining at least 1 million in coverage. You can purchase additional coverage through your employer but it is not typically cost effective in comparison with an individual policy. And just like disability insurance, if you leave your employer, the group policy will go away.
If you are to purchase an individual term-life insurance policy, you want to obtain it when you're younger, as premiums will go up as you age. If you have pre-existing conditions, you might have trouble obtaining life insurance as well, and if you do, your premiums will be much higher. Overall, term-life is much cheaper than disability insurance. Since men are more likely to die earlier than women, premiums are higher for this gender. As discussed previously, you'll want to find an independent insurance agent and therefore be able to receive quotes from various insurance companies. Lastly, stay away from whole life insurance. This product is **not** term-life insurance and the reason you might be recommended to obtain it is because of the extremely high commissions it pays to insurance agents, despite being a bad product for the vast majority of the population. If someone is trying to sell you this, you probably should work with another person.

Student Loans

Student loan management is a complicated topic that requires its own lecture. However, it is important to know some crucial facts. For example, you must know your student loan servicer has no incentive to help you manage your loans (assuming you have not refinanced with a private entity). Afterall, they make more money the longer it takes you to pay them. In addition, their customer service representatives do not have the necessary knowledge to help you figure out the best option for your situation. Oftentimes they offer options such as deferment and forbearance. These might sound good on paper because you want to pay less monthly (or cannot afford the current payment), but will come back to haunt you in the future due to higher interest accrual.
You must be familiar with the details of your student loans and repayment options available. With increasing tuition costs, and therefore increasing debt-to-income ratios, you can't ignore your loans. Public Service Loan Forgiveness (PSFL), pay as you earn (PAYE), revised pay as you earn (REPAYE), income-based repayment (IBR), and income contingent repayment (ICR) are terms to be extremely familiar with. Also important is the difference between loan consolidation and refinancing. The former consists of making all your loans become one, with a single payment. This loan would likely still be eligible for federal repayment plans (you must check!). Refinancing turns many loans into one large loan with a private entity/company. You'll likely have a lower interest rate, but you lose federal features/protections by doing this. None of the above are or should be seen as automatic decisions. Not everyone should refinance. For example, if you're aiming for PSFL, you cannot refinance or you'll become ineligible for this federal program. The federal government has made student loan management so complicated that you might need an advisor that specializes in student loan management alone! Choosing the right repayment plan or electing to refinance or consolidate your loans can save you many thousands of dollars on the long run, so make sure you're up to date on the status of your loans, and make a plan!

Lifestyle Creep

The vast majority of us are basically broke during vet school. If you pursue an internship and residency, there'll be even more time when you're trying to make ends meet. During vet school it's very easy to forget that you're living on borrowed money. You take loans to pay tuition, rent, food, etc. If you're pursuing advanced training, at least you're getting paid a small salary. But remember that for the vast majority of student loans, interest continues to compound during this time. The American Association of Veterinary Medical Colleges (AAVMC) reported the 2019 mean academic internship and residency salaries to be $28,372 and $35,246, respectively. These numbers are far from the mean salary of $61,200 that an MD/DO resident is paid (based on Medscape's Residents Salary and Debt Report 2019).

It's no surprise that when real world (i.e., your first "real" job) salaries come into play, lifestyle creep will occur. Afterall, we just put in years and years of work for that "real" paycheck and we "deserve it!" It's finally time to enjoy the good things in life, right? Increased spending can become so marked that future financial stability might be impacted due to low savings rate, debt numbness, and lack of financial literacy. You've lived so long with student loans that you might not be averse to taking on additional debt. What's an additional $5000 or $10,000 if you already have more than $200,000 of debt? Then comes the mortgage, the fancy car, etc. And we have all felt the pressure to impress our friends or neighbors; keeping up with the Joneses is real.

Instead, how about giving yourself a smaller raise to improve your quality of life, and using the remainder of your higher paycheck to increase wealth (retirements accounts, saving for a down payment on a house, etc.) and pay down debt? If you've just spent many years being "poor," you could probably do it for a few more, with the added benefit of having a bump in living standards.

Written Financial Plan and Budgeting

No matter how good your memory is, you should write down your financial goals. And if it's not just you, you want the whole family to be on the same/similar page so that you can work as a team to reach those goals. While I'm not going to expand on what a financial plan is or could be, it can be as simple or as complicated as you want. You can set simple goals such as "I want to fill my 401(k) every year," or work on more of an overall financial plan with detailed savings, insurance, housing, debt repayment, or spending goals. Remember reevaluate your plan as you go along to ensure you're meeting your goals and to adjust accordingly.

Creating a budget and sticking to it is hard. Plus, everyone dislikes it. However, you need to know what your cash flow is. Cash flow is the amount of money coming in and going out. In other words, you need to know how much money enters your bank account, where that money is going, and how much you're able to save. Then you can start looking in more detail at the different categories where the money is going and write it down. Then track it for a month and at the end of the month decide if there's a category you'd like to spend more or less money on. You can make it as simple or as detailed as you'd like!

If you don't know where you're spending your money, it'll be hard for you to save it. Your savings rate is the most important factor in order to build wealth. It's very easy to start spending more money if you're making more. There's also no guarantee you'll increase how much you're saving, or that you'll put more aside for retirement if your income goes up. Plus, every extra dollar you make will be taxed and, in the end, won't be worth as much as every dollar saved.

Take Home Points

- No one will care more about your money than you. Don't delegate everything and be in the dark about it.
- Don't mix investing and insurance.
- Insure against financial catastrophe.
- Don't trust your student loan provider. They simply don't have enough knowledge (and incentive) to help you.
- Be cautious with lifestyle creep. Live like a resident for as long as you can, which in turn will allow for more wealth accumulation and paying down debt faster.
- Savings rate is more important than increasing your salary. Just because you make more money doesn't mean you'll save more of it.
- Personal finance is personal. No financial plan or advice will be the same and work for everyone—you need your own plan.

References and Resources

1. AAVMC. 2019 internship salaries offered through the VIRMP. www.aavmc.org/public-data/2019-internship-salaries Accessed Feb 12, 2020.
2. AAVMC. 2019 resident salaries offered through the VIRMP. http://www.aavmc.org/public-data/2019-residency-salaries. Accessed Feb 12, 2020.
3. AVMA. 2019 Economic State of the Veterinary Profession. https://www.avma.org/news/press-release/AVMA-2019-Economic-State-of-the-Veterinary-Profession-Report-now-available. Accessed Feb 12, 2020.
4. VIN Foundation Blog. https://vinfoundation.org/blog/. Accessed Feb 12, 2020.
5. VetMed Survival Guide. Financial literacy. https://vetmedsurvivalguide.com/category/financial-literacy/. Accessed Feb 12, 2020.
6. PocketDVM. http://pocketdvm.com/. Accessed Feb 12, 2020.
7. Collins JL. *The Simple Path to Wealth*. https://jlcollinsnh.com/. Accessed Feb 12, 2020.
8. Physician on FIRE. https://www.physicianonfire.com/. Accessed Feb 12, 2020.
9. Taylor Larimore, *et al*. *The Bogleheads' Guide to Investing*.
10. Stanley TJ, Danko WD. *The Millionaire Next Door*.
11. Clements J. *How to Think About Money*.
12. Dahle JM. *The White Coat Investor*.
13. Dahle JM. *The White Coat Investor's Financial Boot Camp*.

AUTHOR INDEX

American College of Veterinary Internal Medicine (ACVIM)
8301 E. Prentice Avenue, Suite 300
Greenwood Village, CO 80111
www.acvim.org

CURRAN ASSOCIATES INC.
proceedings.com

ISBN 978-1-7138-1351-4